Model Test Paper in Agriculture

- Practice papers for students.
- To compete ARS, JRF, SRF, UPCATET and SAU's Exams.
- To compete Civil Services and State PCS exams
- Based on latest/new syllabus of ICAR.
- Short MCQ and answers.

Renu Arya
M.Sc. (Home Science)
Extension Education & Communication
Scientist
Krishi Vigyan Kendra, Bahraich (UP)

Dr. Sonam Arya
M.Sc & Ph.D
Agricultural Biotechnology
Lecturer
College of Social Science & Applied Education, Meerut

Dr. R. L. Arya
M.Sc. & Ph.D.
Agronomy-IARI, Pusa, New Delhi
Ex Senior Scientist
CTRI, Rajahmundry (AP)

Published by

SCIENTIFIC PUBLISHERS (INDIA)
5 A, New Pali Road, P.O. Box 91
JODHPUR 342 001 (India)
E-mail: info@scientificpub.com
Website: www.scientificpubonline.com

ISBN : 978-93-90495-36-8
EISBN : 978-93-90495-46-7

Printed in India.

Preface

Agricultural is the important branch of science which deals with the application of scientific research and new knowledge to agricultural practices through farmer education. Agriculture can help reduce poverty, rise incomes and improve food security for 80% of the world's poor, who live in rural areas and work mainly in farming. The most important aspect of agiculture for human neings is to provide food for people.

The competitive examinations have become a routine procedure of recruitmentand admissions for higher positions and education. Now-a-days a large number of questions appear in the examination. these questions call for quick answering for success wihtin a specified period of time. A subject like agriculture ecompassses a large number of disciplines. A sincere effort has been madeby the authors to present them in most easy, short and understandable language for the benefit of readers. **Model Test Papers in Agriculture** book contains several chapters in agriculture are included in brief and simple language. The book entire book is prepared in most simple, clear and talking language so that the concept of book could be easily understand by the readers.

Model Test Papers in Agriculture is also useful for competitive exams like IAS, IFS, PCS, Banking services, states and national levels of different competition and getting of admission of post doctorate and Ph.D. in agricultural subjects. Finally, I have belief and hope that this book will fulfill the requirement of students of agricultural stream.

Authors shall be grateful to the various books, magazine, newspapers, newsletters, periodicals, journals etc are rendered help for bringing out of this book. The valuable contributions of different persons who are directly or indirectly involved for preparation of this book are hereby gratefully acknowledged. Authors are highly thankful to M/s Scientific Publishers, Jodhpur, Rajasthan and their staff members for printing in the form of book.

Being first edition of book, there could be some shortcomings for which valuable and informative suggestions will be gratefully appreciated, these shortcomings may be incorporated in forthcoming edition so that book may be more useful to the readers. The publisher shall also be grateful for bringing to their notice of any suggestions for further improvement of this book.

— Authors

About the authors

Renu Arya completed her graduation and post graduation in Home Science with specialization in extension education and communication management from C.S.A.U. of Ag. & Tech., Kanpur. Presently she is working as a Subject Matter Specialist (Scientist) at Krishi Vigyan Kendra, Behraich under A.N.D. U. Ag. & Tech., Kumarganj, Ayodhya. She has published many research papers/popular articles in national and international journal of repute. She has authored many books entitled Indira Krishi Pratiyogita Darpan, Indira Agricultural Competition Explorer, Objective in Agricultural Extension, Home Science and Extension education. She is recipient of UGC fellow for doing Ph.D. at PAU, Ludhiana. She is actively involved in extension activities of Home Science at Foot Hills of Himalaya.

Dr. Sonam Arya completed her graduation from College of Biotechnology, CSJM, Kanpur and post graduation in Biotechnology from Babasaheb Bhimrao Ambedker University (Central university), Lucknow. She also obtained M.Phill degree in Life Science from CSJM, Kanpur. She completed Ph.D. in Agricultural Biotechnology, Sardar Ballabh Bhai Patel University of Agriculture and Technology, Modipuram Merut. Presently she is working as a Lecturer in College of Applied education and health Sciences, Merut (UP). She has published several research papers/popular articles in national and International journal of repute. She is author of several books entitled Indira Agricultural Competition Explorer, Indira Competition Explorer, Indira's Objective in Agricultural Biotechnology, Indira's Objective in Agriculture, Home science and Fundamentals of Horticulture. She is awarded Rajiv Gandhi Fellowship from UGC for doing Ph.D.

Dr. R.L. Arya obtained B.Sc. (Ag & AH) degree in 1984 from C.S.A.U. Ag. & Tech., Kanpur and M.Sc. & Ph. D. in Agronomy in 1986 and 1989 from Division of Agronomy, I.A.R.I., Pusa, New Delhi, respectively. He has vast experience in the field of agronomical research in fodder and pulses under dryland/rainfed areas and tobacco production. He has to his credit of more than 250 research publications in national and international journals. He is the author of more than twenty five books in different aspects of agriculture and agronomy. Being a recipient of four ICAR national awards for outstanding contributions in different aspects of agronomy. He was Senior scientist in ICAR and worked CRIDA, Hyderabad, IGFRI, Jhansi, IIPR, Kanpur and CTRI, Rajahmundry. Presently he is working as a extension worker (District Consultant) in centrally sponsored scheme of National Food Security Mission at Kanpur Dehat.

Contents

PART - 3

PART - 1

A AGRICULTURAL EXTENSION

1. Multiple Choice Questions of Agricultural Extension

Practice Paper 1

1. **The term 'Extension' have originated in:**
(A) Latin (B) Persian
(C) Greek (D) Sanskrit

2. **The latin roots for the word 'extension':**
(A) *ex* and *tensio* (B) *ex* and *tension*
(C) *ex* and *luceo* (D) *ex* and *termo*

3. **The word "tentio" in extension is derived from:**
(A) Greek word (B) Latin word
(C) English word (D) None of these

4. **Extension literary means:**
(A) Stretching out (B) Talking with
(C) Deciding for (D) Thinking about

5. **Extension is a/an:**
(A) Informal education
(B) Out of school systems of education
(C) Both (A) & (B)
(D) Formal education

6. **Extension is:**
(A) Education for all village people
(B) Teaching through learning by doing and seeing is believing
(C) Two way channel
(D) All of these

7. **The extension activities was started first time in:**
(A) USA (B) Russia
(C) Germany (D) India

8. **The extension word was used first time in:**
(A) USA (B) India
(C) France (D) USSR

9. **The term 'University Extension' was first used in:**
(A) 1866 (B) 1972
(C) 1901 (D) 1921

10. **The person considered as father of extension:**
(A) James Stewatt (B) Seaman A Knapp
(C) Edgar Dale (D) Paul Leagans

11. **The term 'Extension Education' was first coined in:**
(A) USA (B) UK
(C) Netherlands (D) India

12. **The person considered as father of extension in India:**
(A) Adivi Reddy (B) O P Dahama
(C) K N Singh (D) J B Chitambar

13. **The purpose of introducing agricultural universities in India is to improve the standard of agriculture:**
(A) Extension (B) Teaching
(C) Research (D) All of these

14. **Which extension method is evolved in result demonstration, posters, movies, slides and film strips?**
(A) Spoken (B) Visual
(C) Written (D) None of these

15. **An extension worker, while extending, scientific knowledge to the farmers, starts with:**
(A) Lectures
(B) Practical demonstration
(C) Distribution of leaflet
(D) All of the above

16. **There are how many degrees of evaluation in Extension:**
    (A) 4 (B) 5
    (C) 3 (D) 2
17. **'Vulgarisation' in French is equivalent to the word:**
    (A) Social taboos (B) Education
    (C) Extension (D) Social change
18. **Extension when aims at empowering or uplifting the poor:**
    (A) Emancipatory extension
    (B) Formative extension
    (C) Informative extension
    (D) Persuasive extension
19. **Extension as a science is classified under:**
    (A) Hard Science (B) Soft Science
    (C) Pure Science (D) Physical Science
20. **Extension is not:**
    (A) Voluntary
    (B) Family centered
    (C) Conclusion oriented
    (D) Flexible

Practice Paper 2

1. **Extension can be considered as a:**
   (A) Service (B) Profession
   (C) Discipline (D) All of the above
2. **Extension basically is a two way process between:**
   (A) Farmer and researcher
   (B) Researcher and planner
   (C) Teacher and planner
   (D) Researcher and teacher
3. **Extension that emphasize on human resource development:**
   (A) Emancipatory extension
   (B) Formative extension
   (C) Informative extension
   (D) Persuasive extension
4. **Extension when used as a policy instrument for achieving societal objective:**
   (A) Emancipatory extension
   (B) Formative extension
   (C) Informative extension
   (D) Persuasive extension
5. **Extension that moves beyond the traditional crop production technologies to include all land based activities:**
   (A) Universal extension
   (B) Cyber-extension
   (C) Emancipatory extension
   (D) Broad based extension
6. **Extension as an instrument for helping people to make well-considered choices:**
   (A) Emancipatory extension
   (B) Formative extension
   (C) Informative extension
   (D) Persuasive extension
7. **The instrument used by extension to bring change:**
   (A) Communication (B) Teaching
   (C) Learning (D) Adoption
8. **Extension can be effective only through:**
   (A) Voluntary change
   (B) Democratic change
   (C) Autocratic change
   (D) Persuasive change
9. **The main mode of extension in the farmer first model:**
   (A) Farmer to farmer
   (B) Agent to farmer
   (C) Researcher to agent
   (D) Researcher to farmer
10. **The satellite model concept of extension was formulated by:**
    (A) M S Swaminathan
    (B) G Perumal
    (C) Curt Lewin
    (D) Anil Gupta
11. **The fundamental objective of extension work is:**
    (A) Development of farmer
    (B) Development of individual
    (C) Development of culture
    (D) Development of society
12. **The ultimate objective of extension is:**
    (A) Increase productivity
    (B) Increase income
    (C) Rural development
    (D) Realize ones fullest potential

13. Agricultural extension deals with:
(A) Farmers
(B) Change agents
(C) Farmers and change agents
(D) None of the above

14. An extension worker help farmers to attain desirable changes in his:
(A) Temperament
(B) Behaviour
(C) Health
(D) Wealth

15. Leadership development in extension is:
(A) A means
(B) An end
(C) Both means and end
(D) None of these

16. Most useful leader in agricultural extension is:
(A) Voluntary leader
(B) Bureaucratic leader
(C) Autocratic leader
(D) Democratic leader

17. An extension worker is a:
(A) Professional leader
(B) Lay leader
(C) Local leader
(D) Voluntary leader

18. An extension agent is considered to be a:
(A) Personal cosmopolite
(B) Impersonal cosmopolite
(C) Personal localite
(D) Impersonal localite

19. T & V system is a good example of:
(A) Extension approach
(B) Training approach
(C) Co-operative self help approach
(D) Integrated development approach

20. Which one of the step is not included in programme planning?
(A) Collect facts
(B) Analyze situation
(C) Identify problems
(D) Execute plan

Practice Paper 3

1. .... is the primary institution of society:
(A) Family (B) Village
(C) Individual (D) Block

2. Which of the following extension approaches has got the highest intensity of influence?
(A) Mass approach
(B) Group approach
(C) Community approach
(D) Individual approach

3. Which of the following pair is not matched:
(A) Communication : Berlo
(B) Rural weakers : Small farmers
(C) Mass media : Demonstration
(D) Helping poorest of the poor first : IRD programme

4. General meeting are broadly, the meeting of:
(A) Heterogeneous participation
(B) Homogenuos participation
(C) Community participation
(D) Society participation

5. Campaign is an:
(A) Intensive teaching
(B) Extensive teaching
(C) Invasive teaching
(D) None

6. Programme planning is a procedure of
(A) Working by people
(B) Working for the people
(C) Working with the people
(D) None of these

7. A series of still pictures on one roll is called
(A) Channel (B) Film strip
(C) Mock-up (D) Black board

8. A bulletin should contain
(A) 4-12pages (B) 12-24 pages
(C) 24-48 pages (D) None

9. B.R.Mehta team stimulated an active consideration of .... through democratic bodies in India:

(A) Decentralization (B) Stabilization
(C) Centralization (D) Mobilization

10. Most people retain .... % of what they read:

(A) 5-10 (B) 10-15
(C) 20-25 (D) 30-35

11. Most people kept in mind .... % of what they seen:

(A) 10-15 (B) 20-25
(C) 30-35 (D) 50-55

12. Up to .... % of what is taught is kept in mind by majority of people, if they participate actively, and all the senses (read, hear, and seen) are involved:

(A) 50 (B) 65
(C) 80 (D) 90

13. Symposium is a short series of lectures, usually by speaker:

(A) 1-2 (B) 2-5
(C) 5-7 (D) 7-10

14. Phillips 66 format of hurdle system is related with:

(A) Panel (B) Form
(C) Buzz section (D) Work shop

15. Small group interaction designed to encourage the free introduction of ideas on an unrestricted basis without any limitations to feasibility is known as:

(A) Conference (B) Seminar
(C) Buzz section (D) Brain storming

16. .... essentially follow the seminar method and the focus is on any particular subject or problem:

(A) Work shop (B) Syndicate studies
(C) Institute (D) None

17. News paper articles are included in which method of extension:

(A) Individual (B) Group
(C) Mass (D) None of these

18. .... is essentially a recognizable imitation or replica of the original, whether workable or not:

(A) Model (B) Specimens
(C) Mock-up (D) Objects

19. A working model is known as:

(A) Objects (B) Model
(C) Mock-up (D) Poster

20. "The communication process in Rural development" is written by:

(A) Leagans (B) Berlo
(C) A.A. Reddy (D) Ray

Practice Paper 4

1. .... are expressions of the ends towards which our efforts are directed:

(A) Goal (B) Objectives
(C) Needs (D) Message

2. .... is defined as the distance in any given direction one expects to go during a given period of time:

(A) Goal (B) Objectives
(C) Needs (D) Message

3. The ultimate objective of extension work is the full development of .... :

(A) Individual (B) Group
(C) Mass (D) Village

4. The gap between the situation and objective is the area of .... :

(A) Goal (B) Needs
(C) Interest (D) None

5. In good administration the offices of a .... are arranged in a pyramid of authority and responsibility:

(A) Autocracy (B) Monarchy
(C) Bureaucracy (D) Democracy

6. Which one of the following is a farmer agency?

(A) FCI
(B) IFFCO
(C) Agro service centers
(D) Shree Ram fertilizers

7. Krishi Vigyan Kendra (KVKs) can only be under control of:

(A) Agricultural University
(B) Voluntary organisation
(C) Govt. and Voluntary organization
(D) Agricultural University and Voluntary Organisation

8. **"The Lab to Land" programme also included demonstration but they were different from early demonstrations in terms of:**
   (A) Target groups
   (B) Type of demonstration
   (C) Size of demonstration
   (D) Place of demonstration

9. **Which of the following is not a group approach device:**
   (A) Symposium (B) Film
   (C) Demonstration (D) Workshop

10. **The subject extension is a:**
    (A) Means (B) Discipline
    (C) End (D) Branch of mean

11. **The extension system used to provide precise and reliable scientific information in simple language about a simple practice is called:**
    (A) Leaflet (B) Newspaper
    (C) Literature (D) Circular letter

12. **The benefit of first hand knowledge good will and creating confidence can be developed by:**
    (A) Telephone
    (B) Visiting home and farms
    (C) Organising village fairs
    (D) None of above

13. **Audio-visual aids is:**
    (A) Radio (B) Wireless
    (C) Telephone (D) Television

14. **Which of the following country was first use the term 'extension' to imply extension knowledge from agriculture research institution to the farmers through the process of informal education?**
    (A) U.S.A. (B) U.S.S.R.
    (C) India (D) Japan

15. **Which of the following will give the clear idea of proportion?**
    (A) Model (B) Pie chart
    (C) Radio (D) TV

16. **For providing the precise and reliable scientific information in simple language the extension system used is:**
    (A) Charts (B) Demonstration
    (C) Leaflets (D) None of these

17. **Which of the following visual aids are commonly used by agricultural extension workers in India?**
    (A) Posters, photographs, flash card and fenal graphs
    (B) Slides, film-strips, models and puppets
    (C) Bulletin boards, cultural programmes and black boards
    (D) All of the above

18. **Which of the following is not true?**
    (A) Programme planning is a spontaneous process
    (B) Programme planning is an educational process
    (C) Programme planning is a continuous process
    (D) Programme planning is a co-ordinating process

19. **Which of the following is a function of the RRB's?**
    (A) To develop rural economy
    (B) To provide credit for agriculture and allied activities
    (C) To encourage rural industries
    (D) All of the above

20. **Evaluation done by the agency is known as:**
    (A) Self evaluation
    (B) Internal evaluation
    (C) External evaluation
    (D) None of the above

Practice Paper 5

1. **The all – round development of Indian rural people can be achieved by means of:**
   (A) Extension education
   (B) Adult education
   (C) More money distribution
   (D) All of the above

2. **Slide projector gives the image:**
   (A) Direct (B) Indirect
   (C) Both (A) and (B) (D) Reflected

3. **Puppet show in village as a method for transfer of technology is an example of:**
(A) Mass media (B) Group media
(C) Individual media (D) Tribal media

4. **The main purpose of T and V programme is to:**
(A) Identify research problems
(B) Bring about co-ordination in the administration of agricultural extension
(C) Update the technical knowledge and skills of agricultural extension workers
(D) Help farmers in the adoption of latest innovations in agriculture

5. **Which one of the following is not correctly matched?**

List-I	**List –II**
(A) Mass	Exhibition
(B) Individual	Farm visit
(C) Group	Mela
(D) Mass	Newspaper

6. **Which of the following is the group contact teaching method?**
(A) Farm and home visit
(B) Bulletins
(C) Result demonstration
(D) Method demonstration

7. **The benefit of first hand knowledge, good will and create confidence can be developed by:**
(A) Visiting homes and farm
(B) Telephone
(C) Village fair
(D) All of the above

8. **To create interest in a wider range of people and to influence people to adopt better practices the extension method generally used is:**
(A) Exhibition (B) Tours
(C) Fair (D) Competition

9. **The extension system used to provide precise and reliable scientific information in simple language about a simple practice is called:**
(A) Leaflet (B) Literature
(C) Newspaper (D) Circular letter

10. **Extension of recent agricultural knowledge to rural peoples through trained persons of the same locality is called the principle of:**
(A) Panchayati Raj
(B) Centralized power
(C) Local leadership
(D) None of the above

11. **Effective extension service is a process of:**
(A) Helping the people which helps themselves
(B) Profiting the local leaders
(C) Helping the B.D.O.
(D) All of the above

12. **The concept of demonstration was given by:**
(A) Lewis Jones
(B) O.P. Dahama
(C) Rudramurthy
(D) Dr. Seeman A. Knapp

13. **The functions of pre-extension in the development block was:**
(A) To survey block area
(B) To prepare ground for intensive stage
(C) To conduct agricultural demonstrations
(D) All of the above

14. **The development approach of National Extension Service Block was based mainly on:**
(A) Self help
(B) Govt. help
(C) State Govt. help
(D) Help by village Panchayat

15. **National demonstrations are first line demonstrations which is conducted on the farmer's field by:**
(A) Extension workers
(B) V.L.W.
(C) Research Scientists
(D) B.D.O.

16. **To create interest in a wider range of people to adopt better practices, the extension method generally used is:**
(A) Field trials
(B) Exhibitions
(C) Competition
(D) Fair

17. The elements of teaching learning process are:
(A) Teaching subject matter and learner
(B) Teaching equipments and physical facilities
(C) Both 'A' and 'B'
(D) None of the above

18. "Cone of experience" was devised by:
(A) Rogers (B) O.P. Bhatnagar
(C) A.A. Reddy (D) Edger Dale

19. The programme for transfer of technology started by ICAR at the golden jubilee year occasion was:
(A) Operational Research Project
(B) Lab to Land Programme
(C) National Demonstration
(D) I.R.D.P.

20. Extension workers are:
(A) Professional leaders
(B) Autocratic leader
(C) Lay leaders
(D) None of these

Practice Paper 6

1. VLW (Multi Purpose Worker) concept emerged in which project?
(A) T & V Project (B) NEP
(C) KVK (D) Etawah Pilot Project

2. Which of the extension programmes in India advocated single line of command and strong research-extension linkages?
(A) KVKs (B) T&V
(C) NATP (D) ORP

3. Which of the following programmes is being implemented by SAUs/Institutes to strengthen institute-village linkages?
(A) ATIC (B) NATP
(C) IVLP (D) UPDASP

4. "Farmers Interest Group" concept in agricultural extension is:
(A) A group of farmers meant for cultural activities
(B) A group of farmers in the villages created to watch the interests of the farmers
(C) A group of farmers coming together voluntarily to take up some activities
(D) A group of farmers interested in employment generation for rural youths

5. The internal state which stimulate the individual to carry out certain activities is:
(A) Perception (B) Motivation
(C) Intuition (D) Learning

6. The motivational type most important for extension is:
(A) Extrinsic motivation
(B) Intrinsic motivation
(C) Situational motivation
(D) Classical motivation

7. "If I have to get that gold medal, I should work hard at studies" reflects:
(A) Determination
(B) Intrinsic motivation
(C) Industriousness
(D) Extrinsic motivation

8. Which of wthe following is not an approach to education:
(A) Conventional (B) Prevocational
(C) Progressive (D) Liberating

9. Effective teaching and learning is a matter of:
(A) Chance and luck
(B) Design and effect
(C) Material and non material
(D) Formal and informal

10. Education is shaped by:
(A) Society (B) Teacher
(C) Student (D) Knowledge

11. Cone of experience classify the instructional methods according to:
(A) Experiential correctness
(B) Difficulty in understanding
(C) Ease in use and application
(D) Level of abstraction in thinking

12. The cone of experience was developed by:
(A) Egder Dale (B) Berlo C.K.
(C) Kuldeep Nair (D) S C Parmar

13. The type of experience a farmer have when he saws wheat in lines under the supervision of extension worker:
(A) Symbolic experience
(B) Iconic experience
(C) Supervisory experience
(D) Enactive experience

14. When a person involved in iconic experiences, learning occurs through:
(A) Active participation
(B) Observation
(C) Abstract thinking
(D) Doing

15. The extension worker tells about the varieties characters to a farmer, the farmer learns by:
(A) Abstraction (B) Learning
(C) Intelligence (D) Questioning

16. Classification of extension methods into individual, group and mass is based on:
(A) Foam
(B) Use
(C) Number of people contacted
(D) Periodicity

17. To get first hand knowledge on a farm situation we use:
(A) Village fairs
(B) Farm & home visit
(C) Telephone
(D) Questionnaire

18. The ratio of 'takes' to 'acceptance' is the highest in:
(A) Farm and Home visit
(B) Method demonstration
(C) Result demonstration
(D) Campaign

19. Media forum is classified under:
(A) Individual contact method
(B) Group contact method
(C) Mass contact method
(D) All of the above

20. When the farmer meets the extension worker at his work place:
(A) Farmer's call
(B) Extension campaign
(C) Result demonstration
(D) Office visit

Practice Paper 7

1. The extension method that act as a safety valve in reducing tension:
(A) Group discussion
(B) Method demonstration
(C) Result demonstration
(D) Campaign

2. The best extension method suited to all situation:
(A) Demonstration (B) Campaign
(C) Farm publication (D) None of these

3. Method demonstration aims more at improving ones:
(A) Cognitive skills (B) Perceptual skills
(C) Motor skills (D) Analytical skills

4. Seaman A Knapp conducted result demonstration in:
(A) Cotton (B) Maize
(C) Barley (D) Wheat

5. A method which build the confidence of the researcher, extension worker and the farmer:
(A) Minikit trial
(B) Result demonstration
(C) Method demonstration
(D) All of the above

6. A method of motivating the people to adopt a new practice by showing what has actually been achieved by applying the practice under field conditions:
(A) Campaign
(B) Result demonstration
(C) Field day
(D) Farm bulletin

7. Maintenance of record is absolutely necessary in case of:
(A) Method demonstration (B) Minikit trial
(C) Field day (D) Field trip

8. The method used when the farmer is not there in the field while the extension worker make a visit:
(A) Farm and home visit
(B) Method demonstration
(C) Result demonstration
(D) Flag method

9. The normal group boundaries get obliterated in:
(A) Method demonstration
(B) Farm and home visit
(C) Office calls
(D) Campaign

10. A body of general principles or laws of a field of knowledge is called:
(A) Television (B) Philosophy
(C) Mandate (D) Principle

11. The basic philosophy of extension is to teach people:
(A) How to do (B) What to do
(C) When to do (D) Where to do

12. The difference between what is and what ought to be:
(A) Need (B) Aim
(C) Goal (D) Objective

13. A blue print for action is:
(A) Plan (B) Programme
(C) Objective (D) Project

14. The level of objective operates at field extension work:
(A) Fundamental objective
(B) Working objective
(C) General objective
(D) None of the above

15. Which of the following is not a level of objective:
(A) Basic objective
(B) Fundamental objective
(C) Working objective
(D) General objective

16. Extension work grew out of:
(A) Problem (B) Situation
(C) Condition (D) Participation

17. Measuring performance against predetermined goals is called:
(A) Management (B) Evaluation
(C) Measurement (D) Supervision

18. The element that take time, effort and resource to move from one event to another:
(A) Event (B) Activity
(C) Critical path (D) Slack

19. The largest possible time to perform an activity assuming everything goes well is:
(A) Optimizing time (B) Pessimistic time
(C) Most likely time (D) Expected time

20. SREP is an important component in the functioning of:
(A) NARS (B) ATMA
(C) ATIC (D) KVK

Practice Paper 8

1. Extension teaching at the under graduate level was started by:
(A) Madras University
(B) Meerut University
(C) Calcutta University
(D) Mumbai University

2. The first line system of extension in India is controlled by:
(A) ICAR institutes/SAUs
(B) Ministry of Agriculture
(C) Non Government Organisations
(D) Private Extension Agencies

3. The demonstration in case of AICPND was done by:
(A) Farmer (B) Media personal
(C) Research scientist (D) Extension person

4. The actual physical product of source-encounter is:
(A) Receiver (B) Message
(C) Channel (D) Feedback

5. The three important dimensions of message include:
(A) Code, content, treatment
(B) Source, self, receiver
(C) Encode, decode, feedback
(D) Simplicity and clarity

6. The way in which message is handled before placing in the channel:
(A) Decoding (B) Treatment
(C) Distortion (D) Feedback

7. The link that connects the source with the receiver:
(A) Message
(B) Communication effect
(C) Channel
(D) Treatment

8. **Television is primarily a:**
(A) Media mix (B) Source
(C) Medium (D) Channel

9. **Trustworthiness and competency are elements of:**
(A) Credibility (B) Fidelity
(C) Empathy (D) None of these

10. **Acceptance and rejection are the one of the kinds of:**
(A) Failure of channel
(B) Feedback
(C) Treatment of message
(D) Empathetic feelings

11. **The term class media refers to:**
(A) Mass media
(B) Group media
(C) Interpersonal media
(D) Intra personal media

12. **While communicating, the communicator can write facing the audience in:**
(A) Black board (B) Flash card
(C) Slide (D) OHP

13. **Which of the following statement about audio-visual aids is true:**
(A) Replaces a teacher
(B) Decorates learning situation
(C) Emphasis non-verbal experience
(D) Includes text book

14. **The term electronic chalkboard refers to:**
(A) LCD projector
(B) OHP projector
(C) VHS projector
(D) Beta chrom projector

15. **The film projector commonly used in extension:**
(A) 8 mm (B) 16 mm
(C) 32 mm (D) 70 mm

16. **The object for projection in a slide projector is kept between:**
(A) Reflector and condenser
(B) Condenser and objective
(C) Bulb and condenser
(D) Objective and screen

17. **The principle made used in slide-cum-film projector:**
(A) Direct projection
(B) Indirect projection
(C) Reflected projection
(D) Diffused projection

18. **Which of the following instructional aids essentially involved two sense organs:**
(A) Television (B) Radio
(C) Poster (D) All of the above

19. **Reflected projector are used in:**
(A) OHP (B) Slide projector
(C) Lantern projector (D) Epidiascope

20. **Non-photographic reconstruction of reality is:**
(A) Illustration (B) Pictures
(C) Models (D) Diagrams

Practice Paper 9

1. **The extension system used to provide precise and reliable scientific information in simple language about a simple practice is called:**
(A) Leaflet (B) Literature
(C) News paper (D) Circular letter

2. **The coverage of agricultural news in newspaper is:**
(A) 3 to 5 % (B) 20%
(C) 1 to 2 % (D) 50%

3. **Radio is an example for:**
(A) Hot media (B) Cold media
(C) Optical media (D) Fiber media

4. **The effective rate of delivery for a radio talk is:**
(A) 80-100 words/minute
(B) 120-140 words/minute
(C) 160-180 words/minute
(D) 200-220 words/minute

5. **The best style of presentation for radio talks with the farmers:**
(A) Lecture
(B) Group discussion
(C) Interview
(D) Songs

6. **The knowledge base for a technology usually derives from:**
   (A) Basic research
   (B) Applied research
   (C) Technology testing
   (D) Technology development
7. **Technology development process starts with:**
   (A) Technology testing
   (B) Technology production
   (C) Technology generation
   (D) Technology dissemination
8. **Technology generation mainly happens at:**
   (A) Research station
   (B) Farmers fields
   (C) Training centres
   (D) Educational institutes
9. **The function of extension starts from:**
   (A) Technology testing
   (B) Technology production
   (C) Technology integration
   (D) Technology dissemination
10. **Fine testing and packaging of technologies into specific recommendations is called:**
   (A) Technology testing
   (B) Technology production
   (C) Technology integration
   (D) Technology dissemination
11. **Linkage is the technological term used to denote communication that is:**
   (A) Inter-organizational
   (B) Intra-organizational
   (C) Inter-personal
   (D) Intra-personal
12. **The technology flow in the farmer first model is basically from:**
   (A) Agent to farmer
   (B) Researcher to agent
   (C) Farmer to farmer
   (D) Researcher to farmer
13. **The process of delivery of technology to farmers:**
   (A) Technology testing
   (B) Technology production
   (C) Technology integration
   (D) Technology dissemination
14. **Apart from introduction of technology, to make use of it, extension should concentrate:**
   (A) Adoption of indigenous ideas
   (B) Mobilization of farmers
   (C) Diffusion of technology
   (D) Large farm demonstration
15. **Evaluation of technology means:**
   (A) Value judgment
   (B) Valere
   (C) Positive and negative
   (D) None of these
16. **A technology that represents a major improvement over the existing one embodies:**
   (A) Compatibility
   (B) Visibility
   (C) Utility
   (D) Complexity
17. **The first stage of on-farm research:**
   (A) Technology evaluation
   (B) Technology refinement
   (C) Technology verification
   (D) Technology dissemination
18. **The belief that technology causes changes in the society:**
   (A) Technological fundamentalism
   (B) Technological materialism
   (C) Technological structuralism
   (D) Technological determinism
19. **"If people utilize all the presently available knowledge, there will not be any hunger":**
   (A) Transfer of technology paradigm
   (B) Emancipatory extension paradigm
   (C) Informative extension paradigm
   (D) Persuasive extension paradigm
20. **Transfer of technology is related with:**
   (A) Local idea
   (B) Progressive idea
   (C) Foreign idea
   (D) None of these

Practice Paper 10

1. **Farmers interact with:**
   (A) Many sources
   (B) Few sources
   (C) Intra personal sources
   (D) Information sources

2. **The collectivity of units which are functionally differentiated and engaged in joint problem solving:**
   (A) Social norms (B) Social system
   (C) Social organisation (D) Social unit

3. **Established behaviour patterns for the members of the social system:**
   (A) Norms (B) Values
   (C) Attitude (D) Beliefs

4. **The alterations in the structure and functions of the social system:**
   (A) Social change
   (B) Social movement
   (C) Social alteration
   (D) Social interaction

5. **The social change where in the source of the change is from within the system:**
   (A) Local change
   (B) Immanent change
   (C) Localized change
   (D) Direct change

6. **Microanalysis approach in change analysis refers to:**
   (A) Change at individual level
   (B) Change at family level
   (C) Change at group level
   (D) Change at social system level

7. **Extension agent need to necessitate change initially on:**
   (A) Adoption only
   (B) Knowledge and adoption only
   (C) Belief and value system
   (D) Infrastructural facilities

8. **Voluntary agencies are characterized by:**
   (A) Less acceptability
   (B) Greater continuity
   (C) Less commitment
   (D) Less intensity

9. **Essential characteristics of a primary group is:**
   (A) Face to face contact
   (B) Personal and emotional relationship
   (C) Interpersonal proximity
   (D) Unity in diversity

10. **A person who actually initiates the action in the community is:**
    (A) Professional leader
    (B) Assured leader
    (C) Operational leader
    (D) Popular leader

11. **The leader who is well trained and paid:**
    (A) Autocratic leader
    (B) Democratic leader
    (C) Lay leader
    (D) Professional leader

12. **Contact farmer can be considered as:**
    (A) Opinion leader
    (B) Professional leader
    (C) Official leader
    (D) Lay leader

13. **Which among the following find the greatest application in desire stage of extension teaching:**
    (A) Farm publication (B) Radio
    (C) Exhibition (D) Mass media

14. **Poster is most effective at which stage:**
    (A) Awareness (B) Evaluation
    (C) Affection (D) Persuasion

15. **The extension approach used under IADP was:**
    (A) Proscriptive (B) Perscriptive
    (C) Participatory (D) Commercial

16. **The base level extension functionary in T & V system:**
    (A) BDO (B) VLW
    (C) AEO (D) SDAO

17. **The ratio of extension agent to farmer in a typical T&V system is:**
    (A) 1-250 (B) 1-500
    (C) 1-800 (D) 1-1500

18. **Single line of command is one of the important feature of:**
    (A) T&V (B) CDP
    (C) IAAP (D) IRDP

19. ABC of posters means:

(A) Attractive, Brief and Complete
(B) Attractive, Brief and Clear
(C) Attractive, Best and Creative
(D) None of these

20. A blue print for action is:

(A) Plan
(B) Programme
(C) Project
(D) None of these

Answers Key

■ Practice Paper 1

Q. No.	Ans	Q. No.	Ans	Q. No.	Ans	Q. No.	Ans	Q. No.	Ans	Q. No.	Ans
1.	A	2.	A	3.	B	4.	A	5.	C	6.	D
7.	A	8.	A	9.	A	10.	A	11.	B	12.	C
13.	D	14.	C	15.	B	16.	D	17.	C	18.	A
19.	B	20.	C								

■ Practice Paper 2

Q. No.	Ans	Q. No.	Ans	Q. No.	Ans	Q. No.	Ans	Q. No.	Ans	Q. No.	Ans
1.	D	2.	A	3.	B	4.	D	5.	D	6.	C
7.	A	8.	A	9.	A	10.	B	11.	B	12.	D
13.	A	14.	B	15.	A	16.	A	17.	A	18.	A
19.	B	20.	D								

■ Practice Paper 3

Q. No.	Ans	Q. No.	Ans	Q. No.	Ans	Q. No.	Ans	Q. No.	Ans	Q. No.	Ans
1.	A	2.	D	3.	C	4.	A	5.	A	6.	C
7.	B	8.	C	9.	A	10.	B	11.	C	12.	D
13.	B	14.	C	15.	D	16.	B	17.	C	18.	A
19.	C	20.	A								

■ Practice Paper 4

Q. No.	Ans	Q. No.	Ans	Q. No.	Ans	Q. No.	Ans	Q. No.	Ans	Q. No.	Ans
1.	B	2.	A	3.	A	4.	B	5.	C	6.	B
7.	D	8.	C	9.	B	10.	A	11.	A	12.	B
13.	D	14.	A	15.	B	16.	C	17.	D	18.	A
19.	A	20.	C								

■ Practice Paper 5

Q. No.	Ans	Q. No.	Ans	Q. No.	Ans	Q. No.	Ans	Q. No.	Ans	Q. No.	Ans
1.	D	2.	A	3.	B	4.	C	5.	C	6.	D
7.	A	8.	A	9.	A	10.	C	11.	A	12.	D
13.	D	14.	A	15.	C	16.	B	17.	C	18.	D
19.	B	20.	A								

Practice Paper 6

Q. No.	Ans	Q. No.	Ans	Q. No.	Ans	Q. No.	Ans	Q. No.	Ans	Q. No.	Ans
1.	D	2.	D	3.	C	4.	C	5.	B	6.	B
7.	D	8.	B	9.	B	10.	A	11.	A	12.	A
13.	D	14.	B	15.	A	16.	B	17.	B	18.	A
19.	B	20.	A								

Practice Paper 7

Q. No.	Ans	Q. No.	Ans	Q. No.	Ans	Q. No.	Ans	Q. No.	Ans	Q. No.	Ans
1.	A	2.	D	3.	C	4.	A	5.	A	6.	C
7.	B	8.	D	9.	D	10.	B	11.	A	12.	A
13.	A	14.	B	15.	A	16.	B	17.	B	18.	B
19.	B	20.	B								

Practice Paper 8

Q. No.	Ans	Q. No.	Ans	Q. No.	Ans	Q. No.	Ans	Q. No.	Ans	Q. No.	Ans
1.	C	2.	A	3.	C	4.	B	5.	A	6.	B
7.	C	8.	C	9.	A	10.	B	11.	B	12.	D
13.	C	14.	B	15.	B	16.	B	17.	A	18.	A
19.	D	20.	A								

Practice Paper 9

Q. No.	Ans	Q. No.	Ans	Q. No.	Ans	Q. No.	Ans	Q. No.	Ans	Q. No.	Ans
1.	A	2.	A	3.	A	4.	B	5.	B	6.	A
7.	C	8.	A	9.	A	10.	C	11.	A	12.	C
13.	D	14.	B	15.	A	16.	C	17.	C	18.	D
19.	A	20.	B								

Practice Paper 10

Q. No.	Ans	Q. No.	Ans	Q. No.	Ans	Q. No.	Ans	Q. No.	Ans	Q. No.	Ans
1.	B	2.	B	3.	A	4.	A	5.	B	6.	A
7.	C	8.	B	9.	B	10.	C	11.	D	12.	A
13.	A	14.	A	15.	B	16.	B	17.	C	18.	A
19.	B	20.	A								

2. Multiple Choice Questions Extension Principals, Approaches and Programming

Practice Paper 1

1. **Discouraging the application of chemical fertilizers and advocating organic farming:**
   (A) Principle of transfer
   (B) Principle of set
   (C) Principle of disassociation
   (D) Principle of association
2. **The extension scientific knowledge to villagers by specialists of the same area:**
   (A) Principle of local leadership
   (B) Principle of Panchayati Raj
   (C) Principle of grass root organisation
   (D) Principle of self activity
3. **Learning to use a sprayer correctly requires repeated trying of the instrument:**
   (A) Principle of motivation
   (B) Principle of attitude
   (C) Principle of practice
   (D) Principle of satisfaction
4. **If a farmer has learned the technique of water management in a particular crop, they should be able to use this method in other crops as well:**
   (A) Principle of achievement
   (B) Principle of transfer
   (C) Principle of disassociation
   (D) Principle of association
5. **Issue of advertisement regarding public health and sanitation at the on set of an epidemic outbreak is a classical example of:**
   (A) Principle of set
   (B) Principle of time
   (C) Principle of association
   (D) Principle of transfer
6. **When planting a crop in lines gives better yield, the farmers may be advised not to practice broadcasting:**
   (A) Principle of achievement
   (B) Principle of transfer
   (C) Principle of disassociation
   (D) Principle of association
7. **When farmers are ready to cooperate, with good guidance, they may be able to form a cooperative society:**
   (A) Principle of motivation
   (B) Principle of attitude
   (C) Principle of readiness
   (D) Principle of satisfaction
8. **When a farmer develop a favourable set towards scientific method of cultivation, they shall learn more on the topic:**
   (A) Principle of motivation
   (B) Principle of attitude
   (C) Principle of readiness
   (D) Principle of satisfaction
9. **The attainment of perfection demands that undesirable and useless movements are replaced by desirable and useful movements:**
   (A) Principle of motivation
   (B) Principle of attitude
   (C) Principle of readiness
   (D) Principle of satisfaction
10. **Favourable experience of planting trees motivate tribal farmers to collect samplings from the forest nursery:**
    (A) Principle of motivation
    (B) Principle of attitude
    (C) Principle of readiness
    (D) Principle of satisfaction

11. When insect have appeared or are likely to appear in crops, farmers shall learn about plant protection:
(A) Principle of timing
(B) Principle of adaptation
(C) Principle of maximum response
(D) Principle of satisfaction

12. When villagers assist the extension worker at different levels in implementing a new programme:
(A) Principle of cooperation
(B) Principle of satisfaction
(C) Principle of implementation
(D) Principle of interest

13. When farmers use their crop loan for the very purpose for which the loan is sanctioned:
(A) Principle of achievement
(B) Principle of completion
(C) Principle of clarity of objectives
(D) Principle of satisfaction

14. The principle under which custom religion is findout?
(A) Principle of social change
(B) Principle of cultural differences
(C) Principle of adaptability
(D) Principle of whole family

15. The principle under which children and family are included:
(A) Principle of social change
(B) Principle of cultural differences
(C) Principle of adaptability
(D) Principle of whole family

16. The basis of any extension approach is:
(A) Individual is the end
(B) Individual is a means to an end
(C) Rural development
(D) Transfer of technology

17. Extension approach in India generally follows the norms of:
(A) Cooperation (B) Participation
(C) Prescription (D) Mobilization

18. The most preferred approach in extension work is:
(A) Autocratic (B) Democratic
(C) Lessiz faire (D) None of the above

19. The interactive socio-political process motivated by desire to meet an individuals needs through collective actions on mutual benefit basis:
(A) Participation
(B) Cooperation
(C) Extension
(D) Collaboration

20. People participate by taking initiatives independent of external institutions to change systems:
(A) Self-mobilization
(B) Interactive participation
(C) Functional participation
(D) Passive participation

Practice Paper 2

1. People participation by being told what is going to happen or has already happened:
(A) Self-mobilization
(B) Interactive participation
(C) Functional participation
(D) Passive participation

2. The participatory approach of extension recognizes the importance of:
(A) Exploitation of local resources
(B) Farm planning
(C) Democratic way of problem solving
(D) Learning from farmers

3. Participatory development approach in otherwards is called:
(A) Bottom up approach
(B) Top down approach
(C) Circle centered approach
(D) Centralized approach

4. The extension approach weherein short-run change is the measure of success:
(A) Project approach
(B) Commodity specialized approach
(C) Cost sharing approach
(D) General agricultural extension approach

5. The extension approaches wherein success is measured in terms of farm people's willingness and ability to share some of the cost:

(A) Project approach
(B) Commodity specialized approach
(C) Cost sharing approach
(D) General agricultural extension approach

6. **The privatization approach wherein government grants private entity authority to provide a particular service within a specific geographical area:**
(A) Contracting (B) Fanchise
(C) Vouchers (D) Self help

7. **The privatization approach where in the public pays and private delivers service within a specific geographical area:**
(A) Contracting (B) Fanchise
(C) Vouchers (D) Self help

8. **The privatization approach most commonly used in sugarcane production sector:**
(A) Contracting (B) Fanchise
(C) Vouchers (D) Self help

9. **The extension approaches wherein measure of success is the farming people's attendance at and participation in the school's agricultural extension activities:**
(A) Project approach
(B) Commodity specialized approach
(C) Educational institution approach
(D) General agricultural extension approach

10. **Extension is fairly centralized and is oriented towards one commodity or crop and the agent has many functions:**
(A) Project approach
(B) Commodity specialized approach
(C) Educational institution approach
(D) General agricultural extension approach

11. **The extension approaches wherein the success is measured in terms of production increases of the particular crops covered by the programme:**
(A) Training and visit approach
(B) Commodity specialized approach
(C) Educational institution approach
(D) General agricultural extension approach

12. **A predetermined course of action is called:**
(A) Objective (B) Project
(C) Plan (D) Management

13. **A recognizable imitation or replica of an object whether workable or not is called:**
(A) Chart (B) Specimen
(C) Mock-up (D) Model

14. **A three dimentional recognizable imitation of an object is:**
(A) Illustration (B) Model
(C) Mock up (D) Specimen

15. **A working model is known as:**
(A) Objects (B) Model
(C) Mockup (D) Poster

16. **A visual aid has caption whose letter height is 0.5 inch. It can be visible clearly from:**
(A) 10 Feet (B) 20 Feet
(C) 40 Feet (D) 60 Feet

17. **Affective domain relates to:**
(A) Knowledge (B) Speech
(C) Attitude (D) None of these

18. **After a suitable extension teaching, the villagers attain desirable changes in his:**
(A) Behaviour (B) Health
(C) Habits (D) All of these

19. **Action research programme is considered as an example for:**
(A) Formal training
(B) Non formal training
(C) Direct training
(D) Distant training

20. **Action-reaction interdependence in communication is called:**
(A) Credibility
(B) Feed back
(C) Message treatment
(D) Fidelity

Practice Paper 3

1. **An extension worker while extending scientific knowledge to the farmers starts with:**
(A) Practical demonstration
(B) Lectures (C) Leaflets
(D) All of these

2. **An innovation has relatively high degree of complexity, its adoption will be:**
(A) Quick
(B) Fast
(C) Slow
(D) Cannot be said anything

3. **Aristotle model of communication contains how many ingredients:**
(A) Five (B) Four
(C) Three (D) Two

4. **An individual letter written by the extension worker to a farmer in connection with extension work is called:**
(A) Circular (B) News letter
(C) Personal letter (D) None of these

5. **At which level, communication is fact and problem oriented:**
(A) Explorative level
(B) Participative level
(C) Convenient level
(D) None of these

6. **Behavioral changes in human beings in extension education include:**
(A) Changes in knowledge
(B) Changes in attitude
(C) Changes in skill
(D) All of above

7. **Campaign is an:**
(A) Intensive teaching
(B) Extensive teaching
(C) Invasive teaching
(D) None of these

8. **Communication behaviour essentially consists of:**
(A) Information seeking
(B) Information processing
(C) Information dissemination
(D) All of above

9. **Folkways which deals with moral aspects are:**
(A) Taboos (B) Laws
(C) Mores (D) Norms

10. **Folkways have:**
(A) Moral validity
(B) Conventional validity
(C) Legal validity
(D) All of these

11. **Systems approach is also known as :**
(A) Contingency theory
(B) Integrated theory
(C) Modern theory
(D) Ecological theory

12. **System approach to training has been developed by:**
(A) S. Tyson (B) Alfred York
(C) B. Richard (D) Sameul Paul

13. **The amount of information that can be transmitted per unit of time is called:**
(A) Channel capacity
(B) Rate of transmission
(C) Transmission power
(D) Transmission ratio

14. **The approach to leadership which assumes that " leaders are born and not made" is:**
(A) Innate approach
(B) Trait approach
(C) Contingency approach
(D) Behavioral approach

15. **The communication of innovations:**
(A) Involves a time factor
(B) Does not involve time factor
(C) May or may not involve time factor
(D) Has no relevance

16. **The component of message is/are:**
(A) Code (B) Content
(C) Treatment (D) All of above

17. **The concept of feedback in communication is generally used to reflect:**
(A) Resource orientation
(B) Receiver orientation
(C) Process orientation
(D) All of these

18. The first line extension system mainly comprises of:
(A) ICAR and SAUs
(B) NGO
(C) Ministry of Agriculture and Cooperation
(D) Ministry of Rural Development

19. The social network model ideally suited or imparting knowledge change:
(A) Star (B) Wheel
(C) Circle (D) Slash

20. There are how many basic social processes:
(A) 5 (B) 4
(C) 3 (D) 2

Practice Paper 4

1. Agricultural extension makes available to the farming community to solve the problems of agriculture including:
(A) Animal Husbandry
(B) Gardening
(C) Agricultural Engineering
(D) All of these

2. Agricultural extension is concerned with agricultural education assisting farmers to bring about continuous improvement in their:
(A) Physical
(B) Economic
(C) Social
(D) Physical, economic and social

3. Extension education centres around:
(A) Farmers or rural people
(B) Innovation or inventions
(C) Extension workers
(D) All of these

4. Extension education aims at:
(A) Increasing agricultural production
(B) Increasing milk production
(C) changing knowledge, skills and attitudes of the participants
(D) Changing administrative system

5. The extension approaches wherein success is measured by the extent to which farming people adopt the technologies developed by the programme and continue using them over time:
(A) Commodity specialized approach
(B) General agricultural extension approach
(C) Cost sharing approach
(D) Farming system development approach

6. The farming system development approach essentially aims at:
(A) Development of technology which fits the need
(B) Development of farmer's innovation
(C) Increasing farm production and productivity
(D) Integrated development of the farm

7. "How do I help people achieve what they themselves want to achieve":
(A) "DO TO" strategy
(B) "DO FOR" strategy
(C) "DO WITH" strategy
(D) "DO AGAIST" strategy

8. The communication strategy embedded in the transfer of technology paradigm:
(A) Persuasive + Paternalistic
(B) Persuasive + Participatory
(C) Educational + Paternalistic
(D) Educational + Participatory

9. The strategy that often lead to the so called progressive farmer approach in agricultural extension:
(A) "DO TO" strategy
(B) "DO FOR" strategy
(C) "DO WITH" strategy
(D) "DO AGAIST" strategy

10. "How do I develop an offering which my target clients want" relates with:
(A) "DO TO" strategy
(B) "DO FOR" strategy
(C) "DO WITH" strategy
(D) "DO AGAIST" strategy

11. The most important factor in FSR/E is:
(A) Farmer
(B) Researcher
(C) Extensionist
(D) All of the above

12. Which of the followings is not a factor contributing to the limited application is FSR/E:

(A) Action based approach
(B) Multidisciplinary approach
(C) Resource poor clientale
(D) Limited research output

13. The extension approach wherein success is measured in terms of the rate of take-up of the recommendations, and increases in national production:

(A) Project approach
(B) Commodity specialized approach
(C) Cost sharing approach
(D) General agricultural extension approach

14. Under subsistent farming situation agricultural development is characterized by:

(A) Commercial orientation
(B) Subsidy orientation
(C) Information orientation
(D) Companion-mission orientation

15. Which among the following is a bottom up approach:

(A) Farming system approach
(B) Commodity specialized approach
(C) Training and visit approach
(D) Cost sharing approach

16. .... is a statement of situation, objectives, problems and solutions

(A) A plan of work
(B) Extension programme
(C) A calendar of work
(D) A project

17. .... is a plan of work arranged chronologically.

(A) A project
(B) A calendar of work
(C) A plan of work
(D) Programme planning

18. A sound extension programme building based on

(A) Analysis of the facts in the situation
(B) Select problems based on needs
(C) Determine objective and solutions
(D) All of these

19. First step for making a programme planning includes

(A) Evaluation
(B) Analysis of the situation and determining problems
(C) Deciding on objectives
(D) Teaching

20. The main objective of evaluation in extension is

(A) To facilitate effective decision making directly goes to conclusion
(B) To facilitate effective decision making without jumping to conclusion
(C) To see weak points in programme
(D) None of these

Practice Paper 5

1. When an extension programme is successfully completed and the farmer and his family members are satisfied by solving the problem and getting a new idea, it is called:

(A) The principle of achievement
(B) The principles of completion
(C) The principles of maximum response
(D) The principles of satisfaction

2. The strategy for supporting the rapidly expanding agricultural programme requires:

(A) That cultivators are educated in favour of modern farming
(B) That efforts are made to manufacture more fertilizers
(C) That efforts are made to provide more irrigational facilities
(D) All of the above

3. A good extension programme should be:

(A) Rigid (B) Flexible
(C) Both 'A' and 'B' (D) None of these

4. Determination of extension programme consists of:

(A) Definition and finding solutions
(B) Preparing plan of work and carry out the plan of work

(C) Collection and analysis of data's determination of needs and objectives
(D) Both 'A' and 'B'

5. **ATIC has been initiated with:**
(A) Single window concept of technology dissemination
(B) Diversification of agricultural technology
(C) Intensification of technologies
(D) Intensification and diversification of technologies

6. **PARA extension workers are considered to be:**
(A) The workers working with NGOs
(B) Rural youths trained to take up small extension works on payment basis
(C) Extension workers having specialization in rural areas and enterprises
(D) Workers with knowledge of all the enterprises attached to farm house hold

7. **Which are the main reasons for the limited success of T & V in India:**
(A) Heavy infrastructure investment
(B) Assumptions at different levels
(C) Top to down approach
(D) All the above

8. **Participating extension management envisages:**
(A) Participation of all the stakeholders in the project
(B) Participation of farmers in the project
(C) Technology transfer by extension agencies
(D) None of these

9. **What is "Cyber extension":**
(A) Extension through group approach
(B) Extension through use of television and other mass media
(C) Extension through computers
(D) Extension through NGOs

10. **Gender development in extension programmes is related to:**
(A) Empowering farm women
(B) Development of both genders on equal basis
(C) Development of households with emphasis on gender
(D) None of these

11. **People's participate in an extension programme is significant when:**
(A) Local leaders participate
(B) Literate section of villagers participate
(C) Gram panchayat members participate
(D) Majority of villagers participate

12 **.................. is that of working with rural people through out of school education along with their current interest and needs for getting overall development of rural families:**
(A) Extension education
(B) Extension process (C) Extension jobs
(D) Extension service

13. **The extension programme is a statement of:**
(A) Situation and objectives
(B) Solution and problems
(C) Problems and situation
(D) All of the above

14. **Who, what, when and where are part of:**
(A) Plan of work
(B) Calendar of work
(C) Programme evaluation
(D) Decision making

15. **The plan of work when arranged chronologically is:**
(A) Calendar of work
(B) Calendar of operation
(C) Objective
(D) Programme

16. **A sound programme must have:**
(A) Permanence with rigidity
(B) Permanence with flexibility
(C) Temporarity with rigidity
(D) Temporarity with flexibility

17. **Extension programme planning is a:**
(A) Educational process
(B) Administrative process
(C) Effective process
(D) Uniform process

18. **The first step in extension programme planning at local level is:**
(A) Determination of objectives
(B) Analysis of existing situation
(C) Decision of extension teaching method
(D) Determination of target

19. The participation of people in extension programme is:

(A) Compulsory (B) Voluntary
(C) Involuntary (D) None of these

20. The pilot study in extension programme represent:

(A) Post evaluation
(B) Concurrent evaluation
(C) Pre evaluation
(D) Monitoring

Practice Paper 6

1. Last stage of programme planning process is:

(A) Evaluation (B) Auditing
(C) Supervision (D) Reconsideration

2. Under T & V system what is mainly responsible to make regular contact with farmers?

(A) Deputy Director (Extension)
(B) Kisan Sahayak/Sewak
(C) SDAEO
(D) BDO

3. The extension work must be based on needs and interests of:

(A) People (B) Leader
(C) Officers (D) Government

4. Need based and skill oriented vocational training to practicing farmers and employed extension personnel is provided through:

(A) Training and visit
(B) Krishi Vigyan Kendra
(C) National Demonstration
(D) Lab to Land programme

5. Operational flood programme relate to:

(A) Conserving soil
(B) Controlling floods
(C) Boosting milk supply
(D) Funds for flood victims

6. The first step in extension programme planning is:

(A) Analysis of situation
(B) Fixing objectives
(C) Collection of facts
(D) Determination of goals

7. The best method to convince people is:

(A) Video
(B) Demonstration
(C) Classroom teaching
(D) Group discussion

8. Lab to Land Programme was started by:

(A) Government of India
(B) Government of UP
(C) Smt. Indira Gandhi
(D) ICAR

9. The components of programme planning are:

(A) Simple (B) Complex
(C) Easy (D) None of these

10. Components of programme planning is/are:

(A) People their needs
(B) Useful technology
(C) The educational process
(D) Analysing situation
(E) All of these

11. Principles of programme planning is/are:

(A) It is based on an analysis f the facts of the situation
(B) Based on the needs and interests of the local people
(C) It determines priorities, define objectives and solution which differ satisfaction
(D) All of these

12. The concept of T and V system was evolved in 1973 by:

(A) Daniel Benor (B) Albert Mayer
(C) F. L. Bryne (D) Arthur T. Mosher

13. T and V system firstly evolved in:

(A) France (B) Turkey
(C) Nigeria (D) USA

14. T and V system introduced in India in:

(A) 1973 (B) 1974
(C) 1075 (D) 1976

15. T and V system is introduced in Indian in 1974 in:

(A) Rajasthan canal area
(B) Chambal command area in MP
(C) Both (A) and (B)
(D) None of these

16. The basic objective of T and V system is/ are:

(A) To improve the effectiveness of agricultural extension
(B) Transfer of technical literacy increases agricultural production
(C) Transfer of technical literacy increases agricultural production
(D) To provide training to youth for agricultural and home development

17. Training and visit programme also called:

(A) Intensive agricultural Extension
(B) Sarvodaya programme
(C) Indian Village Scheme
(D) Shatineketan programme

18. The credit for the success of KVK is goes to:

(A) Dr. MS. Swaminathan
(B) Dr. Vergees Kurien
(C) Dr. Chandrika Prasad
(D) Dr. B.P. Pal

19. 4H-Club was established in:

(A) 1900 (B) 1910
(C) 1920 (D 1930

20. 4 H – Club was established in:

(A) Russia (B) Canada
(C) America (D) Japan

Practice Paper 7

1. Basic aim of 4H-Club is:

(A) To improve training at the kendras or out of the centre
(B) Transfer of technical literacy increases agricultural production
(C) To provide training to youth for agricultural and home development
(D) To improve the effectiveness of agricultural extension

2. 4H-Club means Heart, Head, Hand and Health were explained by:

(A) Dr. MS. Swaminathan
(B) Dr. Vergees Kurien
(C) Dr. Chandrika Prasad
(D) A.B. Graham

3. The main objective of KVK is:

(A) To improve training at the kendras or out of the centre
(B) Transfer of technical literacy increases agricultural production
(C) To provide training to youth for agricultural and home development
(D) To improve the effectiveness of agricultural extension

4. First KVK was started at Pandicherry in 1974 with the aim of:

(A) To improve training at the kendras or out of the centre
(B) Transfer of technical literacy increases agricultural production
(C) To provide training to youth for agricultural and home development
(D) To improve the effectiveness of agricultural extension

5. A five minute radio talk has:

(A) 400 words (B) 300 words
(C) 600 words (D) 800 words

6. A new story technology is can be published in:

(A) Extension bulletin (B) Folder
(C) Newspaper (D) Sports magazine

7. A poster has letter height of 22 inches. It can be clearly viewed up to the distance of:

(A) 10 feet (B) 20 feet
(C) 60 feet (D) 80 feet

8. A predetermined course of action is called:

(A) Objective (B) Project
(C) Plan (D) Management

9. A recognizable imitation or replica of an object whether workable or not is called:

(A) Chart (B) Specimen
(C) Mock-up (D) Model

10. A three dimentional recognizable imitation of an object is:

(A) Illustration (B) Model
(C) Mock up (D) Specimen

11. A working model is known as:

(A) Objects (B) Model
(C) Mockup (D) Poster

12. A visual aid has caption whose letter height is 0.5 inch. It can be visible clearly from:

(A) 10 Feet (B) 20 Feet
(C) 40 Feet (D) 60 Feet

13. Affective domain relates to:

(A) Knowledge (B) Speech
(C) Attitude (D) None of these

14. After a suitable extension teaching, the villagers attain desirable changes in his:

(A) Behaviour (B) Health
(C) Habits (D) All of these

15. Action research programme is considered as an example for:

(A) Formal training
(B) Non formal training
(C) Direct training
(D) Distant training

16. Action-reaction interdependence in communication is called:

(A) Credibility
(B) Feed back
(C) Message treatment
(D) Fidelity

17. An extension worker while extending scientific knowledge to the farmers starts with:

(A) Practical demonstration
(B) Lectures
(C) Leaflets
(D) All of these

18. An innovation has relatively high degree of complexity, its adoption will be:

(A) Quick
(B) Fast
(C) Slow
(D) Cannot be said anything

19. Aristotle model of communication contains how many ingredients:

(A) Five (B) Four
(C) Three (D) Two

20. An individual letter written by the extension worker to a farmer in connection with extension work is called:

(A) Circular (B) News letter
(C) Personal letter (D) None of these

Practice Paper 8

1. At which level, communication is fact and problem oriented:

(A) Explorative level
(B) Participative level
(C) Convenient level
(D) None of these

2. Behavioral changes in human beings in extension education include:

(A) Changes in knowledge
(B) Changes in attitude
(C) Changes in skill
(D) All of above

3. Campaign is an:

(A) Intensive teaching
(B) Extensive teaching
(C) Invasive teaching
(D) None of these

4. Communication behaviour essentially consists of:

(A) Information seeking
(B) Information processing
(C) Information dissemination
(D) All of above

5. Folkways which deals with moral aspects are:

(A) Taboos (B) Laws
(C) Mores (D) Norms

6. Folkways have:

(A) Moral validity
(B) Conventional validity
(C) Legal validity
(D) All of these

7. Systems approach is also known as :

(A) Contingency theory
(B) Integrated theory
(C) Modern theory
(D) Ecological theory

8. System approach to training has been developed by:

(A) S. Tyson
(B) Alfred York
(C) B. Richard
(D) Sameul Paul

9. The amount of information that can be transmitted per unit of time is called:

(A) Channel capacity
(B) Rate of transmission
(C) Transmission power
(D) Transmission ratio

10. The approach to leadership which assumes that "leaders are born and not made" is:

(A) Innate approach
(B) Trait approach
(C) Contingency approach
(D) Behavioral approach

11. The communication of innovations:

(A) Involves a time factor
(B) Does not involve time factor
(C) May or may not involve time factor
(D) Has no relevance

12. The component of message is/are:

(A) Code (B) Content
(C) Treatment (D) All of above

13. The concept of feedback in communication is generally used to reflect:

(A) Resource orientation
(B) Receiver orientation
(C) Process orientation
(D) All of these

14. The first line extension system mainly comprises of:

(A) ICAR and SAUs
(B) NGO
(C) Ministry of Agriculture and Cooperation
(D) Ministry of Rural Development

15. The social network model ideally suited or imparting knowledge change:

(A) Star (B) Wheel
(C) Circle (D) Slash

16. There are how many basic social processes:

(A) 5 (B) 4
(C) 3 (D) 2

17. Haryali yojna is related to:

(A) Food grain production
(B) Fertilizer production
(C) Irrigation
(D) Grass management

18. Which one of the following is the correct match?

(A) Etawah pilot project - Albert Mayer (1948)
(B) Grow more food campaign - J.L. Nehru (1947)
(C) Co-operative movement - Nicholas (1904)
(D) Sewagram project - M.K. Gandhi (1927)

19. Which of the following rural development project was launched before independence?

(A) Etawah pilot project
(B) Co-operative movement
(C) Community development
(D) National Extension service

20. Steps in teaching is shortly called as:

(A) AIETA (B) AIDCAS
(C) KPDC (D) AIDACS

Practice Paper 9

1. Effective teaching is:

(A) For informing people
(B) For transforming people
(C) Like filling a bucket
(D) Done by trial and error

2. The basic element of the core of Extension Education is:

(A) Men himself
(B) Man and environment
(C) Man created devices
(D) All of these

3. Principle of extension education is:

(A) Learning by doing
(B) Learning by seeing
(C) Learning by reading
(D) Learning by hearing

4. Goal of extension education is:

(A) To promote income of farmers
(B) To promote production of the crops
(C) To promote new crops
(D) To promote scientific out look

5. **Purpose of extension evaluation is:**
   (A) To identify the weak points
   (B) To identify the strong points
   (C) To identify the gaps and errors
   (D) To identify all the above

6. **The fundamental objective of extension education is:**
   (A) To provide the farmer knowledge and help that will enable him to increase his income
   (B) To encourage the farmer to grow own food, eat well and live well
   (C) The development of the people
   (D) None of these

7. **Extension education is:**
   (A) A discipline (B) A profession
   (C) Discipline and profession both
   (D) None

8. **The objectives of education are decided by:**
   (A) Social sciences
   (B) Physical sciences
   (C) Normative sciences
   (D) Biological sciences

9. **Less intensive and less effective method of extension is:**
   (A) Individual (B) Group
   (C) Mass (D) Both A & B

10. **The goal of extension education is to promote:**
   (A) Income of the farmers
   (B) Area of the new crop
   (C) Scientific outlook
   (D) Production of the crops

11. **Extension education is different from formal education in a sense that:**
   (A) We start from practical and end up with theory
   (B) We start from theory and end up with practical
   (C) We teach practical only
   (D) We teach theory only

12. **Extension education aims at:**
   (A) Increasing agricultural production
   (B) Increasing Milk production
   (C) Changing knowledge, skills and attitudes
   (D) Changing administrative system

13. **In extension education communication is a:**
   (A) Process
   (B) Method
   (C) Idea
   (D) Expression of thoughts

14. **To induce emotional participation of the local community and to create a conductive psychological climate for adoption of a new practice, the method of extension generally applied is:**
   (A) Competition (B) Exhibitions
   (C) Fair (D) Field trails

15. **The method of extension education for effectively impressing the people about a particular recommended practice suitable for a particular area is:**
   (A) Radio (B) Tours
   (C) Charts (D) Demonstrations

16. **The objective of education are decided by:**
   (A) Biological science (B) Physical science
   (C) Normative science (D) Social science

17. **The extension of scientific knowledge to villagers by specialists of the same area is known as the principle of:**
   (A) Panchayat Raj (B) Local leadership
   (C) Modernization (D) None of them

18. **In extension we prefer:**
   (A) Laiser fair approach
   (B) Democrative approach
   (C) Automatic approach
   (D) None of the above

19. **The two important principles of extension education are:**
   (A) Participation and leadership
   (B) Participation and class room teaching
   (C) Leadership and classroom teaching
   (D) Leadership and subject matter specialist

20. **Which of the following extension programmes linked farm plan with credit for farmers?**
   (A) Integrated Rural Development Programme
   (B) Community Development Programme
   (C) Intensive Agricultural Development Programme
   (D) Etawah Pilot Project

Practice Paper 10

1. **Which one of the following categories the discipline of extension education belong?**
   (A) Pure science
   (B) Applied biological science
   (C) Applied behaviour science
   (D) Art

2. **An extension worker practically explains Japanese method of paddy cultivation to a group of farmers and also show its superiority over the traditional method. In the process, which of the following communication methods have been used by them?**
   (1) Method demonstration
   (2) Result demonstration
   (3) Field visit
   (4) Study tour
   Select the correct answer using the codes given below:
   (A) 1, 2, 3 and 4 (B) 1 and 3
   (C) 2 and 3 (D) 1, 2 and 4

3. **Consider the following statements:**
   Extension in the long run means change in:
   (1) Knowledge (2) Skill
   (3) Work (4) Attitude
   Of these statements:
   (A) 1, 2 and 3 are correct
   (B) 1, 3 and 4 are correct
   (C) 1, 2 and 4 are correct
   (D) 2, 3 and 4 are correct

4. **Extension workers should be well versed with economic principles:**
   (A) Never
   (B) Partially
   (C) Both 'A' and 'B'
   (D) Without it they can't give effective advice from business point of view

5. **Extension education focuses upon development of:**
   (A) An individual (B) A family
   (C) A social group (D) A community

6. **Aim of extension work is to bring about the changes:**
   (A) Toward betterment
   (B) In the heart of peoples
   (C) In rural areas
   (D) None of these

7. **The objective of TRYSEM is to provide technical skills to the Rural youth from families below the poverty line. The target group comprises rural youth between the age of:**
   (A) 20 and 25 (B) 18 and 35
   (C) 14 and 35 (D) 14 and 3

8. **An extension education job is:**
   (A) To educate the people
   (B) To advice the Urban people
   (C) To advice the Rural people
   (D) All of the above

9. **Extension work must be based on:**
   (A) The interest of Gram Pradhan
   (B) The interest of BDO
   (C) The needs and interest of the people
   (D) None of the above

10. **In India, the development of poorers from all the sides is possible only by:**
    (A) More money supply
    (B) Making strong rules
    (C) Extension education
    (D) All of the above

11. **In India the development of poorer sections from all the sides possible only by:**
    (A) Extension education
    (B) Making strong rules
    (C) More money supply
    (D) All of the above

12. **The basic element at the core of extension education is:**
    (A) Man itself
    (B) Man and Environment
    (C) Institutions
    (D) Farmers

13. **Extension education is:**
    (A) Formal education
    (B) Informal education
    (C) Non formal education
    (D) All of the above

14. The basic objective of extension education is to create opportunities for:

(A) Greater awareness
(B) Effective learning
(C) Promoting knowledge
(D) Feeling aloof

15. The basic logic applicable in extension education:

(A) Horizontal logic (B) Deductive logic
(C) Inductive logic (D) Lateral logic

16. Education is a process of:

(A) Interaction
(B) Learning
(C) Socialization
(D) Institutionalization

17. Subject and target group specificity are characteristics feature of:

(A) Formal education
(B) Informal education
(C) Non formal education
(D) Elementary education

18. The objectives of education are decided by:

(A) Biological sciences
(B) Physical science
(C) Normative science
(D) Social science

19. The last stage in the extension education process is:

(A) Evaluation
(B) Reconsideration
(C) Adoption
(D) Teaching

20. Education is essentially considered as production of desirable change in:

(A) Knowledge
(B) Attitude
(C) Practice
(D) Behaviour

Answers Key

Practice Paper 1

Q. No.	Ans	Q. No.	Ans	Q. No.	Ans	Q. No.	Ans	Q. No.	Ans	Q. No.	Ans
1.	C	2.	A	3.	C	4.	B	5.	B	6.	C
7.	C	8.	B	9.	C	10.	A	11.	A	12.	A
13.	C	14.	A	15.	D	16.	B	17.	C	18.	B
19.	A	20.	A								

Practice Paper 2

Q. No.	Ans	Q. No.	Ans	Q. No.	Ans	Q. No.	Ans	Q. No.	Ans	Q. No.	Ans
1.	D	2.	D	3.	A	4.	A	5.	C	6.	B
7.	C	8.	A	9.	C	10.	B	11.	A	12.	C
13.	D	14.	B	15.	C	16.	A	17.	C	18.	A
19.	B	20.	B								

Practice Paper 3

Q. No.	Ans	Q. No.	Ans	Q. No.	Ans	Q. No.	Ans	Q. No.	Ans	Q. No.	Ans
1.	B	2.	C	3.	C	4.	C	5.	A	6.	D
7.	A	8.	D	9.	C	10.	B	11.	C	12.	B
13.	A	14.	B	15.	A	16.	D	17.	A	18.	A
19.	A	20.	A								

Practice Paper 4

Q. No.	Ans	Q. No.	Ans	Q. No.	Ans	Q. No.	Ans	Q. No.	Ans	Q. No.	Ans
1.	D	2.	D	3.	D	4.	C	5.	D	6.	A
7.	A	8.	A	9.	A	10.	B	11.	A	12.	C
13.	D	14.	D	15.	A	16.	B	17.	B	18.	D
19.	B	20.	B								

Practice Paper 5

Q. No.	Ans	Q. No.	Ans	Q. No.	Ans	Q. No.	Ans	Q. No.	Ans	Q. No.	Ans
1.	D	2.	D	3.	B	4.	C	5.	A	6.	B
7.	D	8.	A	9.	C	10.	B	11.	D	12.	B
13.	D	14.	A	15.	A	16.	B	17.	A	18.	B
19.	B	20.	C								

■ Practice Paper 6

Q. No.	Ans	Q. No.	Ans	Q. No.	Ans	Q. No.	Ans	Q. No.	Ans	Q. No.	Ans
1.	D	2.	B	3.	A	4.	B	5.	C	6.	C
7.	B	8.	D	9.	B	10.	E	11.	D	12.	A
13.	B	14.	B	15.	C	16.	A	17.	A	18.	C
19.	A	20.	C								

■ Practice Paper 7

Q. No.	Ans	Q. No.	Ans	Q. No.	Ans	Q. No.	Ans	Q. No.	Ans	Q. No.	Ans
1.	B	2.	D	3.	A	4.	B	5.	C	6.	C
7.	C	8.	C	9.	D	10.	B	11.	C	12.	A
13.	C	14.	A	15.	B	16.	B	17.	B	18.	C
19.	C	20.	C								

■ Practice Paper 8

Q. No.	Ans	Q. No.	Ans	Q. No.	Ans	Q. No.	Ans	Q. No.	Ans	Q. No.	Ans
1.	A	2.	D	3.	A	4.	D	5.	C	6.	B
7.	C	8.	B	9.	A	10.	B	11.	A	12.	D
13.	A	14.	A	15.	A	16.	A	17.	C	18.	A
19.	B	20.	B								

■ Practice Paper 9

Q. No.	Ans	Q. No.	Ans	Q. No.	Ans	Q. No.	Ans	Q. No.	Ans	Q. No.	Ans
1.	B	2.	A	3.	A	4.	A	5.	D	6.	C
7.	C	8.	C	9.	C	10.	C	11.	A	12.	C
13.	D	14.	B	15.	D	16.	C	17.	B	18.	B
19.	A	20.	A								

■ Practice Paper 10

Q. No.	Ans	Q. No.	Ans	Q. No.	Ans	Q. No.	Ans	Q. No.	Ans	Q. No.	Ans
1.	C	2.	C	3.	C	4.	D	5.	B	6.	A
7.	B	8.	A	9.	C	10.	C	11.	A	12.	A
13.	C	14.	B	15.	C	16.	C	17.	A	18.	C
19.	B	20.	D								

3. Multiple Choice Questions on Communication and Communication Technology

Practice Paper 1

1. **Linear model of communication was given by:**
(A) Barlo (B) Weaver
(C) Sanane (D) Leagans

2. **Real object taken out of their natural settings is called as:**
(A) Objects (B) Specimens
(C) Model (D) Mock-up

3. **Sound can be recorded in .... ways:**
(A) 1 (B) 2
(C) 3 (D) 4

4. **The disk recording is made by the .... process:**
(A) Mechanical (B) Magnetic
(C) Optical (D) None of these

5. **Tape and wire recording are made by .... process:**
(A) Magnetic (B) Mechanical
(C) Optical (D) None of these

6. **The movie film recording is made by .... process:**
(A) Magnetic (B) Mechanical
(C) Optical (D) None of these

7. **Which one of the following is not a example of still pictures:**
(A) Drawing (B) Painting
(C) Etching (D) None of these

8. **Flash cards are a series of illustrated cards when flashed or presented before a group in proper sequence tell a:**
(A) Part of story (B) Complete story
(C) Incomplete story (D) None of these

9. **A B C of posters means:**
(A) Attractive, Brief and Complete
(B) Attractive, Brief and Clear
(C) Attractive, Brief and Creative
(D) None of these

10. **............... consist of a series of individual charts which are bound together and hung on a supporting stand:**
(A) Bar charts (B) Job charts
(C) Flip charts (D) Tree charts

11. **Communication is more meaningly:**
(A) Extension approach
(B) Subject of broadcasting
(C) Exchange of ideas, facts, feelings etc
(D) Transference of meaning

12. **Which of the following scientist has given the theory of information of communication?**
(A) Shannon and Weaver
(B) Leagon
(C) Berlo
(D) Aristotle

13. **Which one is not the part of communication module?**
(A) Communicator (B) Message
(C) Advice (D) Audience

14. **The number of elements in the communication model of Berlo:**
(A) 4 elements (B) 5 elements
(C) 6 elements (D) 7 elements

15. **The smallest meaningful unit of speech is called:**
(A) Heuristic (B) Productive
(C) Morphemes (D) Phonemes

16. **The communication model better applicable for a person reading newspaper:**
(A) Shannon-Weaver model
(B) Leagan's model
(C) Berlo's model
(D) Barker Wiseman model

17. **The basic elements in Aristotle's model of communication is:**
(A) Speaker (B) Speech
(C) Listener or Audience
(D) All of these

18. The S-M-C-R model of communication is given by:
(A) Berlo (B) Laswell
(C) Rogers (D) Leagans

19. The influence of mass media according to hypodermic model of communication is:
(A) Direct (B) Indirect
(C) Secondary (D) No effect

20. Communication IQ was studied by:
(A) Bhanja (B) Paul Vegans
(C) Hovland (D) None of these

Practice Paper 2

1. Communication competency concept was stressed by:
(A) Wiemann and Backlund
(B) Hymes
(C) Sannob and Weaver
(D) Thayer

2. Source-transmitter-signal-receiver-destination is the model of communication given by:
(A) Aristotle
(B) Shannon and Weaver
(C) Laswel
(D) Westley and Machean

3. The first basic persuasive communication model:
(A) Aristotle
(B) Shannon and Weaver
(C) Laswell
(D) Berlo

4. The communication model that matches well with political symbolism:
(A) Aristotle
(B) Shannon and Weaver
(C) Laswell
(D) Berlo

5. Mathematical theory of communication was put forwarded by:
(A) Aristotle
(B) Shannon and Weaver
(C) Laswell
(D) Berlo

6. The amount of information that can be transmitted per unit of time:
(A) Redundancy (B) Channel capacity
(C) Coupling (D) Rhetoric

7. A communication model developed based on behavioral research:
(A) EM Rogers (B) Lazerfield
(C) D Berlo (D) Stevens

8. Which of the following extension 4 sources of communication do you consider most credible for non progressive/ remote village situation:
(A) Radio (B) Demonstration
(C) Print media (D) Exhibition

9. Which one is not the mode of radio broadcast?
(A) Straight talks (B) Interviews
(C) Songs (D) Lecture

10. Which is the traditional media of communication?
(A) Print (B) Tape recorder
(C) Radio (D) Ram leela

11. Which of the following is a traditional and simple aid of communication?
(A) Model (B) Puppet
(C) Radio (D) TV

12. Attitude of communication source is important from the point of view of:
(A) Code, content, treatment
(B) Source, self, receiver
(C) Encode, decode, feedback
(D) Self, audience, content

13. The degree to which individuals involved in a communication act differ in certain characteristics:
(A) Homophily (B) Credibility
(C) Empathy (D) Heterophily

14. The communication skills that get emphasized on listening and reading:
(A) Attitude (B) Effect
(C) Encoding (D) Decoding

15. Speaking and writing involve which communication skills:
(A) Encoding (B) Decoding
(C) Feedback (D) Knowledge

16. The process of selecting information that is relevant to a problem while ignoring distractions:

(A) Selective encoding
(B) Selective decoding
(C) Selective comparison
(D) None of these

17. Communication behaviour essentially consist of:

(A) Information seeking
(B) Information processing
(C) Information disseminating
(D) All of the above

18. Communication barriers include:

(A) Social (B) Semantic
(C) Technical (D) All of the above

19. The communication barrier wherein people attach different meaning to things:

(A) Semantic barrier (B) Personal barrier
(C) Technical barrier (D) None of these

20. Problem of homogeneity is the communication barrier related to:

(A) Source (B) Message
(C) Channel (D) Receiver

Practice Paper 3

1. Providing too much information is a communication barrier related to:

(A) Communicator (B) Message
(C) Treatment (D) Receiver

2. Which among the following calls for effective communication:

(A) Listening (B) Learning
(C) Training (D) Looking

3. Indirect and inferential feedback is available in:

(A) Group media (B) Mass media
(C) Audio media (D) Visual media

4. The communication type best apply to largely illiterate audience:

(A) Written communication
(B) Informal communication
(C) Oral communication
(D) Formal communication

5. Reading newspaper is:

(A) Intra personal communication
(B) Inter personal communication
(C) Mass communication
(D) Dyadic communication

6. Constant feedback between the sender and the receiver is the characteristic feature of:

(A) Inter personal communication
(B) Mass communication
(C) Written communication
(D) Vertical communication

7. The main problem with mass communication is:

(A) Costly
(B) No local dialect
(C) Same information
(D) Cumbersome

8. Mass communication is generally identified with modern mass media like:

(A) Television and radio
(B) Cinema and news paper
(C) Farm and home visit
(D) All of these

9. An informal communication type involved in the spread of runner:

(A) Horizontal communication
(B) Vertical communication
(C) Diagonal communication
(D) Grapevine communication

10. Cybernetics is the science of:

(A) Decision making
(B) Organizations
(C) Ecology
(D) Communication

11. Communication breakdown is a common feature of:

(A) Ulterior transaction
(B) Cross transaction
(C) Straight transaction
(D) Vertical transaction

12. The art of communication is:

(A) Learning
(B) Listening
(C) Feeling
(D) None of these

13. Which of the following are the elements of communication process?

(A) Source – Message – Channel – Receiver
(B) Production process – Farmers – Farm – Farm business
(C) Research – Education – Extension – Training
(D) Information – Interest – Evaluation – Persuasion

14. is a unit of two or more people in reciprocal communication and interaction with each other:

(A) Family
(B) Group
(C) Society
(D) Social stratification

15. The term community development appears to have originated from in England in 1946:

(A) London (B) France
(C) Cambridge (D) Belfast

16. Among the community, voluntary agencies are characterized by:

(A) Less intensity
(B) Greater continuity
(C) Greater commitment
(D) Less accountability

17. .... is the process by which two or more people exchange ideas, facts, feelings or impressions.

(A) Adoption (B) Teaching
(C) Learning (D) Communication

18. Leagans model of communication process is:

(A) Communication-message-channel-treatment-Audience-Audience response
(B) Sender-encoding-channel-decoding-receiver
(C) Communicationsource-Encoding-Message-Channel-Decoder-Communication receiver
(D) None of these

19. The term 'intercommunication' means:

(A) Communication on any issue
(B) Communication between two persons
(C) Communication between various groups
(D) All of the above

20. The term 'human communication' means:

(A) Exchange of service
(B) Exchange of liabilities
(C) Exchange of ideas
(D) Exchange of ideas with commodity

Practice Paper 4

1. Which one of the sources of the communication is most credible for non-progressive/remote village situation?

(A) Farm radio (B) Demonstration
(C) Printed material (D) Exhibition

2. In extension education exhibition is helpful for:

(A) Individual learning
(B) Group communication
(C) Mass communication
(D) None of the above

3. In extension education communication is a:

(A) Process
(B) Method
(C) Idea
(D) Expression of thoughts

4. The authority of PACS is known as:

(A) President (B) Secretary
(C) General body (D) Executive body

5. Action-reaction inter-dependable communication is termed as:

(A) Barrier (B) Feed back
(C) Fidelity (D) Treatment

6.

List-I	List-II
(a) Demonstration means	(1) Short type demonstration
(b) Method demonstration is	(2) Showing by doing
(c) Result	(3) Long type demonstration
(d) National Demonstration is	(4) Front line extension project

Code	a	b	c	d
(A)	2	1	3	4
(B)	3	4	2	1
(C)	4	3	2	1
(D)	1	2	3	4

7. Under limited resource of a manpower, time and money, we should select:
(A) Individual contact methods
(B) Group contact methods
(C) Mass contact methods
(D) All of the above

8. An extension method that can be considered both spoken and visual:
(A) Posters
(B) Radio
(C) Campaign
(D) Agricultural clinics

9. Flag method is classified under:
(A) Spoken method
(B) Written method
(C) Group method
(D) Mass method

10. Agricultural clinic is classified under:
(A) Group contact method
(B) Individual contact method
(C) Mass contact method
(D) All of the above

11. Educational television is classified under:
(A) Broadcast media
(B) Screen media
(C) Printed media
(D) None of the above

12. Communication is the process by which message are transferred from source to:
(A) Channel (B) Message
(C) Receiver (D) Effect

13. Which of the following is not a method of group communication?
(A) Sabarmati (B) Sevagram
(C) Shriniketan (D) Bhatinda

14. Which of the following is not a method of group communication?
(A) Conference
(B) Field trips
(C) National demonstration
(D) Farm and home visits

15. The most basic function of communication:
(A) Information (B) Instruction
(C) Integration (D) Influence

16. Communication is more meaningfully:
(A) Extension approach
(B) Subset of broadcasting
(C) Exchange of facts and feeling
(D) None of these

17. Level of communication are:
(A) Conventional (B) Exploratory
(C) Participative (D) All of the above

18. On which level, communication is fact and problem oriented:
(A) Explorative level
(B) Participative level
(C) Conventional level
(D) None of the above

19. The essence of communication is:
(A) Sharing information
(B) Imparting knowledge
(C) Sharing understanding
(D) Transmitting information

20. The art of communication is:
(A) Learning (B) Listening
(C) Feeling (D) None of these

Practice Paper 5

1. The most important element in the communication process:
(A) Source (B) Receiver
(C) Channel (D) Message

2. Action-reaction inter-dependence in communication process is:
(A) Feedback (B) Homophily
(C) Competence (D) Fidelity

3. Which of the following is not communication process:
(A) Source (B) Encoding
(C) Transmission (D) Meetings

4. Communication process is stabilized by:
(A) Feedback (B) Credibility
(C) Channel (D) Empathy

5. The type of audience suited for communication process is called:
(A) Active audience
(B) Potential audience
(C) Available audience
(D) None of these

6. The communication channels most active in diffusion process:

(A) Mass media channel
(B) Interpersonal channel
(C) Both A and B
(D) None of these

7. The communication channel most suited to inform the audience about the existence of an innovation:

(A) Mass media channel
(B) Interpersonal channel
(C) Both a and b
(D) None of these

8. Mass communication is intended to be consumed immediately hence it us called:

(A) Rapid (B) Transient
(C) General (D) None of these

9. Feedback is inferential in case of communication made through:

(A) Village Level Worker
(B) Group meeting
(C) Radio
(D) Method Demonstration

10. Mass media had direct, immediate and powerful effects on a mass audience:

(A) Two step flow model
(B) One step flow model
(C) Hypodermic needle model
(D) None of these

11. The controlling of the flow of information through a communication channel:

(A) Gate keeping (B) Gate locking
(C) Gate security (D) Gate guard

12. A system of decision centres interconnected by patterned flows of information:

(A) Communication techniques
(B) Communication network
(C) Communication level
(D) Communication channel

13. The number of communication network links possible in a social system of 100 members:

(A) 1000 (B) 10000
(C) 9876 (D) 4950

14. Two linked individuals in a network having personal communication networks that overlap:

(A) Communication proximity
(B) Communication preference
(C) Communication acceptance
(D) None of these

15. The communication network ideally suited for imparting knowledge change:

(A) Star network (B) Wheel network
(C) Circle network (D) Slash network

16. Which of the following is not a decentralized communication networks:

(A) Star (B) Chain
(C) Circle (D) Both A and B

17. ATIC under NATP functions under:

(A) State Department of Agriculture
(B) Ministry of Agriculture, GOI
(C) Ministry of RD, GOI
(D) ICAR

18. AKIS expands to:

(A) Agricultural Knowledge and Information Syndrome
(B) Agricultural Knowledge and Information Service
(C) Agricultural Knowledge and Information Science
(D) Agricultural Knowledge and Information System

19. Agricultural and rural Industries Information System Network to strengthen Small and Micro Enterprises (SMEs):

(A) AFPINET (B) ARISNET
(C) AGRISNET (D) VISTARNET

20. Land Information System network linking all institutions involved in land and water management for agricultural productivity and production systems:

(A) LISNET
(B) ARISNET
(C) AGRISNET
(D) VISTARNET

Practice Paper 6

1. **Agricultural Research Information System Network:**
(A) HORTNET (B) ARISNET
(C) AGRISNET (D) VISTARNET

2. **Agricultural Extension Information System Network:**
(A) COOPNET (B) FERTNET
(C) AGRISNET (D) VISTARNET

3. **A network that connect agricultural produce wholesale markets and rural markets:**
(A) AFPINET (B) AGMARKNET
(C) HORTNET (D) VISTARNET

4. **An infrastructure network upto block level agricultural offices facilitating agricultural extension services and agribusiness activities to user in rural prosperity:**
(A) AGRISNET (B) ARISNET
(C) HORTNET (D) VISTARNET

5. **The toll free telephone number on kisan call centre is:**
(A) 1551 (B) 1331
(C) 5115 (D) 3113

6. **When started kisan call centre:**
(A) 1970 (B) 1971
(C) 1972 (D) 1973

7. **Warna village network project was started in the state of:**
(A) Uttar Pradesh (B) Maharastra
(C) Tamilnadu (D) Madhya Pradesh

8. **The e-choupal concept in extension delivery is promoted by:**
(A) ITC (B) MSSRF
(C) KINFRA (D) CAPART

9. **"Rice Doctor" an interactive pest diagnostic expert system is developed by:**
(A) ICAR (B) IRRI
(C) ICRISAT (D) INIBAP

10. **Krishi Darshan programme in India is telecasted from:**
(A) Mumbai (B) Kolkata
(C) New Delhi (D) Bangalore

11. **Prasar Bharti came into existence on:**
(A) 15th October, 1995
(B) 18th January, 1996
(C) 20th December, 1996
(D) 23rd January, 1997

12. **The 4-H club colours are:**
(1) Green **(2) Yellow**
(3) White **(4) Red**
(5) Blue

Select the right answer using the codes given below:
(A) 1 and 2 (B) 3 and 4
(C) 1 and 3 (D) 1 and 5

13. **The 4-H club motto is:**
(A) To make the better best
(B) To make the best better
(C) To make the good better
(D) To make the good best

14. **The 4-H clubs are functioning in USA, 4 – H implies:**
(A) Head, Hand, Heart, Health
(B) Head, Home, Health, Hands
(C) Hand, Health, Hope, Heart
(D) Home, Heart, Health, Head

15. **The 4 H of communication are:**
(A) Informal, Instruction, Influence and Integration
(B) Intermediation, Information, Interest and Instruction
(C) Information, Instruction, Influence and Integration
(D) Information, Integration, Influence and Instruction

16. **The abbreviation COIK in COIK fallacy means:**
(A) Clear Only If Known
(B) Clarity Of Informational Knowledge
(C) Capacity Of Indigenous Know-how
(D) Concept of Imaging Knowledge

17. **The COIK fallacy is associated with which of the following:**
(A) Communication channels
(B) Audio visual aids
(C) Communicator
(D) Receiver

18. The 4-H club was forwarded by:

(A) O.H. Benson (B) A.B. Graham
(C) Seaman A. Knapp (D) A.B. Ross

19. The motto of 4 H club is to make:

(A) The poorer rich (B) Better best
(C) Adult motivated (D) Child educated

20. Which one is not the part of communication module:

(A) Communicator (B) Message
(C) Advice (D) Audience

Practice Paper 7

1. Round table discussion is called:

(A) Panel (B) Symposium
(C) Forum (D) Workshop

2. Intelligence Quotient determines:

(A) Capacity age (B) School going age
(C) Chronological age (D) Mental age

3. The extension work must be based on needs and interests of:

(A) People (B) Leader
(C) Worker/Official (D) Government

4. Which of the following is not the element of communication process?

(A) Message (B) Feedback
(C) Channel (D) Fidelity

5. A person receiving favourable response from other is called:

(A) Authoritative
(B) Psychological support
(C) Physical support
(D) Social support

6. The most conspicuous development in the globalization context has been the:

(A) Information technology influx
(B) Communication technology influx
(C) Information and communication technology influx
(D) None of these

7. Developing world is gradually catching up with the technological advancement to solve its:

(A) Sociological problem
(B) Economical problem
(C) Both (A) and (B) (D) None of these

8. The set of activities that facilitates, by electronic means the capturing, storage, processing, transmission and display of information is:

(A) Communication
(B) Information and communication technology
(C) Agricultural extension
(D) Telecommunication technology

9. The objective of Information and Communication Technology is/are:

(A) To elucidate the role of some of the major ICT initiatives in rural development
(B) To study the impact of ICT efforts on agriculture, health, women empowerment and panchayat raj
(C) To print out the impediments for the effective implementation of ICT based project
(D) All of these

10. ICT tackle key problems in mainly:

(A) Agriculture and cooperation sector
(B) Health and education
(C) Panchayat raj and women empowerment
(D) All of these

11. Which programme is designed to assist their development and develop their information, education and communication base of rural people is:

(A) PRA (B) ICT
(C) AV AIDS (D) None of these

12. ICT programme came into existence in India is:

(A) 1982 (B) 1983
(C) 1984 (D) 1985

13. ICT programme was given a vent in 1984 by the:

(A) Congress Govt. (B) BJP Govt
(C) Janta Party Govt. (D) BSP Govt

14. ICT programme was given a vent in 1984 by Rajiv Gandhi and the technologist:

(A) R. S. Paroda (B) Sam Pitroda
(C) Mangla Rai (D) B.P. Pal

15. A high power national task force on information and software development was set up in:

(A) 1996 (B) 1997
(C) 1998 (D) 1999

16. A high power national task force o information technology and software development was set up in 1998 by the:

(A) Congress Govt. (B) BJP Govt
(C) Janta Party Govt. (D) BSP Govt

17. The national task force in information technology prepared the blue print for making the adoption of IT or a national movement by establishing a wide network of empowered task force at:

(A) Government level
(B) Nongovernment level
(C) Both (A) and (B)
(D) None of these

18. The new telecommunication policy was enacted in :

(A) 1999 (B) 2000
(C) 2001 (D) 2002

19. The new telecommunication Policy 1999 was stipulated targets in terms of establishing a telecom network with a view to achieving teledensity by 2005:

(A) 5 per cent (B) 7 per cent
(C) 10 per cent (D) 15 per cent

20. The new telecommunication Policy 1999 was stipulated targets in terms of establishing a telecom network with a view to achieving teledensity by 2010:

(A) 5 per cent (B) 7 per cent
(C) 10 per cent (D) 15 per cent

Practice Paper 8

1. The concept of knowledge society means that it will incorporate new knowledge will take over the role of capital and labour in the economy further leading to:

(A) Structural changes
(B) Functional changes
(C) Both (A) and B)
(D) None of these

2. Sponsoring agency of Cyber Extension Project is:

(A) MANAGE, Hyderabad
(B) ICAR, New Delhi
(C) NAARM, Hyderabad
(D) CSAUA & T, Kanpur

3. Sponsoring agency of Agriculture Technology Information Centre is:

(A) MANAGE, Hyderabad
(B) ICAR, New Delhi
(C) NAARM, Hyderabad
(D) CSAUA&T, Kanpur

4. Sponsoring agency of Agricultural Gateway of India is:

(A) MANAGE, Hyderabad
(B) ICAR, New Delhi
(C) NAARM, Hyderabad
(D) CSAUA&T, Kanpur

5. Sponsoring agency of Helpline Service is:

(A) MANAGE, Hyderabad
(B) ICAR, New Delhi
(C) NAARM, Hyderabad
(D) CSAUA&T, Kanpur

6. Sponsoring agency of Warna Wired Village Project is:

(A) NIC
(B) Warna Cooperative Complex, Warn district Kolhapur (MH)
(C) Both (A) and (B)
(D) CSAUA&T, Kanpur

7. Sponsoring agency of Gyandoot Project is:

(A) MANAGE, Hyderabad
(B) ICAR, New Delhi
(C) CSAUA&T, Kanpur
(D) Gram Panchayat Dhar Dist. MP

8. Sponsoring agency of Interactive Touch Screen Information , KIOSK is:

(A) Dept. of International Development, Rajiv Gandhi College of Animal Sciences (Pondicherry)
(B) University of Reading (UK)
(C) Both (A) and (B)
(D) CSAUA&T, Kanpur

9. **Sponsoring agency of Lokmitra Project is:**
(A) National Information Centre
(B) Gram Panchayat Hamirpur dist HP
(C) Both (A) and (B)
(D) CSAUA&T, Kanpur

10. **Sponsoring agency of E-choupal is:**
(A) Imperial Tobacco Company
(B) EID Parry (Tamilnadu)
(C) Nagarjuna Group
(D) MS Swaminathan Research Foundation, Chennai

11. **Sponsoring agency of Rural Information Kiosks is:**
(A) Imperial Tobacco Company
(B) EID Parry (Tamilnadu)
(C) Nagarjuna Group
(D) MS Swaminathan Research Foundation, Chennai

12. **Sponsoring agency of I-Kisan Portal is:**
(A) Imperial Tobacco Company
(B) EID Parry (Tamilnadu)
(C) Nagarjuna Group
(D) MS Swaminathan Research Foundation, Chennai

13. **Sponsoring agency of Village knowledge Centre is:**
(A) Imperial Tobacco Company
(B) EID Parry (Tamilnadu)
(C) Nagarjuna Group
(D) MS Swaminathan Research Foundation, Chennai

14. **Sponsoring agency of Computer of Wheels is:**
(A) Imperial Tobacco Company
(B) Pingali Rajeswari Enterpreneur, A.P.
(C) Nagarjuna Group
(D) MS Swaminathan Research Foundation, Chennai

15. **Sponsoring agency of www.dctoranywhere.com is:**
(A) Imperial Tobacco Company
(B) EID Parry (Tamilnadu)
(C) Tata Council for Community Initiatives (TCCI), Pune (MS)
(D) MS Swaminathan Research Foundation, Chennai

16. **Important features of Cyber Extension are:**
(A) Village Information Centres
(B) Institutional support to other ICT projects
(C) Both (A) and (B)
(D) None of these

17. **Important features of Agriculture Technology Information Centre is:**
(A) Call centre
(B) Expert system
(C) Researchers and farmers linkage
(D) All of these

18. **Important features of Agricultural Gateway of India are:**
(A) Helpline service through telephones
(B) Researchers-farmers linkage
(C) Both (A) and (B)
(D) None of these

19. **Important features of Helpline Service are:**
(A) Helpline service through telephones
(B) Researchers-farmers linkage
(C) Both (A) and (B)
(D) None of these

20. **Important features of Warna Wired Village Project are:**
(A) Wide Area Network
(B) Computer booths
(C) Both (A) and (B)
(D) None of these

Practice Paper 9

1. **Important features of Gyandoot Project are:**
(A) Soochanalayas
(B) Portal partial recovery
(C) Panchayat community partnership
(D) All of these

2. **Important features of Interactive Touch Screen Information Kiosk are:**
(A) Information Kiosks text
(B) Picture with sound backup
(C) Both (A) and (B)
(D) None of these

3. **Important features of Lokmitra Project are:**
   (A) Lokmitra Soochana Kendra in 25 panchayat
   (B) Panchayat Community partnership
   (C) Researcher-farmers linkage
   (D) None of these

4. **Important features of E-choupal are:**
   (A) Choupal
   (B) Crop specific internet
   (C) Both (A) and (B)
   (D) None of these

5. **Important features of Rural Information Kiosks are:**
   (A) Framers managed Kiosks portal
   (B) Public private partnership in applying ICT improvations and content development
   (C) Both (A) and (B)
   (D) None of these

6. **Important features of I-kisan portal are:**
   (A) Choupal
   (B) Portal in regional language
   (C) Both (A) and (B)
   (D) None of these

7. **Important features of Village Knowledge Centre are:**
   (A) Knowledge centre
   (B) Pro-poor
   (C) Pronature prowomen
   (D) Community ownership
   (E) All of these

8. **Important features of computer on wheels are:**
   (A) Internet on motor bikes
   (B) Portals
   (C) Remote area
   (D) Resource poor
   (E) All of these

9. **Important features of www.doctor anywhere.com are:**
   (A) Telemedicine Service through websites
   (B) Telemedicine Service through computer
   (C) Both (A) and (B)
   (D) None of these

10. **Major thrust or areas or level of cyber extension project are:**
    (A) Post harvest processing
    (B) Women and child welfare
    (C) Both (A) and (B)
    (D) None of these

11. **Major thrust or areas or level of Agricultural Technology Information Centre project are:**
    (A) Agronomic practices
    (B) Technologies inputs
    (C) Both (A) and (B)
    (D) None of these

12. **Major thrust or areas or level of Agricultural Gateway of India are:**
    (A) Cultivation practices
    (B) Plant protection
    (C) New technologies
    (D) All of these

13. **Major thrust or areas or level of Helpline service are:**
    (A) Cultivation practices
    (B) Plant protection
    (C) New technologies
    (D) All of these

14. **Major thrust or areas or level of Warna Wired Village Project are:**
    (A) Cultivation practices
    (B) Pest ad diseases
    (C) Marketing
    (D) All of these

15. **Major thrust or areas or level of Gyandoot Project are:**
    (A) Agricultural produce
    (B) Market intelligence
    (C) Auction rates
    (D) Land records
    (E) All of these

16. **Major thrust or areas or level of Interactive Touch Screen Information Kiosk are:**
    (A) Information about important cattle disease
    (B) Management of cattle
    (C) Both (A) and (B)
    (D) None of these

17. Major thrust or areas or level of Lokmitra Project are:
(A) Classified components
(B) Job vacancies
(C) Market rates
(D) Tenders
(E) All of these

18. Major thrust or areas or level of E-choupal are:
(A) Market price
(B) Cultivation practices
(C) Weather data
(D) All of these

19. Major thrust or areas or level of Rural Information kiosks are:
(A) Farm practices
(B) Farm advisory
(C) Market intelligence
(D) Weather
(E) All of these

20. Major thrust or areas or level of I - Kisan portal are:
(A) Agricultural practices
(B) Plant protection
(C) Animal husbandry
(D) Weather
(E) All of these

Practice Paper 10

1. Major thrust or areas or level of Village Knowledge Centre are:
(A) Agronomic practices
(B) Cattle feed
(C) Weather
(D) All of these

2. Major thrust or areas or level of Computer of wheels are:
(A) Market (B) Weather
(C) Plant protection (D) Animal health
(E) All of these

3. Major thrust or areas or level of www.doctoraywhere.com are:
(A) Medical council
(B) Medical treatment in PHC
(C) Both (A) and (B)
(D) None of these

4. Which ICT project is related to cattle diseases and management of cattle is:
(A) Interactive Touch Screen Information kiosks
(B) E-choupal
(C) Lokmitra project
(D) Gyandoot project

5. Which ICT project is related to medical treatment in PHC is:
(A) Interactive Touch Screen Information kiosks
(B) www.doctoranywhere.com
(C) Lokmitra project
(D) Gyandoot project

6. E-choupal launched by ITC in:
(A) June, 1998 (B) June, 1999
(C) June, 2000 (D) June, 2001

7. ITC's E - choupal is a unique example of using ICT for:
(A) Agricultural development
(B) Animal management
(C) Human health improvement
(D) All of these

8. Which ITC's has already become the largest initiatives among all internet based interventions in rural India is:
(A) E-choupal (B) I-kisan portal
(C) Lokmitra project (D) Gyandoot project

9. E-choupal services render in which crops:
(A) Soybean and Coffee
(B) Wheat and rice
(C) Pulses
(D) All of these

10. E - choupal covers across the:
(A) MP and Karnataka
(B) AP and UP
(C) Maharashtra and Rajasthan
(D) All of these

11. Disseminating various farming information to the farmers at various places through ICT centres by:
(A) I- Kisan portal
(B) E–choupal
(C) Gyandoot project
(D) Lokmitra project

12. **Which ICT's provide strong effective coverage and dissemination of livestock out reach information to the farming community, e-mail conferencing system are:**
    (A) goat-net-cirg.nic.in
    (B) livestock-net@cirg.nic.in
    (C) fishnet@cirg.nic.in
    (D) All of these

13. **Under the animal health project funded by DFID Rajiv Gandhi College of Veterinary and Animal Science, Pondicherry in collaboration with University of Reading, UK has installed a:**
    (A) Interactive Touch Screen Information, Kiosk
    (B) Agricultural Technology Information System
    (C) Warna Wired Village Project
    (D) Agricultural Gateway of India

14. **Specific features of Interactive Touch Screen Information, Kiosk are:**
    (A) Cattle diseases
    (B) Management of cattle
    (C) Method of acquiring information
    (D) All of these

15. **Apollo hospital started telemedicine project on:**
    (A) 1999 (B) 2000
    (C) 2001 (D) 2002

16. **The ISRO and EDCU has initiated several telemedicine project for human health in:**
    (A) Remote areas (B) Rural areas
    (D) Urban areas (D) Tribal areas

17. **Sankara Nethralaya is a super-speciality eye hospital in Chennai launched a telemedicine project through satellite connectivity is:**
    (A) goat-net-cirg.nic.in
    (B) livestock-net@cirg.nic.in
    (C) fishnet@cirg.nic.in
    (D) Tele-ophthalmology

18. **www.doctoranywhere.com is a telemedicine service project with aim that headquarters connected with the PHC's with atleast how many doctors always present at headquarter to give advice and medicine to the rural people:**
    (A) Only one doctor
    (B) Team of two doctors
    (C) Team of 5-6 doctors
    (D) Team of many doctors

19. **Lokmitra project launched in:**
    (A) May, 2000 (B) May, 2001
    (C) May, 2002 (D) May, 2003

20. **For Panchayat raj ICT project are:**
    (A) Lokmitra
    (B) Gyandoot and Saytu
    (C) E-Gram Suvidha
    (D) All of these

Answers Key

Practice Paper 1

Q. No.	Ans	Q. No.	Ans	Q. No.	Ans	Q. No.	Ans	Q. No.	Ans	Q. No.	Ans
1.	D	2.	B	3.	C	4.	A	5.	A	6.	C
7.	D	8.	B	9.	B	10.	C	11.	C	12.	A
13.	C	14.	C	15.	C	16.	D	17.	D	18.	A
19.	A	20.	A								

Practice Paper 2

Q. No.	Ans	Q. No.	Ans	Q. No.	Ans	Q. No.	Ans	Q. No.	Ans	Q. No.	Ans
1.	B	2.	B	3.	A	4.	C	5.	B	6.	A
7.	C	8.	A	9.	D	10.	D	11.	B	12.	D
13.	D	14.	D	15.	A	16.	A	17.	D	18.	D
19.	A	20.	D								

Practice Paper 3

Q. No.	Ans	Q. No.	Ans	Q. No.	Ans	Q. No.	Ans	Q. No.	Ans	Q. No.	Ans
1.	A	2.	B	3.	B	4.	C	5.	A	6.	A
7.	B	8.	A	9.	D	10.	D	11.	B	12.	B
13.	A	14.	B	15.	C	16.	C	17.	D	18.	A
19.	D	20.	C								

Practice Paper 4

Q. No.	Ans	Q. No.	Ans	Q. No.	Ans	Q. No.	Ans	Q. No.	Ans	Q. No.	Ans
1.	B	2.	C	3.	D	4.	B	5.	B	6.	B
7.	C	8.	C	9.	A	10.	B	11.	A	12.	C
13.	C	14.	A	15.	D	16.	C	17.	D	18.	A
19.	C	20.	B								

Practice Paper 5

Q. No.	Ans	Q. No.	Ans	Q. No.	Ans	Q. No.	Ans	Q. No.	Ans	Q. No.	Ans
1.	B	2.	A	3.	D	4.	A	5.	A	6.	C
7.	A	8.	B	9.	B	10.	C	11.	A	12.	B
13.	D	14.	A	15.	A	16.	C	17.	D	18.	D
19.	B	20.	A								

■ Practice Paper 6

Q. No.	Ans	Q. No.	Ans	Q. No.	Ans	Q. No.	Ans	Q. No.	Ans	Q. No.	Ans
1.	B	2.	D	3.	B	4.	A	5.	A	6.	C
7.	B	8.	A	9.	B	10.	C	11.	D	12.	C
13.	B	14.	A	15.	C	16.	A	17.	B	18.	A
19.	B	20.	C								

■ Practice Paper 7

Q. No.	Ans	Q. No.	Ans	Q. No.	Ans	Q. No.	Ans	Q. No.	Ans	Q. No.	Ans
1.	A	2.	D	3.	A	4.	D	5.	C	6.	C
7.	C	8.	B	9.	D	10.	D	11.	B	12.	C
13.	A	14.	B	15.	C	16.	B	17.	C	18.	A
19.	B	20.	D								

■ Practice Paper 8

Q. No.	Ans	Q. No.	Ans	Q. No.	Ans	Q. No.	Ans	Q. No.	Ans	Q. No.	Ans
1.	C	2.	A	3.	B	4.	C	5.	D	6.	C
7.	D	8.	C	9.	C	10.	A	11.	B	12.	C
13.	D	14.	B	15.	C	16.	C	17.	D	18.	C
19.	C	20.	C								

■ Practice Paper 9

Q. No.	Ans	Q. No.	Ans	Q. No.	Ans	Q. No.	Ans	Q. No.	Ans	Q. No.	Ans
1.	D	2.	C	3.	A	4.	C	5.	C	6.	C
7.	E	8.	E	9.	C	10.	C	11.	D	12.	D
13.	D	14.	D	15.	E	16.	C	17.	E	18.	D
19.	E	20.	E								

■ Practice Paper 10

Q. No.	Ans	Q. No.	Ans	Q. No.	Ans	Q. No.	Ans	Q. No.	Ans	Q. No.	Ans
1.	D	2.	E	3.	C	4.	A	5.	B	6.	C
7.	A	8.	A	9.	D	10.	D	11.	A	12.	D
13.	A	14.	D	15.	B	16.	A	17.	D	18.	C
19.	B	20.	D								

4. Multiple Choice Questions on Innovation, Evaluation and Adoption

Practice Paper 1

1. **These are how many degree of evaluation:**
   (A) 4 (B) 3
   (C) 5 (D) 2
2. **Diffusion is the subset of:**
   (A) Communication (B) Teaching
   (C) Learning (D) Motivation
3. **Diffusion has:**
   (A) 3 elements (B) 4 elements
   (C) 6 elements (D) 8 elements
4. **Diffusion and adoption of a new technology are respectively decided by:**
   (A) Society and society
   (B) Individual and Individual
   (C) Individual and Society
   (D) Society and Individual
5. **First step for making a programme planning includes:**
   (A) Evaluation
   (B) Analysis of the situation and determining problems
   (C) Deciding on objectives
   (D) Teaching
6. **At which of the stage of adoption group discussion is most effective:**
   (A) Awareness (B) Evaluation
   (C) Testing (D) Persuasion stage
7. **The correct sequence of innovation - decision process or adoption process is:**
   (A) Interest – awareness – evaluation – trial – adoption
   (B) Evaluation – awareness — interest - trial – adoption
   (C) Awareness -interest — evaluation – adoption - trial
   (D) Awareness – interest – evaluation – trial – adoption
8. **............. are the first farmers to adopt a new idea:**
   (A) Innovators (B) Early adopters
   (C) Early majority (D) Laggards
9. **................. are the last to adopt new idea:**
   (A) Innovators (B) Early adopters
   (C) Laggards (D) Early majority
10. **Which one of the following pair is not correctly matched:**

	Adopter Category	Percentage
(A)	Innovators	2.5
(B)	Early adopters	13.5
(C)	Early majority	40
(D)	Laggards	16

11. **Normal rate of adoption requires how many years from the introduction of the innovation to its adoption throughout the community:**
    (A) 2 – 6 (B) 6 – 10
    (C) 10 – 14 (D) 14 – 18
12. **In case of diffusion of innovation, the overt behavior means:**
    (A) Adoption or rejection
    (B) Remembering or forgetting
    (C) Fooding or lodging
    (D) Mental weighing or field trial
13. **At the adoption stage, information sources is:**
    (A) Village level change agent and to some extent mass media
    (B) Extension agency and other farmers
    (C) Informal personal sources including family members
    (D) Self-experience gained at the trial stage
14. **Adopter category which respect and regarded by many others in the social system as a role model are:**
    (A) Innovators (B) Early adopters
    (C) Late majority (D) Laggards

15. The spread of technology in a social system:

(A) Diffusion (B) Teaching
(C) Learning (D) Adoption

16. Diffusion is essentially concerned with messages that are:

(A) Profitable (B) New ideas
(C) Vital (D) Widely known

17. The elements of diffusion includes all except:

(A) Innovation (B) Social system
(C) Adoption (D) Time

18. The oldest research tradition in diffusion research is from:

(A) Communication (B) Sociology
(C) Education (D) Anthropology

19. The major drawback of past diffusion studies was their inability to take up:

(A) Source research
(B) Channel research
(C) Message research
(D) Audience research

20. Largest number of diffusion studies are conducted in the field of:

(A) Communication (B) Sociology
(C) Education (D) Anthropology

Practice Paper 2

1. The model of diffusion as proposed by Rogers assumes:

(A) The homogeneity of population
(B) The progressiveness of farmers
(C) Availability of infrastructural facilities
(D) The differential farming situation

2. Diffusion and adoption of a new technology are respectively decided by:

(A) Society and society
(B) Individual and Individual
(C) Individual and Society
(D) Society and Individual

3. Diffusion can be considered as a subject of:

(A) Communication (B) Teaching
(C) Learning (D) Motivation

4. The degree to which the results of an innovation may be diffused to others:

(A) Credibility
(B) Compatibility
(C) Communicability
(D) Relative advantage

5. Homophilous diffusion pattern with in a system helps in the spread of ideas:

(A) Horizontally (B) Vertically
(C) Diagonally (D) None of these

6. Which of the following is the first in motivating the farmers:

(A) Food (B) Money
(C) Farm literature (D) All of these

7. CONA expands to:

(A) Career Oriented Need Assessment
(B) Community Oriented Need Assessment
(C) Creative Organizational Need Assessment
(D) Career Oriented Need Analysis

8. The idea which is perceived as new is termed as:

(A) Information (B) Innovation
(C) Perception (D) Invention

9. An innovation may have in general:

(A) One component (B) Two component
(C) Three component (D) Four component

10. The material aspect of an innovation is:

(A) Idea component
(B) Object component
(C) Product component
(D) Relative component

11. Symbolic adoption take place in case of innovation with:

(A) Idea component
(B) Object component
(C) Product component
(D) Relative component

12. To be called as an innovation, an idea, practice or object must be:

(A) Objectively new (B) Subjectively new
(C) All the above (D) None of these

13. The characteristic of an innovation important from extension point of view:

(A) Intrinsic
(B) As people see them
(C) Extraordinary (D) Typological

14. **An innovation can be considered to have originated from:**
(A) Farmer research
(B) Extension programme
(C) Research trials
(D) All of these

15. **An innovation include all except:**
(A) Material products
(B) Ideological belief
(C) Social movement
(D) Traditional technologies

16. **An innovation is liked or disliked based on:**
(A) Innovation effects
(B) Innovation decision
(C) Perceived attributes
(D) Traditional makeup

17. **The perceived newness of the innovation can be expressed in terms of:**
(A) Knowledge
(B) Decision to adopt
(C) Persuasion
(D) All of the above

18. **The perceived attribute of an innovation that reduces the risk component:**
(A) Complexability
(B) Compatibility
(C) Trialability
(D) Absorbability

19. **Providing subsidy to farmers on introduction of farm machinery is an example for:**
(A) Relative advantage
(B) Complexity
(C) Compatibility
(D) Observability

20. **An innovation with low relative advantage may have:**
(A) Slow rate of adoption
(B) High rate of discontinuance
(C) Low return on investment
(D) All of the above

Practice Paper 3

1. **Compatibility is the degree to which an innovation is perceived consistent with:**
(A) Clients needs for innovation
(B) Previously introduced needs
(C) Socio-cultural values and benefits
(D) All of these

2. **The luxuriant growth of the plant on the application of urea relates:**
(A) Compatibility (B) Trialability
(C) Observability (D) Complexity

3. **'Preventive innovations' have less adoption rate mainly due to:**
(A) Low relative advantage
(B) High complexity
(C) Low triability
(D) Low observability

4. **Providing free seed samples to the farmers is closely associated with:**
(A) Relative advantage
(B) Complexity (C) Triability
(D) Observability

5. **The perceived attribute of an innovation that shows a negative correlation with adoption:**
(A) Complexability (B) Compatibility
(C) Triability (D) Observability

6. **A compactable innovation offers the adopter:**
(A) Greater risk (B) Less security
(C) All of these (D) None of these

7. **The attributes of the innovation as matters by whom is most important:**
(A) Research scholars (B) Manufactures
(C) Potential adopters (D) Change agents

8. **Time is very important consideration in the process of:**
(A) Communication (D) Adoption
(C) Diffusion (D) Innovation

9. **Individual adopters in a social system is described in terms of his:**
(A) Economic conditions
(B) Time of adoption
(C) Frequency of adoption
(D) Concurrency ability

10. Frequency of adoptors when plotted over time follows a:
(A) Normal curve (B) Skewed curve
(C) Linear curve (D) Curvilinear shape

11. The rate of adoption of an innovation in a social system over time follows:
(A) Normal curve (B) Skewed curve
(C) Linear curve (D) S- shaped curve

12. The S shape adoption curve increases at an increasing rate for:
(A) Innovators (B) Early adopters
(C) Early majority (D) All of the above

13. The length of time required to pass through the innovation decision process:
(A) Innovation decision matrix
(B) Innovation decision period
(C) Innovation decision index
(D) Innovation decision time

14. Innovation decision period is actually a:
(A) Gestation period
(B) Production period
(C) Mental period
(D) None of these

15. The innovation decision process can be shortly called as:
(A) KPDIC (B) AIDTCA
(C) NAIDTEA (D) AIETA

16. The chief concern of farmer at the persuasion stage is:
(A) How to information
(B) How do information
(C) How much information
(D) None of these

17. The individual become more psychologically involved with the innovation at:
(A) Knowledge stage
(B) Persuasion stage
(C) Decision stage
(D) Confirmation stage

18. The outcome of persuasion leads to subsequent change in:
(A) Covert behaviour
(B) Overt behaviour
(C) Significant behaviour
(D) None of these

19. Persuasion function of innovation decision process is mainly:
(A) Cognitive
(B) Affective
(C) Behavioural
(D) Normative

20. The reinforcement of innovation decision making occurs at:
(A) Persuasion stage
(B) Knowledge stage
(C) Confirmation stage
(D) Decision stage

Practice Paper 4

1. Conflicting information should be strictly avoided during:
(A) Readiness stage
(B) Knowledge stage
(C) Decision stage
(D) Confirmation stage

2. The decision to adopt or reject made after the prior innovation decision is called:
(A) Optimal decision
(B) Contingent decision
(C) Collective decision
(D) Authority decision

3. The basic unit of decision making in the optimal innovation decision model is:
(A) Change agent
(B) Social system
(C) Researcher
(D) Individual

4. The type of innovation decisions that resulted in the emergence of classical diffusion model:
(A) Optimal innovation decisions
(B) Collective innovation decisions
(C) Authority innovation decisions
(D) None of these

5. Generally the fastest rate of adoption of innovation results from:
(A) Optimal innovation decisions
(B) Collective innovation decisions
(C) Authority innovation decisions
(D) Contingent innovation decisions

6. **The unit of decision making and unit of adoption are different in case of:**
   (A) Optimal decision
   (B) Collective decision
   (C) Authority decision
   (D) Mental decision

7. **In collective decision making who screens the idea for approving or puts sanction on it:**
   (A) Stimulators (B) Initiators
   (C) Actors (D) Legitimizers

8. **Which of the following is not an element in the model of innovation decision process:**
   (A) Antecedents (B) Variability
   (C) Consequences (D) Processes

9. **Dissonance does not occur at which of the stages of innovation decision process:**
   (A) Knowledge (B) Persuasion
   (C) Decision (D) Confirmation

10. **The degree to which an innovation is changed or modifies by a user in the process of its adoption:**
   (A) Re-invention (B) Re-engineering
   (C) Re-modeling (D) Re-making

11. **An agricultural innovator is a:**
   (A) Traditional leader
   (B) Opinion leader
   (C) Functional leader
   (D) Formal leader

12. **The decision to make use of a innovation as best course of action:**
   (A) Adoption (B) Diffusion
   (C) Innovation (D) Technology

13. **Adoption is basically a:**
   (A) Mental process
   (B) Social process
   (C) Psychological process
   (D) Physical process

14. **The scientist first to recognize adoption of innovation occurs in stages:**
   (A) Wilkening
   (B) Rogers and Shoemaker
   (C) Johnson and Haver
   (D) Ryan and Gross

15. **The stages of adoption are dynamic and not static according to:**
   (A) Ryan and Gross (B) Wilkening
   (C) Johnson (D) Y P Singh

16. **Who for the first time reported that adoption process involves decision:**
   (A) Pederson (B) Wilkening
   (C) Holmberg (D) Mason

17. **The first stage in the process of adoption according to Rogers is:**
   (A) Attention (B) Interest
   (C) Desire (D) Satisfaction

18. **The stage wherein the individual makes full use of the innovation:**
   (A) Evaluation (B) Adoption
   (C) Awareness (D) Trial

19. **The stage wherein the individual decides to continue the full use of the innovation:**
   (A) Evaluation (B) Adoption
   (C) Awareness (D) Trial

20. **The stage wherein an individual is exposed to the innovation but lacks complete information about it:**
   (A) Attention (B) Interest
   (C) Desire (D) Satisfaction

Practice Paper 5

1. **The most important source of information at the attention stage of adoption:**
   (A) Mass media
   (B) Group media
   (C) Extension worker
   (D) Personal localities

2. **The stage of adoption wherein group discussion is most effective:**
   (A) Awareness (B) Evaluation
   (C) Testing (D) Persuasion

3. **Unfreezing existing behaviour and motivating people to change is expressed at:**
   (A) Desire (B) Interest
   (C) Conviction (D) Attention

4. The extension activity that need to be emphasized in the interest stages of adoption:
(A) Limited trial is good
(B) Rapport building is needed
(C) Group training is needed
(D) Specialists are needed

5. Symbolic adoption is an example for:
(A) Covert behaviour (B) Overt behaviour
(C) Affection (D) Perception

6. Adopter categories are the classification of the members of the social system on the basis of:
(A) Innovativeness
(B) Adoption process
(C) Diffusion of innovation
(D) Innovation decision period

7. The time of adoption of an innovation by all the members in a social system:
(A) Always same (B) Always vary
(C) May vary (D) All of the above

8. The classification of adopters into various categories is based on:
(A) Mean and variance
(B) Mean and standard deviation
(C) Mean and median
(D) Median and standard deviation

9. The adopter category considered as deviant by other members of the society:
(A) Innovators (B) Early majority
(C) Early adopters (D) Laggards

10. The adopter category considered as friends and neighbours by other members in the social system:
(A) Innovators (B) Early majority
(C) Early adopters (D) Laggards

11. The adopter category considered by many as the men to "check with":
(A) Innovators (B) Early majority
(C) Early adopters (D) Laggards

12. Opinion leaders belong to the category of:
(A) Innovators (B) Early majority
(C) Early adopters (D) Late majority

13. The point of reference for laggards is:
(A) Past (B) Present
(C) Future (D) All of the above

14. The adopter categories who is also considered as the custodians of indigenous knowledge:
(A) Innovators (B) Early majority
(C) Early adopters (D) Laggards

15. The person who adopt an innovation before every average member of the society does it:
(A) Innovators (B) Early majority
(C) Early adopters (D) Late adopters

16. Agricultural innovators are generally viewed as:
(A) Opinion leader (B) Social deviants
(C) Skeptical (D) Traditional

17. 'Legitimization' is done by:
(A) Innovators (B) Early adopters
(C) Early majority (D) Late majority

18. Who among the following is more concerned about triability of the innovation:
(A) Early adopters (B) Early majority
(C) Innovators (D) Laggards

19. "Be not the last to lay the old aside nor be the first by which the new is tried" is the motto of:
(A) Innovators (B) Early majority
(C) Early adopters (D) Laggards

20. The percentage of population constituted by the innovators:
(A) 2.5 per cent (B) 7.5 per cent
(C) 16.5 per cent (D) 33 per cent

Practice Paper 6

1. The character best represent an innovator:
(A) Venturesome (B) Skeptical
(C) Traditional (D) Respectfulness

2. The adoption of innovation rather vigourously when experts feel that it should be stopped is called:
(A) Excessive adoption
(B) Over adoption
(C) Vigorous adoption
(D) Overt adoption

3. **A type of rejection which consists of considering adoption of the innovation but then deciding not to adopt it:**
(A) Active rejection
(B) Passive rejection
(C) Pro-active rejection
(D) None of these

4. **Who incorporates the innovations into a specific plan of action adopted to needs of the system:**
(A) Stimulators (B) Initiators
(C) Actors (D) Legitimizers

5. **'Complementary adoption' is usually associated with:**
(A) Agricultural adoption
(B) Technological adoption
(C) Industrial adoption
(D) Both (B) and (C)

6. **The phenomena wherein adoption of one innovation requires the previous adoption of another innovation:**
(A) Interaction effect (B) Diffusion effect
(C) Technology effect (D) Adoption effect

7. **The individuals in the same adopter category have:**
(A) Varied degree of innovativeness
(B) Multitude of innovativeness
(C) Similar degree of innovativeness
(D) None of these

8. **Consequences are the changes that occur is a social system as a result of:**
(A) Adoption of innovation
(B) Rejection of innovation
(C) Both A and B
(D) None of these

9. **Discontinuance occurs only after an innovation has been:**
(A) Fully adopted
(B) Partially adopted
(C) Not at all adopted
(D) None of the above

10. **A decision to reject an idea as a result of dissatisfaction with it's performance:**
(A) Disenchantment discontinuance
(B) Replacement discontinuance
(C) Betterment discontinuance
(D) Distinguishable discontinuance

11. **The decision to reject an idea in order to adopt a better idea:**
(A) Disenchantment discontinuance
(B) Replacement discontinuance
(C) Betterment discontinuance
(D) Distinguishable discontinuance

12. **The dysfunctional effect of an innovation to an individual or to a social system:**
(A) Direct consequence
(B) Indirect consequence
(C) Desirable consequence
(D) Undesirable consequence

13. **The change to an individual or a social system that occur in immediate response to an innovation:**
(A) Direct consequence
(B) Indirect consequence
(C) Desirable consequence
(D) Undesirable consequence

14. **The changes from an innovation that is neither intended nor recognized by the members of a social system:**
(A) Direct consequence
(B) Indirect consequence
(C) Desirable consequence
(D) Unanticipated consequence

15. **Acceptance and rejection are the one of the kinds of:**
(A) Failure of channel
(B) Feed back
(C) Treatment of message
(D) Empathetic feelings

16. **Adoption is a/an:**
(A) Social process
(B) Mental process
(C) Psychological process
(D) Individual process

17. **Serialise the following stages of adoption process:**
(1) Interest (2) Evaluation
(3) Trial (4) Awareness

Pick out the answer from the following
(A) 1, 2, 3, 4 (B) 3, 1, 2, 4
(C) 4, 1, 3, 2 (D) 3, 2, 1, 4

18. Purpose of extension evaluation is:

(A) To identify the weak points
(B) To identify the strong points
(C) To identify the gaps and errors
(D) To identify all the above

19. Read the following statements and mark your answer according to the code given below:

I. Extension evaluation is a planned process
II. Extension evaluation is spontaneous
III. Extension evaluation is indivisible portion of teaching
IV. Extension evaluation is beginning process

Answer code:

(A) I and III are correct
(B) II and III are correct
(C) I and IV are correct
(D) II and IV are correct

20. In case of diffusion of innovation, the overt behavior means:

(A) Adoption or rejection
(B) Remembering or forgetting
(C) Fooding or lodging
(D) Mental weighing or field trial

Practice Paper 7

1. The collection of data in PRA is done by:

(A) Local Extension agents
(B) Progressive farmers
(C) Research assistants
(D) A multidisciplinary team

2. For a study of farming system, the best PRA exercise is:

(A) Transect walk
(B) Resource mapping
(C) Time line
(D) Chapatti diagram

3. PRA is known for:

(A) People Resource Appraisal
(B) Public Resource Appraisal
(C) Participatory Rural Appraisal
(D) People Rural Agency

4. The concept of PRA was given by:

(A) Henry Ford (B) Robert Chambers
(C) Neils Roling (D) Nila Mukerjee

5. The person considered as father of PRA:

(A) Neils Rolling (B) Robert Chambers
(C) Van den Ban (D) Paul Leagans

6. The term PRA was first used in:

(A) Kenya (B) USA
(C) UK (D) Israel

7. PRA was introduced in India by:

(A) Government departments
(B) ICAR
(C) NGO's
(D) Ford foundation

8. PRA had it's origin during late:

(A) Sixties (B) Seventies
(C) Eighties (D) Nineties

9. The main innovation in PRA is:

(A) Methods (B) Behaviour
(C) Process (D) Actors

10. The role of extension personnel's in PRA is:

(A) Elective
(B) Extractive
(C) Participatory
(D) None of the above

11. Which of the following is not the characteristic of PRA:

(A) Quick (B) Rigid
(C) Interactive (D) Collective

12. The collection of data in PRA is based on:

(A) Extension workers
(B) Multi disciplinary team
(C) Research scientists
(D) Progressive farmers

13. The type of PRA used to understand broad patterns and trends in a rural society:

(A) Exploratory (B) Topical
(C) Participatory (D) Monitoring

14. Which among the following is not a principle of PRA:

(A) Optimal ignorance
(B) Setting uniformity
(C) Offsetting bias
(D) Triangulation

15. The PRA technique that shows cause effect relationship:

(A) Venn diagrams
(B) Flow diagram
(C) Social mapping
(D) Livelihood analysis

16. The PRA technique wherein rural elders narrates life histories:

(A) Time line (B) Village transits
(C) History sheets (D) Ranking

17. Social calendar is a tool most commonly used in:

(A) PRA (B) RRA
(C) LFA (D) FSR

18. Which of the following is not a tool for conducting PRA:

(A) Transect walk (B) Social map
(C) Time line analysis (D) Role play

19. The PRA technique that represent the degree of importance of individuals and institutions in decision making:

(A) Social mapping
(B) Venn diagram
(C) Semi structured interview
(D) None of the above

20. People's participation in an extension programme is significant when:

(A) Literate section off villagers participate
(B) Gram panchayat members participate
(C) Majority of villagers participate
(D) Local leader participate

Practice Paper 8

1. Consider the following statements:

Assertion (A) : Farmers posses keen interest in training programme at block headquarter

Reason (R) : Farmers enjoy free meal during training period.

Select the correct answer from the code given below

(A) A is true but R is false
(B) R is true but A is false
(C) Both A and R are correct and R is the correct explanation of A
(D) Both A and R are true but R is not the correct explanation of A

2. Participatory rural appraisal is directly related to:

(A) Soil science
(B) Entomology
(C) Agricultural economics
(D) Agricultural extension

3. Systematic process of involvement of people for the analysis of their own situations and finding out solution for improving their present condition is called:

(A) Participatory research
(B) Explorative research
(C) Ameliorative research
(D) None of these

4. An effort to introduce innovation not known to social system is called:

(A) Participatory research
(B) Explorative research
(C) Ameliorative research
(D) None of these

5. It emphasizes in the improvement of what people are doing (at present). It is an effort to make the good better (Compatible) is:

(A) Participatory research
(B) Explorative research
(C) Ameliorative research
(D) None of these

6. The characteristics of participatory research is/are:

(A) An activity carried out b group of people to understand their own situation for their own benefit
(B) It is a collective effort to bring desirable results taking into consideration the various aspects of problem in a collective manner
(C) It is an analysis that lead to action
(D) Understanding of personal realities and larger social realities it will facilitate

awareness raising about the connections between micro-aspects of their life and the real macro-aspects of society at large

(E) All of these

7. **Participatory research is a:**
(A) Cyclic process
(B) Noncyclic process
(C) Both of these
(D) None of these

8. **The limitations of the cyclic process of participatory research y be the:**
(A) More participation of practioners
(B) Least participation of practioners
(C) Either (A) or (B)
(D) None of these

9. **Steps in conducting a participatory research are:**
(A) To know the problem
(B) Solution is needed or not
(C) To know the present status and additional information needed for the problem to be tackled
(D) All of these

10. **A well known method of data collection is to carry out a:**
(A) Big story (B) Medium story
(C) Small story (D) None of these

11. **A small number of people (10-25) come together to start experiences and decide for any action through:**
(A) Individual discussion
(B) Group discussion
(C) Individual interview
(D) Group interview

12. **A well known method of data collection is to carry out a:**
(A) Small survey
(B) Big survey
(C) Reconscence survey
(D) None of these

13. **A small number of people (10-25) come together to store experiences and decide for any action through:**
(A) Individual discussion
(B) Group discussion
(C) Individual interview
(D) Group interview

14. **Sometimes the group engaged in participatory research may find itself uncomfortable with:**
(A) Written method
(B) Oral method
(C) Written and oral method both
(D) None of these

15. **Women can be engaged to draw their perceptions of the problem on:**
(A) Blackboard
(B) Paper or cloth
(C) Black and white photographs
(D) Cutting from old magazines
(E) All of these

16. **A group discussion or individual interview can be taped and then can be:**
(A) Replayed for analysis
(B) Data collection
(C) Questionnaire
(D) None of these

17. **A group of people can investigate a particular problem by using .... to known the opinions of a larger number of people:**
(A) Discussion (B) Survey
(C) Questionnaire (D) None of these

18. **The questionnaire should be single have only questions?**
(A) 4-6 (B) 6-8
(C) 8-10 (D) 10-12

19. **Each question of questionnaire should be:**
(A) Short complex and straight forward
(B) Long, simple and straight forward
(C) Short complex and straight forward
(D) Long, complex and straight forward

20. **A questionnaire of self survey should possess:**
(A) Should have few (8-10) questions
(B) Should be short simple and straight forward
(C) Should have easy in conversation
(D) All of these

Practice Paper 9

1. **Group discussion is one of the most popular and easy method of combining:**
(A) Data collection (B) Analysis
(C) Action planning (D) All of these

2. **The important consideration of group discussion is to insure the:**
(A) Uncomfortable atmosphere
(B) Comfortable atmosphere
(C) Positive atmosphere
(D) None of these

3. **Which method can easily be combined with other methods is/are:**
(A) Individual discussion
(B) Group discussion
(C) Individual interview
(D) Group interview

4. **The important characteristics of peoples camp is/are:**
(A) May be one or two days programme
(B) Common lodging ad boarding arrangement
(C) Peoples comes from longer distance
(D) All of these

5. **Which method can provide an informal and free atmosphere:**
(A) People's camp (B) Exhibition
(C) Campaign (D) Seminar

6. **In a people's camp how many peoples can participate easy:**
(A) 30-40 (B) 50-100
(C) 100-200 (D) Unlimited

7. **It is always beneficial to visit an area where a similar problem has been solved for which the group is interested is:**
(A) Demonstrations (B) Group discussion
(C) Exhibition (D) Campaign

8. **Indian folk theatre has a important role to play in the process of:**
(A) Social changes
(B) Political changes
(C) Economic changes
(D) Socio-economic changes both

9. **Folk theatre media are:**
(A) Low cost
(B) Indigenous
(C) May provide better for focusing and solving rural problems
(D) All of these

10. **Rural poor can communicate better and convince people by making their:**
(A) Own theatre (B) Own house
(C) Leased house (D) None of these

11. **Any inanimate thing animated on stage to communicate any subject is called:**
(A) Puppets (B) Drama
(C) Kisan mela (D) All of these

12. **All the instincts basic to human learning are stimulated and get satisfied through puppet show:**
(A) Communicative dramatic
(B) Artistic musical
(C) Inquisitive constructive
(D) All of these

13. **The science that deals with adult learning is:**
(A) Androgogy (B) Pedagogy
(C) Pedology (D) Edaohology

14. **The emphasis of androgogical method in comparison with pedogological method is on:**
(A) Content and Process
(B) Content
(C) Process
(D) None of the above

15. **Androgogy literally means:**
(A) Adult leading (B) Adult learning
(C) Child leading (D) Child learning

16. **Adult learning is:**
(A) Problem centered (B) Subject centered
(C) Instructor centered (D) Institute centered

17. **The central problem in adult learning is:**
(A) Age (B) Literacy
(C) Sex difference (D) Motivation

18. **Adult literacy programme can be considered as:**
(A) Formal education
(B) Informal education
(C) Non-formal education
(D) All of the above

19. **"Learning the rope" refers to:**
    (A) Training
    (B) Socialization
    (C) Both A and B
    (D) None of the above
20. **Learning is facilitated mostly by:**
    (A) Taste (B) Hearing
    (C) Touch (D) Sight

Practice Paper 10

1. **Learning is motivated by:**
   (A) Self (B) Group
   (C) Environment (D) All of these
2. **Learning is an internal process mainly controlled by:**
   (A) The learner
   (B) The teacher
   (C) The subject matter
   (D) The learning situation
3. **Learning the rope refers to:**
   (A) Training (B) Socialization
   (C) Both a and b (D) None of these
4. **Learning become the most rapid by maximizing concentration on:**
   (A) One seasons at a time
   (B) Two seasons at a time
   (C) Three seasons at a time
   (D) All the seasons simultaneously
5. **Learning is a process which continues:**
   (A) Upto college education
   (B) Upto employment
   (C) Life long
   (D) Occasionally
6. **Which is not a principle of learning:**
   (A) Learning is the growth like and continuous
   (B) Learning should be meaningful
   (C) Learning should be challenging and satisfying
   (D) Learning should not develop functional understanding of learners
7. **Which one of the following is not an element of good learning situation:**
   (A) Physical facilities (B) Extension worker
   (C) Subject matter (D) Transfer process
8. **The significant aspect of non formal learners:**
   (A) Undirectional flow of knowledge
   (B) Flexible curriculum
   (B) Hierarchically structure system
   (D) Homogenous learners
9. **Learning is a internal process mainly controlled by:**
   (A) Teacher
   (B) Learner
   (C) Subject matter
   (D) Teaching materials
10. **The process of acquiring or imparting the ability to perform a behavioral pattern through experience and practice:**
    (A) Communication (B) Learning
    (C) Perception (D) Adoption
11. **Learning become the most rapid by maximizing concentration on:**
    (A) One sense at a time
    (B) Two sense at a time
    (C) Three sense at a time
    (D) All the sense simultaneously
12. **"People learn more rapidly and permanently when the learning experience is pleasant or enjoyable" relate to:**
    (A) Law of readiness (B) Law of exercise
    (C) Law of effect (D) Law of belonging
13. **To Do, By Doing and For Doing are the fundamental objectives of:**
    (A) Learning (B) Teaching
    (C) Training (D) Perception
14. **The learning curve in teaching follows a:**
    (A) S-Shape (B) U-Shape
    (C) V-Shape (D) ∩ - Shape
15. **The experimental technique of learning emphasis on:**
    (A) Use of psychoanalysis
    (B) Use of simulation games
    (C) Use of psychiatry
    (D) Use of learner groups
16. **Progressive behaviour adaptation of an individual to a stimulus is called:**
    (A) Learning (B) Teaching
    (C) Perceiving (D) Listening

17. Learning that suddenly appears when a reward or incentives for performance is given:

(A) Discovery learning
(B) Individualized learning
(C) Rote learning
(D) Latent learning

18. The general approach in progressive mode of learning is:

(A) Authoritarian (B) Paternalistic
(C) Humanitarian (D) Democratic

19. New learning must be related to the old:

(A) Principle of readiness
(B) Principle of association
(C) Principle of practice
(D) Principle of timing

20. Life long process of learning happening through daily experience and exposure:

(A) Informal education
(B) Non formal education
(C) Formal education
(D) Professional education

Answers Key

■ Practice Paper 1

Q. No.	Ans	Q. No.	Ans	Q. No.	Ans	Q. No.	Ans	Q. No.	Ans	Q. No.	Ans
1.	C	2.	A	3.	C	4.	D	5.	B	6.	B
7.	D	8.	A	9.	C	10.	C	11.	B	12.	A
13.	D	14.	B	15.	A	16.	B	17.	C	18.	D
19.	A	20.	B								

■ Practice Paper 2

Q. No.	Ans	Q. No.	Ans	Q. No.	Ans	Q. No.	Ans	Q. No.	Ans	Q. No.	Ans
1.	A	2.	D	3.	A	4.	C	5.	A	6.	D
7.	B	8.	B	9.	B	10.	B	11.	A	12.	A
13.	B	14.	D	15.	B	16.	C	17.	D	18.	C
19.	A	20.	D								

■ Practice Paper 3

Q. No.	Ans	Q. No.	Ans	Q. No.	Ans	Q. No.	Ans	Q. No.	Ans	Q. No.	Ans
1.	D	2.	C	3.	D	4.	C	5.	A	6.	D
7.	C	8.	C	9.	B	10.	A	11.	D	12.	D
13.	B	14.	A	15.	A	16.	A	17.	B	18.	B
19.	B	20.	C								

■ Practice Paper 4

Q. No.	Ans	Q. No.	Ans	Q. No.	Ans	Q. No.	Ans	Q. No.	Ans	Q. No.	Ans
1.	D	2.	B	3.	D	4.	A	5.	C	6.	C
7.	D	8.	B	9.	B	10.	A	11.	C	12.	A
13.	A	14.	D	15.	D	16.	B	17.	A	18.	D
19.	B	20.	A								

■ Practice Paper 5

Q. No.	Ans	Q. No.	Ans	Q. No.	Ans	Q. No.	Ans	Q. No.	Ans	Q. No.	Ans
1.	A	2.	B	3.	A	4.	B	5.	A	6.	A
7.	B	8.	B	9.	A	10.	B	11.	C	12.	C
13.	A	14.	D	15.	B	16.	B	17.	B	18.	A
19.	B	20.	A								

■ Practice Paper 6

Q. No.	Ans	Q. No.	Ans	Q. No.	Ans	Q. No.	Ans	Q. No.	Ans	Q. No.	Ans
1.	A	2.	B	3.	A	4.	B	5.	A	6.	A
7.	C	8.	C	9.	A	10.	A	11.	B	12.	D
13.	A	14.	D	15.	B	16.	B	17.	C	18.	D
19.	C	20.	A								

■ Practice Paper 7

Q. No.	Ans	Q. No.	Ans	Q. No.	Ans	Q. No.	Ans	Q. No.	Ans	Q. No.	Ans
1.	D	2.	B	3.	C	4.	A	5.	B	6.	A
7.	C	8.	C	9.	B	10.	C	11.	B	12.	B
13.	C	14.	B	15.	B	16.	A	17.	A	18.	D
19.	B	20.	A								

■ Practice Paper 8

Q. No.	Ans	Q. No.	Ans	Q. No.	Ans	Q. No.	Ans	Q. No.	Ans	Q. No.	Ans
1.	D	2.	D	3.	A	4.	B	5.	C	6.	E
7.	A	8.	B	9.	D	10.	C	11.	B	12.	A
13.	B	14.	A	15.	E	16.	A	17.	C	18.	C
19.	A	20.	D								

■ Practice Paper 9

Q. No.	Ans	Q. No.	Ans	Q. No.	Ans	Q. No.	Ans	Q. No.	Ans	Q. No.	Ans
1.	D	2.	B	3.	B	4.	D	5.	A	6.	B
7.	A	8.	A	9.	D	10.	A	11.	A	12.	D
13.	A	14.	C	15.	A	16.	A	17.	D	18.	C
19.	B	20.	D								

■ Practice Paper 10

Q. No.	Ans	Q. No.	Ans	Q. No.	Ans	Q. No.	Ans	Q. No.	Ans	Q. No.	Ans
1.	A	2.	A	3.	B	4.	A	5.	C	6.	D
7.	D	8.	B	9.	B	10.	B	11.	A	12.	C
13.	A	14.	A	15.	B	16.	A	17.	D	18.	B
19.	B	20.	A								

5. Multiple Choice Questions on Participatory Rural Appraisal and Audio-Visual Aids

Practice Paper 1

1. **Groups of local people analyzing their own conditions and choosing their own means of improving them is:**
   (A) Participatory Rural Appraisal
   (B) Rapid Rural Appraisal
   (C) Both (A) and (B)
   (D) None of these

2. **The technique developed for analyzing the needs of rural community quickly, cost effectively and with little destruction to every day life is:**
   (A) Participatory Rural Appraisal
   (B) Rapid Rural Appraisal
   (C) Both (A) and (B)
   (D) None of these

3. **The techniques are a range of survey methods to be used principally in the extraction of qualitative data is:**
   (A) Participatory Rural Appraisal
   (B) Rapid Rural Appraisal
   (C) Both (A) and (B)
   (D) None of these

4. **Rapid Rural Appraisal (RRA) evolved in the :**
   (A) 1979 (B) 1980
   (C) 1981 (D) 1982

5. **PRA which began in the late:**
   (A) 1970s (B) 1975s
   (C) 1980s (D) 1990s

6. **Which method emphasized multi-disciplinary teams, careful observation, semi-structured interviewing and focus group in:**
   (A) Participatory Rural Appraisal
   (B) Rapid Rural Appraisal
   (C) Both (A) and (B)
   (D) None of these

7. **The method which involve group rather than individuals and visual representations rather than solely verbal communication in:**
   (A) Participatory Rural Appraisal
   (B) Rapid Rural Appraisal
   (C) Both (A) and (B)
   (D) None of these

8. **Key issue of Participatory Rural Appraisal is:**
   (A) To highlight the different PRA techniques available for understanding the access strengths and constraints of the rural poor
   (B) To describe the best condition for use of PRA techniques
   (C) To highlights the constraints of using PRA techniques
   (D) All of these

9. **PRA, a field based methodology which has evolved and spread in the:**
   (A) 1980s
   (B) 1985s
   (C) 1990s
   (D) 2000s

10. **The primary focus of PANR is an local natural resources from:**
    (A) Local community
    (B) Urban community
    (C) Rural community
    (D) None of these

11. **The conventional techniques help the experts to arrive at numerical value for appraising:**
    (A) Manmade resources
    (B) Artificial resources
    (C) Natural resources
    (D) None of these

12. PRA is a technique for involving people to work for their:
(A) Other development
(B) Own development
(C) Rural development
(D) Urban development

13. All principles of PRA are framed on the:
(A) NGO involvement
(B) Government involvement
(C) People's involvement
(D) None of these

14. A method of linking different survey methods is order to cross-check the information collected from each method is:
(A) Triangulation (B) Pilot experiment
(C) Action research (D) Survey design

15. Triangulation is of particular use when employing:
(A) Mapping
(B) Ranking and scoring
(C) Flow and venn diagrams
(D) All of these

16. One of the strengths of PRA is that many of the methods are visual and therefore, accessible to a:
(A) Longer group of people
(B) Small group of peoples
(C) Peoples are not required
(D) None of these

17. Participatory methods have contributed to adjusting the interview to make it more conversational, while still:
(A) Controlled
(B) Structured
(C) Controlled and structured both
(D) None of these

18. What are the characteristics of informal conversational interview:
(A) Questions emerge from the immediate context and are asked in the natural course of things
(B) There is no predetermination of question topics or wording
(C) Either (A) or (B)
(D) Both (A) and (B)

19. What are the strengths of informal conversational interview:
(A) Increases the salience and relevance of questions
(B) Interview the built on and emerge from observations
(C) The interview can be matched to individuals and circumstances
(D) All of these

20. Key interview methods in PRA system are:
(A) Informal conversational interview
(B) Interview guide approach
(C) Standardised open-ended interview
(D) Closed quantitative interviews
(E) All of these

Practice Paper 2

1. Assumptions of PRA are:
(A) The environmental problem is a common concern, the solution of which requires the active ad responsible involvement of the entire community
(B) The collective action is possible only when a stakeholders of a community develop a clear common understanding about the issue
(C) Both (A) and (B)
(D) None of these

2. Rapid Rural Appraisal is a:
(A) Extractive in nature
(B) Elicited
(C) Information owned by outsiders
(D) All of these

3. Participatory Rural Appraisal is a:
(A) Learning with local people
(B) Information owned by rural people but shared with outsiders
(C) Both (A) and (B)
(D) None of these

4. Principles of PRA is/are:
(A) Using optimal ignorance
(B) Offsetting biases
(C) Triangulation
(D) Learning from and with rural people
(E) Learning rapidly and progressively
(F) All of these

5. **Good features of PRA is/are:**
(A) Iterative (B) Innovative
(C) Interactive (D) Informal
(E) In the community (F) All of these

6. **The menu of PRA is/are:**
(A) Secondary data review
(B) Semi structured interviews
(C) Analytical games
(D) Stories ad portraits
(E) Diagrams
(F) Workshop
(G) All of these

7. **Helpers of PRA is/are:**
(A) What and Why (B) Who and How
(C) Where and When (D) All of these

8. **Important tips of PRA is/are:**
(A) Look (B) Listen
(C) Lear (D) All of these

9. **Participatory approaches are not substitutes for , but are rather a integral part of:**
(A) Long term dialogue
(B) Sustained interaction
(C) Both (A) and (B)
(D) None of these

10. **A single, brief participatory exercise with a group of local people will not lead to:**
(A) Positive and lasting change
(B) Negative and lasting change
(C) Positive and integral change
(D) Negative and integral change

11. **PRA are not a panacea of:**
(A) Qualitative surveying
(B) Quantitative surveying
(C) Either A or B
(D) Neither A or B

12. **PRA work must effectively where they are carried out over a sufficient:**
(A) Length of time (B) Money
(C) Research (D) None of these

13. **Primary importance of PRA is:**
(A) Behaviour (B) Attitudes
(C) Longevity (D) All of these

14. **Television, fairs, radios and exhibition are the media of:**
(A) Group communication
(B) Individual communication
(C) Commodity communication
(D) Mass communication

15. **Powerful audio-visual media for group communication is:**
(A) Television
(B) Film
(C) Puppet
(D) None of these

16. **Puppet can be made out of easily available and cheap material in the community like:**
(A) Paper mach and cloth
(B) Cardboard and mud
(C) Saw dust and hay
(D) All of these

17. **Puppetry in an 'INDIAN' medium of communication which attracts people of:**
(A) Young (B) Old
(C) Children (D) All age group

18. **Video tapes can be used for promoting audio-visual literacy for:**
(A) Motivation and attitudinal change
(B) Behaviour reinforcement
(C) Community participation and entertainment
(D) All of these

19. **Indian folk theatre is full of exploitative stories of kings and priests tuned into Gods, picked up from different Hindu epics like:**
(A) Ramayana
(B) Mahabharat
(C) Ramayana and Mahabharat
(D) None of these

20. **Like video, similar results even with more clarity and immediacy can be now used through:**
(A) Computer system
(B) Television system
(C) Gramophone system
(D) All of these.

Practice Paper 3

1. **What are the weaknesses of informal conversational interview are:**
   (A) Different information collected from different people with different questions
   (B) Less systematic and comprehensive
   (C) Data organization and analysis can be quite difficult
   (D) Requires maximum attention by interviewer
   (E) All of these

2. **What are important characteristics of interview guide approach:**
   (A) Topics and issues to be covered are specified in advance, in out line form
   (B) Interviewer decide sequence and working of questions in the course of the interview
   (C) Both of these
   (D) None of these

3. **What are the strengths of Interview guide approach:**
   (A) The outline increases the comprehensiveness of the data
   (B) Makes data collection systematic for each respondent
   (C) Interviews remain fairly conversational and situational
   (D) All of these

4. **What are the weaknesses of Interview guide approach:**
   (A) Important and salient topics may be inadventently omitted
   (B) Interviewer flexibility in working questions can result in incomparatively of responses
   (C) Both (A) and (B)
   (D) None of these

5. **Important characteristics of Standardised open-ended interview:**
   (A) The exact wording and sequence of questions are determined in advance
   (B) Al interviewees are asked the same basic questions in the same order
   (C) Both (A) and (B)
   (D) None of these

6. **Important strengths of standardized open-ended interview:**
   (A) Respondents answer the same questions, thus increasing comparability of response
   (B) Reduces interviewer bias when several interviewers are used
   (C) Facilitates organization and analyses of data
   (D) All of these

7. **What are the weakness of standardized open - ended interview:**
   (A) Little flexibility in rilchig the interview to particular individuals and circumstances
   (B) Requires maximum attention by interviewer
   (C) Data organization and analysis can be quite difficult
   (D) All of these

8. **Important characteristics of closed quantitative interviews:**
   (A) Questions and response categories are determined in advance
   (B) Responses are fixed
   (C) Respondent chooses from among these fixed responses
   (D) All of these

9. **Important strengths of closed quantitative interviews:"**
   (A) Data analysis is simple
   (B) Responses can be directly compared and easily aggregate
   (C) Man questions can be asked in a short time
   (D) All of these

10. **Important weaknesses of closed quantitative interviews:**
    (A) Respondents must fit their experience and fallings into the researchers categories
    (B) May be perceived as impersonal, irrelevant and mechanistic
    (C) Can distort what respondents really mean or experienced
    (D) All of these

11. Participatory methods tend to make use if the:

(A) Informal conversational interview
(B) Inter view guide approach
(C) Both (A) and (B)
(D) None of these

12. Maps can be used to locate houses services and infrastructure within an area is called:

(A) Participatory mapping
(B) Social mapping
(C) Contour mapping
(D) Flow diagram

13. Flow diagram can act as a basis for discussing the relationship between:

(A) Different groups
(B) Individuals or issues
(C) Can demonstrate potential multiplier effects
(D) All of these

14. The diagrams are used to depict key institutions, organizations and individuals, and their interaction with the local community is called:

(A) Venn diagram (B) Flow diagram
(C) Both of these (D) None of these

15. In which communication system of farmers works:

(A) Knowledge, Generating and System
(B) Knowledge, Disseminating and System
(C) Knowledge, Consuming and System
(D) None of these

16. In the rural transport context, venn diagram can be used to demonstrate the interaction between:

(A) Local villagers
(B) Transport operators
(C) Local government
(D) All of these

17. The process of making the community inventory is sometimes called:

(A) Semi-structured interviewing
(B) High-structured interviewing
(C) Both of these
(D) None of these

18. The techniques widely used in recent years in most of the developmental programme:

(A) Participatory Rural Appraisal
(B) Rapid Rural Appraisal
(C) Participatory Resource Appraisal
(D) All of these

19. A process that empowers local people to change their own condition and situation:

(A) Participatory Rural Appraisal
(B) Rapid Rural Appraisal
(C) Participatory Resource Appraisal
(D) All of these

20. A way of learning for local community about their natural resources is called:

(A) Participatory Rural Appraisal
(B) Rapid Rural Appraisal
(C) Participatory Resource Appraisal
(D) Participatory Appraisal of Natural Resources

Practice Paper 4

1. The important principles which PRA is governed are:

(A) Listening and learning
(B) Offsetting biases
(C) Proper utilization of precious community time
(D) Seeking diversity
(E) Cross checking
(F) All of these

2. Components of PRA are:

(A) Attitude and Behaviour
(B) Process of participation
(C) Methods
(D) Sharing
(E) Facilitation
(F) All of these

3. PRA is a:

(A) Unidimensional approach
(B) Multidimensional approach
(C) Straight dimensional approach
(D) All of these

4. Sharing of information and knowledge is integral in the process of:

(A) Participation (B) Communication
(C) Learning (D Adoption

5. **Attitude and behavior of the agent constitute the base for approaching and establishing links with:**
(A) Local communities
(B) Outside communities
(C) Urban communities
(D) None of these

6. **PRA is an essential component of:**
(A) Rural developmental projects
(B) Rural developmental schemes
(C) Both (A) and (B)
(D) None of these

7. **Advantages of PRA is/are:**
(A) Empowering the poor and weak
(B) Diversity
(C) Research priority
(D) Organizational changes
(E) All of these

8. **PRA has extensive range of application in research and planning for:**
(A) Land tenure
(B) Urban planning
(C) Natural resource management and domestic violence
(D) All of these

9. **Sharing of information and knowledge has magic effort in the process of :**
(A) Participation (B) Communication
(C) Learning (D Adoption

10. **Successful training requires encouraging new ways of learning:**
(A) Within oragnisation
(B) Outside the organization
(C) Both of these
(D) None of these

11. **In which communication system of mass media works:**
(A) Knowledge, Generating and System
(B) Knowledge, Disseminating and System
(C) Knowledge, Consuming and System
(D) None of these

12. **The single practice demonstrations is:**
(A) Method demonstration
(B) Result demonstration
(C) Frontline demonstration
(D) Single line demonstration

13. **Father of method demonstration is:**
(A) Mr Arthur T. Mosher
(B) F. L. Bryne
(C) Dr. Seeman A Knapp
(D) Mr Albert Mayer

14. **In which demonstration process, extension worker may take help of audio-visual aids:**
(A) Method demonstration
(B) Result demonstration
(C) Frontline demonstration
(D) Single line demonstration

15. **Communication is a :**
(A) One way process
(B) Two way process
(C) Three way process
(D) Multi way process

16. **In which demonstration cannot compare between old and new techniques or skill:**
(A) Method demonstration
(B) Result demonstration
(C) Frontline demonstration
(D) Single line demonstration

17. **The two important principles of extension education are:**
(A) Participation and leadership
(B) Participation and classroom teaching
(C) Leadership and classroom teaching
(D) Leadership and subject knowledge

18. **The method of extension education for effectively impressing the people about a particular recommended practice suitable for a particular area is:**
(A) Radio (B) Demonstration
(C) Tours (D) None of these

19. **To prove the worth of a new practice suited to a particular area, the best method of extension education is:**
(A) Method demonstration
(B) Result demonstration
(C) Field trails
(D) Farm visits

20. **The steps in extension teaching was put forwarded by:**
(A) Paul Leagans (B) Curl Lewin
(C) Ensminger (D) Wilson & Gallup

Practice Paper 5

1. **For effective Extension teaching, one must use:**
   (A) Combination of appropriate aids
   (B) Only visual aids
   (C) Only audio aids
   (D) Printed literature

2. **Correct sequence of steps in extension teaching is:**
   (A) Attention – Interest – Desire – Conviction – Action - Satisfaction
   (B) Desire – Interest – Attention – Conviction – Action - Satisfaction
   (C) Interest – Attention – Desire – Conviction – Action - Satisfaction
   (D) Action – Attention – Interest – Conviction – Desire - Satisfaction

3. **Which methods of extension teaching is a way of showing people the value of an improved practice:**
   (A) Result demonstration
   (B) Method demonstration
   (C) Newspaper
   (D) Circular letter

4. **Leaflet is a single sheet of paper folded to make a .... page piece of printed matter:**
   (A) 4 (B) 6
   (C) 8 (D) 10

5. **For one talk, how many flash cards should be used?**
   (A) 10 – 12 (B) 12 – 14
   (C) 14 – 18 (D) 20 – 22

6. **Steps of extension teaching are:**
   (A) Source, Message, Channel, Recommendation etc.
   (B) Awareness, Interest, Trial, Evaluation
   (C) Attention, Interest, Desire, Conviction, Action and Satisfaction
   (D) None of these

7. **The process of arranging situation in which the things to be learned are called to the attention of the learners, their interest developed, desire aroused and action promoted:**
   (A) Communication (B) Training
   (C) Teaching (D) Learning

8. **Steps in teaching is shortly called as:**
   (A) AIETA (B) AIDCAS
   (C) KPDC (D) AIDACS

9. **In extension teaching the teacher should first know:**
   (A) Subject matter
   (B) Teaching aids
   (C) Place of teaching
   (D) Attitude of the learner

10. **Effective teaching is:**
    (A) For informing people
    (B) For transforming people
    (C) Like filling a bucket
    (D) Done by trial and error

11. **When an extension worker approaches large number of people, spread over large areas for disseminating new improved production technology, it is called:**
    (A) Mass Communication
    (B) Undirectional communication
    (C) Public communication
    (D) Group communication

12. **Written form of extension teaching methods does not include:**
    (A) Bulletins (B) News articles
    (C) Circular letters (D) None of these

13. **To show the relative worth of a new practice, the extension method best suited is:**
    (A) Result demonstration
    (B) Method demonstration
    (C) Campaign
    (D) Group discussion

14. **Result demonstration is conducted by:**
    (A) Researcher (B) Extension worker
    (C) Farmer (D) All of the above

15. **Method demonstration works on the principle of:**
    (A) Learning is believing
    (B) Learning by teaching
    (C) Seeing and doing
    (D) Learning by doing

16. **Result demonstration is done in:**
    (A) Research station
    (B) Farmer's field
    (C) District farm
    (D) None of the above

17. Result demonstration works on the principle of:
(A) Learning is believing
(B) Learning by teaching
(C) Seeing is believing
(D) Learning by doing

18. An extension method that can also be used as a group teaching method:
(A) Campaign
(B) Method demonstration
(C) Office calls
(D) Farm and home visits

19. Education cause:
(A) Change in knowledge
(B) Change in attitude
(C) Change in skills
(D) All of these

20. Which extension method is evolved in result demonstration, posters and movies?
(A) Spoken (B) Visual
(C) Written (D) None of these

Practice Paper 6

1. The term extension was first used by:
(A) USA (B) England
(C) India (D) China

2. The basic unit of I.R.D.P. programme is:
(A) A village
(B) A community development block
(C) A family (D) A district

3. Extension education is:
(A) A discipline (B) A profession
(C) Discipline ad profession both
(D) None o these

4. Principles of extension education is:
(A) Learning by hearing
(B) Learning by seeing
(C) Learning by reading
(D) Learning by doing

5. Goal of extension education is:
(A) To promote income of the farmers
(B) To promote production of the crops
(C) To promote new crops
(D) To promote scientific out look

6. Extension education aims at:
(A) Increasing agricultural production
(B) Increasing milk production
(C) Changing knowledge, skills and attitudes of the participants
(D) Changing administrative system

7. The work at Shantineketan with the objectives of:
(A) To create a real interest in people for rural improvement
(B) To study rural problems
(C) To improve village sanitation
(D) All of these

8. Sh. Rabindra Nath Tagore established a rural reconstruction institute in Shanti-neketan in:
(A) 1921 (B) 1922
(C) 1923 (D) 1924

9. The main principles of extension education is:
(A) Principles of interest and need
(B) Principles of cultural change
(C) Principles of learning by doing
(D) Principles of leadership
(E) All of these

10. Rural welfare movement on a mass scale was:
(A) Work at Shantineketan
(B) Gurgaon experiment
(C) Marthndam Programme
(D) Indian Village Service

11. Self help with internate expert counsel is the technique of:
(A) Work at Shantineketan
(B) Gurgaon experiment
(C) Marthndam Programme
(D) Indian Village Service

12. The aim of Etawah Pilot Project was:
(A) Introducing intensive work on the rural recontraction front
(B) Promotion of cottage industries and spread of literacy
(C) The development of village panchayat
(D) None of these

13. **Who is the founder of Indian Village Service (IVS):**
(A) Arthor T. Mosher
(B) B.N. Gupta
(C) Both (A) and (B)
(D) None of these

14. **The main features of Sarvodaya Programme were:**
(A) Simplicity and non violence
(B) Holiness of laboures
(C) Reconstruction of human value
(D) All of these

15. **The intensive Agricultural District Programme was popularly known as:**
(A) Package programme
(B) Knowledge programme
(C) Innovative Programme
(D) None fo these

16. **The main aims of Firka development scheme of Madras were:**
(A) Educational and economic improvements
(B) Sanitary improvement
(C) Making the people self confident
(D) All of these

17. **The package programme name was given of I.A.D.P. due to:**
(A) Collective application of all improved practices
(B) Simultaneous application of all improved practices
(C) Collective and simultaneous application of all improved practices
(D) None of these

18. **IADP programme was started in 1960 in the month of:**
(A) July (B) August
(C) September (D) October

19. **Etawah Pilot Project was started in 1948 in which month:**
(A) July (B) August
(C) September (D) October

20. **Extension education centres around:**
(A) Farmers or rural peoples
(B) Innovation or inventions
(C) Extension workers
(D) All of these

Practice Paper 7

1. **The technology development of education can be related to various field of extension like:**
(A) Agriculture
(B) Home Science
(C) Agriculture and Home Science
(D) None of these

2. **Agricultural extension is concerned with agricultural education assisting farmers to bring about continuous improvement in their:**
(A) Physical
(B) Economic
(C) Social
(D) Physical, economic and social

3. **Agricultural extension makes available to the farming community to solve the problems of agriculture including:**
(A) Animal husbandry
(B) Gardening
(C) Agricultural engineering
(D) All of these

4. **Flash cards, pull charts , slides and film strips are the type of aids:**
(A) Audio
(B) Visual
(C) Audio-visual
(D) None of these

5. **Flash cards are a series of illustrated cards when flashed or presented before a group in proper sequence tell a:**
(A) Part of story
(B) Complete story
(C) Incomplete story
(D) None of these

6. **Television broadcast for rural development in India, started in:**
(A) 1947 (B) 1957
(C) 1967 (D) 1977

7. **Radio mass medium is characterized by:**
(A) One way with immediate feedback
(B) One way without instant audience response
(C) One way and colourful
(D) One way and timeless

8. The important source of information to village women is:
(A) Barber's wife (B) Dai
(C) Washer women (D) Bangle sellers
(E) All of these

9. Which one of the following is not correctly matched?

Approach	Technique
(A) Mass	Television
(B) Mass	Film
(C) Mass	Tape recorder
(D) Mass	Demonstration

10. Which source of interview is the best in a rural society?
(A) Key informant (B) Group
(C) Focused group (D) None of the above

11. Method of mass communication is:
(A) Demonstration (B) Farm visit
(C) Group discussion (D) News paper

12. Which one is not the part of communication module?
(A) Communicator (B) Message
(C) Advice (D) Audience

13. Which one of the following is not a method of mass contacts:
(A) Circular letters
(B) Television
(C) Leaflets
(D) Result demonstration

14. Which one of the following is a method of contact group:
(A) Farms and home visits
(B) Office calls
(C) Personal letters
(D) Tours

15. Which one of the following is an audio - visual method of extension teaching:
(A) Drama (B) Recordings
(C) Flash board (D) Tape recorders

16. Which one is a non - projected aid:
(A) Cinema
(B) Slide
(C) Over head projector
(D) Pictures

17. Puppet shows are included in:
(A) Audio aids
(B) Visual aids
(C) Audio – visual aids
(D) None of these

18. Flash cards, pull charts, slides and films strips are the type of aids:
(A) Audio
(B) Visual
(C) Audio – visual aids
(D) None of these

19. Visual teaching aid flannel graph is also known as:
(A) Chalk board (B) Bulletin board
(C) Khadder graph (D) None of these

20. Television, Fairs, Radios and Exhibitions are media of:
(A) Group communication
(B) Individual communication
(C) Community communication
(D) Mass communication

Practice Paper 8

1. If we want to impress the people about a particular recommended practice suitable for a specific area, then effective extension method is:
(A) TV (B) Radio
(C) Demonstration (D) Pumphlets

2. Radio is the medium of mass communication in which, use is made up of:
(A) Audio device (B) Visual device
(C) both 'A' and 'B' (D) None of these

3. Bulletins, leaflets, circular letters, posters, fairs, exhibitions, radio and television are the media of:
(A) Public communication
(B) Official communication
(C) Group communication
(D) Mass communication

4. Puppet, Drama, Television and Cinema are example of:
(A) Audio-aids (B) Visual – aids
(C) Audio-visual aids (D) None of these

5. **Radio and tape recorder is the examples of:**
(A) Audio-aids (B) Visual – aids
(C) Audio-visual aids (D) All of these

6. **The fastest method of delivering of message in the village is:**
(A) Newspaper (B) Radio
(C) Television (D) Group discussion

7. **Less intensive and less effective method of extension is**
(A) Individual (B) Group
(C) Both (A) & (B) (D) Mas

8. **Which one of the following is not correctly matched?**

	Approach	Technique
(A)	Mass	Television
(B)	Mass	Demonstration
(C)	Mass	Tape recorder
(D)	Mass	Film

9. **Which one is related to mass approach method?**
(A) Film, Demonstration and Radio
(B) Radio, TV and Lecture
(C) Phone call, Radio and TV
(D) Exhibition, Newspaper and Radio

10. **"Stay at home" type of people can be contacted by which method of extension:**
(A) Individual (B) Group
(C) Mass (D) None of these

11. **Which one of the sources of communication is most credible for non-progressive/remote village situation?**
(A) Farm radio (B) Demonstration
(C) Printed material (D) Exhibition

12. **Method of mass communication is:**
(A) Demonstration (B) Farm visit
(C) Group discussion (D) Newspaper

13. **Best method for showing the value of new practice is:**
(A) Result demonstration
(B) Exhibition
(C) Farmer's fair
(D) Meeting

14. **In a good chart facts are presented:**
(A) Only in words
(B) Only in figures
(C) Only in photographs
(D) Both in figures and photographs

15. **Non-institutional communication method is:**
(A) Exhibition
(B) Family member
(C) Extension literature
(D) Radio

16. **Which one of the following aids involves two sensory organs simultaneously?**
(A) Poster (B) Model
(C) Radio (D) Television

17. **Identify the projected audio-visual aid:**
(A) Puppet (B) Film
(C) Drama (D) Folk song

18. **Which one is not the example of nonverbal communication?**
(A) Ram Leela
(B) Movements of arms
(C) Facial expression
(D) Gestures

19. **We learn maximum through:**
(A) Taste (B) Touch
(C) Hearing (D) Sight

20. **The extension service is called "Co-operative Extension Service" in:**
(A) USA (B) India
(C) Mexico (D) China

Practice Paper 9

1. **The main functions of an agricultural university in India are:**
(A) Technology generation, technology transfer and technology adoption
(B) Drawing attention, creating interest and taking action
(C) Teaching, research and extension
(D) Preparation, deliberation and action

2. **General meeting are broadly the meeting of:**
   (A) Heterogeneous participation
   (B) Homogeneous participation
   (C) Community participation
   (D) Social participation

3. **A photograph is called:**
   (A) A diagnostic representation of facts or ideas
   (B) A meaningful exchange of ideas
   (C) An exact visual recording of a thing
   (D) An act of influencing others

4. **Identify one which is not correctly matched in the following pairs:**
   (A) Method demonstration : Group approach
   (B) Circular letter : Mass approach
   (C) Radio : Individual approach
   (D) Campaign : Mass approach

5. **Non projected teaching method is:**
   (A) Cinema (B) Slides
   (C) Puppets (D) 3-D film

6. **Which one is not the group approach device?**
   (A) Demonstration (B) Symposium
   (C) Workshop (D) Film

7. **Learning must result in functional understanding:**
   (A) Principle of readiness
   (B) Principle of association
   (C) Principle of practice
   (D) Principle of timing

8. **Learning must make sense to the learner:**
   (A) Principle of readiness
   (B) Principle of clarity
   (C) Principle of practice
   (D) Principle of timing

9. **A relatively enduring change in overt and covert response as a result of perceived stimulus is called:**
   (A) Endurance (B) Attribute
   (C) Learning (D) Perception

10. **Adult learning is:**
   (A) Problem centered
   (B) Subject centered
   (C) Instructor centered
   (D) Institute centered

11. **Adult literacy programme can be considered:**
   (A) Formal education
   (B) Informal education
   (C) Non-formal education
   (D) All of the above

12. **Which of the following statements represents true concepts of extension evaluation?**
   (A) Extension evaluation begins before the start of an extension programme
   (B) Extension evaluation begins at the mid step of the programme
   (C) Extension evaluation begins at the last step of the programme
   (D) Extension evaluation begins inbuilt in each step of programme

13. **Important element of evolution process is:**
   (A) Observations, criteria to observations and making some decisions
   (B) Observations and criteria to observations
   (C) Observations and collection of some information and final decision
   (D) All of them

14. **Pamphlets contain number of pages:**
   (A) 4-12 (B) 12-24
   (C) 24-48 (D) Unlimited

15. **Bulletin contain number of pages:**
   (A) 4-12 (B) 12-24
   (C) 24-48 (D) Unlimited

16. **Quick and cheap method of written communication and proves very effective if done well is:**
   (A) Leaflet (B) Pamphlet
   (C) Bulletin (D) Circular letter

17. **Which written method of communication can teach villagers easily:**
   (A) Circular letters (B) Leaf lets
   (C) Pamphlets (D) Bulletin

18. Brainstorming is basically a:
(A) Creative technique
(B) Motivation technique
(C) Evaluation technique
(D) Reasoning technique

19. Which type of communication is called as public relations:
(A) Across communication
(B) Interpersonal communication
(C) Internal communication
(D) External communication

20. An official communication either oral or written is called:
(A) Informal communication
(B) Formal communication
(C) Both (A) and (B)
(D) None of these

Practice Paper 10

1. A formal communication may be:
(A) Mandatory
(B) Indicative
(C) Both (A) and (B)
(D) None of these

2. A mandatory communication is mostly:
(A) Vertical (B) Horizontal
(C) Both (A) and (B) (D) None of these

3. The indicative or explanatory communication may exist:
(A) Diagonal (B) Horizontal
(C) Vertical (D) All of these

4. The communication type which provides instructions, informal and clarification is called:
(A) Informal communication
(B) Formal communication
(C) Administrative communication
(D) Both (A) and (B)

5. The communication that take place between two or more functionaries at the same level under the same supervisors:
(A) Downward communication
(B) Upward communication
(C) Horizontal communication
(D) Grapevine communication

6. Which one of the following is not an element of communication:
(A) Consistency (B) Uniformity
(C) Rigidity (D) Adequacy

7. A large number of communication models exist because of:
(A) Contextual variation
(B) Behavioural variation
(C) Sequential variation
(D) Geographical variation

8. 'Communicator' is the characteristic element in the model of:
(A) Leagans (B) Lazwell
(C) Berlo (D) Rogers

9. Which is correct:
(1) Method demonstration is the short type demonstration
(2) In method demonstration cannot compare between old and new techniques
(3) The result demonstration practices are shown in comparison to existing practices
(4) In result demonstration process, the extension worker may take the help of audio-visual aids

Options
(A) 1, 2 and 4 (B) 1, 3 and 4
(C) 1, 3 and 4 (D) 1, 2, 3 and 4

10. The important points of result demonstration is/are:
(A) Never try to discover new truths
(B) Do not repeat demonstration repeatedly
(C) Use audio-visual aids to support the result demonstration
(D) Better to conduct more than one demonstration in the same village on the same subject
(E) All of these

11. There are methods of recording the sound:
(A) Mechanical process
(B) Magnetic process
(C) Optical process
(D) All of these

12. Disc recording is done by:
(A) Mechanical process
(B) Magnetic process
(C) Optical process
(D) All of these

13. Tape and wire recording is done by:
(A) Mechanical process
(B) Magnetic process
(C) Optical process
(D) All of these

14. Recording of sound on movie film by the:
(A) Mechanical process
(B) Magnetic process
(C) Optical process
(D) All of these

15. In India and England, projector is also called as Epidiascope:
(A) Opaque projector
(B) Slide projector
(C) Overhead projector
(D) Movie projector

16. The projector are based on principle of direct projection:
(A) Opaque projector
(B) Slide projector
(C) Overhead projector
(D) Movie projector

17. The projector are based on principle of indirect projection:
(A) Opaque projector
(B) Slide projector
(C) Overhead projector
(D) Movie projector

18. Using the principle of magnetic process recording can be done on tapes made of:
(A) Paper (B) Plastic
(C) Metal (D) All of these

19. The disc or gramophone records, the disc is made of:
(A) Lac (B) Plastic
(C) Both (A) and (B) (D) None of these

20. Leaflets contain number of pages:
(A) 4-12 (B) 12-24
(C) 24-48 (D) Unlimited

Answers Key

Practice Paper 1

Q. No.	Ans	Q. No.	Ans	Q. No.	Ans	Q. No.	Ans	Q. No.	Ans	Q. No.	Ans
1.	A	2.	B	3.	A	4.	B	5.	C	6.	B
7.	A	8.	D	9.	C	10.	A	11.	C	12.	B
13.	C	14.	A	15.	D	16.	A	17.	C	18.	D
19.	D	20.	E								

Practice Paper 2

Q. No.	Ans	Q. No.	Ans	Q. No.	Ans	Q. No.	Ans	Q. No.	Ans	Q. No.	Ans
1.	C	2.	D	3.	C	4.	F	5.	F	6.	G
7.	D	8.	D	9.	C	10.	A	11.	A	12.	A
13.	D	14.	D	15.	C	16.	D	17.	D	18.	D
19.	C	20.	A								

Practice Paper 3

Q. No.	Ans	Q. No.	Ans	Q. No.	Ans	Q. No.	Ans	Q. No.	Ans	Q. No.	Ans
1.	E	2.	D	3.	D	4.	C	5.	C	6.	D
7.	D	8.	D	9.	D	10.	D	11.	C	12.	A
13.	D	14.	A	15.	C	16.	D	17.	A	18.	D
19.	A	20.	D								

Practice Paper 4

Q. No.	Ans	Q. No.	Ans	Q. No.	Ans	Q. No.	Ans	Q. No.	Ans	Q. No.	Ans
1.	F	2.	F	3.	B	4.	C	5.	A	6.	C
7.	E	8.	D	9.	A	10.	A	11.	B	12.	A
13.	C	14.	B	15.	B	16.	A	17.	A	18.	B
19.	B	20.	D								

Practice Paper 5

Q. No.	Ans	Q. No.	Ans	Q. No.	Ans	Q. No.	Ans	Q. No.	Ans	Q. No.	Ans
1.	A	2.	A	3.	A	4.	A	5.	A	6.	C
7.	C	8.	B	9.	D	10.	B	11.	A	12.	A
13.	A	14.	C	15.	D	16.	B	17.	C	18.	B
19.	D	20.	C								

■ Practice Paper 6

Q. No.	Ans	Q. No.	Ans	Q. No.	Ans	Q. No.	Ans	Q. No.	Ans	Q. No.	Ans
1.	B	2.	C	3.	C	4.	D	5.	D	6.	C
7.	D	8.	A	9.	E	10.	B	11.	C	12.	A
13.	C	14.	D	15.	A	16.	D	17.	C	18.	A
19.	C	20.	D								

■ Practice Paper 7

Q. No.	Ans	Q. No.	Ans	Q. No.	Ans	Q. No.	Ans	Q. No.	Ans	Q. No.	Ans
1.	C	2.	D	3.	D	4.	C	5.	B	6.	C
7.	B	8.	D	9.	D	10.	C	11.	D	12.	C
13.	D	14.	D	15.	A	16.	D	17.	C	18.	B
19.	C	20.	D								

■ Practice Paper 8

Q. No.	Ans	Q. No.	Ans	Q. No.	Ans	Q. No.	Ans	Q. No.	Ans	Q. No.	Ans
1.	C	2.	A	3.	D	4.	C	5.	A	6.	B
7.	D	8.	B	9.	D	10.	A	11.	A	12.	D
13.	A	14.	D	15.	D	16.	D	17.	B	18.	A
19.	D	20.	A								

■ Practice Paper 9

Q. No.	Ans	Q. No.	Ans	Q. No.	Ans	Q. No.	Ans	Q. No.	Ans	Q. No.	Ans
1.	C	2.	B	3.	C	4.	C	5.	C	6.	D
7.	C	8.	B	9.	C	10.	A	11.	C	12.	B
13.	A	14.	B	15.	C	16.	D	17.	A	18.	A
19.	D	20.	B								

■ Practice Paper 10

Q. No.	Ans	Q. No.	Ans	Q. No.	Ans	Q. No.	Ans	Q. No.	Ans	Q. No.	Ans
1.	C	2.	C	3.	B	4.	C	5.	C	6.	C
7.	A	8.	A	9.	D	10.	E	11.	A	12.	D
13.	B	14.	C	15.	A	16.	B	17.	C	18.	D
19.	C	20.	A								

B | AGRONOMY

6. Multiple Choice Questions of History and Concept of Agronomy

Practice Paper 1

1. **Agronomy is originated from:**
   (A) Greek words (B) Latin words
   (C) German words (D) English words
2. **Agronomy is made of two words namely:**
   (A) Agro + Nomous (B) Agros + Nomous
   (C) Agros + Nomos (D) None of these
3. **Literary meaning of agronomy is:**
   (A) Nutrients management
   (B) Soil management
   (C) Crop management
   (D) All the above
4. **The branch of agriculture which deals with the principles and practices of crop production and field management is called:**
   (A) Agronomy (B) Agroforestry
   (C) Agriculture (D) Sericulture
5. **First agronomic research was conducted in:**
   (A) 1832 (B) 1834
   (C) 1932 (D) 1934
6. **First agronomic research was conducted at:**
   (A) Assells (B) Brussels
   (C) USA (D) India
7. **The separate branch of agronomy was originated from agriculture in:**
   (A) 1900 (B) 1901
   (C) 1902 (D) 1903
8. **Agronomy is a:**
   (A) Applied science (B) Basic science
   (C) Informative science (D) All of these
9. **The expansion of area for increasing total agricultural production is called:**
   (A) Horizontal growth
   (B) Vertical growth
   (C) Linear growth
   (D) Cubical growth
10. **The increase in production of per unit time and area from intensive cropping system is called:**
    (A) Horizontal growth
    (B) Vertical growth
    (C) Linear growth
    (D) Cubical growth
11. **The food requirement of ever increasing population can be met out by:**
    (A) Horizontal growth (B) Vertical growth
    (C) Linear growth (D) Cubical growth
12. **Sustainable agriculture can be achieved by:**
    (A) More resources
    (B) More subsidiers
    (C) More technological support
    (D) Better prices of the production
13. **Cultivation of hybrid rice is most popular in:**
    (A) Japan (B) Korea
    (C) China (D) India
14. **Green revolution in wheat is basically attributed to:**
    (A) Greater use of seeds of HYV
    (B) More use of NPK fertilizer
    (C) Greater availability of irrigation
    (D) All of these
15. **The farming is a more precise farm management practices taking into account the variability in soil and micro climatic conditions that occurs within the field:**

(A) Precision farming
(B) Multistoried farming
(C) Conservation agriculture
(D) All of these

16. The cultivation practices is based on GIS, GPS and RSS is called:
(A) Precision farming
(B) Multistoried farming
(C) Conservation agriculture
(D) All of these

17. To study the relation of agricultural crops and environment is called:
(A) Agro-ecology
(B) Urban agriculture
(C) Per urban agriculture
(D) Conservation agriculture

18. The practice of cultivating, processing and distributing food in or around (peri-urban), village, town or city is called:
(A) Urban agriculture
(B) Peri urbon agriculture
(C) Both (A) and (B)
(D) Conservation agriculture

19. Urban agriculture contributes to:
(A) Food security
(B) Food safety
(C) Both of these
(D) None of these

20. The ability of a farm to produce food in definitely without causing severe or irregular damage to eco-system health is called:
(A) Sustainable agriculture
(B) Conservation agriculture
(C) Organic agriculture
(D) Urban agriculture

Practice Paper 2

1. The production system that sustain the health of soils, ecosystems and peoples is called:
(A) Sustainable agriculture
(B) Conservation agriculture
(C) Organic agriculture
(D) Urban agriculture

2. The concept for resource-saving agricultural crop productive that strives to achieve acceptable profits together with high and sustained production levels while concurrently conserving the environment is referred as:
(A) Sustainable agriculture
(B) Conservation agriculture
(C) Organic agriculture
(D) Urban agriculture

3. Sustainable agriculture integrates main goals are:
(A) Environmental stewardship
(B) Farm profitability
(C) Prosperous farming communities
(D) All of these

4. New technology that allows farmers to look at their fields more site specifically than before and apply inputs in a manner more specific than a blanket application is called:
(A) Industrial agriculture
(B) Precision agriculture
(C) Bio-dynamic agriculture
(D) Community supported agriculture

5. Cutting and burning of forests or woodlands to create field for agriculture or pasture for livestock, or for a variety of other purposes is called:
(A) Industrial agriculture
(B) Slash and Burn agriculture
(C) Bio-dynamic agriculture
(D) Community supported agriculture

6. Principles of conservation agriculture is/are:
(A) Minimal disturbances of the soil
(B) Permanent soil cover with organic materials
(C) Precision land leveling
(D) All of these

7. The harmful constituents in tobacco are:
(A) Nicotine and Tar
(B) Carbon mono oxide and Nitric oxides
(C) Nitrosamines and benzopyrenes
(D) All of these

8. The harmful constituents in tobacco can be minimized by application of:

(A) Potassic fertilizers
(B) Phosphatic fertilizers
(C) Nitrogenous fertilizers
(D) None of these

9. By which process the harvested tobacco leaf is made ready for the market?

(A) Bulking
(B) Curing
(C) Loading in the barn
(D) None of these

10. Nicotine sulphate and solanesol can be obtained from:

(A) Tobacco (B) Tea
(C) Coffee (D) None of these

11. Nicotine based pesticides are:

(A) Non persistent
(B) Non systematic
(C) Contact insecticide
(D) All of these

12. The processes like steeping, retting and extraction are the important processes in:

(A) Tobacco (B) Jute
(C) Sugarcane (D) Rice

13. The process of dipping of jute bundles in water is called:

(A) Steeping (B) Retting
(C) Extraction (D) None of these

14. *Triticum aestivum* is a variety of wheat is known as:

(A) Bread wheat (B) Hard wheat
(C) Emmer wheat (D) None of these

15. *Triticum duram* is a variety of wheat is known as:

(A) Bread wheat (B) Hard wheat
(C) Emmer wheat (D) None of these

16. *Triticum dicocum* is a variety of wheat is known as:

(A) Bread wheat (B) Hard wheat
(C) Emmer wheat (D) None of these

17. *Oryza japonica* is mainly grown in:

(A) Japan (B) Indonesia
(C) Western Africa (D) None of these

18. *Oryza javanica* is mainly grown in:

(A) Japan (B) Indonesia
(C) Western Africa (D) None of these

19. *Oryza glaberrima* is mainly grown in:

(A) Japan (B) Indonesia
(C) Western Africa (D) None of these

20. The most common and largely grown species of jute is:

(A) *Corchorus antichorous*
(B) *Corchorus acutangulus*
(C) *Corchorus capsularis*
(D) *Corchorus olitorius*

Practice Paper 3

1. Which jute species are called Tosha pat, Sona pat and Meena pat is:

(A) *Corchorus antichorous*
(B) *Corchorus acutangulus*
(C) *Corchorus capsularis*
(D) Corchorus olitorius

2. Which jute species are called Kakya Bombay, deshi jute, safed jute, Sada pat, Virajganj and dhaka jute is:

(A) *Corchorus antichorous*
(B) *Corchorus acutangulus*
(C) *Corchorus capsularis*
(D) *Corchorus olitorius*

3. Soil amendment used to reclaim usar land is:

(A) Zypsum
(B) Sodium sulphate
(C) Potassium sulphate
(D) Calcium carbonate

4. Which microorganism is present in large quantities in soil:

(A) Algae (B) Bacteria
(C) Fungi (D) Protozoa

5. Which one is a insect trap crop is:

(A) Arhar (B) Paddy
(C) Bhindi (D) Potato

6. Winter maize is recommended to be sown on slopes of ridges in:

(A) Southern part of India
(B) Northern part of India
(C) Eastern part of India
(D) Western part of India

7. American society of Agronomy is established in year:

(A) 1831 (B) 1873
(C) 1908 (D) 1941

8. Indian Society of Agronomy came into existence in years:

(A) 1908 (B) 1941
(C) 1955 (D) 1971

9. First agronomic research was started by:

(A) J.B. Bossinggault (B) Leibig
(C) Borlaug (D) None of these

10. Agronomy journal is published from:

(A) India (B) China
(C) USA (D) UK

11. The land management system in which agriculture and forest crops are grown in combined form on same unit of the land is called:

(A) Conservation farming system
(B) Ley farming system
(C) Alley cropping system
(D) Agroforestry system

12. The land on which production of crop is economical and practical both is known as:

(A) Productive land (B) Economical land
(C) Conserved land (D) Arable land

13. For which type of fertilizer India depends on imports?

(A) Nitrogenous fertilizers
(B) Phosphatic fertilizers
(C) Potassic fertilizers
(D) All of these

14. Active factor of production is:

(A) Land (B) Labour
(C) Capital (D) All of these

15. Total cropped area includes:

(A) Net area sown+area sown more than once
(B) Net area sown + fallow land
(C) Net area sown + fallow land + area under forest
(D) None of these

16. The total dry matter produced by a crop is known as:

(A) Dry matter yield (B) Biological yield
(C) Crop yield (D) Economic yield

17. The maximum dry weight in plants is formed by:

(A) Carbon and Oxygen
(B) Carbon and Hydrogen
(C) Nitrogen and Carbon
(D) Hydrogen and Oxygen

18. Hay is prepared from crops rich in:

(A) Sugar (B) Protein
(C) Fat (D) Crude fibre

19. The hilly soils of Uttar Pradesh are formed by:

(A) Granite and quartzite rocks
(B) Schist and biotite rocks
(C) Sand stone
(D) Lime stone

20. The process of topping is practiced in tobacco to remove:

(A) Top leaves
(B) Additional branches
(C) Inflorescence
(D) None of these

Practice Paper 4

1. Pegging is useful phenomenon in:

(A) Topioca
(B) Groundnut
(C) Sweet potato
(D) Sugarcane

2. Arrowing is known as:

(A) Tillering in sugarcane
(B) Emergence of inflorescence in sugarcane
(C) Arrow like shape of sugarcane leaf
(D) None of these

3. Decomposition of organic matter in submerged soil is carried out by:

(A) Bacteria
(B) Actinomycetes
(C) Fungi
(D) Earthworm

4. Cultivation of duram wheat is primarily confined to:

(A) Central India
(B) Southern India
(C) Central and Southern India
(D) Uttarakhand hills

5. Tick the correct agro-climatic zone of UP which has highest average population density per km^2?

(A) Bundelkhand zone
(B) Central plain zone
(C) Hill zone
(D) North-eastern plains zone

6. The crop grown in Uttar Pradesh during all the three seasons is:

(A) Urd (B) Sorghum
(C) Moong (D) Maize

7. The ginning percentage in cotton can be worked out by the formula:

(A) $\frac{\text{Weight of cotton seed}}{\text{Weight of lint}} \times 100$ (B) $\frac{\text{Weight of lint}}{\text{Weight of cotton seed}} \times 100$

(C) $\frac{\text{Weight of lint}}{\text{Weight of seed cotton}} \times 100$ (D) $\frac{\text{Weight of seed cotton}}{\text{Weight of lint}} \times 100$

8. The lint index of cotton can be worked out by the formula:

(A) $\frac{\text{Weight of 100 seed} \times \text{Ginning percentage}}{100 - \text{Ginning percentage}} \times 100$

(B) $\frac{\text{Weight of lint}}{\text{Weight of cotton seed}} \times 100$

(C) $\frac{\text{Weight of lint} \times \text{Ginning percentage}}{\text{Weight of seed cotton}} \times 100$

(D) $\frac{\text{Weight of seed cotton x Ginning percentage}}{100 - \text{Ginning percentage}} \times 100$

9. Punjab and Haryana have high percentage of net sown area under farming because of:

(A) Fertile soil
(B) Perennial rivers and canals
(C) Modern farming method
(D) All of these

10. Sustainable agriculture can be achieved by:

(A) More resources
(B) More subsidies
(C) More technological support
(D) Better prices of the produce

11. Green revolution in wheat is basically attributed to:

(A) Greater use of seeds of high yielding varieties
(B) Greater availability of irrigation
(C) Higher use of NPK fertilizer
(D) Greater use of plant protection chemicals

12. Coarse grained crops are grown in:

(A) High rainfed areas
(B) Waterlogged areas
(C) Lowland areas
(D) Upland areas

13. Hay is prepared from crops rich in:

(A) Sugar (B) Protein
(C) Fat (D) Crude fibre

14. Diara lands are:

(A) Areas flooded by sea
(B) Areas created by sand storms
(C) Areas on both sides of canal
(D) Areas located on either side of river with get flooded every year

15. Which of the following is commonly used for green manuring in India?

(A) Dhaincha
(B) Lentil
(C) Pea
(D) Potato

16. Following are the millet's grown in India:
(A) Jowar, bajra, rice, maize
(B) Bajra, wheat, jowar, rice
(C) Jowar, bajra, ragi, maize
(D) Jowar, bajra, ragi, kodon

17. The leading producing states of jowar, bajra and ragi respectively are:
(A) Maharastra, Rajasthan and Karnataka
(B) Karnataka, Rajasthan and Maharastra
(C) Karnataka, Rajasthan and Maharastra
(D) Karnataka, Rajasthan and Maharastra

18. The millet is gown both as fodder and food crops:
(A) Wheat (B) Rice
(C) Maize (D) Bajra

19. Indian is a major consumer of pulses because:
(A) Pulses are rich in dietary fibre
(B) Pulses are the major source of proteins
(C) Pulses are tasty to eat
(D) Pulses are inexpensive

20. Which of the following oilseeds are grown both as rabi and kharif crops?
(A) Mustard and groundnut
(B) Sunflower and groundnut
(C) Caster and sesame
(D) Soyabean and cotton seeds

Practice Paper 5

1. Why should the pulses be grown in rotation with other crops:
(A) They increase the taste of pulses
(B) Being leguminous crop, pulses help in restoring soil fertility
(C) They require less water for irrigation if grown in rotation with ccrops
(D) None of these

2. Millets are important for us to grow for the following reasons:
(A) Coarse grains
(B) Inexpensive
(C) Rich in dietary fibre
(D) Rich in nutrients
(E) All of these

3. Which one of the following groups of food grains is an appropriate indicator of the first three major food grains grown in India?
(A) Wheat, jowar and rice
(B) Rice, wheat and jowar
(C) Rice, wheat and bajra
(D) Wheat, bajra and rice

4. In which of the following states, three crops of paddy in Aus, Aman and Boro are grown in India:
(A) Assam, West Bengal and Orissa
(B) Punjab, Haryana and Uttar Pradesh
(C) Bihar, Assam and Tamilnadu
(D) Punjab, Assam and Orissa

5. The four major fibre crops grown in India are:
(A) Rope, Coffee, Silk and Wool
(B) Wool, Silk, Tea and Rice
(C) Cotton, Jute, Hemp and Dhaincha
(D) Cotton, Silk, Jute and Rope

6. Which of the following reforms in agriculture do not pertain to technological reforms?
(A) Use of HYV's for higher production
(B) Introduction of Kissan Credit Card and Personal Accident Insurance Scheme
(C) Application of fertilizer and manures for raising crop production
(D) Mechanization in agriculture and pest control measures

7. Development in agriculture is decelerating not due to one of the following reasons:
(A) Reduction in public investment in agriculture
(B) Reduction in import duties on agriculture products
(C) Rise in subsidy on fertilizers, power and irrigation
(D) Rise in the cost of production

8. Which one of the following has not contributed to the development and mechanization of agriculture?
(A) Establishment of agriculture universities and research farms
(B) Modern technologies inputs

(C) Development of agriculture infrastructures
(D) Reduced import duties and decrease in subsidies

9. Bloodless revolution pertains to which one of the following?
(A) Land donations
(B) Green revolution
(C) Agriculture reforms
(D) White revolution

10. Profit making is the sole aim of this type of farming:
(A) Subsistence farming
(B) Intensive subsistence farming
(C) Shifting cultivation
(D) Commercial farming

11. The success of rabi crops depends on:
(A) Heavy cyclonic rainfall
(B) Intense irrigation
(C) Light winter rainfall
(D) All of these

12. Which one of the following is not the characteristics of commercial farming?
(A) Crop is grown for self-consumption
(B) Modern inputs of farming are used
(C) Monoculture is practiced
(D) Crops are grown with the purpose of selling in market

13. Hoe, das and digging sticks are used in which of the following farming types?
(A) Subsistence farming
(B) Intensive subsistence farming
(C) Shifting cultivation
(D) Commercial farming

14. Farm productivity is low in shifting cultivation because of which one of the following reasons?
(A) Size of farm is small
(B) Farmers uses fertilizers irrationally
(C) He does not use modern inputs in farming
(D) Mechanisation in agriculture

15. Which one of the following describes a system of agriculture where a single crop is grown on a large area?
(A) Shifting agriculture
(B) Plantation agriculture
(C) Horticulture
(D) Intensive agriculture

16. Which one of the following is a oilseeds crops?
(A) Niger (B) Cowpea
(C) Jute (D) Sunhemp

17. Which one of the following is a sugar crop?
(A) Sugarbeet (B) Sorghum
(C) Clusterbean (D) Sweet potato

18. Which one of the following is a narcotic crop?
(A) Tea (B) Coffee
(C) Poppy (D) All of these

19. Which one of the following prices is announce by the government in support of a crop?
(A) Maximum Support Price
(B) Minimum Support Price
(C) Moderate Support Price
(D) Influential Support Price

20. Which of the following about pulses is not true?
(A) Help fixing nitrogen in soil
(B) Protein rich
(C) Restore soil fertility
(D) A major import item

Practice Paper 6

1. Limited crops are grown in:
(A) Kharif season
(B) Rabi season
(C) Zaid season
(D) None of these

2. Jhumming is a form of shifting cultivation practiced in:
(A) North-Eastern states
(B) South India
(C) Western India
(D) North India

3. 'Slash and burn' agriculture is also termed as in North-East India is:
(A) Jhumming (B) Kuruwa
(C) Dahiya (D) Bringa

4. Empirical formula of Crop Equivalent Yield is:

(A) $\sum_{i=1}^{n}(Yi-ei)$

(B) $La+Lb=\frac{Ya}{Sa}+\frac{Yb}{Sb}$

(C) $\frac{\text{Mixture yield of 'a'}}{\text{Pure stand yield of 'a'}-\text{mixture yield of 'a'}}$

(D) $\frac{\text{Mixture yield of 'a'}}{\text{Pure stand yield of 'a'}-\text{mixture yield of 'a'}}\times 100$

5. Empirical formula of Land Equivalent Ratio is:

(A) $\sum_{i=1}^{n}(Yi-ei)$

(B) $La+Lb=\frac{Ya}{Sa}+\frac{Yb}{Sb}$

(C) $\frac{\text{Mixture yield of 'a'}}{\text{Pure stand yield of 'a'}-\text{mixture yield of 'a'}}$

(D) $\frac{\text{Mixture yield of 'a'}}{\text{Pure stand yield of 'a'}-\text{mixture yield of 'a'}}\times 100$

6. Empirical formula of Relative Crowding Coefficient, if mixture is 50:50 is:

(A) $\sum_{i=1}^{n}(Yi-ei)$

(B) $La+Lb=\frac{Ya}{Sa}+\frac{Yb}{Sb}$

(C) $\frac{\text{Mixture yield of 'a'}}{\text{Pure stand yield of 'a'}-\text{mixture yield of 'a'}}$

(D) $\frac{\text{Mixture yield of 'a'}}{\text{Pure stand yield of 'a'}-\text{mixture yield of 'a'}}\times 100$

7. Empirical formula of Relative Crowding Coefficient for a mixing differing from 50:50 is:

(A) $\frac{Yab \times Zba}{(Yaa - Yab) \times (Zab)}$

(B) $La+Lb=\frac{Ya}{Sa}+\frac{Yb}{Sb}$

(C) $\frac{\text{Mixture yield of 'a'}}{\text{Pure stand yield of 'a'}-\text{mixture yield of 'a'}}$

(D) $\frac{\text{Mixture yield of 'a'}}{\text{Pure stand yield of 'a'}-\text{mixture yield of 'a'}}\times 100$

8. Empirical formula of Aggressivity is:

(A) $\sum_{i=1}^{n}(Yi-ei)$

(B) $La+Lb=\frac{Ya}{Sa}+\frac{Yb}{Sb}$

(C) $\frac{\text{Mixture yield of 'a'}}{\text{Pure stand yield of 'a'}-\text{mixture yield of 'a'}}$

(D) $\frac{\text{Mixture yield of 'a'}}{\text{Pure stand yield of 'a'}-\text{mixture yield of 'a'}}\times 100$

9. Empirical formula of Competition Index is:

(A) $\frac{(Yab - Yaa) \times (Yba - Ybb)}{(Yaa - Ybb)}$

(B) $\frac{Yab}{(Yaa - Zab)} \div \frac{Yba}{(Ybb - Zba)}$

(C) $\frac{\text{Relative crowding coefficient of a specific species}}{\text{Total relative crowding coefficient of all crops in the mixture}}\times 100$

(D) Value of combined intercrop yield $\times \frac{LER - 1}{LER}$

10. Empirical formula of Competition Ratio is:

(A) $\frac{(Yab - Yaa) \times (Yba - Ybb)}{(Yaa - Ybb)}$ (B) $\frac{Yab}{(Yaa - Zab)} \div \frac{Yba}{(Ybb - Zba)}$

(C) $\frac{\text{Relative crowding coefficient of a specific species}}{\text{Total relative crowding coefficient of all crops in the mixture}} \times 100$

(D) Value of combined intercrop yield × $\frac{\text{LER -1}}{\text{LER}}$

11. Empirical formula of Competition Coefficient is:

(A) $\frac{(Yab - Yaa) \times (Yba - Ybb)}{(Yaa - Ybb)}$

(B) $\frac{Yab}{(Yaa - Zab)} \div \frac{Yba}{(Ybb - Zba)}$

(C) $\frac{\text{Relative crowding coefficient of a specific species}}{\text{Total relative crowding coefficient of all crops in the mixture}} \times 100$

(D) Value of combined intercrop yield × $\frac{\text{LER -1}}{\text{LER}}$

12. Empirical formula of Monetary Advatage is:

(A) $\frac{(Yab - Yaa) \times (Yba - Ybb)}{(Yaa - Ybb)}$

(B) $\frac{Yab}{(Yaa - Zab)} \div \frac{Yba}{(Ybb - Zba)}$

(C) $\frac{\text{Relative crowding coefficient of a specific species}}{\text{Total relative crowding coefficient of all crops in the mixture}} \times 100$

(D) Value of combined intercrop yield × $\frac{\text{LER -1}}{\text{LER}}$

13. In a intercropping system, the chickpea and mustard yield was obtained 1800 and 800 kg/ha, where sale price was Rs. 12 and 15/kg. Findout chickpea equivalent yield.

(A) 2800 (B) 2900
(C) 3000 (D) 3100

14. The pure crop yield of chickpea and mustard is 2000 and 1800 kg/ha, respectively. When these crops grown in intercropping the chickpea and mustard yield was obtained 1500 and 1200 kg/ha respectively. Findout the LER:

(A) 0.42 (B) 1.42
(C) 2.42 (D) 3.42

15. Which agronomic assessment is used for calculation of relative crowding coefficient of all crops in the mixture, thereby maximum production can be obtained without leaving any harmful effect.

(A) Competition Index
(B) Competition coefficient
(C) Competition ratio
(D) Monetary advantage

16. If the competition index is more than one (>1) it means that:

(A) No effect in mixing
(B) Less advantage in mixing
(C) Advantage in mixing
(D) Unknown

17. Relative Crowding Index was proposed by:

(A) Hall (1974)
(B) De Wit (1960)
(C) Blackman (1919)
(D) Turners (1986)

18. Relative Crowding Index was examined by:

(A) Hall (1974)
(B) De Wit (1960)
(C) Blackman (1919)
(D) Turners (1986)

19. Relative Growth Rate was proposed by:

(A) Hall (1974)
(B) De Wit (1960)
(C) Blackman (1919)
(D) Turners (1986)

20. Relative Water Content was proposed by:

(A) Hall (1974)
(B) De Wit (1960)
(C) Blackman (1919)
(D) Turners (1986)

Practice Paper 7

1. Net Assimilation Rate is expressed as:

(A) $g/m^2/day$
(B) kg/kg nutrient applied
(C) day or week
(D) cm^2/g

2. Agronomic Efficiency is expressed as:

(A) $g/m^2/day$
(B) kg/kg nutrient applied
(C) day or week
(D) cm^2/g

3. Leaf Area Duration is expressed as:

(A) $g/m^2/day$
(B) kg/kg nutrient applied
(C) day or week
(D) cm^2/g

4. Specific Leaf Area is expressed as:

(A) $g/m^2/day$
(B) kg/kg nutrient applied
(C) day or week
(D) cm^2/g

5. An index of the amount of growing material per unit dry weight of plant per unit time is called:

(A) Relative Growth Rate
(B) Specific Leaf Area
(C) Net assimilation Rate
(D) Lead Area Duration

6. The ratio of area of the leaf to the weight of the leaf is called:

(A) Relative Growth Rate
(B) Specific Leaf Area
(C) Net assimilation Rate
(D) Lead Area Duration

7. Relative Yield Total was proposed by:

(A) Hall (1974)
(B) De Wit (1960)
(C) De Wit and Van Deb Burgh
(D) Turners (1986)

8. Empirical formula of Multiple Cropping Index is:

(A) $\frac{\text{Total cropped area in a year}}{\text{Net cultivated area}} \times 100$

(B) $\frac{\text{Total number of crops} \times \text{Total area under crops}}{\text{Total land available for cropping}} \times 100$

(C) $\frac{\text{Area under Kharif + Rabi + Zaid}}{\text{Area under actual cultivation}} \times 100$

(D) $\frac{\text{Yield obtained in fertilized plot} - \text{Yield obtained in unfertilized plot}}{\text{Quantity of nutrient applied in plot}}$

9. Empirical formula of Cropping Intensity Index is:

(A) $\dfrac{\sum_{i=1}^{n} ai}{A} \times 100$

(B) $\dfrac{\sum_{i=1}^{n} ai\ ti}{AoT + \sum_{j=1}^{m}}$

(C) $\dfrac{\text{Area under Kharif + Rabi + Zaid}}{\text{Area under actual cultivation}} \times 100$

(D) $\dfrac{\text{Yield obtained in fertilized plot} - \text{Yield obtained in unfertilized plot}}{\text{Quantity of nutrient applied in plot}}$

10. Empirical formula of Cropping Intensity is:

(A) $\dfrac{\sum_{i=1}^{n} ai}{A} \times 100$

(B) $\dfrac{\sum_{i=1}^{n} ai\ ti}{AoT + \sum_{j=1}^{m}}$

(C) $\dfrac{\text{Area under Kharif + Rabi + Zaid}}{\text{Area under actual cultivation}} \times 100$

(D) $\dfrac{\text{Yield obtained in fertilized plot} - \text{Yield obtained in unfertilized plot}}{\text{Quantity of nutrient applied in plot}}$

11. Empirical formula of Agronomic Efficiency is:

(A) $\dfrac{\sum_{i=1}^{n} ai}{A} \times 100$

(B) $\dfrac{\sum_{i=1}^{n} ai\ ti}{AoT + \sum_{j=1}^{m}}$

(C) $\dfrac{\text{Area under Kharif + Rabi + Zaid}}{\text{Area under actual cultivation}} \times 100$

(D) $\dfrac{\text{Yield obtained in fertilized plot} - \text{Yield obtained in unfertilized plot}}{\text{Quantity of nutrient applied in plot}}$

12. Empirical formula of Crop Growth Rate:

(A) $\dfrac{W_2 - W_1}{t_2 - t_1}$ (B) $\dfrac{Log\ W_2 - Log\ W_1}{t_2 - t_1}$ (C) $\dfrac{\text{Total leaf area}}{\text{Ground area}}$ (D) $LAI \times M$

13. Empirical formula of Relative Growth Rate:

(A) $\dfrac{W_2 - W_1}{t_2 - t_1}$ (B) $\dfrac{Log\ W_2 - Log\ W_1}{t_2 - t_1}$ (C) $\dfrac{\text{Total leaf area}}{\text{Ground area}}$ (D) $LAI \times M$

14. Empirical formula of Leaf Area Index:

(A) $\frac{W_2 - W_1}{t_2 - t_1}$ (B) $\frac{Log\ W_2 - Log\ W_1}{t_2 - t_1}$ (C) $\frac{\text{Total leaf area}}{\text{Ground area}}$ (D) $LAI \times M$

15. Empirical formula of Leaf Area Duration:

(A) $\frac{W_2 - W_1}{t_2 - t_1}$ (B) $\frac{Log\ W_2 - Log\ W_1}{t_2 - t_1}$ (C) $\frac{\text{Total leaf area}}{\text{Ground area}}$ (D) $LAI \times M$

16. Empirical formula of Leaf Area Ratio:

(A) $\frac{L_2 - L_1}{Log\ L_2 - Log\ L_1} \times \frac{Log\ W_2 - Log\ W_1}{W_2 - W_1}$ (B) $\frac{Area\ of\ the\ leaf\ (\text{cm}^2)}{Weight\ of\ the\ leaf\ (\text{g})}$

(C) $\frac{(W_2 - W_1)\ (Log\ L_2 - Log\ L_1)}{(T_2 - T_1)\ (L_2 - L_1)}$ (D) $\frac{FW - DW}{TW - DW} \times 100$

17. Empirical formula of Specific Leaf Area:

(A) $\frac{L_2 - L_1}{Log\ L_2 - Log\ L_1} \times \frac{Log\ W_2 - Log\ W_1}{W_2 - W_1}$ (B) $\frac{Area\ of\ the\ leaf\ (\text{cm}^2)}{Weight\ of\ the\ leaf\ (\text{g})}$

(C) $\frac{(W_2 - W_1)\ (Log\ L_2 - Log\ L_1)}{(T_2 - T_1)\ (L_2 - L_1)}$ (D) $\frac{FW - DW}{TW - DW} \times 100$

18. Empirical formula of Net Assimilation Rate:

(A) $\frac{L_2 - L_1}{Log\ L_2 - Log\ L_1} \times \frac{Log\ W_2 - Log\ W_1}{W_2 - W_1}$ (B) $\frac{Area\ of\ the\ leaf\ (\text{cm}^2)}{Weight\ of\ the\ leaf\ (\text{g})}$

(C) $\frac{(W_2 - W_1)\ (Log\ L_2 - Log\ L_1)}{(T_2 - T_1)\ (L_2 - L_1)}$ (D) $\frac{FW - DW}{TW - DW} \times 100$

19. Empirical formula of Relative Water Content:

(A) $\frac{L_2 - L_1}{Log\ L_2 - Log\ L_1} \times \frac{Log\ W_2 - Log\ W_1}{W_2 - W_1}$ (B) $\frac{Area\ of\ the\ leaf\ (\text{cm}^2)}{Weight\ of\ the\ leaf\ (\text{g})}$

(C) $\frac{(W_2 - W_1)\ (Log\ L_2 - Log\ L_1)}{(T_2 - T_1)\ (L_2 - L_1)}$ (D) $\frac{FW - DW}{TW - DW} \times 100$

20. The information technology and mechotronics have provided a strong sophisticate technology base to support precision farming such as:

(A) Global Positioning System (GPS)
(B) Geographical Information System (GIS)
(C) Simulation Modeling for Decision Support System (DSS)
(D) All of these

Practice Paper 8

1. **Which system provides the accurate positional information, which will be more useful in location the spatial variability with accuracy.**
   (A) GPS (B) GIS
   (C) DSS (D) All of these
2. **The inherent accuracy of the GPS is about 20 m, which is based on a 95% probability that the position given will be within:**
   (A) 10 m (B) 20 m
   (C) 30 m (D) 40 m
3. **Development of precision GIS/GPS auto navigation systems increased the efficiency of the field operations in**
   (A) Precision Farming
   (B) Conservation agriculture
   (C) Multistory farming
   (D) Cooperative farming
4. **In precision farming a software system, designed to provide the tools to manipulate and display spatial data:**
   (A) Global Positioning System (GPS)
   (B) Global Information system (GIS)
   (C) Simulation Modeling for Decision Support System (DSS)
   (D) All of these
5. **In precision farming a deterministic model is needed that reflects different soil conditions within fields rather than a breed regional based generalized capacity models:**
   (A) Global Positioning System (GPS)
   (B) Geographical Information System (GIS)
   (C) Simulation Modeling
   (D) Decision Support System (DSS)
6. **A precision farming decision support system could link:**
   (A) Site specific attribute data models
   (B) GIS technology
   (C) Both (A) and (B)
   (D) None of these
7. **DSS integrate information from many sources such as:**
   (A) Synthesize management plans
   (B) Evaluates them
   (C) Both (A) and (B)
   (D) None of these
8. **Growing trees along with agricultural crops is known as:**
   (A) Silviculture (B) Horticulture
   (C) Social forestry (D) Agroforestry
9. **"Block wood" is obtained from:**
   (A) Acacia (B) Pongamia
   (C) Dalbergia (D) Albizia
10. **Type of nursery used in paddy trans-plantor is:**
    (A) Protray (B) Mat
    (C) Conventional (D) Raised bed
11. **Arrowing is known as:**
    (A) Tillering in sugarcane
    (B) Emergence of inflorescence in sugarcane
    (C) Arrow like shape of sugarcane leaf
    (D) None of these
12. **Land area having definite boundary and contributing runoff to be a single point is known as:**
    (A) Watershed (B) Catchment area
    (C) Command area (D) A field unit
13. **Rotational intensity of maize-barley-late potato-mungbean (first year) and rice-potato-wheat-blackgram (second year) calculated as:**
    (A) 800% (B) 600%
    (C) 400% (D) 200%
14. **The unit of crop growth rate (CGR):**
    (A) g/m^2 (leaf area)/day
    (B) g/g/day
    (C) g/m^2 (land area)/day
    (D) g/day
15. **Crop intensification in both time and space dimension observed in:**
    (A) Multiple cropping
    (B) Double cropping
    (C) Intercropping
    (D) Intensive cropping
16. **Which od the following indices is used to assess the intercropping system:**
    (A) Leaf area index
    (B) Land equivalent ratio
    (C) Land utilization index
    (D) Multiple cropping

17. Study of agronomy dealing with the relationship of yield to the quantity of an added/available fertilizer element is called:

(A) Soil fertility
(B) Agrobiology
(C) Agrostology
(D) Agrology

18. Agronomic efficiency is expressed as:

(A) kg/kg
(B) kg/ha
(C) kg/ha/day
(D) kg/day

19. Growing of annual crops in between the perennial crops is:

(A) Relay cropping
(B) Intercropping
(C) Alley cropping
(D) Multiple cropping

20. An intercropping can not be beneficial if it has LER:

(A) = 1.0 (B) < 1.0
(C) > 1.0 (D) Zero

Practice Paper 9

1. Annidation is:

(A) Legume effect
(B) Allelopathy
(C) Competitive interaction
(D) Complementary interaction

2. Which one of the following planting geomentry is recommended to optimizing system productivity in intercropping system in semi-arid and sub-humid ecosystem?

(A) Slip row planting
(B) Three rows of main crop and two rows of intercrops
(C) Four rows of main crop and one row of intercrop
(D) Normal planting of intercrops between two rows of normal planted crop

3. The formula LER – 1/LER × value of combined intercrop represents:

(A) MAI (B) LER
(C) LAI (D) IER

4. If L1 and L2 are leaf area and W1 and W2 are dry weight of leaves at time t1 and t2, respectively, then {(W – W1) (Log L2 – Log L1)}/{(t2-t1) (L2 – L1)} is

(A) RGR
(B) NAR
(C) LAR
(D) CGR

5. If an a 20 acres farm, crops are raised on 15 acres in rabi, 15 acres in kharif and 20 acres in zaid in a year, what shall be the cropping intensity:

(A) 150%
(B) 250%
(C) 300%
(D) 400%

6. Productivity of an intercrop per unit area of ground compared with that expected from sole crop sown in the same proportion is termed as:

(A) Land equivalent ratio
(B) Competitive ratio
(C) Land equivalent coefficient
(D) Crop performance

7. Which one of the following crop rotation is the best for maintaining soil fertility:

(A) Maize-toria-wheat
(B) Paddy-wheat-cowpea
(C) Paddy-potato-greengram
(D) Soybean-wheat-greengram

8. The ratio between gross and net cropped area is called an index of:

(A) Cropping index
(B) Multiple cropping
(C) High intensity cropping
(D) None of these

9. Repetitive growing of same sole crop on the same field is known as:

(A) Monocropping
(B) Sequential cropping
(C) Relay cropping
(D) Multiple cropping

10. In intercrop situations when subsidiary crops are usually of a shorter duration in a long duration main crop is called:

(A) Companian crop (B) Riparian crops
(C) Smother crops (D) Soiling crop

11. Crops grown alongwith irrigation or drainage channels or waterbodies is known as:

(A) Companian crop (B) Riparian crops
(C) Smother crops (D) Soiling crop

12. Such crops have ability to smother (suppress) the growth and population of weeds by providing dense foliage and quick growing capacity is known as:

(A) Companian crop (B) Riparian crops
(C) Smother crops (D) Soiling crop

13. Crops grown to harvest while they are still green and fed fresh to livestock in stall feeding is known as:

(A) Companian crop (B) Riparian crops
(C) Smother crops (D) Soiling crop

14. Those crops are grown to break the continuity of the agroecological situation of the field under multiple cropping systems is known as:

(A) Brake crops (B) Restorative crops
(C) Fouling crops (D) Ratoon crop

15. Crops provide a good harvest along with enrichment or restoration or amelioration of soil is known as:

(A) Brake crops (B) Restorative crops
(C) Fouling crops (D) Ratoon crop

16. Whose cultural practices allows the infestation of weeds intensively is known as:

(A) Brake crops (B) Restorative crops
(C) Fouling crops (D) Ratoon crop

17. In perennial or multicut crop plants, this refers to the subsequent harvests taken from the regrowth of the root stocks, stubble and stumps after harvest is known as:

(A) Brake crops (B) Restorative crops
(C) Fouling crops (D) Ratoon crop

18. Ratational growing of gasses and legume fodder with cereal crops are known as:

(A) Ley farming (B) Alley cropping
(C) Relay cropping (D) Utera cultivation

19. Growing of lentil before harvesting of rice is known as:

(A) Ley farming (B) Alley cropping
(C) Relay cropping (D) Utera cultivation

20. Utera cultivation is generally practices in area of:

(A) Tal area of Mokama in Bihar
(B) Chattisgarh
(C) Jharkhand
(D) All of these

Practice Paper 10

1. The practices of stacking, tying and wrapping are important phenomenon of:

(A) Tea (B) Tobacco
(C) Cotton (D) Sugarcane

2. The practices of topping, desuckering, primining and curing are important phenomenon of:

(A) Tea (B) Tobacco
(C) Cotton (D) Sugarcane

3. The practices of primining , curling, twisting and curing are important phenomenon of:

(A) Tea (B) Tobacco
(C) Cotton (D) Sugarcane

4. The most common and largely grown species of jute is:

(A) *Corchorus antichorous*
(B) *Corchorus acutangulus*
(C) *Corchorus capsularis*
(D) *Corchorus olitorius*

5. Jute cann't be grown in any soil other than this type of soil:

(A) Bangar of alluvial soils
(B) Black soils of Deccan
(C) Soils of flood plains
(D) Sandy soils of Rajasthan

6. Profit making is the sole aim of this type of farming:

(A) Subsistence farming
(B) Intensive subsistence farming
(C) Shifting cultivation
(D) Commercial farming

7. The major consumer of rubber in India is:

(A) Latex form
(B) Auto tyres and tubes
(C) Foot wear
(D) Cycle tyres

8. What are the most important beverage crops in India:

(A) Tea (B) Coffee
(C) Both (A) and (B) (D) None of these

9. Which one of the following is a leguminous crop?

(A) Niger (B) Millet
(C) Jowar (D) Groundnut

10. Winter maize is recommended to be sown on slopes of ridges in:

(A) Southern part of India
(B) Northern part of India
(C) Eastern part of India
(D) Western part of India

11. Ratio of grain to stalk in pigeonpea is:

(A) 1 : 2 (B) 1 : 3
(C) 1 : 6-8 (D) 1 : 10

12. Ratio of grain to stalk in urdbean is:

(A) 1 : 1.2 (B) 1 : 3
(C) 1 : 6-8 (D) 1 : 10

13. Ratio of grain to stalk in chickpea is:

(A) 1 : 2 (B) 1 : 5
(C) 1 : 6-8 (D) 1 : 10

14. Ratio of grain to stalk in wheat is:

(A) 1 :1.25 (B) 1 : 3
(C) 1 : 6-8 (D) 1 : 10

15. Ratio of grain to stalk in jowar is:

(A) 1 : 2 (B) 1 : 3
(C) 1 : 6-8 (D) 1 : 10

16. Ratio of grain to stalk in bajra is:

(A) 1 : 2 (B) 1 : 3
(C) 1 : 8 (D) 1 : 10

17. Coarse grained crops are grown in:

(A) High rainfed areas
(B) Waterlogged areas
(C) Lowland areas
(D) Upland areas

18. Development in agriculture is decelerating not due to one of the following reasons:

(A) Reduction in public investment in agriculture
(B) Reduction in import duties on agriculture products
(C) Rise in subsidy on fertilizers, power and irrigation
(D) Rise in the cost of production

19. Which one of the following has not contributed to the development and mechanization of agriculture?

(A) Establishment of agriculture universities and research farms
(B) Modern technologies inputs
(C) Development of Agriculture infrastructures
(D) Reduced import duties and decrease in subsidies

20. In Punjab, rice is grown as an irrigated crop, but in West Bengal it is a:

(A Irrigated
(B) Rainfed
(C) Both (A) and (B)
(D) None of these

Answers Key

■ Practice Paper 1

Q. No.	Ans	Q. No.	Ans	Q. No.	Ans	Q. No.	Ans	Q. No.	Ans	Q. No.	Ans
1.	A	2.	C	3.	D	4.	A	5.	B	6.	A
7.	A	8.	A	9.	A	10.	B	11.	B	12.	C
13.	C	14.	D	15.	A	16.	A	17.	A	18.	C
19.	C	20.	A								

■ Practice Paper 2

Q. No.	Ans	Q. No.	Ans	Q. No.	Ans	Q. No.	Ans	Q. No.	Ans	Q. No.	Ans
1.	C	2.	B	3.	D	4.	B	5.	B	6.	D
7.	D	8.	A	9.	B	10.	A	11.	D	12.	B
13.	A	14.	A	15.	B	16.	C	17.	A	18.	B
19.	C	20.	C								

■ Practice Paper 3

Q. No.	Ans	Q. No.	Ans	Q. No.	Ans	Q. No.	Ans	Q. No.	Ans	Q. No.	Ans
1.	D	2.	C	3.	A	4.	B	5.	C	6.	C
7.	C	8.	C	9.	A	10.	C	11.	D	12.	A
13.	B	14.	D	15.	A	16.	B	17.	C	18.	D
19.	B	20.	A								

■ Practice Paper 4

Q. No.	Ans	Q. No.	Ans	Q. No.	Ans	Q. No.	Ans	Q. No.	Ans	Q. No.	Ans
1.	B	2.	B	3.	A	4.	C	5.	B	6.	D
7.	B	8.	A	9.	D	10.	C	11.	A	12.	D
13.	D	14.	D	15.	A	16.	D	17.	A	18.	D
19.	B	20.	B								

■ Practice Paper 5

Q. No.	Ans	Q. No.	Ans	Q. No.	Ans	Q. No.	Ans	Q. No.	Ans	Q. No.	Ans
1.	B	2.	E	3.	B	4.	A	5.	C	6.	B
7.	C	8.	D	9.	A	10.	D	11.	C	12.	A
13.	C	14.	C	15.	B	16.	A	17.	A	18.	D
19.	B	20.	D								

■ Practice Paper 6

Q. No.	Ans	Q. No.	Ans	Q. No.	Ans	Q. No.	Ans	Q. No.	Ans	Q. No.	Ans
1.	C	2.	A	3.	A	4.	A	5.	B	6.	C
7.	A	8.	D	9.	A	10.	B	11.	C	12.	D
13.	A	14.	B	15.	B	16.	C	17.	B	18.	A
19.	C	20.	D								

■ Practice Paper 7

Q. No.	Ans	Q. No.	Ans	Q. No.	Ans	Q. No.	Ans	Q. No.	Ans	Q. No.	Ans
1.	A	2.	B	3.	C	4.	D	5.	A	6.	B
7.	C	8.	B	9.	B	10.	C	11.	D	12.	A
13.	B	14.	C	15.	D	16.	A	17.	B	18.	C
19.	D	20.	D								

■ Practice Paper 8

Q. No.	Ans	Q. No.	Ans	Q. No.	Ans	Q. No.	Ans	Q. No.	Ans	Q. No.	Ans
1.	A	2.	B	3.	A	4.	B	5.	C	6.	C
7.	C	8.	D	9.	A	10.	B	11.	B	12.	B
13.	C	14.	C	15.	C	16.	B	17.	A	18.	A
19.	C	20.	B								

■ Practice Paper 9

Q. No.	Ans	Q. No.	Ans	Q. No.	Ans	Q. No.	Ans	Q. No.	Ans	Q. No.	Ans
1.	B	2.	A	3.	A	4.	B	5.	B	6.	A
7.	D	8.	A	9.	A	10.	A	11.	B	12.	C
13.	D	14.	A	15.	B	16.	C	17.	D	18.	A
19.	D	20.	D								

■ Practice Paper 10

Q. No.	Ans	Q. No.	Ans	Q. No.	Ans	Q. No.	Ans	Q. No.	Ans	Q. No.	Ans
1.	D	2.	B	3.	A	4.	C	5.	C	6.	D
7.	B	8.	C	9.	D	10.	C	11.	D	12.	A
13.	B	14.	A	15.	D	16.	C	17.	D	18.	C
19.	D	20.	B								

7. Multiple Choice Questions of Cropping System

Practice Paper 1

1. **The rotation intensity of maize - wheat + gram - moong is:**
(A) 200% (B) 250%
(C) 300% (D) 400%

2. **Intercropping of mustard with potato is recommended in:**
(A) Replacement series
(B) Additive series
(C) Replacement cum additive series
(D) None of these

3. **Which one of the following is the example of parallel intercropping:**
(A) Maize+soybean
(B) Sugarcane + mustard
(C) Potato + mustard
(D) Sugarcane + potato

4. **Maize – toria – wheat - moong rotation is termed as a system of:**
(A) Mixed cropping (B) Relay cropping
(C) Multiple cropping (D) Intercropping

5. **Each species has its own coefficient (K) which gives a measure of whether the species has produce more or less yield than expected is called:**
(A) Crop Equivalent Yield
(B) Land Equivalent Ratio
(C) Relative Crowding Coefficient
(D) Monetary Advantage

6. **The common practice of examining intercropping advantages by expressing yields in monetary term is called:**
(A) Crop Equivalent Yield
(B) Land Equivalent Ratio
(C) Relative Crowding Coefficient
(D) Monetary Advantage

7. **If LER is 1.15, it mean the increase in yield by:**
(A) 10 % (B) 15 %
(C) 20 % (D) 25 %

8. **Criteria for assessing yield advantages are:**
(A) Where intercropping must give full yield of a main crop and some yield of a second crop
(B) Where the combined intercrop yield must exceed the higher sole crop yield
(C) Where the combined intercrop yield must exceed a combined sole crop yield
(D) All of these

9. **If LER is less than 1.0 it mean that the intercropping is:**
(A) Profitable (B) Not profitable
(C) No change (D) None of these

10. **Relative crowding coefficient assumes that mixture treatments from a:**
(A) Replacement series
(B) Additive series
(C) Both of these
(D) None of these

11. **Relative crowding coefficient is less than one, it means it has produced:**
(A) Less yield (B) More yield
(C) Same yield (D) Unknown

12. **Relative crowding coefficient is more than one, it means it has produced:**
(A) Less yield (B) More yield
(C) Same yield (D) Unknown

13. **Relative crowding coefficient is equal, it means it has produced:**
(A) Less yield (B) More yield
(C) Same yield (D) Unknown

14. **It assume that mixture from a replacement series and it gives a simple measure of how much the relative yield increase in species a is greater than that for species b is known as:**
(A) Aggressivity
(B) Land Equivalent Ratio
(C) Relative Crowding Coefficient
(D) Monetary Advantage

15. **An aggressivity value is zero indicates that the component species are:**
    (A) Not competitive
    (B) Equally competitive
    (C) Less competitive
    (D) None of these
16. **Land Equivalent Ratio (LER) can be used in:**
    (A) Replacement series
    (B) Additive series
    (C) Both of these
    (D) None of these
17. **Aggressivity can be used in:**
    (A) Replacement series
    (B) Additive series
    (C) Both of these
    (D) None of these
18. **The index is the product of the two equivalent factors is called:**
    (A) Competition index
    (B) Monetary advantage
    (C) Multiple cropping index
    (D) Intensity index
19. **The sum of area planted to different crops and harvested in single year divided by the total cultivated area times 100 is called:**
    (A) Competition index
    (B) Monetary advantage
    (C) Multiple cropping index
    (D) Intensity index
20. **The ratio between total cropped area and actual net cultivated area is called:**
    (A) Cropping intensity
    (B) Intensity index
    (C) Both of these
    (D) None of these
21. **The competition index is more than 1 (>1) there has been an:**
    (A) Advantage of mixing
    (B) No advantage in mixing
    (C) No change in mixing
    (D) Unknown

Practice Paper 2

1. **The conversion of crop yield into one form to compare the crops grown in mixed/ inter/ sequentially cropped is called:**
   (A) Crop Equivalent Yield
   (B) Land Equivalent Ratio
   (C) Relative Crowding Coefficient
   (D) Monetary Advantage
2. **The ratio of relative crowding coefficient of a specific and total relative crowding coefficient of all crops in the mixture is called:**
   (A) Competition coefficient
   (B) Competition ratio
   (C) Both of these
   (D) None of these
3. **Competition coefficient is used for calculation of relative crowding coefficient of all crops in the mixture, thereby maximum production can be obtained without leaving any:**
   (A) Harmful effect
   (B) Beneficial effect
   (C) Equal
   (D) Unknown
4. **The amount of area-time devoted to each crop or group of crops compared to total area time available to the farms for crop production is termed as:**
   (A) Specific cropping index
   (B) Diversity index
   (C) Relative yield index
   (D) Relative spread index
5. **The mean yield for a crop in a region divided by the mean of all India yields of the same crop times hundred is known as:**
   (A) Specific cropping index
   (B) Diversity index
   (C) Relative yield index
   (D) Relative spread index

6. **The multiplicity of farm products which are planted by computing the reciprocal of sum of share of gross revenue received from each individual farm enterprises in a single year is called:**
   (A) Specific cropping index
   (B) Diversity index
   (C) Relative yield index
   (D) Relative spread index

7. **An area of the crop expressed as percentage of the total cultivated area in the zone divided by the area of the crop expressed as percentage of the total cultivated area in the country, times hundred is called:**
   (A) Specific cropping index
   (B) Diversity index
   (C) Relative yield index
   (D) Relative spread index

8. **The relative yields, the value in a total which has been called:**
   (A) Specific cropping index
   (B) Diversity index
   (C) Relative yield total
   (D) Relative spread index

9. **The amount of area time allocated to one crop or a group of crops relative to the area - time actually used in production of all crops is called:**
   (A) Relative crop intensity index
   (B) Crop growth rate
   (C) Relative growth rate
   (D) Net assimilation rate

10. **The dry weight gained by a unit area of crop in a unit time is called:**
    (A) Relative crop intensity index
    (B) Crop growth rate
    (C) Relative growth rate
    (D) Net assimilation rate

11. **The increase in dry weight in unit time over unit original weight of the plant is called:**
    (A) Relative crop intensity index
    (B) Crop growth rate
    (C) Relative growth rate
    (D) Net assimilation rate

12. **The increase in dry weight of plant per unit leaf area per unit time is called:**
    (A) Relative crop intensity index
    (B) Crop growth rate
    (C) Relative growth rate
    (D) Net assimilation rate

13. **Crop growth rate expressed as:**
    (A) $g/day/m^2$ (B) g/g/day
    (C) cm^2/g (D) mm/day

14. **Relative growth rate expressed as:**
    (A) $g/day/m^2$ (B) g/g/day
    (C) cm^2/g (D) mm/day

15. **Leaf area ratio expressed as:**
    (A) $g/day/m^2$ (B) g/g/day
    (C) cm^2/g (D) mm/day

16. **The ratio of leaf area to covered ground area is called:**
    (A) Leaf Area Index
    (B) Leaf Area Ratio
    (C) Leaf Area Duration
    (D) None of these

17. **The ability of the plant produce and maintain leaf area and is obtained by integrating the leaf area index over crop growth period is called:**
    (A) Leaf Area Index
    (B) Leaf Area Ratio
    (C) Leaf Area Duration
    (D) None of these

18. **The ratio of leaf area to dry weight of plant is called:**
    (A) Leaf Area Index
    (B) Leaf Area Ratio
    (C) Leaf Area Duration
    (D) None of these

19. **Which one of the following is the example of parallel cropping:**
    (A) Maize + Soybean
    (B) Sugarcane + Mustard
    (C) Potato + Mustard
    (D) Sugarcane + Potato

20. **Maize-toria-wheat-moong rotation is termed as a system of:**
    (A) Mixed cropping
    (B) Relay cropping
    (C) Multiple cropping
    (D) Intercropping

Practice Paper 3

1. **Growing plants of different heights in the same field at the same time is known as:**
   (A) Multiple cropping
   (B) Multistoreyed cropping
   (C) Relay cropping
   (D) Mixed cropping

2. **Planting of succeeding crop before harvest of standing crop is known as:**
   (A) Mixed cropping (B) Relay copping
   (C) Inter cropping (D) Multiple cropping

3. **Soil fertility is reduced due to:**
   (A) Continuous cropping
   (B) Imbalanced use of fertilizers
   (C) Over irrigation
   (D) Poor drainage

4. **In sugarcane + wheat intercropping, sugarcane is planted in the month of:**
   (A) October (B) February
   (C) March (D) April

5. **The relative land area under sole crops that is required to produce the yields achieved in intercropping is called:**
   (A) Crop Equivalent Yield
   (B) Land Equivalent Ratio
   (C) Relative Crowding Coefficient
   (D) Monetary Advantage

6. **The intensity of following one year crop rotation Maize-potato-onion is:**
   (A) 100% (B) 150%
   (C) 200% (D) 300%

7. **Intercropping refers to:**
   (A) Growing two or more crops simultaneously and intermingled
   (B) Growing two or more crops simultaneously in definite row proportions
   (C) Growing more than one crop on same land in one year
   (D) None of these

8. **Practice of raising forest or fruit trees on the society land to reduce atmospheric pollution is known as:**
   (A) Farm forestry (B) Silvipasture
   (C) Agroforestry (D) Social forestry

9. **Concept of agro-climatic zoning was developed by:**
   (A) World bank (B) USA
   (C) FAO (D) ICAR

10. **Lay farming refers:**
    (A) Growing of field crops in rotation with grasses and fodder legumes
    (B) Growing trees in rotation with grasses and fodder legumes
    (C) Growing of pastures in rotation with grasses and fodder legumes
    (D) Growing of grasses and fodder legumes in forest areas

11. **Alley cropping means:**
    (A) Growing of pastures in between two widely spaced rows of fast growing trees
    (B) Growing of field crops in between two widely spaced rows of fast growing trees
    (C) Growing of only short duration crops in between two widely spaced rows of fast growing tree
    (D) Growing of only fodder crops in between two widely spaced rows of fast growing trees

12. **The intensity of groundnut + arhar - sugarcane crop rotation would be:**
    (A) 100 % (B) 150 %
    (C) 200 % (D) 250 %

13. **The intensity of paddy - pea -sugarcane-fallow-lobia - wheat is:**
    (A) 100 % (B) 150 %
    (C) 166.6 % (D) 200 %

14. **The intensity of jowar + arhar - fallow - wheat is:**
    (A) 100 % (B) 150 %
    (C) 200 % (D) 250 %

15. **The intensity of fallow - sugarcane - rotoon is:**
    (A) 66.6 % (B) 100 %
    (C) 125 % (D) 150 %

16. **The intensity of maize (cob) -early potato - late potato - sugarcane- ratoon - berseem is:**
    (A) 100 % (B) 150 %
    (C) 200 % (D) 250 %

17. The intensity of jowar + guar – berseem + oat is:

(A) 100 % (B) 150 %
(C) 200 % (D) 250 %

18. The intensity of maize-potato-onion is:

(A) 100 % (B) 150 %
(C) 200 % (D) 300 %

19. The intensity of sunhemp (GM) - sugarcane is:

(A) 100 % (B) 25 %
(C) 50 % (D) 75 %

20. Multiple cropping index proposed by:

(A) Dalrimpal (1971)
(B) Rao and Willey (1980)
(C) Monge and Associates (1978)
(D) Monge (1978)

Practice Paper 4

1. Competition ratio proposed by:

(A) Dalrimpal (1971)
(B) Rao and Willey (1980)
(C) Monge and Associates (1978)
(D) Monge (1978)

2. Cropping intensity index proposed by:

(A) Dalrimpal (1971)
(B) Rao and Willey (1980)
(C) Monge and Associates (1978)
(D) Monge (1978)

3. Specific cropping intensity index proposed by:

(A) Dalrimpal (1971)
(B) Rao and Willey (1980)
(C) Monge and Associates (1978)
(D) Monge (1978)

4. Cultivated land utilization index proposed by:

(A) Choung (1973)
(B) Wong and You and Strout (1978)
(C) Donald (1963)
(D) Mc Gilchrist (1965)

5. Competition index proposed by:

(A) Choung (1973)
(B) Wong and You and Strout (1978)
(C) Donald (1963)
(D) Mc Gilchrist (1965)

6. Diversity index proposed by:

(A) Choung (1973)
(B) Wong and You and Strout (1978)
(C) Donald (1963)
(D) Mc Gilchrist (1965)

7. Aggressivity proposed by:

(A) Choung (1973)
(B) Wong and You and Strout (1978)
(C) Donald (1963)
(D) Mc Gilchrist (1965)

8. Net assimilation rate proposed by:

(A) Dalrimpal (1971)
(B) Rao and Willey (1980)
(C) Evans (1982)
(D) Monge (1978)

9. Leaf area index proposed by:

(A) Choung (1973)
(B) Radford (1967)
(C) Donald (1963)
(D) Mc Gilchrist (1965)

10. Growing of crops one after the other without season fallowing is called:

(A) Continuous cropping
(B) Contingency cropping
(C) Contour cropping
(D) Conservation cropping

11. Growing of crops in aberrant situations like drought and floods is called:

(A) Continuous cropping
(B) Contingency cropping
(C) Contour cropping
(D) Conservation cropping

12. The cultivation of crops along the contours of a slope is called:

(A) Continuous cropping
(B) Contingency cropping
(C) Contour cropping
(D) Conservation cropping

13. A way of farming that aims to maximize the protection against erosion that can be achieved through soil and crop management for sustained farm productivity is called:

(A) Continuous cropping
(B) Contingency cropping
(C) Contour cropping
(D) Conservation cropping

14. **As the ability to regenerate by shoot or root suckers, or a forest so established is called:**
(A) Coppice farming (B) Cover crop
(C) Crop competition (D) Crop canopy

15. **The practice of intensive exploitation of land capability by regular harvesting of such trees which produce numerous shoots from the stump after cutting is called:**
(A) Coppice farming (B) Cover crop
(C) Crop competition (D) Crop canopy

16. **Crops which are grown primarily to cover the soil and to reduce the loss of moisture due to leaching and erosion by wind and water is called:**
(A) Coppice farming (B) Cover crop
(C) Crop competition (D) Crop canopy

17. **Two plants are in competition with each other when the growth of either one or both of them is reduced or their form modified as compared with their growth or form in isolation is called:**
(A) Coppice farming (B) Cover crop
(C) Crop competition (D) Crop canopy

18. **It is the structures of aerial vegetative parts with special reference to the size, orientation, depth, density and arrangement of leaves influencing the penetration and interception of radiant energy is called:**
(A) Coppice farming (B) Cover crop
(C) Crop competition (D) Crop canopy

19. **The concept, approach method and process of growing more than one crop per year in the same field by increasing cropping intensity is called:**
(A) Crop intensification
(B) Crop modeling
(C) Cropping pattern
(D) Cropping scheme

20. **How inputs or variable enter, move, interact and are controlled within the system and how the output of the system is generated and affected by other inputs and controlled variables is called:**
(A) Crop intensification
(B) Crop modeling
(C) Cropping pattern
(D) Cropping scheme

Practice Paper 5

1. **Yearly sequence and spatial arrangement of crops and fallow, on a given area is called:**
(A) Crop intensification
(B) Crop modeling
(C) Cropping pattern
(D) Cropping scheme

2. **The plan according to which crops are raised on individual plots of a farm with an object of getting the maximum returns from each crops without impairing fertility of the soil is called:**
(A) Crop intensification
(B) Crop modeling
(C) Cropping pattern
(D) Cropping scheme

3. **Pattern of crops taken up for a given piece of land, or other in which the crops are cultivated on piece of land over a fixed period and other interaction with farm resources and other farm enterprises and available technology which determine their make up is called:**
(A) Cropping system
(B) Crop production
(C) Crop productivity
(D) Sequential cropping

4. **The exploitation of plant morphological and plant physiological responses within a soil and atmospheric environment to produce a high yield per unit area of land is called:**
(A) Cropping system
(B) Crop production
(C) Crop productivity
(D) Sequential cropping

5. **Production of plant product of economic importance, expressed in standard units per unit area is called:**
(A) Cropping system
(B) Crop production
(C) Crop productivity
(D) Sequential cropping

6. **The growing of different crops on a piece of land in a pre-planned succession is called:**
(A) Cropping system
(B) Crop production
(C) Crop productivity
(D) Crop rotation/Sequential cropping

7. **Comprises all components required for the production of a particular crop and the inter-relationship between them and environment is called:**
(A) Crop system
(B) Crop technology
(C) Diversified cropping
(D) Diversified farm

8. **The theoretical and practical approach to crop production in which concepts of applied science and advanced techniques are related to the financial consequences of their use is called:**
(A) Crop system
(B) Crop technology
(C) Diversified cropping
(D) Diversified farm

9. **A cropping plan in which no single crop contributes 50 % or more towards the total crop production or monetary income annually is called:**
(A) Crop system
(B) Crop technology
(C) Diversified cropping
(D) Diversified farm

10. **A farm on which no single product or source of income equals as much as 50% of the total receipt and on such farm for farmer depends on several sources of income is called:**
(A) Crop system
(B) Crop technology
(C) Diversified cropping
(D) Diversified farm

11. **A cropping system in which the same crop if grown year after year in the same field is known as:**
(A) Mono cropping
(B) Double cropping
(C) Triple cropping
(D) Quandruple cropping

12. **Taking two crops in a year in sequence from the same piece of land is called:**
(A) Mono cropping (B) Double cropping
(C) Triple cropping (D) Quandruple cropping

13. **Growing three crops in a year in sequence is called:**
(A) Mono cropping (B) Double cropping
(C) Triple cropping
(D) Quandruple cropping

14. **Growing four crops in a year in sequence is called:**
(A) Mono cropping
(B) Double cropping
(C) Triple cropping
(D) Quandruple cropping

15. **Growing two or more crops consecutively on the same field in the same year is called:**
(A) Multiple cropping
(B) Diversified cropping
(C) Relay cropping
(D) Opportunity cropping

16. **Measures the sum of area planted to different crops and harvested in single year divided by the total cultivated area times 100 is known as:**
(A) Multiple cropping index
(B) Diversified index
(C) Competition index
(D) Leaf area index

17. **System of growing together, crops of different heights at the same time on the same piece of land and thus using land, water and space most efficiently and economically is known as:**
(A) One tier system (B) Two tier system
(C) Three tier system (D) Multi tier system

18. **A companion crop which nourishes the main crop by way of nitrogen fixation and/or adding the organic matter in to the soil is called:**
(A) Nurse crop (B) Truck crop
(C) Catch crop (D) Guard crop

19. **Growing main crop in the centre surrounded by hardy and thorny crop such as safflower around pea, mesta**

around sugarcane and sorghum round maize is called:

(A) Nurse crop (B) Truck crop
(C) Catch crop (D) Guard crop

20. The practice of placing primary emphasis on the use of stored soil moisture while determining whether or not to establish a crop is called:

(A) Multiple cropping
(B) Diversified cropping
(C) Relay cropping
(D) Opportunity cropping

Practice Paper 6

1. The range of environmental factors over which an organism or a physiological process can function is called:

(A) Ecological amplitudes
(B) Ecological equivalence
(C) Ecological niche
(D) Eco-type

2. The situation in which two or more species can replace one another in the same ecological niche because of the similarity in their ecological amplitude is called:

(A) Ecological amplitudes
(B) Ecological equivalence
(C) Ecological niche
(D) Eco-type

3. The position or functional status of an organism within its community and ecosystem, resulting in organisms structural adaptation, physiological response and specific behaviour is called:

(A) Ecological amplitudes
(B) Ecological equivalence
(C) Ecological niche
(D) Eco-type

4. Plant type or strain within a species, resulting from exposure to a particular environment is called:

(A) Ecological amplitudes
(B) Ecological equivalence
(C) Ecological niche
(D) Eco-type

5. It is a cropping zone in which relative yield index is 200 % and relative spread indices vary from 90-200 % is called:

(A) Efficient cropping zone
(B) Relative yield index
(C) Relative spread index
(D) None of these

6. A concept involving the farming of fast growing plants/trees for the purpose of providing biomass that can be used directly as fuel or converted in to other forms of fuel or energy products is called:

(A) Energy farming (B) Organic farming
(C) Diversified farming
(D) Energy productivity ratio

7. Ratios showing food yield per unit energy inputs and are most often used in evaluating the energetics of agricultural systems is called:

(A) Energy farming (B) Organic farming
(C) Diversified farming
(D) Energy productivity ratio

8. It is the practice of allowing crop land to lie idle during a growing season to build up the soil moisture and fertility content so that a better crop can be produced the following year is called:

(A) Fallow (B) Fallow land
(C) Fallow system (D) None of these

9. Land left unfarmed for one or more growing seasons to kill weeds, make the soil richer is called:

(A) Fallow (B) Fallow land
(C) Fallow system (D) None of these

10. The sequence of crop years with fallow years is called:

(A) Fallow (B) Fallow land
(C) Fallow system (D) None of these

11. Extensive fallow systems are:

(A) Shifting cultivation system
(B) Bush fallow or grass fallow systems
(C) Fallow system
(D) None of these

12. Intensive fallow systems are:

(A) Shifting cultivation system
(B) Bush fallow or grass fallow systems
(C) Fallow system (D) None of these

13. The crops are grown in parallel strips laid out across the general slope but which do not fallow the contour is known as:

(A) Field strip cropping
(B) Contour strip cropping
(C) Both of these
(D) None of these

14. Yield of the plant parts of economic interest (economic yield) as percentage of total biological yield in terms of dry matter is called:

(A) Harvest index (B) Harvest ratio
(C) Both of these (D) None of these

15. Maximum use of the land by means of frequent succession of harvested crop is called:

(A) Intensive cropping
(B) Intercropping
(C) Mixed cropping
(D) Mixed inter cropping

16. Growing two or more generally dissimilar crop simultaneously on the same piece of land, base crop necessarily in distinct row arrangement is called:

(A) Intensive cropping
(B) Intercropping
(C) Mixed cropping
(D) Mixed inter cropping

17. Growing of two or more crops simultaneously on the same piece of land, without any definite row arrangement is called:

(A) Intensive cropping
(B) Intercropping
(C) Mixed cropping
(D) Mixed inter cropping

18. Growing two or more crops simultaneously with no distinct row management is called:

(A) Intensive cropping
(B) Intercropping
(C) Mixed cropping
(D) Mixed inter cropping

19. The crops which are grown around the field boundaries in narrow strips is called:

(A) Outer crop (B) Guard crop
(C) Border crop (D) All of these

20. Plants tolerant to high acidic soil conditions is called:

(A) Oxylophytes (B) Acedophytes
(C) Basilophytes (D) None of these

Practice Paper 7

1. Crop sown broad cast in the standing crop of low land rice before its harvest where the residual moisture is used for the establishment of sown crop is called:

(A) Paira crop
(B) Utera crop
(C) Both of these
(D) None of these

2. A rotation of arable crops requiring annual cultivation and artificial pastures occupying field for two years or longer is called:

(A) Ley farming
(B) Alley cropping
(C) Silvi pastoral system
(D) None of these

3. The practice of placing primary emphasis on the use of stored soil moisture while determining whether or not to establish a crop is called:

(A) Multiple cropping
(B) Diversified cropping
(C) Relay cropping
(D) Opportunity cropping

4. Crops which are grown during the rainy season is called:

(A) Kharif crops (B) Rabi crops
(C) Zaid crops (D) Perennial crops

5. Crops which are grown during the winter season is called:

(A) Kharif crops (B) Rabi crops
(C) Zaid crops (D) Perennial crops

6. Crops which are grown during the summer season is called:

(A) Kharif crops (B) Rabi crops
(C) Zaid crops (D) Perennial crops

7. Crops which are grown during the two years is called:

(A) Kharif crops (B) Rabi crops
(C) Zaid crops (D) Binnial crops

8. **Crops which are grown during the more than two years is called:**
   (A) Kharif crops
   (B) Rabi crops
   (C) Zaid crops
   (D) Perennial crops

9. **The emergence of new tillers from the root/crown of a plant specially as in sugarcane, sorghum etc. is called:**
   (A) Ratoon
   (B) Ratoon cropping
   (C) Rangeland management
   (D) Recession farming

10. **The cultivation of an additional crop from the regrowth of stubbles of previous main crop after its harvest, thereby avoiding reseeding/replanting is called:**
   (A) Ratoon
   (B) Ratoon cropping
   (C) Rangeland management
   (D) Precession farming

11. **Management of a range to produce maximum forage for sustained use without jeopardizing the other resources or uses of land is called:**
   (A) Ratoon
   (B) Ratoon cropping
   (C) Rangeland management
   (D) Recession farming

12. **System in which crops are planted in flooded areas as the rainy season ends and water recedes is called:**
   (A) Recession farming
   (B) Diara land farming
   (C) Both of these
   (D) None of these

13. **The population of a particular species expressed as a percentage of all the species in a habitat is called:**
   (A) Relative dominance
   (B) Relative density
   (C) Relative spread index
   (D) Relative yield index

14. **The actual area occupied by a species expressed as a percentage of total area occupied by all species in a given land area is called:**
   (A) Relative dominance
   (B) Relative density
   (C) Relative spread index
   (D) Relative yield index

15. **LAI refers to:**
   (A) Lower Available Index
   (B) Light Available Intensity
   (C) Leaf Area Index
   (D) Lux Area Index

16. **The rotation intensity of maize-potato-sugarcane-sunflower is:**
   (A) 100% (B) 150%
   (C) 200% (D) 250%

17. **The mean of plant relative yield (ratio of the per plant yields in a mixture) to that in monoculture will have a value close to unity is called:**
   (A) Relative Yield total
   (B) Relative crop intensity index
   (C) Relative spread index
   (D) Relative yield index

18. **Land Equivalent ratio (LER) is required in:**
   (A) Inter-cropping system
   (B) Sequential cropping system
   (C) Mixed cropping system
   (D) Sole cropping system

19. **The leaf area per unit leaf weight expressed as cm^2g^{-1} is called:**
   (A) Relative Yield total
   (B) Specific leaf area
   (C) Relative spread index
   (D) Relative yield index

20. **Growing two or more crops simultaneously where one or more crops are planted in rows is called:**
   (A) Row intercropping
   (B) Row mixed cropping
   (C) Strip cropping
   (D) None of these

Practice Paper 8

1. **The practice of cultivating clearings scattered in the reservoir of natural vegetation and of abandoning them as soon as the soil is exhausted; and this includes the practice of shifting homesteads in order to fallow the cultivator's search for new fertile land is called:**
   (A) Row intercropping
   (B) Shifting cultivation
   (C) Strip cropping
   (D) None of these
2. **These crops which are preserved in a succulent condition by partial fermentation in a tight receptacle is called:**
   (A) Silage crop (B) Sod culture
   (C) Soilage crop (D) None of these
3. **A system of soil management wherein the plants are grown in permanent grass without tillage is called:**
   (A) Silage crop (B) Sod culture
   (C) Soilage crop (D) None of these
4. **Forage cut green and fed to livestock while in the fresh conditions are known as:**
   (A) Silage crop (B) Sod culture
   (C) Soilage crop (D) None of these
5. **A crop grown in pure stand at optimum population is called:**
   (A) Sole crop (B) Intercrop
   (C) Strip crop (D) None of these
6. **Growing soil covering and soil depleting crops in alternate strips running perpendicular to the slope of the land or to the direction of prevailing winds for the purpose of reducing erosion is called:**
   (A) Sole cropping (B) Intercropping
   (C) Strip cropping (D) None of these
7. **Growing two or more crops simultaneously in different strips wide enough to permit independent cultivation but narrow enough for the crop to interact agronomically is called:**
   (A) Sole cropping (B) Intercropping
   (C) Strip intercropping (D) None of these
8. **The crop which is grown under certain problematic conditions when no other crop can be grown as floating rice in flood prone areas is called:**
   (A) Subsistence crop
   (B) Subsistence farming
   (C) Substitution farming
   (D) None of these
9. **It is farming enterprise which provides food and commodities just sufficient for the farming family, and there are no surplus to sell is called:**
   (A) Subsistence crop
   (B) Subsistence farming
   (C) Substitution farming
   (D) None of these
10. **Substitution of existing inefficient crop with an identified efficient crop in a zone or region is called:**
    (A) Subsistence crop
    (B) Subsistence farming
    (C) Substitution farming
    (D) None of these
11. **The crops grown for the purpose of producing sugar from them is called:**
    (A) Sugar crop
    (B) Support crop
    (C) Nurse crop
    (D) Guard crop
12. **Certain fast growing crops which work as support to vine crop is called:**
    (A) Sugar crop (B) Support crop
    (C) Nurse crop (D) Guard crop
13. **Crops which yield in tones and grown for different market requiring heavy transport is called:**
    (A) Truck crops (B) Guard crops
    (C) Nurse crops (D) None of these
14. **A type of agro-forestry system in which trees are planted with food crops which serve as a nursery for the young tree seedlings is called:**
    (A) Taungya cultivation
    (B) Shifting cultivation
    (C) Both of these
    (D) None of these

15. Growing of two crops in a definite row proportion on the same piece of land is called:

(A) Parallel cropping (B) Inter-cropping
(C) Mixed cropping (D) Sole cropping

16. Which one of the following can be expressed by using equation

$$\frac{\text{Harvest Diversity Index} \times 10000}{\text{Multiple cropping index}}$$

(A) Diversity index
(B) Multiple cropping index
(C) Simultaneous cropping index
(D) Cropping intensity index

17. The best example of synergistic cropping system is:

(A) Pigeonpea + Green gram
(B) Potato + Wheat
(C) Autumn planted sugarcane + mustard
(D) Autumn planted sugarcane + potato

18. Multi-intercropping includes:

(A) 2 crops in intercropping
(B) Relay cropping
(C) Mixed cropping
(D) More than 2 crops in intercropping

19. The plant geometry of wheat crop is:

(A) Square
(B) Spherical
(C) Extreme rectangular
(D) None of these

20. Number of steps taken in preparation of cropping scheme are:

(A) 2 (B) 4
(C) 6 (D) 8

Practice Paper 9

1. Read the following crop rotation and mark the correct answer:

(A) Arhar (Type 7) – Wheat (K 8020)
(B) Arhar (Bahar) – Wheat (K 8020)
(C) Arhar (Type 17) – Wheat (HD 2285)
(D) Arhar (UPAS 120) – Wheat (Golden Halna)

2. Yearly sequence and spatial arrangement of crops on a given land area is called:

(A) Cropping system
(B) Cropping pattern
(C) Crop rotation
(D) Sequential cropping

3. Number of crops taken in relay cropping should be:

(A) 1 (B) 2
(C) 3 (D) 4

4. Evaluation of intercropping is done by:

(A) Rotational intensity
(B) Cropping intensity
(C) Land Equivalent ratio
(D) None of the above

5. The recommended row ratio for intercropping of wheat + mustard is:

(A) 3 : 1 (B) 5 : 1
(C) 9 : 1 (D) 7 : 1

6. The soybean crop can be grown with:

(A) Wheat (B) Chickpea
(C) Maize (D) Tobacco

7. Safflower can be grown as an intercrop with:

(A) Wheat (B) Urdbean
(C) Maize (D) Sorghum

8. Sesamum is mainly grown in UP as a:

(A) Paira crop (B) Mixed crop
(C) Inter crop (D) Pure crop

9. The example of intercropping in replacement series is:

(A) Sugarcane + Potato
(B) Sugarcane + Wheat
(C) Pigeonpea + Urdbean
(D) Potato + Mustard

10. Multiple cropping increases over single cropping:

(A) Soil fertility
(B) Production of crops
(C) Autotoxins (D) Soil productivity

11. Main pre-requisite for adoption of multiple cropping is:

(A) Longer duration varieties
(B) Soil conservation
(C) Irrigation facility
(D) Weed control chemicals

12. Crop rotation intensity is equal to:

(A) $\frac{\text{No. of crops in a rotation}}{\text{Duration of crop rotation}} \times 100$

(B) $\frac{\text{No. of crops in a rotation}}{\text{Number of crop rotation}} \times 100$

(C) $\frac{\text{Duration of rotation}}{\text{Number of crops in rotation}} \times 100$

(D) None of these

13. Cropping intensity is equal to:

(A) $\frac{\text{Total croped area}}{\text{Net available area}} \times 100$

(B) $\frac{\text{No. of crop sown}}{\text{Net available area}} \times 100$

(C) $\frac{\text{Total croped area}}{\text{Duration of rotation}} \times 100$

(D) None of these

14. Growing plants of different heights in the same field at the same time is known as:

(A) Multiple cropping
(B) Multi-storied cropping
(C) Relay cropping
(D) Mixed cropping

15. In an intercropping system, if increase in the output of one crop help increase in the output of other crop, it is known as:

(A) Competitive
(B) Complementary
(C) Supplementary
(D) Mutual inhibitive

16. Cropping intensity of Paddy - Wheat - Urd bean rotation is:

(A) 100% (B) 200%
(C) 300% (D) 400%

17. Presently the cropping intensity of India is:

(A) 100% (B) 127%
(C) 137% (D) 150%

18. Which one of the following is the example of parallel intercropping?

(A) Sugarcane + Potato
(B) Sugarcane + Mustard
(C) Sugarcane + Soybean
(D) Potato + Mustard

19. Monoculture is always harmful due to:

(A) Development of harmful soil organisms
(B) Reduction in soil fertility
(C) Accumulation of allelochemicals
(D) All of these

20. Planting of succeeding crop before harvest of standing crop is known as:

(A) Mixed cropping
(B) Relay cropping
(C) Inter cropping
(D) Multiple cropping

Practice Paper 10

1. Multiple cropping means:

(A) Growing > 2 crops in a year
(B) Growing > 3 crops in a year
(C) Growing > 1 crops in a year
(D) All of these

2. For getting 10% advantage, the LER should be:

(A) 1.0 (B) 0.90
(C) 1.10 (D) 110

3. Cropping intensity of the rotation chari - Berseem - Rice - Pea is:

(A) 100 % (B) 200 %
(C) 300 % (D) 400 %

4. Cropping intensity of the rotation green manure-wheat is:

(A) 50 % (B) 100 %
(C) 150 % (D) 200 %

5. Rotation Rice - Pea - Sugarcane-Ratoon is of:

(A) 1 year (B) 2 year
(C) 3 year (D) 4 year

6. Mixed cropping refers to:

(A) Growing of two crops together
(B) Growing of two crops together in different lines
(C) Sowing of mixed seeds of two crops
(D) All of these

7. In sugarcane + wheat intercropping, sugarcane is planted in the month of:

(A) October (B) February
(C) March (D) April

8. Which of the following cropping system is most prevalent in India:

(A) Rice – Wheat
(B) Rice – Soybean
(C) Maize – Gram
(D) Maize-Sugarcane-Ratoon-Wheat

9. Intercropping refers to:

(A) Growing of 2 crops together
(B) Growing of 2 crops together in different lines with least competition
(C) Sowing of mixed seeds of 2 crops
(D) All of these

10. Which one of the example of relay cropping?

(A) Paddy – Wheat
(B) Soybean – Wheat
(C) Maize – Wheat
(D) Maize – Potato – Cucurbit

11. In wheat, mustard may be grown as inter crop in the row ratio of:

(A) 6 : 1 (six rows of Wheat + One row of Mustard)
(B) 9 : 1 (Nine rows of Wheat + One row of Mustard)
(C) 12 : 2 (Twelve rows of Wheat + Two rows of Mustard)
(D) 18 : 3 (Eighteen rows of Wheat + Three rows of Mustard)

12. Rice - Rice - Rice rotation indicates:

(A) Monoculture
(B) Duoculture
(C) Polyculture
(D) None of these

13. Sowing of toria in standing crop of maize is:

(A) Mixed cropping
(B) Mixed intercropping
(C) Alley cropping
(D) Relay cropping

14. The highest yield of potato is obtained when cultivated in the rotation:

(A) Maize – Potato – Wheat
(B) Maize – Potato – Sunflower
(C) Dhaincha – Potato – Sunflower
(D) Moong – Potato – Sunflower

15. Which value can be assessed by using equation

$$\frac{\text{Total number of crops} \times \text{Their respective area}}{\text{Total area available for cropping pattern}} \times 100$$

(A) Diversity index
(B) Simultaneous cropping index
(C) Multiple cropping index
(D) Cropping intensity index

16. Growing plants of different heights in the same field at the same time is known as:

(A) Multiple cropping
(B) Multi-storeyed cropping
(C) Relay cropping
(D) Mixed cropping

17. The intensity of following one year crop rotation 'Maize - potato-Onion' is:

(A) 100% (B) 150%
(C) 200% (D) 300%

18. Relay cropping means:

(A) Growing of more than one crop on the same land in a year
(B) Growing of four crops on the same land in a year
(C) Growing of four crops on the same land in a year in such a manner that the following crop is sown before the harvest of preceding crop
(D) None of these

19. A short duration crop in between two main seasonal crops is termed as:

(A) Cash crop
(B) Inter crop
(C) Companion crop
(D) Catch crop

20. Which of the following intercropping system has synergetic effect on component crop:

(A) Autumn planted sugarcane + mustard
(B) Autumn planted sugarcane + chickpea
(C) Autumn planted sugarcane + wheat
(D) Autumn planted sugarcane + potato

Answers Key

Practice Paper 1

Q. No.	Ans	Q. No.	Ans	Q. No.	Ans	Q. No.	Ans	Q. No.	Ans	Q. No.	Ans
1.	D	2.	A	3.	A	4.	B	5.	C	6.	D
7.	B	8.	D	9.	B	10.	A	11.	A	12.	B
13.	C	14.	A	15.	B	16.	C	17.	A	18.	A
19.	C	20.	C	21.	A						

Practice Paper 2

Q. No.	Ans	Q. No.	Ans	Q. No.	Ans	Q. No.	Ans	Q. No.	Ans	Q. No.	Ans
1.	A	2.	A	3.	A	4.	A	5.	C	6.	B
7.	D	8.	C	9.	A	10.	B	11.	C	12.	D
13.	A	14.	B	15.	C	16.	A	17.	C	18.	B
19.	C	20.	C								

Practice Paper 3

Q. No.	Ans	Q. No.	Ans	Q. No.	Ans	Q. No.	Ans	Q. No.	Ans	Q. No.	Ans
1.	B	2.	B	3.	A	4.	A	5.	B	6.	D
7.	B	8.	D	9.	D	10.	A	11.	B	12.	B
13.	C	14.	B	15.	A	16.	C	17.	C	18.	D
19.	C	20.	A								

Practice Paper 4

Q. No.	Ans	Q. No.	Ans	Q. No.	Ans	Q. No.	Ans	Q. No.	Ans	Q. No.	Ans
1.	B	2.	C	3.	D	4.	A	5.	C	6.	B
7.	D	8.	C	9.	B	10.	A	11.	B	12.	C
13.	D	14.	A	15.	A	16.	B	17.	C	18.	D
19.	A	20.	B								

Practice Paper 5

Q. No.	Ans	Q. No.	Ans	Q. No.	Ans	Q. No.	Ans	Q. No.	Ans	Q. No.	Ans
1.	C	2.	D	3.	A	4.	B	5.	C	6.	D
7.	A	8.	B	9.	C	10.	D	11.	A	12.	B
13.	C	14.	D	15.	A	16.	A	17.	D	18.	A
19.	D	20.	D								

■ Practice Paper 6

Q. No.	Ans	Q. No.	Ans	Q. No.	Ans	Q. No.	Ans	Q. No.	Ans	Q. No.	Ans
1.	A	2.	B	3.	C	4.	D	5.	A	6.	A
7.	D	8.	A	9.	B	10.	C	11.	A	12.	B
13.	A	14.	A	15.	A	16.	B	17.	C	18.	D
19.	D	20.	A								

■ Practice Paper 7

Q. No.	Ans	Q. No.	Ans	Q. No.	Ans	Q. No.	Ans	Q. No.	Ans	Q. No.	Ans
1.	C	2.	A	3.	D	4.	A	5.	B	6.	C
7.	D	8.	D	9.	A	10.	B	11.	C	12.	C
13.	B	14.	A	15.	C	16.	C	17.	A	18.	A
19.	B	20.	A								

■ Practice Paper 8

Q. No.	Ans	Q. No.	Ans	Q. No.	Ans	Q. No.	Ans	Q. No.	Ans	Q. No.	Ans
1.	B	2.	A	3.	B	4.	C	5.	A	6.	C
7.	C	8.	A	9.	B	10.	C	11.	A	12.	B
13.	A	14.	A	15.	B	16.	C	17.	D	18.	D
19.	C	20.	C								

■ Practice Paper 9

Q. No.	Ans	Q. No.	Ans	Q. No.	Ans	Q. No.	Ans	Q. No.	Ans	Q. No.	Ans
1.	D	2.	D	3.	D	4.	C	5.	C	6.	C
7.	A	8.	D	9.	D	10.	B	11.	C	12.	A
13.	A	14.	B	15.	B	16.	C	17.	C	18.	C
19.	B	20.	B								

■ Practice Paper 10

Q. No.	Ans	Q. No.	Ans	Q. No.	Ans	Q. No.	Ans	Q. No.	Ans	Q. No.	Ans
1.	B	2.	C	3.	B	4.	B	5.	C	6.	C
7.	A	8.	A	9.	B	10.	D	11.	B	12.	A
13.	D	14.	C	15.	C	16.	B	17.	D	18.	C
19.	D	20.	D								

8. Multiple Choice Questions of Weed Management

Practice Paper 1

1. **The practice of following a systematic, rotational sequence of herbicide in the same field is called:**
(A) Herbicide rotation
(B) Crop rotation
(C) Herbicide sequence
(D) None of these

2. **The physically homogenous mixture of herbicide concentrate and the carrier is called:**
(A) Herbicide solution
(B) Herbicide emulsion
(C) Herbicide suspension
(D) All of these

3. **Herbicide concentration dispersed in carrier (and vice-versa), where each component maintain its original identity as minute globules is called:**
(A) Herbicide solution
(B) Herbicide emulsion
(C) Herbicide suspension
(D) All of these

4. **The fine particles of the wattable powder are dispersed in a suitable carrier is called:**
(A) Herbicide solution
(B) Herbicide emulsion
(C) Herbicide suspension
(D) All of these

5. **The movement of a herbicide from the surface into the plant body in a limited zone is called:**
(A) Absorption (B) Adsorption
(C) Translocation (D) None of these

6. **The transfer of a herbicide, food, or other material from one part of the plant to another is called:**
(A) Absorption (B) Adsorption
(C) Translocation (D) None of these

7. **The molecules with electro positive and electro negative regions and they tend to ionize is called:**
(A) Polar compound
(B) Apolar compound
(C) Molecules
(D) All of these

8. **The braodspectrum systematic herbicide glyphosate kills the weeds by inhibiting the biosynthesis of:**
(A) Phenylaqlanine (B) Alanine
(C) Glutamine (D) Cysteine

9. **What kind of parasitic weed, the Orobanche is?**
(A) Holo-stem (B) Hemi-root
(C) Holo-stem (D) Hemi-stem

10. ***Chenopodium album* is a pollution indicator plants for:**
(A) H_2S (B) SO_2
(C) Both H_2S and SO_2 (D) None of these

11. **Molecules with no surface distribution of charge are said to be :**
(A) Polar compound
(B) Apolar compound
(C) Molecules
(D) All of these

12. **Substance capable of antagonizing specific herbicides phytotoxicity to plant is called:**
(A) Safner (B) Antidote
(C) Antagonists (D) All of these

13. **Sprayable bioagent, presently the fungal pores of their extracts, applied like herbicides to kill the existing growth of weeds is called:**
(A) Bioherbicide (B) Mycoherbicide
(C) Both of these (D) None of these

14. **Chemical capable of improving the herbicidal effect in some way is known as:**
(A) Adjuvants (B) Safners
(C) Antidotes (D) All of these

15. **The adjuvants are:**
    (A) Triton & APSA 80
    (B) Agrocer & Active 80
    (C) AG – F & Luffmul
    (D) All of these

16. **Materials which facilitated and accentuates the emulsifying, dispersing, spreading, wetting and other surface modifying proportion of herbicide formulation is called:**
    (A) Surfactants
    (B) Stabilizers
    (C) Herbicide extenders
    (D) None of these

17. **Chemical for obtaining stable emulsion and suspension in the spray tank, individually known as emulsifier and dispersant, respectively known as:**
    (A) Surfactants
    (B) Stabilizers
    (C) Herbicide extenders
    (D) None of these

18. **The non-phytotoxic, specific microbial activity inhibitors delay the biodegradation of soil applied herbicides and thus maintain then weed control activity for some extended period, same time by a week or more is called:**
    (A) Surfactants
    (B) Stabilizers
    (C) Herbicide extenders
    (D) None of these

19. **Coupling of herbicide molecules to some plant constituent, with consequent herbicide deactivation is called:**
    (A) Conjugation
    (B) Coupling agents
    (C) Compatibility agent
    (D) All of these

20. **Isoproturon kills Phalaris minor but does not affect wheat because of the presence of ?**
    (A) Cytochrome P-450 enzyme
    (B) Cytochrome A-370 enzyme
    (C) Cytochrome enzyme
    (D) Cytokinin enzyme

Practice Paper 2

1. **Solvants used to solublize parent herbicide compound to form either its soluble concentrate or emulsifiable concentrate is called:**
   (A) Conjugation
   (B) Coupling agents
   (C) Compatibility agent
   (D) All of these

2. **The measurement of the potency and stimulus by a chemical by means of reach in which it produces in living matter is called:**
   (A) Bioassey (B) Spray drift
   (C) Fate of herbicide (D) None of these

3. **The diluting the herbicide in soil by its transfer or decomposition is known as:**
   (A) Bioassey (B) Spray drift
   (C) Fate of herbicide (D) None of these

4. **A herbicide that kills the target plant by way of inducing excessive production of singlet oxygen in its green tissue cells is called:**
   (A) Bioassey (B) Lesser herbicide
   (C) Fate of herbicide (D) None of these

5. **The tendency of a compound to undergo redox reaction is called:**
   (A) Bioassey (B) Spray drift
   (C) Redox reaction (D) None of these

6. **Weeds that has become an integral part of a crop ecosystem is called:**
   (A) Satellite weed (B) Allien weed
   (C) Facultative weed (D) Obligate weed

7. **The ratio of ED 50 of test crop to ED 50 of associated weeds is called:**
   (A) Selectivity index (B) Crop index
   (C) Both of these (D) None of these

8. **Grain yield in crop yield due to weeding as percentage of yield of the weed free crop is called:**
   (A) Weed Competition Index
   (B) Crop-Weed Competition
   (C) Weed Control Efficiency
   (D) None of these

9. The complete elimination of all live plants, plant parts and seeds of a from an area is called:

(A) Weed eradication (B) Weed prevention
(C) Weed control (D) None of these

10. Chemical employed to keep a herbicide moist longer on the leaf surface is called:

(A) Humicant
(B) Hygroscopic agent
(C) Both of these
(D) None of these

11. A component which causes the leaves or foliage to drop from the plant is called:

(A) Defoliant (B) Defoliator
(C) Both of these (D) None of these

12. The time needed for half the pesticide to disappear from the soil is known as:

(A) Half time (B) Full time
(C) Both of these (D) None of these

13. All measure which deny the entry and establishment of weeds in an area is called:

(A) Weed eradication (B) Weed prevention
(C) Weed control (D) None of these

14. Recommended dose of sulfosulfuron is:

(A) 15-20 g/ha (B) 25-50 g/ha
(C) 50-75 g/ha (D) 75-100 g/ha

15. Which herbicide applied by soil injection method is:

(A) Ethylene
(B) Carbon bi sulphide
(C) Vernolate
(D) All of these

16. The application of herbicides over an entire area is called:

(A) Pre-plant
(B) Pre-emergence
(C) Blanket application
(D) Direct contact application

17. Pre-emergence herbicides are applied in a restricted area over the crop row only is called:

(A) Pre-plant
(B) Band application
(C) Blanket application
(D) Direct contact application

18. Certain non-selective or partially selective herbicides can be used to control weed in the inter rows by directing the spray on weeds is called:

(A) Direct application
(B) Band application
(C) Blanket application
(D) Direct contact application

19. Non-selective herbicides can be employed to control weeds in widely spaced vegetables or ornamentals by covering the crop plants with plastic of metallic covers is called:

(A) Direct application
(B) Band application
(C) Protected spraying
(D) Direct contact application

20. Application of herbicides to small patches of weeds leaving the weed free gaps unsprayed is called:

(A) Direct application
(B) Spot application
(C) Blanket application
(D) Direct contact application

Practice Paper 3

1. Spot application practices in treating patches of:

(A) Annual weeds
(B) Biennial weed
(C) Perennial weeds
(D) Noxious perennial weeds

2. Application of herbicides after the last cultivated in crops is called:

(A) Direct application
(B) Lay by application
(C) Protected application
(D) Band application

3. Highly selective herbicides are suitable for which application:

(A) Blanket application
(B) Lay by application
(C) Protected application
(D) Band application

4. **By which method of application to prevent crop phytotoxicity caused by spray drift that occurs with conventional spray application is:**
   (A) Spraying (B) Broadcast
   (C) Rope-wick (D) None of these

5. **Ideal method of controlling perennial weeds (*Cyperus rotundus*) in standing crop is:**
   (A) Spraying (B) Broadcast
   (C) Rope-wick (D) None of these

6. **Optimum soil moisture is pre-requisite for which type of herbicides:**
   (A) Pre-emergence (B) Post emergence
   (C) Both of these (D) None of these

7. **At what time of herbicide spraying gave maximum benefit:**
   (A) Sunny day (B) Cloudy day
   (C) Night (D) None of these

8. **Untimely rain or heavy irrigation may accentuate:**
   (A) Crop injury (B) Bo effect
   (C) Unknown (D) None of these

9. **The spray tank should be atleast full:**
   (A) One-third (B) One-fourth
   (C) Half (D) Two-third

10. **Which type of spray nozzle should be used for spraying the herbicides:**
    (A) Flood jet nozzle
    (B) Flat fan nozzle
    (C) Both of these
    (D) None of these

11. **Most herbicides are formulated as:**
    (A) Wettable powder (WP)
    (B) Emulsifiable concentration (EC)
    (C) Both of these
    (D) All of these

12. **Under submerged condition which type of herbicides formulation is used:**
    (A) Wettable powder (WP)
    (B) Emulsifiable concentration (EC)
    (C) Granular formulation (G)
    (D) All of these

13. **Application of herbicides through irrigation water is called:**
    (A) Herbigation (B) Herbicultre
    (C) Herbilogy (D) None of these

14. **Herbigation method of herbicide application mostly used in:**
    (A) Maharastra (B) Gujarat
    (C) Rajasthan (D) Punjab

15. **Commercial product of herbicide can be computed by:**

(A) Commercial product = $\frac{\text{Dose in kg a.i./ha}}{\text{\% a.i. in the product}} \times 100$

(B) Commercial product = $\frac{\text{Dose in kg a.i./ha}}{\text{\% a.i. in the product}}$

(C) Commercial product = $\frac{\text{Dose in kg/ha}}{\text{\% a.i. in the product}} \times 100$

(D) Commercial product = $\frac{\text{Dose in kg a.i./ha}}{\text{a.i. in the product}} \times 100$

16. **Isoproturon is available as 75% WP. If the recommended rate of application is 0.75 kg a.i./ha then the amount of commercial product required is:**
    (A) 0.75 kg/ha (B) 1.00 kg/ha
    (C) 1.25 kg/ha (D) 1.50 kg/ha

17. **Isoproturon is available as 50%WP. If the recommended rate of application is 0.75 kg a.i./ha then the amount of commercial product required is:**
    (A) 0.75 kg/ha (B) 1.00 kg/ha
    (C) 1.25 kg/ha (D) 1.50 kg/ha

18. **Paraquat is to be applied at 0.5 kg a.i./ha. The herbicide is available as paraquat which contains 25% paraquat is:**
    (A) 1.0 kg/ha (B) 1.50 kg/ha
    (C) 2.0 kg/ha (D) 2.50 kg/ha

19. Herbicide used to clear the vegetation and as a substitute for tillage is:

(A) Glyphosate (B) Atrazine
(C) Simazine (D) Pendimethalin

20. Excellent control of *Phalaris minor* by:

(A) Cyhalofop-butyl (B) Sulfosulfuron
(C) Isoproturon (D) Metribuzin

Practice Paper 4

1. When chickpea intercropped with mustard which herbicide to be spray for controlling of broad leaved weeds is:

(A) Pendimethalin (B) Fluchloralin
(C) Both of these (D) None of these

2. *Cuscuta* spp is a:

(A) Perrenial (B) Rabi
(C) Kharif (D) Annual

3. Yield is the trade name of:

(A) 2,4,5-T
(B) Methabenzthiazuron
(C) Simazine
(D) Monuron

4. Imparting resistance to normally herbicide susceptible crops to produce:

(A) Herbicide resistant crop
(B) Genetically modified crops
(C) Both of these
(D) None of these

5. A chemical in commercial product that is directly responsible for its herbicidal effect is known as:

(A) Active ingredient (B) Acid equivalent
(C) Both in these (D) None of these

6. The theoretical yield of parent acid from such a herbicide formulation is called:

(A) Active ingredient (B) Acid equivalent
(C) Both in these (D) None of these

7. Long term use of herbicide can lead to the development in weeds:

(A) Resistance (B) Persistence
(C) Both of these (D) None of these

8. In rice-wheat cropping system, which herbicide is more effective for control of *Eichinoclova colonum* is:

(A) Butachlor (B) Metholachlor
(C) Alachlor (D) Isoproturon

9. Prominent bioagent (*Zygogramma bicolorata*) is effective against is:

(A) *Chondrilla juncea*
(B) *Cirsium arvense*
(C) *Cyperus rotundus*
(D) *Parthenium hysterophorus*

10. Prominent bioagent (*Crocidosema lantana*) is effective against is:

(A) *Chondrilla juncea*
(B) *Cirsium arvense*
(C) *Lantana camara*
(D) *Echinoclova spp.*

11. Which species of fish consume large quantity of aquatic weeds:

(A) Chinese grass carp (B) Carp
(C) Both of these (D) None of these

12. Many weeds in India originated in some other part of the world is called:

(A) Alien weeds (B) Native weeds
(C) Facultative weeds (D) Noxicious weeds

13. Weeds occur within the geographical limits of their origin is called:

(A) Alien weeds (B) Native weeds
(C) Facultative weeds (D) Noxicious weeds

14. Weeds grow in highly acidic soils are called:

(A) Acidophills (B) Basiphills
(C) Neutrophills (D) None of these

15. Weeds grow in saline and alkali soils are called:

(A) Acidophills (B) Basiphills
(C) Neutrophills (D) None of these

16. Weeds grow in neutral soils are called:

(A) Acidophills (B) Basiphills
(C) Neutrophills (D) None of these

17. Those weed species that grow primarily in wild commodities but often escape to cultivated field, associated themselves closely with men's affairs is called:

(A) Alien weeds (B) Native weeds
(C) Facultative weeds (D) Noxicious weeds

18. Facultative weeds is also called:

(A) Alien weeds (B) Native weeds
(C) Facultative weeds (D) Apophytes

19. The weeds occur only in the cultivated or otherwise disturbed land is called:
(A) Obligate weeds (B) Native weeds
(C) Facultative weeds (D) Noxicious weeds

20. The weeds undesirable, troublesome and difficult to control is called:
(A) Alien weeds (B) Native weeds
(C) Facultative weeds (D) Noxicious weeds

Practice Paper 5

1. The noxious weeds whose seeds are difficult to separate, once mixed with crop stand is called:
(A) Alien weeds
(B) Objectionable weeds
(C) Facultative weeds
(D) Noxicious weeds

2. Difficult perennial weed are also called:
(A) Alien weeds (B) Native weeds
(C) Pernicious weeds (D) Noxicious weeds

3. Some weeds survive because of their similarity in morphology with the host crop is called:
(A) Mimicry
(B) Native weeds
(C) Facultative weeds
(D) Noxicious weeds

4. Out of the total losses due to various pests, weeds account for nearly:
(A) One-third (B) Half
(C) Two-third (D) Three-fourth

5. The undesirable, injurious, unsightly and troublesome plant, which interfere with cultivated crops and affect human affairs is called:
(A) Weeds (B) Insect
(C) Crop (D) None of these

6. The fibre obtained from weed is:
(A) *Sida cardifolia* (B) *Abutilon indicum*
(C) *Corchorus spp* (D) All of these

7. The industrial effluents absorb to heavy metal by:
(A) Water hyacinth (B) Hydrilla
(C) Pimpernells (D) All of these

8. The natural agencies for dissemination of weeds is:
(A) Wind (B) Water
(C) Animals (D) All of these

9. Dissemination by *Xanthium strumarium* mainly by:
(A) Wind (B) Water
(C) Animals (D) All of these

10. Weeds endanger biodiversity is:
(A) *Mimosa rubicaulis*
(B) *Mikania micrantha*
(C) Both of these
(D) All of these

11. Management of an alien weeds is a great challenge to the weed research scientists in the country is:
(A) *Ipomoea cernua*
(B) *Chenopodium album*
(C) *Avena fatua*
(D) *Convolvulus arvensis*

12. Recently resitance to isoproturon is developed in:
(A) *Phalaris minor*
(B) *Avena fatua*
(C) *Chenopodium album*
(D) None of these

13. New herbicide successfully controlled *Phalaris minor* is:
(A) Cladinafop (B) Fenoxaprop
(C) Sulfosulfuron (D) All of these

14. Post emergence herbicide mentribuzin is recommended for control of broad leaves weeds in:
(A) Chickpea (B) Lentil
(C) Field pea (D) Lathyrus

15. Post emergence herbicide pursuit is recommended for control of broad leaves weeds in:
(A) Pigeonpea (B) Lentil
(C) Field pea (D) Black gram

16. Orobanche is a:
(A) Total root parasite
(B) Total stem parasite
(C) Semi root parasite
(D) Semi stem parasite

17. In pigeonpea + soybean intercropping following herbicide can be used for successful weed control:

(A) 2,4-D (B) Pendimethalin
(C) Isoproturon (D) Butachlor

18. Herbicide used for weed control in sorghum is:

(A) Atrazine at 0.5-1.0 kg a.i./ha
(B) Atrazine at 1.25-1.50 kg a.i./ha
(C) Basalin at 0.5-1.0 kg a.i./ha
(D) Basalin at 1.25-1.50 kg a.i./ha

19. Herbicide has been recommended for the control of weeds in mungbean:

(A) Basalin @ 1.0 kg a.i./ha
(B) 2,4-D @ 1 litre a.i./ha
(C) Simazine @ 1 kg a.i./ha
(D) Atrazine @ 1 kg a.i./ha

20. It has been found that the incorporation of basalin provides relatively more effective control of weeds in mungbean:

(A) Pre-planting spray
(B) Post-emergence (20 DAS)
(C) Pre-emergence
(D) Post emergence (40 DAS)

Practice Paper 6

1. Weeds which complete their life cycle more than two years is called:

(A) Annual (B) Binnial
(C) Perennial (D) None of these

2. Binnial weeds is:

(A) *Daucos carota*
(B) *Trianthema portulacastrum*
(C) *Agropyron repens*
(D) All of these

3. Perennial weeds is:

(A) *Daucos carota*
(B) *Trianthema portulacastrum*
(C) *Agropyron repens*
(D) All of these

4. Many weeds in India originated in some other part of the world is called:

(A) Native weeds (B) Allien weeds
(C) Facultative weeds (D) Obligate weeds

5. Weeds occur within the geographical limits of their origin is called:

(A) Native weeds (B) Allien weeds
(C) Facultative weeds (D) Obligate weeds

6. The weed species that grow primarily is called:

(A) Native weeds (B) Allien weeds
(C) Facultative weeds (D) Obligate weeds

7. Weeds of crop land is capable of surviving is a wild community is called:

(A) Native weeds (B) Allien weeds
(C) Facultative weeds (D) Obligate weeds

8. Facultative weeds are also called:

(A) Apopytes (B) Symplasts
(C) Apoplasts (D) None of these

9. Alien weeds are:

(A) *Eichhornia crasiipus*
(B) *Lantana camara*
(C) *Argemone mexicana*
(D) All of these

10. Weeds that undesirable, troublesome and difficult to control is called:

(A) Native weeds
(B) Noxious weeds
(C) Facultative weeds
(D) Obligate weeds

11. Noxious weeds whose seeds are difficult to separate, when once mixed with crop stand is called:

(A) Objectionable weeds
(B) Allien weeds
(C) Facultative weeds
(D) Obligate weeds

12. Difficult perennial weeds are also called:

(A) Native weeds (B) Allien weeds
(C) Pernicious (D) Obligate weeds

13. Those weeds are similar in morphology with the host crop is called:

(A) Mimicry (B) Allien weeds
(C) Facultative weeds (D) Obligate weeds

14. Which one is a troublesome weeds of agricultural fields?

(A) *Chenopodim album*
(B) *Cyperus rotundus*
(C) *Partthenium hysterophorus*
(D) *Eichhornia crassipus*

15. Congress grass is:

(A) *Cynodon dactylon*
(B) *Digitaria purpurea*
(C) *Cymbopogon citrates*
(D) *Parthenium hysterophorus*

16. Troublesome American water weed found in India is:

(A) *Eichhornia* (B) *Typha*
(C) *Cyprus* (D) *Trapa*

17. The herbicide applied in pulse crops is:

(A) Isoproturon (B) Pendimethalin
(C) 2, 4-D (D) Atrazine

18. Pendimethalin is pulses is applied as:

(A) Pre-planting
(B) Pre-emergence
(C) Post emergence
(D) None of the above

19. Which weedicide and its dose is correct for groundnut is:

(A) Isoproturon @ 1.5 kg/ha
(B) Glyphoset @ 4.00 lit/ha
(C) Pendimethalin @ 2.2 lit/ha
(D) Pendimethalin @ 3.3 lit/ha

20. The spray drift of 2,4-D (ester from) can cause considerable damage in:

(A) Maize (B) Rice
(C) Sorghum (D) Cotton

Practice Paper 7

1. Dispersing agents reduce .... between similar particles:

(A) Adhesion (B) Cohesion
(C) Tension (D) None of these

2. Addition of ammonium sulphate to a herbicidal spray solution enhances the .... against *Agropyron repens*:

(A) Glyphosate (B) Amylase
(C) Reductase (D) Oxi-reductase

3. Propanil, a post-emergence herbicide is readily absorbed by the :

(A) Stem (B) Root
(C) Leaf (D) None of these

4. Once the phenoxy herbicide penetrates the cuticle, it moves inside the plant:

(A) Only apoplastically
(B) Only symplastically
(C) By both mechanisms
(D) None of these

5. Thiocarbamates *viz.*, a soil-incorporated herbicide is readily absorbed by roots and coleoptile and .... translocated:

(A) Only upward (B) Only downward
(C) Both of these (D) None of these

6. It has been observed that propanil (a herbicide) interfred with ATP generation and electron transport in .... mitochondria:

(A) Wheat (B) Barley
(C) Srghum (D) Greengram

7. Triazines and phenoxy acids are subject to ring cleavage, triazine, heterocyclic ring can be cleaved resulting in liberation of:

(A) Chlorine (B) Oxygen
(C) (D)

8. Urea herbicides undergo conjugation with:

(A) Lipids (B) Carbohydrates
(C) Proteins (D) All of these

9. Glyphosate has been reported to be metabolized to aminomethylphosphoric acid and glyphosate in:

(A) Maize (B) Sugarcane
(C) Sunflower (D) Safflower

10. The .... structure of the herbicide also has a great influence on its volatility through adsorption to soil:

(A) Biological
(B) Chemical
(C) Physical
(D) Biphysical

11. .... comprises an important mechanism of herbicide detoxification in soil under field conditions especially surface incorporation subsequent to rainfall or irrigation:

(A) Thermodecomposition
(B) Biodecomposition
(C) Chemodecomposition
(D) Photodecomposition

12. Photodecomposition plays an important role in the loss of urea herbicides from the soil surface in which areas:

(A) Moist (B) Wet
(C) Dry (D) None of these

13. .... degradation plays a major role in the persistence and behaviour of herbicides in the soil:

(A) Microbial (B) Physical
(C) Chemical (D) Thermal

14. The organisms involved in herbicide detoxification include:

(A) Earthworms (B) Nematodes
(C) Algae (D) None of these

15. If the weedicide kills the plants after reaching in to the system, the weedicide is a:

(A) Contact type
(B) Translocated type
(C) Selective type
(D) None-selective type

16. The herbicide that kills the contact part of the weeds is known as:

(A) Contact type
(B) Translocated type
(C) Selective type
(D) None-selective type

17. 2,4-D kills only:

(A) Broad leaved weeds
(B) Narrow leaved weeds
(C) Both of the above
(D) None of these

18. Prickly pear can be killed biologically by:

(A) *Dactylopius tomentosus*
(B) *Lemna minor*
(C) *Hydrilla*
(D) None of these

19. Application of propanil after .... in rice causes prohibitive vegetative injury:

(A) DDT (B) Endosulfan
(C) Carbofuran (D) Sevin

20. The *Phalaris minor* can be controlled by using the weedicide called .... at 0.75-1.25 kg a.i./ha depending upon soil type:

(A) 2,4-D (B) Isoproturon
(C) Basalin (D) Simazine

Practice Paper 8

1. The rate of application of atrazine in maize under heavy soils varies from:

(A) 0.75 kg a.i./ha
(B) 1.25-1.50 kg a.i./ha
(C) 1.75-2.00 kg a.i./ha
(D) 1.00-1.25 kg a.i./ha

2. Soybean crop should be free from weed competition during first .... days for its good productivity:

(A) 10-20 (B) 20-30
(C) 30-40 (D) 40-50

3. Fluchloralin can be used for weed control in soybean as:

(A) Pre-emergence
(B) Pre-plant incorporation
(C) Post-emergence
(D) Through herbigation

4. The weeds can be controlled in linseed by the application of MCPB at:

(A) 0.5 kg a.i./ha (B) 1.5 kg a.i./ha
(C) 1.0 kg a.i./ha (D) 2.0 kg a.i./ha

5. The critical period of weed control in upland rice varies between:

(A) 15-20 DAS (B) 20-25 DAS
(C) 25-30 DAS (D) 30-35 DAS

6. For mungbean the period between .... days after sowing has been regarded as critical for weed control:

(A) 0 and 30 (B) 5 and 25
(C) 15 and 40 (D) 25 and 45

7. *Parthenium hysterophorus* may be controlled biologically by:

(A) *Zygogramma biocolorata*
(B) *Delias hyperata*
(C) *Melanogromyra* spp.
(D) All of these

8. *Sorghum helepense* may be controlled biologically by:

(A) Geese (B) Hydrilla
(C) Verticillata (D) None of these

9. Which of the following tillage system is useful to control *Phalaris minor*?

(A) Reduced tillage (B) FIRBS
(C) Zero tillage (D) None of these

10. Which one is pre plant applied herbicide?

(A) Diquat (B) Fluchloralin
(C) Alachlor (D) B and C

11. If farmer wants to apply Atrazine 80% @ 0.5 kg a.i./ha, the quantity of atrazine would be:

(A) 0.40 kg (B) 0.48 kg
(C) 0.52 kg (D) 0.62 kg

12. Important monocot weed found in most of the wheat field is:

(A) *Cyperus rotundus*
(B) *Phalaris minor*
(C) *Avena fatua*
(D) *Cynodon dactylon*

13. The weed that causes Asthama is:

(A) Dommon grass weed
(B) Kunch weed
(D) Parthenium
(D) Calotrapis

14. Weedicide 2,4-D is not recommended in the crop:

(A) Sugarcane (B) Maize
(C) Moong (D) Potato

15. Glyphosate is a:

(A) Selective translocated herbicide
(B) Non selective translocated herbicide
(C) Non selective, non translocated herbicide
(D) Contact type herbicide

16. Triazines are mostly used in:

(A) Barley (B) Wheat
(C) Pulses (D) Maize

17. Montha belongs to the family:

(A) Gramineae (B) Malvaceae
(C) Cyperaceae (D) Liliaceae

18. Isoproturon resistance in *Phalaris minor* was reported by:

(A) Singh and Mishra
(B) Malik and Singh
(C) Singh and Singh
(D) Singh and Prasad

19. Which of the following is a selective contact herbicide:

(A) 2,4-D (B) Propanil
(C) Butachlor (D) Bromoxynil

20. Butachlor is an ideal weedicide for:

(A) Pre-emergence application in direct seeded rice
(B) Soon after the transplanting of rice
(C) Before the transplantation of rice
(D) None of these

Practice Paper 9

1. Which weedicide is used to kill weeds of broad leaves in legumes?

(A) 2, 4-D (B) 2, 4, 5-T
(C) 2, 4-DB (D) None of these

2. Sprayable bioagent, presently the fungal pores of their extracts, applied like herbicides to kill the existing growth of weeds is called:

(A) Bioherbicide (B) Bioinsecticide
(C) Biopesticide (D) None of these

3. Which type of herbicides also known as Mycoherbicides:

(A) Bioherbicide (B) Bioinsecticide
(C) Biopesticide (D) None of these

4. The plant pathogen *Cercospora rodmanii* is used to control the weeds of :

(A) *Abutilon theophorsii*
(B) *Aeschynomene virginiea*
(C) *Echhornia crassipes*
(D) *Malva pustilla*

5. Application of herbicides to small patches of weeds leaving the weed free-gaps unsprayed. This is mostly practices in treating patches of noxious perennial weeds with potential herbicides.

(A) Protected spraying
(B) Rope Wick application
(C) Blanket application
(D) Spot application

6. Application of herbicides after the last cultivation in crops, for example after ridging in sugarcane or cotton, after earthing - up in potato, etc. Usually soil-acting herbicides are used, which will prevent further emergence of weeds is known as:

(A) Lay by application
(B) Rope Wick application
(C) Blanket application
(D) Band application

7. **Technique is used in the application of non-selective herbicides particularly the translocative type (e.g. glyphosate) is known as:**
(A) Protected spraying
(B) Rope Wick application
(C) Blanket application
(D) Band application

8. **Non-selective herbicides can be employed to control weeds in widely spaced vegetables or ornamentals by covering the crop plants with plastic of metallic covers. This is practicable only in high value plants or crops is known as:**
(A) Protected spraying
(B) Rope Wick application
(C) Blanket application
(D) Band application

9. **Application of herbicides over an entire area. In case of post-emergence applications this includes crop and weeds and hence only highly selective herbicides are suitable for such applications is known as:**
(A) Protected spraying
(B) Rope Wick application
(C) Blanket application
(D) Band application

10. **Pre-emergence herbicides are applied in a restricted are over the crop row only is known as:**
(A) Protected spraying
(B) Rope Wick application
(C) Blanket application
(D) Band application

11. **Isoproturon is available as 50% WP. If the recommended rate of application is 0.75 kg a.i./ha then the amount of commercial product required is:**
(A) 1.0 kg/ha (B) 1.5 kg/ha
(C) 2.0 kg/ha (D) 2.5 kg/ha

12. **Isoproturon is available as 75% WP. If the recommended rate of application is 0.75 kg a.i./ha then the amount of commercial product required is:**
(A) 1.0 kg/ha (B) 1.5 kg/ha
(C) 2.0 kg/ha (D) 2.5 kg/ha

13. **Paraquat is to be applied at 0,5 kg a.i./ha. The herbicide is available as Gramoxone (commercial name of paraquat) which contains 25% paraquat. The quantity of Gramoxone required is:**
(A) 1.5 kg/ha (B) 2.0 kg/ha
(C) 2.5 kg/ha (D) 3.0 kg/ha

14. **Application of herbicide in the form of dust not recommended due to:**
(A) More hazards (B) Less hazards
(C) Drift hazards (D) None of these

15. **A herbicide formulation which is a clear solution turns milky while when mixed with water is:**
(A) Soluble concentrate
(B) Suspension concentrate
(C) Aqueous concentrate
(D) Emulsifiable concentrate

16. **Butachlor is recommended for control of weeds in:**
(A) Maize (B) Wheat
(C) Transplanted rice (D) Sugarcane

17. **Successful control of Kans (*Saccharum spontaneum*) is done by:**
(A) Glyphosate 41/SL 2-3 litres/ha
(B) Glyphosate 41/SL 3-4 litre/ha
(C) Atrazine 50% w.p. 3 litre/ha
(D) Pendimethalin 30 E.C. 3.3 litre/ha

18. ***Orobanche* sp. is an associated weed of:**
(A) Potato (B) Sugarcane
(C) Tobacco (D) Berseem

19. **The herbicides containing carbon and hydrogen in their molecules are called:**
(A) Arsenic (B) Organic herbicide
(C) Salt (D) Acid

20. **Which of the following is now more suitable for control of *Phalaris minor* in wheat field:**
(A) Sulfo-sulphuran
(B) Pendimethalin
(C) Isoproturon
(D) Methabenzthiazuron

Practice Paper 10

1. **Which of the following herbicide is more suitable for control of broad leave weeds in pulse crop as a pre-emergence application:**
   (A) Pendimethaline (B) Methabenthiazuran
   (C) Metribuzin (D) Basalin

2. **Which of the following herbicide is more suitable for control of broad leave weeds in pulse crop as a pre-plant application:**
   (A) Pendimethaline
   (B) Methabenthiazuran
   (C) Metribuzin
   (D) Basalin

3. **Which of the following herbicide is more suitable for control of broad leave weeds in urd crop as a standing crop:**
   (A) Pendimethaline
   (B) Methabenthiazuran
   (C) Metribuzin
   (D) Basalin

4. **Which of the following herbicide is more suitable for control of broad leave weeds in field pea crop as a standing crop:**
   (A) Pendimethaline
   (B) Methabenthiazuran
   (C) Metribuzin
   (D) Basalin

5. **Which one of the following is not narrow leaved weeds?**
   (A) *Cynodon dactylon*
   (B) *Cyperus rotundus*
   (C) Setaria glouca
   (D) Melilotus indica

6. **Weedicide used for control of broadleaf weeds in pulse crops is:**
   (A) 2, 4-D (B) Basaline
   (C) Atrazine (D) Butachor

7. **Propanil herbicide is readily absorbed by:**
   (A) Stem (B) Root
   (C) Leaf (D) None of these

8. **Read the following and mark the correct answer:**

Cropping system	**Herbicide**
(A) Potato + wheat	Pendimethalin
(B) Potato + mustard	Pendimethalin
(C) Chickpea + Linseed	Pendimethalin
(D) Sugarcane + Potato	Pendimethalin

9. **Impirical formula of weed control efficiency:**

   (A) $\frac{\text{Dry weight of weeds in control plot} - \text{Dry weight of weeds in treated plot}}{\text{Dry weight of weeds in control plot}} \times 100$

   (B) $\frac{\text{Dry matter weight of weeds in control plot} - \text{Drymatter weight of weeds in treated plot}}{\text{Dry matter weight of weeds in treated plot}} \times 100$

   (C) $\frac{\text{Dry matter weight of weeds in control plot} - \text{Drymatter weight of weeds in treated plot}}{\text{Dry matter weight of weeds in control plot}}$

   (D) $\frac{\text{Dry matter weight of weeds in control plot} - \text{Drymatter weight of weeds in treated plot}}{\text{Dry matter weight of weeds in treated plot}}$

10. **Impirical formula of weed index:**

   (A) $\frac{\text{Yield from weed free plots} - \text{Yield from treated plots}}{\text{Yield from treated plots}} \times 100$

   (B) $\frac{\textit{Yield from weed free plots} - \textit{Yield from treated plots}}{\textit{Yield from weed free plots}} \times 100$

(C) $\frac{\text{Yield from weed free plots} - \text{Yield from treated plots}}{\text{Yield from weed free plots}}$

(D) $\frac{\text{Yield from weed free plots} - \text{Yield from treated plots}}{\text{Yield from treated plots}}$

11. Impirical formula of weed density:

(A) $\frac{\text{No. of individuals in all quadrates}}{\text{No. of all species in all quadrates}}$

(B) $\frac{\text{No. of all species in all quadrates}}{\text{No. of individuals in all quadrates}} \times 100$

(C) $\frac{\text{No. of individuals in all quadrates}}{\text{No. of all species in all quadrates}} \times 100$

(D) $\frac{\text{No. of al species in all quadrates}}{\text{No. of individuals in all quadrates}}$

12. Impirical formula of weed per cent occurrence:

(A) $\frac{\text{No. of occurrences of a species in a district}}{\text{Total no. of observations recorded in a district}}$

(B) $\frac{\text{No. of occurrences of a species in a district}}{\text{Total no. of observations recorded in a district}}$

(C) $\frac{\text{Total no. of observations recorded in a district}}{\text{No. of occurrences of a species in a district}} \times 100$

(D) $\frac{\text{No. of occurrences of a species in a district}}{\text{Total no. of observations recorded in a district}} \times 100$

13. Weed control efficiency formula has been suggested by:

(A) Patil and Patil (1983)
(B) Gill and Kumar (1969)
(C) Misra (1968)
(D) Raju (1977)

14. Weed index formula has been suggested by:

(A) Patil and Patil (1983)
(B) Gill and Kumar (1969)
(C) Misra (1968)
(D) Raju (1977)

15. Weed density formula has been suggested by:

(A) Patil and Patil (1983)
(B) Gill and Kumar (1969)
(C) Misra (1968)
(D) Raju (1977)

16. Weed per cent occurrence formula has been suggested by:

(A) Patil and Patil (1983)
(B) Gill and Kumar (1969)
(C) Misra (1968)
(D) Raju (1977)

17. Paraquat kills plant by:

(A) Inhibiting respiration
(B) Inhibiting Pigment System-II
(C) Inhibiting Pigment System-I
(D) Inhibiting dark reaction

18. Glyphosate is a broad spectrum herbicide. Upon application to leaves, it is translocated to meristematic areas and underground rhizomes by:

(A) Phloem (B) Xylem
(C) Border piths (D) Tracheids

19. Atrazine effectively control grasses in maize field without causing harm to the crop plants, as degradation of the chemical in maize plant occurs due to the presence of:

(A) R-Q enzyme (B) Amylases
(C) Aryl Acyl amidase (D) GSH enzyme

20. A narrow leaf dicotyledonous weed is:

(A) *Sphenoclea zeylanica*
(B) *Spergula arvensis*
(C) *Conyza canadiensis*
(D) Asphodelus tennifolius

Answers Key

■ Practice Paper 1

Q. No.	Ans	Q. No.	Ans	Q. No.	Ans	Q. No.	Ans	Q. No.	Ans	Q. No.	Ans
1.	A	2.	A	3.	B	4.	C	5.	A	6.	C
7.	A	8.	A	9.	A	10.	B	11.	B	12.	D
13.	C	14.	A	15.	D	16.	A	17.	B	18.	C
19.	A	20.	A								

■ Practice Paper 2

Q. No.	Ans	Q. No.	Ans	Q. No.	Ans	Q. No.	Ans	Q. No.	Ans	Q. No.	Ans
1.	B	2.	A	3.	C	4.	B	5.	C	6.	A
7.	A	8.	A	9.	A	10.	C	11.	C	12.	A
13.	B	14.	B	15.	D	16.	C	17.	B	18.	A
19.	C	20.	B								

■ Practice Paper 3

Q. No.	Ans	Q. No.	Ans	Q. No.	Ans	Q. No.	Ans	Q. No.	Ans	Q. No.	Ans
1.	D	2.	B	3.	A	4.	C	5.	C	6.	A
7.	A	8.	A	9.	A	10.	C	11.	C	12.	C
13.	A	14.	B	15.	A	16.	B	17.	D	18.	C
19.	A	20.	B								

■ Practice Paper 4

Q. No.	Ans	Q. No.	Ans	Q. No.	Ans	Q. No.	Ans	Q. No.	Ans	Q. No.	Ans
1.	C	2.	A	3.	B	4.	C	5.	A	6.	B
7.	A	8.	A	9.	D	10.	C	11.	A	12.	A
13.	B	14.	A	15.	B	16.	C	17.	C	18.	D
19.	A	20.	D								

■ Practice Paper 5

Q. No.	Ans	Q. No.	Ans	Q. No.	Ans	Q. No.	Ans	Q. No.	Ans	Q. No.	Ans
1.	B	2.	C	3.	A	4.	A	5.	A	6.	D
7.	A	8.	D	9.	C	10.	C	11.	A	12.	A
13.	D	14.	C	15.	D	16.	A	17.	B	18.	A
19.	A	20.	A								

■ Practice Paper 6

Q. No.	Ans	Q. No.	Ans	Q. No.	Ans	Q. No.	Ans	Q. No.	Ans	Q. No.	Ans
1.	C	2.	A	3.	C	4.	B	5.	A	6.	C
7.	D	8.	A	9.	D	10.	B	11.	A	12.	C
13.	A	14.	C	15.	D	16.	A	17.	B	18.	B
19.	D	20.	D								

■ Practice Paper 7

Q. No.	Ans	Q. No.	Ans	Q. No.	Ans	Q. No.	Ans	Q. No.	Ans	Q. No.	Ans
1.	B	2.	A	3.	C	4.	C	5.	C	6.	D
7.	B	8.	C	9.	A	10.	B	11.	D	12.	C
13.	A	14.	C	15.	B	16.	A	17.	A	18.	C
19.	C	20.	B								

■ Practice Paper 8

Q. No.	Ans	Q. No.	Ans	Q. No.	Ans	Q. No.	Ans	Q. No.	Ans	Q. No.	Ans
1.	B	2.	C	3.	B	4.	A	5.	C	6.	B
7.	A	8.	C	9.	C	10.	D	11.	D	12.	B
13.	D	14.	B	15.	B	16.	D	17.	C	18.	B
19.	B	20.	B								

■ Practice Paper 9

Q. No.	Ans	Q. No.	Ans	Q. No.	Ans	Q. No.	Ans	Q. No.	Ans	Q. No.	Ans
1.	C	2.	A	3.	A	4.	C	5.	D	6.	A
7.	B	8.	A	9.	C	10.	D	11.	B	12.	A
13.	B	14.	C	15.	D	16.	C	17.	B	18.	C
19.	B	20.	A								

■ Practice Paper 10

Q. No.	Ans	Q. No.	Ans	Q. No.	Ans	Q. No.	Ans	Q. No.	Ans	Q. No.	Ans
1.	A	2.	D	3.	B	4.	C	5.	D	6.	B
7.	C	8.	C	9.	A	10.	B	11.	C	12.	D
13.	A	14.	B	15.	C	16.	D	17.	C	18.	A
19.	D	20.	D								

9. Multiple Choice Questions of Water Management

Practice Paper 1

1. **Prediction of that the share of water for agriculture will reduce by 2025 AD is:**
(A) 78% (B) 73%
(C) 70% (D) 65%

2. **On domestic sector front, the present per capita annual water availability of 2100 m^3 is likely to decrease to about 1700 m^3 by:**
(A) 2000 AD (B) 2025 AD
(C) 2030 AD (D) 2035 AD

3. **In India, agriculture sector consumes around .... of total water use is:**
(A) 89 % (B) 80%
(C) 75% (D) 70%

4. **Indian, domestic sector consumes around .... of total water use is:**
(A) 7% (B) 8 %
(C) 9% (D) 10%

5. **Indian, industrial sector consumes around .... of total water use is:**
(A) 3% (B) 4 %
(C) 5% (D) 6%

6. **Artificial application of water to soil to escape stress of plant is called:**
(A) Irrigation (B) Drainage
(C) Both of these (D) None of these

7. **The main objectives of irrigation is/are:**
(A) To make soil surface soft
(B) For proper seed germination
(C) Proper growth of plant
(D) All of these

8. **Water available in the soil at standard crop growth- stages is called:**
(A) Soil moisture constant
(B) Water holding capacity
(C) Field capacity
(D) None of these

9. **When all the pores of the soil are filled with water, the soil said to be:**
(A) Saturation
(B) Water holding capacity
(C) Both of these
(D) None of these

10. **For production of 1 kg dry matter, how much water is needed:**
(A) 300-400 litres (B) 400-500 litres
(C) 500-600 litres (D) 600-700 litres

11. **Loss of cell turgor and wilting cessation of cell enlargement, closure of stomata, reduction of photosynthesis and interference with many basic metabolic processes is accompanied by:**
(A) Decreasing water content
(B) Increasing water content
(C) Increasing water pressure
(D) Decreasing water pressure

12. **The apical portion of roots and stem contain water:**
(A) More than 90% (B) 80 %
(C) 70 % (D) Less than 60%

13. **The wood of large trees may contain water:**
(A) More than 90% (B) 80 %
(C) 70 % (D) 50- 60%

14. **The stem of barley, wheat and sorghum contain water:**
(A) More than 90% (B) 80 %
(C) 60-70 % (D) Less than 50%

15. **The freshly harvested grains of most crop contain water:**
(A) More than 30% (B) 20-25 %
(C) 10-15 % (D) Less than 10%

16. **The moisture content in the grains which determines that:**
(A) Storage life
(B) Viability
(C) Germinability
(D) All of these

17. Water is needed mainly to meet the demands of evaporation, transpiration and metabolic needs of plants, all together is known as:
(A) Consumptive use
(B) Irrigation requirement
(C) Evapo-transpiration
(D) Water requirement

18 At maximum water holding capacity, water held at:
(A) Less than 1 bar (B) 1 bar
(C) More than 1 bar (D) None of these

19. Water stored in soil than field capacity is:
(A) Same (B) Double
(C) Triple (D) Any amount

20. The maximum amount of moisture that will be held in the pores of a saturated soil is known as:
(A) Saturation
(B) Water holding capacity
(C) Field capacity
(D) All of these

Practice Paper 2

1. To constitute field capacity together with:
(A) Capillary water
(B) Hygroscopic water
(C) Both of these
(D) None of these

2. The moisture constant of a soil at field capacity is taken as:
(A) 25 % (B) 50 %
(C) 75 % (D) 100 %

3. Field capacity generally comes after:
(A) 10-12 hrs (B) 12-24 hrs
(C) 24-48 hrs (D) 48-72 hrs

4. At field capacity water held between:
(A) 0.1 to 0.3 bar (B) less than 0.1 bar
(C) 15 bar (D) 31 bars

5. The moisture level at which plants will no longer get enough moisture for their survival is called:
(A) Wilting point
(B) Filed capacity
(C) Saturation
(D) Water holding capacity

6. Wilting point in most soils is in the region of field capacity is:
(A) 25 % (B) 50 %
(C) 75 % (D) 100 %

7. Permanent wilting point is also known as:
(A) Wilting coefficient
(B) Hygroscopic coefficient
(C) Moisture equivalent
(D) None of these

8. Moisture held at permanent wilting point is:
(A) 0.1 to 0.3 bar (B) less than 0.1 bar
(C) 15 bar (D) 31 bars

9. Soil moisture between field capacity and permanent wilting point is referred as:
(A) Available soil moisture
(B) Field capacity
(C) Saturation
(D) Water holding capacity

10. Air dried soil at atmosphere contains tightly bound water with soil particles and water of macro and micro pores are completely replaced by air is called:
(A) Wilting coefficient
(B) Hygroscopic coefficient
(C) Moisture equivalent
(D) Sticky point

11. The amount of water retained by a sample of initially saturated soil material after being subjected to a centrifugal force of 1000 times that of gravity for a definite period of time, usually half a hour is called:
(A) Wilting coefficient
(B) Hygroscopic coefficient
(C) Moisture equivalent
(D) Sticky point

12. Hygroscopic coefficient is mainly depend upon:
(A) Colloidal phase of the soil
(B) Available soil moisture of the soil
(C) Field capacity of the soil
(D) None of these

13. Clay soil contains more colloidal materials and hygroscopic water than:
(A) Sandy soil (B) Clay loam
(C) Sandy loam (D) All of these

14. Moisture equivalent is measures in laboratory, generally used in place of:

(A) Field capacity
(B) Permanent wilting point
(C) Temporary wilting point
(D) All of these

15. Moisture equivalent is also used for available moisture in:

(A) Micro capillary pores
(B) Macro capillary pores
(C) Both of these
(D) None of these

16. In medium textured soil moisture content equivalent to:

(A) Field capacity
(B) Permanent wilting point
(C) Temporary wilting point
(D) All of these

17. In clay soil the field capacity is .... than moisture equivalent:

(A) Less than (B) More than
(C) Same (D) Unknown

18. In sandy soil the field capacity is .... than moisture equivalent:

(A) Less than (B) More than
(C) Same (D) Unknown

19. The performance of soil moisture of hygroscopic water is:

(A) 31-10000 atm (B) 31 atm
(C) 15 atm (D) 0.3-31 atm

20. The performance of soil moisture of hygroscopic coefficient is:

(A) 31-10000 atm (B) 31 atm
(C) 15 atm (D) 0.3-31 atm

Practice Paper 3

1. The performance of soil moisture of wilting coefficient is:

(A) 31-10000 atm (B) 31 atm
(C) 15 atm (D) 0.3-31 atm

2. The performance of soil moisture of capillary water is:

(A) 31-10000 atm (B) 31 atm
(C) 15 atm (D) 0.3-31 atm

3. The performance of soil moisture of moisture equivalent is:

(A) 0.3-1 atm (B) 0.3 atm
(C) less than 0.3 atm (D) 0.3-31 atm

4. Poiseulle's law describes as:

(A) Volume of water flow is proportional to radius of the tube
(B) Volume of water flow is proportional to pressure difference
(C) Volume of water flow is inversely related to viscosity of liquid
(D) All of these

5. Total fresh water available in the world on the earth is:

(A) 2.5% (B) 2.6%
(C) 2.7% (D) 2.9%

6. Wells and ponds irrigation is known as:

(A) Macro irrigation
(B) Minor irrigation
(C) Both (A) and (B)
(D) None of these

7. Water movement in saturated soil is measured by:

(A) Poiseuille's law (B) Brown's law
(C) Newton's Law (D) None of these

8. The performance of soil moisture of field capacity is:

(A) 0.3-1 atm (B) 0.3 atm
(C) less than 0.3 atm (D) 0.3-31 atm

9. The performance of soil moisture of sticky point is:

(A) 0.3-1 atm (B) 0.3 atm
(C) less than 0.3 atm (D) 0.3-31 atm

10. The performance of soil moisture of gravitational water is:

(A) 0.3-1 atm (B) 0.3 atm
(C) less than 0.3 atm (D) 0.3-31 atm

11. Moisture content of a soil at which it does not sticks with other materials is called:

(A) Wilting coefficient
(B) Hygroscopic coefficient
(C) Moisture equivalent
(D) Sticky point

12. Sticky point occurs at:

(A) 1/3 atm (B) ½ atm
(C) 1 atm (D) 2/3 atm

13. After the soil is completely saturated the excess infiltration that starts moving downwards due to the force of gravity to finally join the ground water reservoir is known as:

(A) Gravitational water
(B) Capillary water
(C) Hygroscopic water
(D) Crystalline water

14. Movement of gravitational water down the soil profile depends upon the:

(A) Permeability
(B) Infiltration
(C) Seepage
(B) None of these

15. Gravitational water is also called:

(A) Free water
(B) Joint water
(C) Combined water
(D) None of these

16. Gravitational water retained in the soil at:

(A) Below 1/3 atm
(B) at 1/3 atm
(C) Above 1/3 atm
(D) None of these

17. Gravitational water remain in the soil too long, serious damage may take place due to:

(A) Lack of O_2 (B) Lack of CO_2
(C) Excess of O_2 (D) Excess of CO_2

18. The amount of water will remain held within the soil pores due to capillary force is called:

(A) Gravitational water
(B) Capillary water
(C) Hygroscopic water
(D) Crystalline water

19. Capillary water behave according to:

(A) Law of gravity
(B) Law of capillary
(C) Law of atmosphere
(D) None of these

20. Capillary water retained in the soil at tension of:

(A) 0.3-1 atm (B) 0.3 atm
(C) less than 0.3 atm (D) 0.3-31 atm

Practice Paper 4

1. Capillary water held at tension of 15-31 atm is known as:

(A) Intercapillary water
(B) Intra capillary water
(C) Both of these
(D) None of these

2. Availability of intercapillary water to the plants is:

(A) Never available (B) Hardly available
(C) Not available (D) More available

3. The part of the water which a dry soil will absorbed in order to reach a static state with the atmosphere water is known as:

(A) Gravitational water
(B) Capillary water
(C) Hygroscopic water
(D) Crystalline water

4. Hygroscopic water retained in the soil at tension of:

(A) 0.3-31 atm (B) 31 atm
(C) less than 31 atm (D) More than 31 atm

5. Hygroscopic water held by the force of molecular attraction and can be removed only by:

(A) Heating (B) Condensation
(C) Both of these (D) None of these

6. Availability of Hygroscopic water to the plants is:

(A) Never available (B) Hardly available
(C) Not available (D) More available

7. The water is held in the soil particles with the chemical forces is called:

(A) Gravitational water
(B) Capillary water
(C) Combined water
(D) Crystalline water

8. The water is present in intra spaces of minerals and binds with minerals is called:

(A) Gravitational water
(B) Capillary water
(C) Combined water
(D) Crystalline water

9. **The water is present in soil in the form of vapour specially at higher temperature is known as:**
(A) Vapour water
(B) Capillary water
(C) Combined water
(D) Crystalline water

10. **Which form of water is not available to the plants is:**
(A) Vapour water
(B) Crystalline water
(C) Combined water
(D) All of these

11. **The quantity of water, which is needed for the maximum yield of crop is called:**
(A) Water Requirement
(B) Irrigation Requirement
(C) Consumptive Use of water
(D) None of these

12. **Quantity of surface and ground water absorbed by crops at transpired or use directly in the building of plant tissues together with that evaporated from the cropped area is called:**
(A) Water Requirement
(B) Irrigation Requirement
(C) Consumptive Use of water
(D) None of these

13. **Total water required of a crop, excluding of effective rainfall and soil moisture stored in the root zone or that contributed from the shallow ground water table is known as:**
(A) Water Requirement
(B) Irrigation Requirement
(C) Consumptive Use of water
(D) None of these

14. **Downward movement of water through saturated or nearly saturated soil in response to the force of gravity is known as:**
(A) Percolation (B) Seepage
(C) Infiltration (D) Permeability

15. **Lateral movement of water is called:**
(A) Percolation (B) Seepage
(C) Infiltration (D) Permeability

16. **The ratio of flow of water through a porus medium as it directly proportional to the cros sectional area of the flow and the hydraulic gradient under which the flow moves is known as:**
(A) Percolation
(B) Seepage
(C) Infiltration
(D) Permeability

17. **Flow of ground water through a porus medium is controlled by:**
(A) Law of gravity
(B) Law of capillary
(C) Law of atmosphere
(D) Darcy's law

18. **The rate of flow of water through a unit cross-sectional area under a unit hydraulic gradient is known as:**
(A) Coefficient of permeability
(B) Hygroscopic coefficient
(C) Wilting coefficient
(D) None of these

19. **The ratio of irrigation water transpired by the crop on irrigation farm or project during their growth period over the water diverted from a river or other natural source into the farm or project canal or canals during the same period is known as:**
(A) Irrigation application efficiency
(B) Project efficiency
(C) Water conveyance efficiency
(D) Water storage efficiency

20. **The ratio of volume of irrigation water consumptively used in a specified period to the total volume of irrigation water that has been diverted from the capture system is called:**
(A) Irrigation application efficiency
(B) Project efficiency
(C) Water conveyance efficiency
(D) Water storage efficiency

Practice Paper 5

1. **The ratio of water diverted to the farm to water diverted from the river or reservoir is called:**
   (A) Irrigation application efficiency
   (B) Project efficiency
   (C) Water conveyance efficiency
   (D) Water storage efficiency

2. **The ratio of water stored in the root zone during the irrigation to the water needed in the root zone prior to the irrigation is called:**
   (A) Irrigation application efficiency
   (B) Project efficiency
   (C) Water conveyance efficiency
   (D) Water storage efficiency

3. **In drought areas appears in a field which is not irrigated uniformly unless excess water is applied, which in turn results in a waste of water is termed as:**
   (A) Irrigation application efficiency
   (B) Water distribution efficiency
   (C) Water conveyance efficiency
   (D) Water storage efficiency

4. **The proportional of normal consumptive use of water to the net amount of water depleted from root zone is known as:**
   (A) Consumptive use efficiency
   (B) Project efficiency
   (C) Water conveyance efficiency
   (D) Water storage efficiency

5. **The natural process by which water continue to change into the vapour from its free surface known as:**
   (A) Evaporation
   (B) Transpiration
   (C) Evapo-transpiration
   (D) None of these

6. **Evaporation is measured by:**
   (A) US Open evaporimeter
   (B) Lysimeter
   (C) Potometer
   (D) Pressure membrane apparatus

7. **The average annual potential evaporation in different agro-climatic zones ranges from:**
   (A) Less than 1000 mm per annum
   (B) 1000-3500 mm per annum
   (C) 3500-5000 mm per annum
   (D) More than 5000 mm per annum

8. **Water entering the soil from the ground surface is referred to:**
   (A) Evaporation (B) Transpiration
   (C) Evapo-transpiration
   (D) None of these

9. **The infiltration rate is 12.5 to 12.5 mm/hr categorized as:**
   (A) Low (B) Medium
   (C) High (D) Very high

10. **The infiltration rate is 12.5-25.0 mm/hr categorized as:**
    (A) Low (B) Medium
    (C) High (D) Very high

11. **The infiltration rate is more than 25.0 mm/hr categorized as:**
    (A) Low (B) Medium
    (C) High (D) Very high

12. **Water losses through combined effect of evaporation and transpiration is called:**
    (A) Evapo-transpiration
    (B) Consumptive use
    (C) Both of these
    (D) None of these

13. **Water held tightly by soil colloids through gravitational force is called:**
    (A) Capillary water
    (B) Hygroscopic water
    (C) Gravitational water
    (D) Free water

14. **Availability of gravitational water to the plant is:**
    (A) More (B) Less
    (C) Limited (D) Unknown

15. **Furrow length for irrigation should be kept under normal soil:**
    (A) Less than 100 meter
    (B) 100-300 meter
    (C) 500-1000 meter
    (D) More than 1000 meter

16. **Depth of furrow for irrigation should be:**
    (A) 5-6 inches (B) 8-12 inches
    (C) 12-15 inches (D) more than 15 inches

17. Water Use Efficiency highest in case of:
(A) Flood method
(B) Border method
(C) Sprinkler method
(D) Drip method

18. Among the following types of soil-water which one is the most useful for the plants:
(A) Capillary water
(B) Hygroscopic water
(C) Gravitational water
(D) Combined water

19. Near the canal banks soils become:
(A) Alkaline (B) Acidic
(C) Neutral (D) Well drained

20. The weight of a half mole of water is:
(A) 0.5 g (B) 1.0 g
(C) 9.0 g (D) 18.0

Practice Paper 6

1. Water requirement of mentha crop is:
(A) High
(B) Less
(C) Same as other crop
(D) Drought crop

2. What will be the moisture content of soil at 50% ASM, if it has field capacity and permanent wilting point values of 20% and 8 %, respectively:
(A) 18 % (B)
(C) 12 % (D) 10 %

3. A device for measuring percolation and leaching losses from a column of soil under controlled conditions is called:
(A) Lysimeter
(B) USDA Open pan evoporimeter
(C) Parshal flume
(D) Permeameter

4. A calibrated device for measuring the flow of water in open conduits is:
(A) Lysimeter
(B) USDA Open pan evoporimeter
(C) Parshal flume
(D) Permeameter

5. A device for measuring the permeability of soils or other materials is:
(A) Lysimeter
(B) USDA Open pan evoporimeter
(C) Parshal flume
(D) Permeameter

6. Instrument used to measure the humidity of the atmosphere is:
(A) Psychrometer
(B) USDA Open pan evaporimeter
(C) Parshal flume
(D) Permeameter

7. During absorption of water by roots, the water potential of cell sap is lower than that of:
(A) Pure water and soil solution
(B) Neither pure water nor soil solution
(C) Pure water but higher than that of soil solution
(D) Soil solution but higher than that of pure water

8. Cotton fibre dipped in water absorb water through:
(A) Endosmosis (B) Exosmosis
(C) Capillarity (D) Imbibition

9. The term water potential was coined by:
(A) Sayre (B) Von Mohl
(C) Lloyd (D) Slater and Taylor

10. Water potential of a solution is depicted by:
(A) ψ_p (B) ψ_s
(C) ψ_x (D) $\Delta\psi$

11. A soil which has its interspaces or void spaces completely filled with water to the point where runoff occurs:
(A) Saturated soil
(B) Saturated deficit
(C) Saturation Extract
(D) Saturation index

12. The amount of water vapours required to bring non saturated air at a given temperature and pressure to the point of saturation is called:
(A) Saturated soil
(B) Saturated deficit
(C) Saturation Extract
(D) Saturation index

13. The solution obtained by pressure or vacuum filtration of a soil paste that has been made up to a saturated condition by adding water while stirring is called:

(A) Saturated soil (B) Saturated deficit
(C) Saturation Extract (D) Saturation index

14. An estimate of carbonate precipitation from irrigation water as a function of the degree of calcium carbonate saturation of the soil solution is called:

(A) Saturated soil (B) Saturated deficit
(C) Saturation Extract (D) Saturation index

15. The irrigation projects having culturable command area of more than 10 thousand hectare is:

(A) Major irrigation
(B) Medium irrigation
(C) Minor irrigation
(D) None of these

16. The irrigation projects having culturable command area of less than 10 thousand but more than 1 thousand hectare is:

(A) Major irrigation
(B) Medium irrigation
(C) Minor irrigation
(D) None of these

17. The irrigation projects having culturable command area of less than 1 thousand hectare is:

(A) Major irrigation
(B) Medium irrigation
(C) Minor irrigation
(D) None of these

18. Irrigation done through tanks, surface percolation, wells, tube wells and fluxial water is:

(A) Major irrigation
(B) Medium irrigation
(C) Minor irrigation
(D) None of these

19. Which furrow method of irrigation is suitable where rate of infiltration is minimum :

(A) Flat furrow
(B) Alternate furrow
(C) Contour furrow
(D) Corrugated furrow

20. Under saline-alkali soil and water scarce areas, suitable furrow method of irrigation is :

(A) Flat furrow
(B) Alternate furrow
(C) Contour furrow
(D) Corrugated furrow

Practice Paper 7

1. When furrow are prepared across the slope, the method of irrigation is :

(A) Flat furrow
(B) Alternate furrow
(C) Contour furrow
(D) Corrugated furrow

2. When furrow are prepared to the same direction of slop, suitable furrow method of irrigation is :

(A) Flat furrow
(B) Alternate furrow
(C) Sloppy furrow
(D) Corrugated furrow

3. Furrow method of irrigation is suitable in:

(A) Potato (B) Sugarcane
(C) Tobacco (D) All of these

4. Which is the suitable irrigation method where root rot is a severe problem:

(A) Furrow method (B) Basin method
(C) Ring method (D) Surface flooding

5. Sprinkler irrigation is also called:

(A) Over head irrigation
(B) Trickle irrigation
(C) Under head irrigation
(D) Gravity irrigation

6. Surface irrigation method is also called:

(A) Over head irrigation
(B) Trickle irrigation
(C) Under head irrigation
(D) Gravity irrigation

7. The suitable sprinkler system of irrigation at low pressure (0.5-25 kg/cm^2) :

(A) Rotating head system
(B) Perforated pipe system
(C) Both of these (D) None of these

8. Component of sprinkler system is:
(A) Pumping set or water lifting unit
(B) Pipe line
(C) Sprinkler head
(D) All of these

9. Under steep sloppy areas which irrigation system is more suitable:
(A) Furrow method
(B) Border method
(C) Sprinkler method
(D) Drip method

10. The most suitable method of irrigation in horticultural and vegetable crops is:
(A) Furrow method
(B) Border method
(C) Sprinkler method
(D) Drip method

11. Flood method of irrigation is suitable when:
(A) High water requiring crop
(B) Water is not limiting factor
(C) Less erosion affected areas
(D) All of these

12. The practice of irrigation method is mainly adopted in India is:
(A) Surface (B) Sub surface
(C) Aerial (D) All of these

13. Border method of irrigation is suitable under:
(A) Permeability of soil should be less
(B) Soil erosion should be less
(C) Light slope should be one direction
(D) All of these

14. Basin method of irrigation is more suitable in:
(A) Orchard (B) Field crops
(C) Fodder crops (D) Cash crops

15. Ring basin method of irrigation is more suitable in:
(A) Mango (B) Jackfruit
(C) Jamun (D) All of these

16. Water use efficiency is maximum in:
(A) Sprinkler system
(B) Drip irrigation
(C) Surface method
(D) None of these

17. Main feature of drip irrigation system is:
(A) Less infiltration (B) Less runoff
(C) Less evaporation (D) All of these

18. Under undulated soil and high infiltration rate, which method of irrigation is more suitable:
(A) Sprinkler (B) Drip
(C) Surface (D) Subsurface

19. When moisture movement is in two direction, which method of irrigation is more suitable:
(A) Sprinkler (B) Drip
(C) Surface (D) Subsurface

20. Removal of excess water form the soil or crop by artificial means to maintain water table and salt concentration in the soil to increase crop production is called:
(A) Irrigation (B) Drainage
(C) Percolation (D) None of these

Practice Paper 8

1. Spacing between the sprinklers at high wind velocity is:
(A) ½ of wetting diameter
(B) 2/3 of wetting diameter
(C) 3/4 of wetting diameter
(D) Equal to wetting diameter

2. Sprinkler irrigation stopped at wind velocity of:
(A) <10 km/hr (B) >10 km/hr
(C) <15 km/hr (D) >15 km/hr

3. If EC of drainage water is 8 mhos/cm and EC of irrigation water is 1 mhos/cm, the leaching requirement will be equal to:
(A) 10% (B) 11%
(C) 13% (D) 15%

4. Entry of water into soil is known as:
(A) Infilteration (B) Percolation
(C) Waterlogging (D) Evaporation

5. Rice can be survive under water logged conditions due to:
(A) Absence of aerenchymatous tissues
(B) Presence of aerenchymatous tissues
(C) Rice plant likes water
(D) Presence of parenchymatous tissue

6. **Darcy's law is related to:**
   (A) Water flow (B) Infilterability
   (C) Permeability (D) Water potential

7. **Darcy's law states that:**
   (A) Velocity of water flow is directly proportional to the difference of hydraulic heads and flow length
   (B) Velocity of water flow is directly proportional to the difference of hydraulic heads and inversely proportional to the flow length
   (C) Both of the above
   (D) None of these

8. **Alternate wetting and drying in soil change the:**
   (A) Soil texture (B) Soil structure
   (C) Soil colour (D) Soil temperature

9. **Percentage of water available at wilting point:**
   (A) 0 % (B) 2 %
   (C) 10 % (D) 20 %

10. **Moisture content at field capacity in comparison to percent of moisture content at saturation, is approximately:**
    (A) 100 % (B) 50 %
    (C) 75 % (D) 30 %

11. **Water left in soil after 24-72 hrs of irrigation is known as:**
    (A) Available water (B) Field capacity
    (C) PWP (D) TWP

12. **Plant show physiological drought when:**
    (A) Increase in salt concentration
    (B) Decrease salt conecentration
    (C) Both of these
    (D) None of these

13. **The yield of marketable crop produced per unit of water used in evapotraspiration is known as:**
    (A) Water distribution efficiency
    (B) Water use efficiency
    (C) Water application efficiency
    (D) Conveyence efficiency

14. **At what stages, wheat crop should be irrigated if two irrigations are available?**
    (A) Crown root and tillering
    (B) Tillering and flowering
    (C) Flowering and milking
    (D) Crown root and flowering

15. **What would happen when pulse crops is irrigated during flowering stage?**
    (A) More flowering
    (B) Dropping of flowers and reduction in yield
    (C) No favourable effect or adverse effect
    (D) None of these

16. **Which soil has higher water holding capacity?**
    (A) Sandy soil (B) Sandy loam soil
    (C) Loam soil (D) Clay loam soil

17. **Water requirement of crop is equal to:**
    (A) CUW + Application losses
    (B) CUW
    (C) CUW + Application losses + water needed for special operations
    (D) All of the above

18. **Irrigation requirement is equal to:**
    (A) Water requirement – (Effective rainfall + Soil water contribution + Ground water contribution)
    (B) Water requirement – CUW
    (C) CUW
    (D) None of these

19. **CUW is equal to:**
    (A) ET
    (B) ET + $\leq$ 1% for plant metabolite activities
    (C) E + T
    (D) ET + Effective rainfall

20. **The amount of water consumptively used in 24 hours period is called:**
    (A) CUW
    (B) Daily consumptive use
    (C) Seasonal consumptive use
    (D) All the above

Practice Paper 9

1. **Actual crop ET (Et crop) and reference crop ET (ETo) has the relation ship equal to:**
   (A) CT crop = Ki. ETo
   (B) ETo = Kc. ET crop
   (C) Kc = ETo/ET crop
   (D) None of these

2. **Crop ET can be estimated by:**
   (A) Lysimeter method
   (B) Field Experiment method
   (C) Soil water depletion method
   (D) All the above

3. **Reference crop ET was proposed by:**
   (A) Doorenbos and Pruitt
   (B) Blaney –Criddle
   (C) Tanner
   (D) Sharma and Dastane

4. **Weighing and non-weighing type lysimeter were suggested by:**
   (A) Harrold and Dreibelbis
   (B) Slatyer and Mellroy
   (C) Pelton
   (D) None of these

5. **PET can be calculated as:**
   (A) PET = Epan × crop factor
   (B) $\frac{\text{Epan}}{\text{Crop factor}}$
   (C) PET = Epan
   (D) None

6. **Duty of water refers to:**
   (A) Number of hectares of a crop successfully raised with a constant flow of one cusec of water throughout the growth period
   (B) Total area irrigated
   (C) Total cropped area irrigated
   (D) None of these

7. **The amount of water consumptively used by a crop during the entire growing season or crop period is called as:**
   (A) Daily consumptive use
   (B) Seasonal consumptive use
   (C) CUW
   (D) All of the above

8. **The application of water divided in strips along with contour is known as:**
   (A) Strip irrigation
   (B) Contour irrigation
   (C) Control irrigation
   (D) Basin irrigation

9. **Specific heat of soil water is:**
   (A) 1.0 calories/g
   (B) 0.40 calories/g
   (C) 0.5 calories/g
   (D) 0.22 calories/g

10. **The ratio of irrigation water transpired by the crops on an irrigation farm or project during their growth period, over the water diverted from a river or other natural source in to the farm or project canal or canals during the same period of time is called:**
    (A) Irrigation application efficiency
    (B) Project efficiency
    (C) Water conveyance efficiency
    (D) Water use efficiency

11. **The ratio of volume of irrigation water consumptively used in a specified period to the total volume of irrigation water that has been diverted from the capture system is called:**
    (A) Irrigation application efficiency
    (B) Project efficiency
    (C) Water conveyance efficiency
    (D) Water use efficiency

12 **Most irrigation water that come from diversion from streams or reservoirs. Losses which occurred while conveying water were often excessive is known as:**
    (A) Irrigation application efficiency
    (B) Project efficiency
    (C) Water conveyance efficiency
    (D) Water use efficiency

13. **What proportion of the water delivered was beneficially used on the project, farm, or field is called:**
    (A) Irrigation application efficiency
    (B) Project efficiency
    (C) Water conveyance efficiency
    (D) Water use efficiency

14. Empirical formula of Irrigation Application Efficiency is:

(A) $\frac{\text{Average depth of irrigation water contributing to target}}{\text{Average depth of irrigation water applied}} \times 100$

(B) $\frac{\text{Volume of irrigation water consumptively used}}{\text{Volume of irrigation water diverted to the project area}} \times 100$

(C) $\frac{\text{Water diverted to the farm}}{\text{Water diverted from the river or reservoir}} \times 100$

(D) $\frac{\text{Water beneficially used}}{\text{Water delivered}} \times 100$

15. Empirical formula of Project Efficiency is:

(A) $\frac{\text{Average depth of irrigation water contributing to target}}{\text{Average depth of irrigation water applied}} \times 100$

(B) $\frac{\text{Volume of irrigation water consumptively used}}{\text{Volume of irrigation water diverted to the project area}} \times 100$

(C) $\frac{\text{Water diverted to the farm}}{\text{Water diverted from the river or reservoir}} \times 100$

(D) $\frac{\text{Water beneficially used}}{\text{Water delivered}} \times 100$

16. Empirical formula of Water Conveyance Efficiency is:

(A) $\frac{\text{Average depth of irrigation water contributing to target}}{\text{Average depth of irrigation water applied}} \times 100$

(B) $\frac{\text{Volume of irrigation water consumptively used}}{\text{Volume of irrigation water diverted to the project area}} \times 100$

(C) $\frac{\text{Water diverted to the farm}}{\text{Water diverted from the river or reservoir}} \times 100$

(D) $\frac{\text{Water beneficially used}}{\text{Water delivered}} \times 100$

17. Empirical formula of Water Use Efficiency is:

(A) $\frac{\text{Average depth of irrigation water contributing to target}}{\text{Average depth of irrigation water applied}} \times 100$

(B) $\frac{\text{Volume of irrigation water consumptively used}}{\text{Volume of irrigation water diverted to the project area}} \times 100$

(C) $\frac{\text{Water diverted to the farm}}{\text{Water diverted from the river or reservoir}} \times 100$

(D) $\frac{\text{Water beneficially used}}{\text{Water delivered}} \times 100$

18. Empirical formula of Water Storage Efficiency of:

(A) $\frac{\text{Average depth of irrigation water contributing to target}}{\text{Average depth of irrigation water applied}} \times 100$

(B) $\frac{\text{Water stored in the root zone during the irrigation}}{\text{Water needed in the root zone prior to the irrigation}} \times 100$

(C) $\left[1-\frac{y}{D}\right] \times 100$

(D) $\frac{\text{Normal consumptive use of water}}{\text{Net amount of water depleted from root zone}} \times 100$

19. Empirical formula of Water Distribution Efficiency:

(A) $\frac{\text{Average depth of irrigation water contributing to target}}{\text{Average depth of irrigation water applied}} \times 100$

(B) $\frac{\text{Water stored in the root zone during the irrigation}}{\text{Water needed in the root zone prior to the irrigation}} \times 100$

(C) $\left[1-\frac{y}{D}\right] \times 100$

(D) $\frac{\text{Normal consumptive use of water}}{\text{Net amount of water depleted from root zone}} \times 100$

20. Empirical formula of Consumptive Use Efficiency:

(A) $\frac{\text{Average depth of irrigation water contributing to target}}{\text{Average depth of irrigation water applied}} \times 100$

(B) $\frac{\text{Water stored in the root zone during the irrigation}}{\text{Water needed in the root zone prior to the irrigation}} \times 100$

(C) $\left[1-\frac{y}{D}\right] \times 100$

(D) $\frac{\text{Normal consumptive use of water}}{\text{Net amount of water depleted from root zone}} \times 100$

Practice Paper 10

1. World water day celebrated on:

(A) 22 March (B) 22 April

(C) 22 May (D) 22 June

2. Main source of irrigation in India is:

(A) Canal

(B) Tubewells

(C) Wells

(D) Ponds

3. Surge irrigation is useful mainly for:

(A) Reducing percolation loss

(B) Increased water use efficiency

(C) Infiltration increased

(D) All the above

4. Total soil water potential is equal to:

(A) $\psi g + \psi p + \psi o$

(B) $\psi g + \psi p$

(C) $\psi g + \psi o$

(D) $\psi g + \psi p - \psi o$

5. **pF is:**
   (A) Logarithm of water column height in cm
   (B) Logarithm of H+
   (C) Equal to pH
   (D) None of these
6. **Which irrigation system gives the highest irrigation efficiency:**
   (A) Surface method
   (B) Subsurface method
   (C) Sprinkler method
   (D) Trickle method
7. **Discharge of water from trapezoidal weir is equal to:**
   (A) $186\times10^{-4}LH^{3/2}$ (B) $184\times10^{-4}LH^{3/2}$
   (C) $138\times10^{-4}LH^{5/2}$ (D) None of these
8. **Discharge of water from V-notch will be is equal to:**
   (A) $186\times10^{-4}LH^{3/2}$ (B) $184\times10^{-4}LH^{3/2}$
   (C) $138\times10^{-4}LH^{5/2}$ (D) None of these
9. **Discharge of water from rectangular weir is equal to:**
   (A) $186\times10^{-4}LH^{3/2}$ (B) $184\times10^{-4}LH^{3/2}$
   (C) $138\times10^{-4}LH^{5/2}$ (D) None of these
10. **Discharge from an orifice is calculated as:**
    (A) $Q=61\times10^{-5}A\sqrt{2gh}$
    (B) $Q=6.1\times10^{-5}A\sqrt{2gh}$
    (C) $Q=184\times10^{-4}LH^{3/2}$
    (D) $Q=138\times10^{-4}LH^{5/2}$
11. **If the ratio of Ha/Hb of a parshall flume is < 0.5, the flow will be:**
    (A) Free flow (B) Submerged flow
    (C) Over flow (D) Under flow
12. **Field water use efficiency is equal to:**
    (A) $\frac{Y}{WR}$ (B) $\frac{WR}{Y}$
    (C) $\frac{Y}{ET}$ (D) $\frac{ET}{Y}$
13. **Crop water use efficiency is equal to:**
    (A) $\frac{Y}{WR}$ (B) $\frac{WR}{Y}$
    (C) $\frac{Y}{ET}$ (D) $\frac{ET}{Y}$
14. **Optimum soil water regime for rice crop is:**
    (A) 50% ASW
    (B) 75% ASW
    (C) Submergence to 4 cm water height
    (D) None of these
15. **Major irrigation project should have total cultural command area:**
    (A) 500 ha (B) 1000 ha
    (C) 10000 ha (D) >10000 ha
16. **Minor irrigation project should have total cultural command area:**
    (A) 2000 ha (B) <1000 ha
    (C) 10000 ha (D) 2000 - 10000 ha
17. **Medium irrigation project should have total cultural command area:**
    (A) 2000 ha (B) 1000 - 10000 ha
    (C) >10000 ha (D) None of these
18. **National Water Policy was adopted in year:**
    (A) 1975 (B) 1981
    (C) 1987 (D) 1965
19. **Indicator plants can be used for:**
    (A) Scheduling irrigation
    (B) Identifying nutritional deficiency symptoms
    (C) Soil water variations
    (D) All of these
20. **Tensiometer was first used for scheduling irrigation by:**
    (A) Richards (B) Taylor
    (C) Miller (D) All the above

Answers Key

■ Practice Paper 1

Q. No.	Ans	Q. No.	Ans	Q. No.	Ans	Q. No.	Ans	Q. No.	Ans	Q. No.	Ans
1.	A	2.	C	3.	A	4.	B	5.	A	6.	A
7.	D	8.	A	9.	C	10.	B	11.	A	12.	A
13.	D	14.	C	15.	C	16.	D	17.	A	18.	A
19.	B	20.	C								

■ Practice Paper 2

Q. No.	Ans	Q. No.	Ans	Q. No.	Ans	Q. No.	Ans	Q. No.	Ans	Q. No.	Ans
1.	C	2.	D	3.	D	4.	A	5.	A	6.	B
7.	A	8.	C	9.	A	10.	B	11.	C	12.	A
13.	D	14.	A	15.	A	16.	A	17.	A	18.	B
19.	A	20.	B								

■ Practice Paper 3

Q. No.	Ans	Q. No.	Ans	Q. No.	Ans	Q. No.	Ans	Q. No.	Ans	Q. No.	Ans
1.	C	2.	D	3.	A	4.	D	5.	B	6.	B
7.	A	8.	B	9.	B	10.	C	11.	D	12.	A
13.	A	14.	A	15.	A	16.	A	17.	A	18.	B
19.	B	20.	D								

■ Practice Paper 4

Q. No.	Ans	Q. No.	Ans	Q. No.	Ans	Q. No.	Ans	Q. No.	Ans	Q. No.	Ans
1.	A	2.	B	3.	C	4.	D	5.	A	6.	C
7.	C	8.	D	9.	A	10.	D	11.	A	12.	C
13.	A	14.	A	15.	B	16.	D	17.	D	18.	A
19.	A	20.	B								

■ Practice Paper 5

Q. No.	Ans	Q. No.	Ans	Q. No.	Ans	Q. No.	Ans	Q. No.	Ans	Q. No.	Ans
1.	C	2.	D	3.	B	4.	A	5.	A	6.	A
7.	B	8.	B	9.	A	10.	B	11.	C	12.	C
13.	C	14.	C	15.	B	16.	B	17.	D	18.	A
19.	A	20.	C								

■ Practice Paper 6

Q. No.	Ans	Q. No.	Ans	Q. No.	Ans	Q. No.	Ans	Q. No.	Ans	Q. No.	Ans
1.	A	2.	B	3.	A	4.	C	5.	D	6.	A
7.	A	8.	C	9.	D	10.	B	11.	A	12.	A
13.	C	14.	D	15.	A	16.	B	17.	C	18.	C
19.	A	20.	B								

■ Practice Paper 7

Q. No.	Ans	Q. No.	Ans	Q. No.	Ans	Q. No.	Ans	Q. No.	Ans	Q. No.	Ans
1.	C	2.	C	3.	D	4.	A	5.	A	6.	D
7.	B	8.	D	9.	C	10.	D	11.	D	12.	A
13.	D	14.	A	15.	D	16.	B	17.	D	18.	B
19.	B	20.	B								

■ Practice Paper 8

Q. No.	Ans	Q. No.	Ans	Q. No.	Ans	Q. No.	Ans	Q. No.	Ans	Q. No.	Ans
1.	A	2.	D	3.	C	4.	A	5.	B	6.	A
7.	B	8.	C	9.	C	10.	C	11.	B	12.	A
13.	B	14.	D	15.	B	16.	D	17.	A	18.	A
19.	B	20.	B								

■ Practice Paper 9

Q. No.	Ans	Q. No.	Ans	Q. No.	Ans	Q. No.	Ans	Q. No.	Ans	Q. No.	Ans
1.	A	2.	D	3.	A	4.	A	5.	A	6.	A
7.	B	8.	A	9.	A	10.	A	11.	B	12.	C
13.	D	14.	A	15.	B	16.	C	17.	D	18.	B
19.	C	20.	D								

■ Practice Paper 10

Q. No.	Ans	Q. No.	Ans	Q. No.	Ans	Q. No.	Ans	Q. No.	Ans	Q. No.	Ans
1.	A	2.	A	3.	D	4.	A	5.	A	6.	D
7.	C	8.	A	9.	C	10.	A	11.	A	12.	A
13.	C	14.	C	15.	D	16.	B	17.	B	18.	B
19.	D	20.	A								

10. Multiple Choice Questions on Dryland Agriculture

Practice Paper 1

1. **The practice of crop production entirely with rain water received during the crop season or on conserved soil moisture in low rainfall (<750 mm) areas of arid and semi-arid climate and the crop may face mild to very severe moisture stress during their life cycle is called:**
   (A) Dry farming (B) Dry season
   (C) Drought year (D) Rainfed farming
2. **Growing of field crops entirely with rain water received during the crop season (rainfall usually > 750 mm) under humid or subhumid climate and the crop may face little or no moisture stress during their life cycle is called:**
   (A) Dry farming (B) Dry season
   (C) Drought year (D) Rainfed farming
3. **A period during which water deficiency occurs as stored water is used for evapotranspiration and actual evapotranspiration falls below potential evapotranspiration is called:**
   (A) Dry farming (B) Dry season
   (C) Drought year (D) Drought prone
4. **Usually crop face little or no moisture stress during their life cycle in:**
   (A) Dry farming (B) Dry season
   (C) Drought year (D) Rainfed farming
5. **Usually crop face moisture stress during their life cycle in:**
   (A) Dry farming (B) Dry season
   (C) Drought year (D) Rainfed farming
6. **Rainfall received less than 750 mm during their life cycle in:**
   (A) Dry farming
   (B) Dry season
   (C) Drought year
   (D) Rainfed farming
7. **Rainfall received more than 750 mm during their life cycle in:**
   (A) Dry farming (B) Dry season
   (C) Drought year (D) Rainfed farming
8. **Supplemental irrigation can be given to save the crop in:**
   (A) Dry farming (B) Dry season
   (C) Drought year (D) Rainfed farming
9. **No supplemental irrigation are required to save the crop in:**
   (A) Dry farming (B) Dry season
   (C) Drought year (D) Rainfed farming
10. **Drier part of Rajasthan, Gujarat, Maharashtra etc. comes under:**
    (A) Dry farming (B) Dry season
    (C) Drought year (D) Rainfed farming
11. **Humid and sub humid part of North-eastern states comes under:**
    (A) Dry farming (B) Dry season
    (C) Drought year (D) Rainfed farming
12. **Permits a crop to grow for a longer time in a given moisture stress conditions is called:**
    (A) Drought avoidance
    (B) Drought resistance
    (C) Drought escape
    (D) Drought prone
13. **Relative ability to maintain growth and yield with little or no major set back under moisture stress conditions is called:**
    (A) Drought avoidance
    (B) Drought resistance
    (C) Drought escape
    (D) Drought prone
14. **Which one is drought resistant crop?**
    (A) Paddy
    (B) Sorghum
    (C) Chickpea
    (D) Wheat

15. Physiological drought is caused by:

(A) Excess water
(B) Salinity and alkalinity
(C) Diseases
(D) All of these

16. The areas where drought has occurred in .... per cent of years during the period are considered drought areas:

(A) 10 (B) 20
(C) 30 (D) 40

17. The area where drought has occurred in more than .... per cent of years are called chronic drought areas:

(A) 30 (B) 40
(C) 50 (D) 60

18. The year in which rainfall of particular place is short by more than twice the standard deviation is called:

(A) Dry farming
(B) Dry season
(C) Drought year
(D) Drought prone

19 A period of atleast 15 consecutive days, during which not a single wetting rain has fallen is called:

(A) Dry farming
(B) Dry season
(C) Dry spell
(D) Drought prone

20. In hilly areas and in Western Himalayas people have constructed diversion channels is called:

(A) Guls (B) Kuls
(C) Both (A) and (B) (D) None of these

Practice Paper 2

1. The International Centre for Agricultural Research in Dry Areas (ICARDA), Syria was established in:

(A) 1957 (B) 1967
(C) 1977 (D) 1987

2. Under rainfed conditions one of the suitable varieties of wheat is for central India:

(A) Mukta (B) Bijaga red
(C) VL-421 (D) K-65

3. Mulching is helpful in:

(A) Increasing soil fertility
(B) Improvement of soil structure
(C) Moisture conservation
(D) Better crops

4. The objective of tillage is:

(A) Seed bed preparation
(B) Weed control
(C) Soil and water conservation
(D) All of these

5. According to CRIDA, Hyderabad, 15-20 cm deep ploughing is known as:

(A) Shallow (B) Medium
(C) Deep (D) None of these

6. Best suited soil for rainfed cotton crop is:

(A) Sandy loam (B) Clay loam
(C) Silty loam (D) Black soil

7. Which of the following is not a tillage conservation technique?

(A) Zero tillage (B) Reduced tillage
(C) Minimum tillage (D) All of these

8. The chemical which induces the stomata closing is:

(A) PMA (B) 2,4-D
(C) Kaoline (D) Cycocil

9. Maximum wind soil erosion occurs in which state:

(A) Uttar Pradesh (B) West Bengal
(C) Rajasthan (D) Himachal Pradesh

10. Tilt angle of disc furrow is:

(A) 20° (B) 5°
(C) 45° (D) 0°

11. Contour bunding and terracing are the measures of:

(A) Agronomical measures
(B) Mechanical measures
(C) Biological measures
(D) None of these

12. Name the first phase of water erosion is:

(A) Sheet erosion
(B) Rill erosion
(C) Splash erosion
(D) None of these

13. Geological erosion is also known as:
(A) Accelerated erosion
(B) Natural erosion
(C) Soil erosion
(D) Water erosion

14. Wind breaks are planted in orchard towards:
(A) South-West (B) North-West
(C) North-East (D) East-South

15. Mulching is related to:
(A) Check percolation of moisture from the soil
(B) Check evaporation of moisture from the soil
(C) Check loss of any type of moisture from the soil
(D) None of these

16. Zero tillage system was first used successfully in 1950 in pasture renovation in:
(A) Germany (B) Japan
(C) United Kingdom (D) USA

17. Though tilth is dynamic in nature, can be measured by:
(A) Aggregate analysis
(B) Chemical analysis
(C) Biological analysis
(D) None of these

18. The Central Arid Zone Research Institute, Jodhpur (Rajasthan) was established in:
(A) 1952 (B) 1957
(C) 1959 (D) 1961

19. The lack of resistance of a soil to soil erosion is called:
(A) Erodibility
(B) Erosion
(C) Erosion accelerated
(D) Erosion resistant

20. The watering away of the land surface by water, winds or other forces is called:
(A) Erodibility
(B) Erosion
(C) Erosion accelerated
(D) Erosion resistant

Practice Paper 3

1. Abnormally rapid erosion in an environment disturbed by animal life, chiefly man is called:
(A) Erodibility
(B) Erosion
(C) Erosion accelerated
(D) Erosion resistant

2. A crop, because of its dense foliage, root system etc., provides effective protection against soil loss by erosion is called:
(A) Erodibility
(B) Erosion
(C) Erosion accelerated
(D) Erosion resistant crops

3. A crop, which are increased the soil losses by erosion is called:
(A) Erosion permitting crops
(B) Erosion
(C) Erosion accelerated
(D) Erosion resistant crops

4. Water erosion producing very small and numerous channels is called:
(A) Splash erosion (B) Sheet erosion
(C) Rill erosion (D) Gully erosion

5. A form of soil erosion resulting from soil splash under the impact of falling rain drops is called:
(A) Splash erosion (B) Sheet erosion
(C) Rill erosion (D) Gully erosion

6. Erosion of a fairly uniform layer of material from the land surface, often scarcely perceptible especially when caused by wind is called:
(A) Splash erosion (B) Sheet erosion
(C) Rill erosion (D) Gully erosion

7. Removal of soil by running water, with formation of channels, that cannot be smothered out completely by normal cultivation is called:
(A) Splash erosion (B) Sheet erosion
(C) Rill erosion (D) Gully erosion

8. Climate under crop canopy is called:
(A) Eco climate (B) Crop climate
(C) Both of these (D) None of these

9. **The fraction of total precipitation which forms a part of the crop consumptive use or that becomes utilizable for crop production is called:**
   (A) Effective rainfall
   (B) Effective irrigation
   (C) Effective climate
   (D) None of these

10. **Controlled and uniform application of water to crop land in required amount at required time, with minimum cost to produce optimum yields without waste of water and adverse effect on soil in form of soil salinity and water-logging problems is called:**
   (A) Effective rainfall
   (B) Effective irrigation
   (C) Effective climate
   (D) None of these

11. **A small body of water retained behind a small dam or held in a pit dug in the ground by water harvesting techniques is called:**
   (A) Farm pond (B) Gully plug
   (C) Both of these (D) None of these

12. **The amount of water vapour in the air is called:**
   (A) Humidity (B) Frost
   (C) Fog (D) Mist

13. **Leaf and canopy temperature can be measured by:**
   (A) Infra red thermometer
   (B) Barometer
   (C) Lysimeter
   (D) Dry bulb thermometer

14. **Organic substances which in minute amounts may participate in the control of growth processes is called:**
   (A) Growth regulator (B) Growth retardant
   (C) Growth substances (D) None of these

15. **A chemical that selectivity interferes with normal hormonal production of growth but without appreciable toxic effects is called:**
   (A) Growth regulator
   (B) Growth retardant
   (C) Growth substances
   (D) None of these

16. **A growth regulating chemical produced by a plant of a synthetic one with similar activity even at very low concentration is called:**
   (A) Growth regulator
   (B) Growth retardant
   (C) Growth substances
   (D) None of these

17. **Chemical such as auxins, gibberellins, kinins etc. are the example of:**
   (A) Growth regulator
   (B) Growth retardant
   (C) Growth substances
   (D) None of these

18. **A chemical such as CCC is the example of:**
   (A) Growth regulator
   (B) Growth retardant
   (C) Growth substances
   (D) None of these

19. **Tillage operation which more soil to create desired soil configurations is called:**
   (A) Land forming (B) Land grading
   (C) Land leveling (D) Land planning

20. **Reshaping of the field surface to a desired elevation and slope is called:**
   (A) Land forming (B) Land grading
   (C) Land leveling (D) Land planning

Practice Paper 4

1. **The reshaping of the land surface to facilitate a more uniform application of irrigation water or for preventing soil erosion is called:**
   (A) Land forming (B) Land grading
   (C) Land leveling (D) Land planning

2. **Tillage operation that cuts and moves small layers of soil to provide a smooth and refined surface is called:**
   (A) Land forming (B) Land grading
   (C) Land leveling (D) Land planning

3. **Measuring depth of water from precipitation is:**
   (A) Raingauge (B) Anemometer
   (C) Sunshine recorder (D) None of these

4. **The period of the year in which monthly rainfall exceeds 10 per cent of the monthly evapotranspiration in:**
   (A) Rainy season (B) Dry season
   (C) Wet season (D) None of these
5. **A long, low ridge of earth with gently sloping sides and a shallow channel along the upper side, to control erosion by divesting surface runoff across the slope instead of permitting it to flow uninterrupted down to slope is called:**
   (A) Ridge terrace (B) Ridge planting
   (C) Rotary tillage (D) None of these
6. **Planting on ridges is called:**
   (A) Ridge terrace (B) Ridge planting
   (C) Rotary tillage (D) None of these
7. **Tillage operation employing rotary action to cut, break and mix soil is called:**
   (A) Ridge terrace (B) Ridge planting
   (C) Rotary tillage (D) None of these
8. **A kind of drought which occurs during distinct annual periods of dry weather is called:**
   (A) Seasonal drought (B) Recurrent drought
   (C) Both of these (D) None of these
9. **Tillage operations which constitute the initial major soil working operation is called:**
   (A) Primary tillage (B) Secondary tillage
   (C) Preparatory tillage (D) Reduced tillage
10. **The tillage operations following primary tillage to create a good seed bed for proper seeding/planting is called:**
   (A) Primary tillage
   (B) Secondary tillage
   (C) Preparatory tillage
   (D) Reduced tillage
11. **Various tillage operations carried out on the farm before the crops are sown to make a fine seedbed is called:**
   (A) Primary tillage
   (B) Secondary tillage
   (C) Preparatory tillage
   (D) Reduced tillage
12. **A tillage system in which the primary tillage operation is performed in conjunction with special planting procedure in order to reduce or eliminate secondary tillage operations is called:**
   (A) Primary tillage
   (B) Secondary tillage
   (C) Preparatory tillage
   (D) Reduced tillage
13. **Reduction in the specific volume of soil by means of mechanical manipulation is called:**
   (A) Primary tillage
   (B) Secondary tillage
   (C) Preparatory tillage
   (D) Soil compaction
14. **A land or climate where the rainfall is greater many crops. Irrigation or dryland farming methods are commonly required for crop production is called:**
   (A) Arid (B) Semi arid
   (C) Humid (D) Sub humid
15. **The zone delineated by Thornthwaite's moisture indices varying from -20 to -40 is:**
   (A) Semi-arid (B) Arid
   (C) Severe arid (D) Humid
16. **The zone delineated by Thornthwaite's moisture indices varying from -60 to -80 is:**
   (A) Semi-arid (B) Arid
   (C) Severe arid (D) Humid
17. **The zone delineated by Thornthwaite's moisture indices varying from less than -80 is:**
   (A) Semi-arid (B) Arid
   (C) Severe arid (D) Humid
18. **The zone delineated by Thornthwaite's moisture indices varying from 20 to 50 is:**
   (A) Semi-arid (B) Arid
   (C) Severe arid (D) Humid
19. **The zone delineated by Thornthwaite's moisture indices varying from 50 to 100 is:**
   (A) Semi-arid (B) Humid
   (C) Severe humid (D) Extreme Humid
20. **The zone delineated by Thornthwaite's moisture indices varying from more than 100 is:**
   (A) Semi-arid (B) Humid
   (C) Severe humid (D) Extreme Humid

Practice Paper 5

1. **Thornthwaite's moisture index can be calculated by:**

(A) $MI = \frac{P-PE}{PE} \times 100$ (B) $MI = \frac{PE-P}{PE} \times 100$

(C) $MI = \frac{P-PE}{P} \times 100$ (D) $MI = \frac{P-PE}{PE}$

2. **A dry soil layer at the surface may be created *in situ* by loosening the surface soil and hastening the drying of the upper soil is known as:**

(A) Soil mulch (B) Straw mulch
(C) Plastic mulch (D) Stuble mulch

3. **Mulching with crop residue like straw, husk, dry leaves, twings etc helps to reduce surface runoff is known as:**

(A) Soil mulch (B) Straw mulch
(C) Plastic mulch (D) Stuble mulch

4. **Mulching with costly plastic films and paper is practices to reduce moisture losses from soil surface is called:**

(A) Soil mulch (B) Straw mulch
(C) Plastic mulch (D) Stuble mulch

5. **Which is mulching is generally practiced in high value crops is:**

(A) Soil mulch (B) Straw mulch
(C) Plastic mulch (D) Stuble mulch

6. **Which type of plastic is more effective in mulching:**

(A) Black plastic
(B) White plastic
(C) Transparent plastic
(D) None of these

7. **Plastic mulching is also act as:**

(A) Reduces weed infestation
(B) Increases seed germination
(C) Both of these
(D) None of these

8. **Mechanical manipulations of the soil that are used to provide necessary soil conditions favourable for the growth of crop is called:**

(A) Tillage (B) Hoeing
(C) Sowing (D) All of these

9. **Soil are tilled to create soil physical conditions which facilitates:**

(A) Seed germination
(B) Seedling emergence
(C) Development of rooting
(D) All of these

10. **Loosing of the top soil hastens the development of a dry layer at the surface which reduces further loss of water from soil is known as:**

(A) Shallow tillage
(B) Traditional tillage
(C) Conventional tillage
(D) Minimum tillage

11. **The tillage practices is also associated with faster decomposition of soil organic matter, greater erosion risk and degradation of soil structure is known as:**

(A) Shallow tillage
(B) Traditional tillage
(C) Conventional tillage
(D) Minimum tillage

12. **The tillage envisages minimum disturbance of soil, residues are left in the field and weeds are killed with herbicides is called:**

(A) Shallow tillage
(B) Traditional tillage
(C) Conventional tillage
(D) Minimum tillage

13. **Tillage operations are considerably reduced to the minimum required for good seed bed preparation is called:**

(A) Shallow tillage
(B) Traditional tillage
(C) Conventional tillage
(D) Minimum tillage

14. **Ultimate stage of minimum tillage is called:**

(A) Zero tillage
(B) Traditional tillage
(C) Conventional tillage
(D) Minimum tillage

15. **Sowing of without any tillage operation is called:**

(A) Shallow tillage
(B) Zero tillage
(C) Conventional tillage
(D) Minimum tillage

16. The tillage practices combined with crop residues, dry weeds and other types of vegetative material spread on the soil surface is called:

(A) Shallow tillage
(B) Traditional tillage
(C) Mulch tillage
(D) Minimum tillage

17. Which method of harvested water is directed to the crop is known as:

(A) Inter plot water harvesting
(B) Inter row water harvesting
(C) Water harvesting in pods
(D) None of these

18. There may not be enough rain to support a crop in some areas and therefore by conserving more water in furrows and planting the crop in furrows may give some yields is called:

(A) Inter plot water harvesting
(B) Inter row water harvesting
(C) Water harvesting in pods
(D) None of these

19. A portion of the excess water after allowing maximum *in-situ* moisture conservation in collected in:

(A) Inter plot water harvesting
(B) Inter row water harvesting
(C) Water harvesting in ponds
(D) None of these

20. A tillage system in which only isolated bunds of soil are tilled is called:

(A) Strip tillage
(B) Sub soiling
(C) Sub soil placement
(D) Surface tillage

Practice Paper 6

1. Breaking of compact sub soils without inverting them with a special narrow cultivator, showed or chisel at depth usually from 30-60 cm from the surface is called:

(A) Strip tillage (B) Sub soiling
(C) Sub soil placement (D) Surface tillage

2. Placement of fertilizers in the subsoil with thc help of heavy power machinery usually followed in humid and sub humid region is known as:

(A) Strip tillage
(B) Sub soiling
(C) Sub soil placement
(D) Surface tillage

3. To till the surface without turning the tilled layers is called:

(A) Strip tillage
(B) Sub soiling
(C) Sub soil placement
(D) Surface tillage

4. To plough or cultivate soil for seeding is called:

(A) Strip tillage (B) Sub soiling
(C) Till (D) Surface tillage

5. The action of a tillage tool in executing a specific form of soil manipulation such as soil cutting, pulverizing or inversion is called:

(A) Tillage action (B) Tillage depth
(C) Tillability (D) Vertical mulching

6. Vertical distance from the initial soil surface to a specific point of penetrations of the tool is called:

(A) Tillage action (B) Tillage depth
(C) Tillability (D) Vertical mulching

7. The degree of ease with which a soil may be manipulated for a specific purpose is called:

(A) Tillage action (B) Tillage depth
(C) Tillability (D) Vertical mulching

8. Incorporation of vegetative mulch in a bund in the soil for the purpose of harvesting rain water and conserving soil and water is called:

(A) Tillage action (B) Tillage depth
(C) Tillability (D) Vertical mulching

9. Conservation of rain water under unirrigated conditions, by collecting runoff of precipitation in order to supplement soil moisture in an adjacent area is called:

(A) Water harvesting (B) Soil conservation
(C) Runoff harvesting (D) None of these

10. A combination of all management and land use methods which safe guard the soil against depletion or deterioration by natural or by man induced factors is called:

(A) Water harvesting
(B) Soil conservation
(C) Runoff harvesting
(D) None of these

11. The area above a given point on a stream that contributes water to the flow at that point is called:

(A) Water harvesting (B) Soil conservation
(C) Runoff harvesting (D) Watershed

12. Machine cuts the soil, pulverizes the soil to powder and spreads uniformly to level the field is:

(A) Rotavator
(B) Ridge and furrow makers
(C) Harrow
(D) None of these

13. A machine is used for making irrigation channel and boundaries of a field is:

(A) Rotavator
(B) Ridge and furrow makers
(C) Harrow
(D) None of these

14. The speed of rotavator is:

(A) 100-200 rpm (B) 200-300 rpm
(C) 300-400 rpm (D) 400-500 rpm

15. Rotavator work with the help of:

(A) Hydraulic Pressure
(B) PTO shaft
(C) Both (A) and (B)
(D) None of these

16. Harrows are tillage implement of:

(A) Primary tillage (B) Secondary tillage
(C) Both of these (D) None of these

17. A ridger plough consists of:

(A) V - shaped (B) U - shaped
(C) L - shaped (D) M - shaped

18. An implement used for puddling the paddy field before transplanting is:

(A) Rotavotor
(B) Ridge and furrow makers
(C) Harrow
(D) Puddler

19. Water absorbing chemicals are:

(A) Grow soak
(B) Agrolite-400
(C) Jalshakti
(D) All of these

20. Bund construction across the slopes are known as:

(A) Gulley plugs (B) Terraces
(C) Field bunds (D) Contour bunds

Practice Paper 7

1. Series of flat plateform constructed across a slopping land which give rise to the formation of a step or bench like feature is known as:

(A) Gulley plugs (B) Terraces
(C) Field bunds (D) Contour bunds

2. Small earthen barriers constructed in field is known as:

(A) Gulley plugs (B) Terraces
(C) Field bunds (D) Contour bunds

3. Construction of field bunds along the contour line is known as:

(A) Gulley plugs (B) Terraces
(C) Field bunds (D) Contour bunds

4. Over flow weir is also known as:

(A) Check dams (B) Field bunds
(C) Gully plugs (D) Terraces

5. Soil erosion can be caused by:

(A) High plant density
(B) Overgrazing
(C) Terracing
(D) Strip cropping

6. Soil temperature can be regulated by use of straw mulch both in summer and winter season because:

(A) It reduces evaporation
(B) It lowers the soil temperature
(C) It keeps the temperature of soil high
(D) It provides insulating effect

7. Chemical which regulates opening and closing of stomata is:

(A) Kaoline (B) 2, 4-D
(C) PMA (D) Paraquat

8. Gully erosion of soil is mainly caused by:
(A) Stream formed by rain water
(B) Wind
(C) Terrace farming
(D) Sea waves

9. Which one of the following can lead to soil erosion?
(A) Contour ploughing
(B) Terrace farming
(C) Deforestation
(D) Mixed cropping

10. Soil erosion is the state of Punjab is due to which one of the following reasons?
(A) Excessive cultivation
(B) Irrigation
(C) Overgrazing
(D) Deforestation

11. In which one of the following states is the terrace cultivation practice?
(A) Punjab
(B) Haryana
(C) Plains of Uttar Pradesh
(D) Uttarakhand

12. Name the first phase of water erosion:
(A) Sheet erosion (B) Rill erosion
(C) Splash erosion (D) None of these

13. Geological erosion is also known as:
(A) Accelerated erosion
(B) Natural erosion
(C) Soil erosion
(D) Water erosion

14. Area under dryland is:
(A) 33.5 mha (B) 34.5 mha
(C) 35.5 mha (D) 36.5 mha

15. Share of dryland area in total agriculture land is:
(A) 15.0% (B) 16.5%
(C) 17.0% (D) 18.5%

16. Share of total food grain production under dryland/rainfed area is:
(A) 35% (B) 44%
(C) 45% (D) 50%

17. Drought affected area in the country is:
(A) 107 mha (B) 108 mha
(C) 109 mha (D) 110 mha

18. When precipitation received less than 50% is called:
(A) Severe drought (B) Medium drought
(C) Less drought (D) Drought

19. Maximum evapotranspiration occurs in:
(A) Jalgaon
(B) Ahmedabad
(C) Jalgaon and Ahmedabad
(D) None of these

20. World average rainfall is:
(A) 890 mm (B) 990 mm
(C) 1090 mm (D) 1190 mm

Practice Paper 8

1. Average rainfall in India is:
(A) 950 mm (B) 1195 mm
(C) 1260 mm (D) 1350 mm

2. Indian agriculture is gamble of:
(A) Monsoon (B) Indian economy
(C) Politics (D) Social

3. Share of human population under rainfed agriculture is:
(A) 30% (B) 40%
(C) 50% (D) 60%

4. Per cent area under rainfed agriculture is:
(A) 60% (B) 65%
(C) 70% (D) 75%

5. Per cent livestock population under rainfed agriculture is:
(A) 60% (B) 65%
(C) 70% (D) 75%

6. Per cent area in pulses and millets under rainfed conditions is:
(A) 87% (B) 77%
(C) 66% (D) 50%

7. Per cent area in oilseeds under rainfed conditions is:
(A) 87% (B) 77%
(C) 66% (D) 50%

8. Per cent area in cereals under rainfed conditions is:
(A) 87% (B) 77%
(C) 66% (D) 50%

9. Per cent area in cotton under rainfed conditions is:

(A) 87% (B) 77%
(C) 66% (D) 50%

10. Area under rainfed condition in India is:

(A) 80 mha (B) 84 mha
(C) 52 mha (D) 50 mha

11. Area under irrigated condition in India is:

(A) 80 mha (B) 82 mha
(C) 52 mha (D) 50 mha

12. Wilting of plants occurs when:

(A) Lateral roots are removed
(B) Cortex is injured
(C) Phloem is blocked
(D) Xylem is blocked

13. Mulching is a process that helps mainly in:

(A) Improvement of soil
(B) Moisture conservation
(C) Weed control
(D) Soil drainage

14. A two bladed flywheel type forage cutter rotates at 60 rrpm. At a feed rate of 1-26 meters per minutes, theoretical length of cut of the forage will be:

(A) 21.0 mm (B) 42.0 mm
(C) 5.3 mm (D) 10.5 mm

15. Contour trenching is preferred to:

(i) Control of landslides

(ii) Conserve the moisture

(A) (i) is false and (ii) is true
(B) Both (i) and (ii) are false
(C) Both (i) and (ii) are true
(D) (i) is true and (ii) is false

16. Major limiting factor in drylands are:

(A) Nutrients (B) Moisture
(C) Both (A) and (B) (D) None of these

17. Phenyl mercuric acetate (PMA) is a chemical used in agricultural crops in order to:

(A) Increase CO_2 uptake
(B) Reduced respiration
(C) Reduce transpiration
(D) Increase respiration

18. Cotour bunds are effective when slope of the land is:

(A) upto 6% (B) upto 16%
(C) upto 20% (D) upto 50%

19. A film forming anti-transpirant used in dryland agriculture is:

(A) Kaolin (B) Atrazin
(C) Mobileaf (D) PMA

20. In universal soil loss equation, what does K refer to ?

(A) Cropping management factor
(B) Slope gradient factor
(C) Soil erodibility factor
(D) Rainfall factor

Practice Paper 9

1. Which of the following tillage operations is particularly beneficial under dry farming situations?

(A) Planking (B) Harrowing
(C) Dust mulching (D) Zero tillage

2. An imaginary line drawn on a map connecting the point of equal rainfall is called:

(A) Isotherm (B) Isobath
(C) Isohyt (D) Isobar

3. Semi-arid climate has a moisture index of:

(A) -80 to -100 (B) -40 to-80
(C) -33.3 to -66.7 (D) 0 to 20

4. Activities of which hormone increases with moisture stress:

(A) Gibberellic acid (B) Indoleacetic acid
(C) Abscisic acid (D) Cytokinins

5. Drought avoidance mechanism is found in:

(A) Barley (B) Pearlmillet
(C) Sorghum (D) Maize

6. The size of large watershed is:

(A) More than 2000 ha
(B) 1000-10000 ha
(C) 100- 1000 ha
(D) 10-100 ha

7. The size of mini watershed is:

(A) 10000-20000 ha (B) 500-2000 ha
(C) 100- 1000 ha (D) 10-100 ha

8. The size of micro watershed is:
(A) 10000-20000 ha (B) 1000-10000 ha
(C) 100- 1000 ha (D) 10-100 ha

9. CAM system is prevalent in:
(A) Arid legumes (B) Cactus
(C) Pearlmillet (D) Sorghum

10. In India, first Dry Farming Research Station, Manjri was started in:
(A) 1923 (B) 1933
(C) 1934 (D) 1936

11. Under stress condition, which amino acid accumulated in crop plants:
(A) Methionine (B) Tryptophan
(C) Proline (D) Phenylalanine

12. Sorghum crop is considered as camel crop because of:
(A) Deep root system
(B) Resistant to drought
(C) Shallow root system
(D) Nutrient exhaustiveness

13. Soil mulch is useful in:
(A) Improving aeration
(B) Improving drainage
(C) Minimizes evaporation losses
(D) Removing weeds

14. When rainfall departure is 11 to 25% from normal rainfall is called:
(A) Slight drought (B) Moderate drought
(C) Severe drought (D) Drought year

15. When rainfall departure is 26 to 50% from normal rainfall is called:
(A) Slight drought (B) Moderate drought
(C) Severe drought (D) Drought year

16. When rainfall departure is more than 50% from normal rainfall is called:
(A) Slight drought (B) Moderate drought
(C) Severe drought (D) Drought year

17. The year is considered drought when less than 75% of the normal rainfall is received is called:
(A) Slight drought (B) Moderate drought
(C) Severe drought (D) Drought year

18. A period during which stream flows are inadequate to supply established used under given water management system is called:
(A) Hydrological drought
(B) Agricultural drought
(C) Socio-economic drought
(D) Meteorological drought

19. As the situation when actual rainfall is less than 75% of the normal rainfall over as area is called:
(A) Hydrological drought
(B) Agricultural drought
(C) Socio-economic drought
(D) Meteorological drought

20. The shortage of water harmful to agricultural activities is called:
(A) Hydrological drought
(B) Agricultural drought
(C) Socio-economic drought
(D) Meteorological drought

Practice Paper 10

1. In the regions with clearly defined rainy (wet) and dry climates is called:
(A) Seasonal drought
(B) Permanent drought
(C) Contingent drought
(D) Meteorological drought

2. Drought which results due to irregular and variability of rainfall especially in humid and sub humid regions is called:
(A) Seasonal drought
(B) Permanent drought
(C) Contingent drought
(D) Meteorological drought

3. The area having permanent dry arid or desert regions is called:
(A) Seasonal drought
(B) Permanent drought
(C) Contingent drought
(D) Meteorological drought

4. The condition when soil moisture depletes and falls short to meet the potential Evapotranspiration (PET) of crop is called:
(A) Seasonal drought
(B) Soil drought
(C) Contingent drought
(D) Meteorological drought

5. Drought which results for low humidity, dry and hot winds and causes desiccations of plants is called:

(A) Seasonal drought
(B) Permanent drought
(C) Contingent drought
(D) Atmospheric drought

6. Survival under drought conditions is to escape drought:

(A) Drought Escape
(B) Drought Avoidance
(C) Drought Tolerance
(D) Drought Resistance

7. The ability of the plant to maintain a favourable internal water balance under moisture stress is called:

(A) Drought Escape
(B) Drought Avoidance
(C) Drought Tolerance
(D) Drought Resistance

8. The ability of crop plants to withstand low tissue water content is referred to:

(A) Drought Escape
(B) Drought Avoidance
(C) Drought Tolerance
(D) Drought Resistance

9. The sum of drought avoidance and drought tolerance is knownas:

(A) Drought Escape
(B) Drought Avoidance
(C) Drought Tolerance
(D) Drought Resistance

10. Roof top rain water harvesting was commonly practices to store drinking water in:

(A) Rajasthan
(B) Tamilnadu
(C) West Bengal
(D) Assam

11. The mulch offers potential benefits for moisture conservation in many rainfed agricultural systems of the arid regions are:

(A) Plastic mulch (B) Stone mulch
(C) Stubble mulch (D) Dust mulch

12. Which is the best practice in terms of water conservation?

(A) Minimum tillage (B) Antitranspirants
(C) Mulches (D) Subsoiling

13. Improvement in the drought tolerance ability of a plant is known as:

(A) Drought Escape
(B) Drought Avoidance
(C) Drought hardening
(D) Drought Resistance

14. A favourable water balance under drought conditions can be achieved by transpiration before as soon as stress is experienced. These are called:

(A) Water savers (B) Water spenders
(C) Water harvesting (D) Watershed

15. Accelerating water uptake sufficiently so as to replenish the lost water called as:

(A) Water savers (B) Water spenders
(C) Water harvesting (D) Watershed

16. Conservation of rain water under unirrigated conditions, by collecting runoff of precipitation in order to supplement soil moisture in an adjacent area is called;

(A) Water savers (B) Water spenders
(C) Water harvesting (D) Watershed

17. The area above a given point on a stream that contributes water to the flow at that point is called:

(A) Water savers (B) Water spenders
(C) Water harvesting (D) Watershed

18. Gadhak ki Baoli in Mehrauli complex in Delhi built in 14th century during the reign of:

(A) Akbar (B) Shahjahan
(C) Iltutmish (D) Sikandar Lodi

19. Rajaon ki Bais or Sukhi Baoli (15th -16th century) built during:

(A) Akbar (B) Shahjahan
(C) Iltutmish (D) Sikandar Lodi

20. Step well at Patan in Gujarat built 11th century also called:

(A) Rani ki vav (B) Kund
(C) Khadin (D) Johads

Answers Key

■ Practice Paper 1

Q. No.	Ans	Q. No.	Ans	Q. No.	Ans	Q. No.	Ans	Q. No.	Ans	Q. No.	Ans
1.	A	2.	D	3.	B	4.	D	5.	A	6.	A
7.	D	8.	A	9.	D	10.	A	11.	D	12.	A
13.	B	14.	B	15.	B	16.	B	17.	B	18.	C
19.	C	20.	C								

■ Practice Paper 2

Q. No.	Ans	Q. No.	Ans	Q. No.	Ans	Q. No.	Ans	Q. No.	Ans	Q. No.	Ans
1.	C	2.	A	3.	C	4.	D	5.	B	6.	D
7.	A	8.	A	9.	C	10.	A	11.	B	12.	C
13.	B	14.	A	15.	B	16.	D	17.	A	18.	C
19.	A	20.	B								

■ Practice Paper 3

Q. No.	Ans	Q. No.	Ans	Q. No.	Ans	Q. No.	Ans	Q. No.	Ans	Q. No.	Ans
1.	C	2.	D	3.	A	4.	C	5.	A	6.	B
7.	D	8.	A	9.	A	10.	B	11.	A	12.	A
13.	A	14.	A	15.	B	16.	C	17.	A	18.	B
19.	A	20.	B								

■ Practice Paper 4

Q. No.	Ans	Q. No.	Ans	Q. No.	Ans	Q. No.	Ans	Q. No.	Ans	Q. No.	Ans
1.	C	2.	D	3.	A	4.	A	5.	A	6.	B
7.	C	8.	A	9.	A	10.	B	11.	C	12.	D
13.	D	14.	B	15.	A	16.	B	17.	C	18.	D
19.	C	20.	D								

■ Practice Paper 5

Q. No.	Ans	Q. No.	Ans	Q. No.	Ans	Q. No.	Ans	Q. No.	Ans	Q. No.	Ans
1.	A	2.	A	3.	B	4.	C	5.	C	6.	A
7.	C	8.	A	9.	D	10.	A	11.	B	12.	C
13.	D	14.	A	15.	B	16.	C	17.	A	18.	B
19.	C	20.	A								

■ Practice Paper 6

Q. No.	Ans	Q. No.	Ans	Q. No.	Ans	Q. No.	Ans	Q. No.	Ans	Q. No.	Ans
1.	B	2.	C	3.	D	4.	C	5.	A	6.	B
7.	C	8.	D	9.	A	10.	B	11.	D	12.	A
13.	B	14.	B	15.	B	16.	B	17.	A	18.	D
19.	D	20.	A								

■ Practice Paper 7

Q. No.	Ans	Q. No.	Ans	Q. No.	Ans	Q. No.	Ans	Q. No.	Ans	Q. No.	Ans
1.	B	2.	C	3.	D	4.	A	5.	B	6.	D
7.	C	8.	A	9.	C	10.	C	11.	D	12.	C
13.	B	14.	B	15.	C	16.	B	17.	C	18.	A
19.	C	20.	B								

■ Practice Paper 8

Q. No.	Ans	Q. No.	Ans	Q. No.	Ans	Q. No.	Ans	Q. No.	Ans	Q. No.	Ans
1.	B	2.	A	3.	B	4.	B	5.	A	6.	A
7.	B	8.	D	9.	C	10.	B	11.	C	12.	D
13.	B	14.	D	15.	A	16.	C	17.	C	18.	A
19.	C	20.	C								

■ Practice Paper 9

Q. No.	Ans	Q. No.	Ans	Q. No.	Ans	Q. No.	Ans	Q. No.	Ans	Q. No.	Ans
1.	C	2.	D	3.	C	4.	C	5.	C	6.	A
7.	B	8.	C	9.	B	10.	A	11.	C	12.	B
13.	C	14.	A	15.	B	16.	C	17.	D	18.	A
19.	D	20.	B								

■ Practice Paper 10

Q. No.	Ans	Q. No.	Ans	Q. No.	Ans	Q. No.	Ans	Q. No.	Ans	Q. No.	Ans
1.	A	2.	C	3.	B	4.	B	5.	D	6.	A
7.	B	8.	C	9.	D	10.	A	11.	B	12.	D
13.	C	14.	A	15.	B	16.	C	17.	D	18.	C
19.	D	20.	A								

PART - 2

11. Multiple Choice Questions of General Agriculture

Practice Paper 1

1. **The practice of raising plants and animals is called:**
(A) Agriculture (B) Horticulture
(C) Agronomy (D) Silviculture

2. **Agriculture word derived from:**
(A) Latin (B) French
(C) Greek (D) English

3. **Agriculture denotes:**
(A) Agri + cultura (B) Ager + cultura
(C) Agri + culture (D) Agris + cultura

4. **Agriculture is the branch of:**
(A) Social science (B) Basic science
(C) Applied science (D) None of these

5. **Agriculture originated about:**
(A) 2300-3000 years back
(B) 20000-30000 years back
(C) 7000-13000 years back
(D) 90000-100000 years back

6. **Kung tribe is:**
(A) Primitive agriculturist
(B) Hunter-gatherer
(C) Advanced agriculturist
(D) Completely dependent on horticulture

7. **In various parts of the world, agriculture:**
(A) Came from Asia Minor
(B) Has spread from Yellow Rive Valley
(C) Originated independently
(D) Has developed to various degree

8. **Fire helped in the cultivation of:**
(A) Fruit plants
(B) Fibre Plant
(C) Drug yielding plants
(D) Grain yielding plants

9. **Early human civilization was based on:**
(A) Development of fibre crops
(B) Fire
(C) Agriculture
(D) Fire and Agriculture

10. **A nonagricultural society living today is:**
(A) Kung tribe (B) Urban Society
(C) Hintza tribe (D) Kuntizha tribe

11. **Early man preferred plants:**
(A) Grain yielding (B) Fruit yielding
(C) Multipurpose (D) Fibre yielding

12. **Cabbage originated in:**
(A) Mediterranean
(B) Mexico and Central America
(C) Peruvian Andes
(D) South west Asia

13. **Maize evolved in:**
(A) Brazil
(B) Mexico and Central America
(C) USA
(D) Asia minor and Afganistan

14. **The origin home of rice is:**
(A) South West Asia
(B) Mediterranean
(C) South East Asia
(D) China

15. **Peruvian Andes are believed to be centre of origin of:**
(A) Tomato
(B) Potato
(C) Tomato and Potato
(D) Tomato, Potato and Chillies

16. **The origin of sunflower is believed to be in:**
(A) Mexico and Central America
(B) Brazil
(C) USA
(D) Peruvian Andes

17. **South East Asia is thought to be centre for the origin of:**
    (A) Rice
    (B) Rice and Sugarcane
    (C) Rice, Sugarcane and Mango
    (D) Rice, Sugarcane, Mango and Banana

18. **The famous cultivated plant which developed in China is:**
    (A) Tea (B) Coffee
    (C) Cotton (D) Orange

19. **The centre of origin of Wheat is:**
    (A) South West Asia
    (B) South East Asia
    (C) Asia minor and Afganistan
    (D) Mexico

20. **Which is not a new world crop:**
    (A) Tapioca (B) Peanut
    (C) Cashewnut (D) Poppy

Practice Paper 2

1. **Which one belong to old world:**
   (A) Squash (B) Pumpkin
   (C) Water Melon (D) Guava

2. **Select a crop which belongs to new world:**
   (A) Jackfruit (B) Papaya
   (C) Spinach (D) Pear

3. **Which crop belong to new world:**
   (A) Red Pepper (B) Black pepper
   (C) Onion (D) Garlic

4. **Which is the crop which originated in Peru and Brazil but is now mainly grown in India is:**
   (A) Groundnut (B) Potato
   (C) Maize (D) None of these

5. **The crop which originated in Peru but is most extensively grown in East Europe is:**
   (A) Coffee (B) Potato
   (C) Rubber (D) Maize

6. **Name the crop which had its birthplace in Tropical America but the centre of production is now Midwest USA:**
   (A) Pineapple (B) Oil palm
   (C) Cocoa (D) Maize

7. **Christopher Columbus landed in island of Bahmas (Carribean sea) in:**
   (A) 1692 (B) 1592
   (C) 1492 (D) 1792

8. **Tropical forests are being lost at the annual rate of:**
   (A) 11 million ha/yr
   (B) 110 million ha/yr
   (C) 1 million ha/yr
   (D) 1100 million ha/yr

9. **Post Vavilov megadiversity centre is:**
   (A) Australia (B) Siberia
   (C) Andamana (D) Tibet

10. **The number of centres of origin of crop plants recognized by Vavilove is:**
    (A) 8 (B) 11
    (C) 7 (D) 21

11. **Ethopia is the origin home of:**
    (A) Coffee (B) Maize
    (C) Rice (D) Cabbage

12. **Potato famine of Ireland occurred in:**
    (A) 1645 (B) 1745
    (C) 1845 (D) 1854

13. **Potato famine of Ireland was caused by the attack of fungus:**
    (A) *Phytophthora palmivora*
    (B) *Plasmopora viticola*
    (C) *Puccinia graminis*
    (D) *Phytophthora infestans* (Late blight disease of potato)

14. **Natural resistance to late blight of potato is present in:**
    (A) *Solanum tuberosum*
    (B) *Solanum acaule*
    (C) *S. demissum*
    (D) *S stoloniferum*

15. **Gene bank are part of:**
    (A) *Ex situ* conservation
    (B) *In situ* conservation
    (C) Both *ex situ* and *in situ* conservation
    (D) Tribal diet

16. ***In situ* conservation are carried out through:**
    (A) Natural parks
    (B) Wildlife santuaries
    (C) Biosphere reserves
    (D) All the above

17. ***Ex situ* conservation is carried out through:**
    (A) Funneling the threatened species into trade
    (B) Offsite collection
    (C) Gene banks
    (D) All the above

18. **Gene banks comprise:**
    (A) Seed banks, orchards, tissue culture and cryopreservation
    (B) Seed banks, tissue culture and cryopreservation
    (C) Seed banks and cryopreservation
    (D) Tissue culture and cryopreservation

19. **Orthodox seeds can be preserved for long periods at a temperature of:**
    (A) -196 °C (B) -10 to -20 °C
    (C) 0 – 4 °C (D) 10 to 15 °C

20. **Which one will be scarched for obtaining maximum genetic diversity:**
    (A) Recent varieties (B) Old varieties
    (C) Wild relatives (D) All the above

Practice Paper 3

1. **Paraguay Tea is got from:**
   (A) Leaves of *Ilex*
   (B) Seeds of *Ilex*
   (C) Leaves of *Paullinia*
   (D) Paste of crushed kernels of *Paullinia*

2. **Arabian Tea is:**
   (A) Leaves and buds of *Paullinia*
   (B) Seed of Ilex
   (C) Leaves and buds of *Catha edulis*
   (D) Seeds of Catha

3. **Cryopreservation is carried out at:**
   (A) -10 to -20 °C (B) -50 to -60 °C
   (C) -196 °C (D) -100 °C

4. **Co-varieties of sugarcane obtained red rot resistance from:**
   (A) *Saccharum munja*
   (B) *S. spontaneum*
   (C) *S. arundinacearum*
   (D) *S. edula*

5. **A calcitrant seed will be killed in:**
   (A) Low temperature
   (B) Dry air
   (C) Further dehydration
   (D) All the above

6. **A gift from new world to old world is:**
   (A) Barley (B) Rice
   (C) Wheat (D) Maize

7. **Which plant was introduced from old to new world:**
   (A) Potato (B) Rice
   (C) Corn (D) Sweet potato

8. **Dwarf wheat varieties brought into India from Mexico were:**
   (A) Sonora-64 and Sonalika
   (B) Sharbati Sonora and Pusa Lerma
   (C) Sonalika and Sharbati Sonora
   (D) Sonora-64 and Lerma Rojo-64

9. **Vavilov centres of origin of crop plants are located in:**
   (A) Tropical plains
   (B) Mountains of tropical areas
   (C) Mountains of temperate areas
   (D) Both (B) and (C)

10. **An under utilized oil crop is:**
    (A) *Brassica campestris*
    (B) *Cocos nucifera*
    (C) *Helianthus annus*
    (D) *Azadiracta indica*

11. **Growing degree days concept is useful in forecasting:**
    (A) Rainfall (B) Drought
    (C) Crop harvest dates (D) Frost

12. **Wheat varieties collected by Vavilov were:**
    (A) 26000 (B) 2600
    (C) 4600 (D) 260

13. **Hevea Rubber belongs to:**
    (A) Malayasia (B) Peruvial Andes
    (C) Brazil (D) South east Asia

14. **Germplasm is contained in:**
    (A) Somatic cells
    (B) Growing points
    (C) Reproductive cells
    (D) Both (A) and (C)

15. Which one will be best for germplasm collection:

(A) Britain (B) France
(C) Portugal (D) Peru

16. Dwarf wheat were evolved by:

(A) Swaminathan (B) Borlaug
(C) Khush (D) Vavilov

17. In a gene bank, genetic material is stored as:

(A) Genes
(B) Seeds
(C) Vegetative material/tissue culture
(D) Both (B) and (C)

18. Genetic erosion is due to:

(A) Deforestation
(B) Shifting cultivation
(C) Adoption of genetically uniform varieties
(D) All the above

19. Exchange of germplasm is carried mostly through shoot tips because it is:

(A) Cheap (B) Small and handy
(C) Virus free (D) Genetically stable

20. Tomato and potato belong to:

(A) North America (B) South America
(C) Asia minor (D) Ethiopia

Practice Paper 4

1. A plant introduced from new world to old world is:

(A) Rice (B) Wheat
(C) Sugarcane (D) Potato

2. In India, the largest herbarium is located at:

(A) Dehradun (B) Delhi
(C) Lucknow (D) Calcutta

3. Which of the following countries is the centre of origin of rice:

(A) India
(B) China
(C) Indo-Malayan region
(D) Russia

4. A new world spice that has become an essential part of Indian cuisine is:

(A) Red pepper (B) Black Pepper
(C) Cardamom (D) Ginger

5. A plant species threatened with extinction can be preserved in:

(A) Herbarium (B) National park
(C) Botanical Garden (D) Gene Bank

6. A place where endangered genetic materials of plants are kept is:

(A) Gene library (B) Genebank
(C) Gene pool (D) Herbarium

7. Wild varieties of plants must be conserved to:

(A) Feeding wild animals
(B) Maintain ecosystem
(C) Future evolution
(D) Incorporate useful traits in future crop varieties

8. Which is the most important crop in the world:

(A) Maize (B) Sorghum
(C) Barley (D) Wheat

9. The great popularity of wheat is on accounts of:

(A) Well balanced carbohydrates and proteins
(B) Inability for complete mechanized production
(C) Short storage period of grain and flour storage
(D) Limited to narrow range of climate

10. According to Vavilov , the common bread wheat originated in :

(A) The Caucasus-Turkey-Iraq area
(B) Temperate zone of America
(C) Naturally fertile regions of France
(D) Cool areas of China

11. Commercial plantation farming comprises an:

(A) Import-oriented system
(B) Export-oriented system
(C) Indigenous use-oriented system
(D) None of these

12. First State Agricultural University of India is:

(A) Gujarat Agricultural University
(B) Punjab Agricultural University
(C) G. B. Pant University of Agricultural & Technology
(D) Chaudhary Charan Singh Hisar Agricultural University

13. Marketing and Designing centre is situated at:
(A) Pune
(B) Faridabad
(C) Nagpur
(D) New Delhi

14. Central Agmark Laboratory is situated at:
(A) Nagpur
(B) New Delhi
(C) Bangalore
(D) Kanpur

15. Indian Standard Institute is situated at:
(A) Kolkata
(B) New Delhi
(C) Madras
(D) Hyderabad

16. APEDA refers to:
(A) Agricultural and Processed Food Products Export Development Authority
(B) Agricultural Production and Export Development Authority
(C) Agricultural Production and Export Development Agency
(D) Asian Policy and Export Development Association

17. Agriculture originated about
(A) 2300-3000 years back
(B) 20000-30000 years back
(C) 7000-13000 years back
(D) 90000-100000 years back

18. Kung tribe is
(A) Primitive agriculturist
(B) Hunter-gatherer
(C) Advanced agriculturist
(D) Completely dependent on horticulture

19. Fire helped in the cultivation of
(A) Fruit plants
(B) Fibre Plants
(C) Drug yielding plants
(D) Grain yielding plants

20. Early human civilization was based on
(A) Development of fibre crops
(B) Fire
(C) Agriculture
(D) Fire and Agriculture

Practice Paper 5

1. A nonagricultural society living today is
(A) Kung tribe
(B) Urban Society
(C) Hintza tribe
(D) Kuntizha tribe

2. Early man preferred plants
(A) Grain yielding (B) Fruit yielding
(C) Multipurpose (D) Fibre yielding

3. The main source of irrigation in India is:
(A) Canals (B) Tubewells
(C) Wells (D) Tanks

4. ICAR was born in:
(A) 1908 (B) 1910
(C) 1929 (D) 1949

5. What was the name of ICAR in initial years of its establishment?
(A) Indian Council of Agricultural Research
(B) Imperial Council of Agricultural Research
(C) Imperial Centre of agricultural Research
(D) Indian Centre of Agricultural Research

6. The establishment if ICAR is based on the recommendation of which commission?
(A) Royal Commission on Agriculture
(B) Famine Commission
(C) Grow More Food Commission
(D) Balwant Rai Mehta Commission

7. ICAR was renamed Indian Council of Agricultural Research in which year?
(A) 1929 (B) 1946
(C) 1949 (D) 1952

8. What is the main mandate of ICAR?
(A) To plan aid, undertake, promote and coordinate agriculture research in the country
(B) To plan, aid undertake, promote and coordinate agricultural research in the country
(C) To plan, aid undertake, promote and coordinate agricultural education and research in the country
(D) To plan, aid undertake, promote and coordinate agricultural research, education and extension in the country

9. **Who works as President of ICAR?**
   (A) The Prime Minister of India
   (B) The Union Agriculture Minister
   (C) The State Minister of Agriculture
   (D) Director General, ICAR

10. **The growth of ICAR started with the establishment of which of the following institutes, being the first one?**
   (A). Imperial Bacteriological Laboratory, Pune
   (B) Agricultural Research Institute, Pusa, Bihar
   (C) Indian Agricultural Research Institute, New Delhi
   (D) Agriculture College, Kanpur

11. **Which of the following was the first institute/university to be establishment only land Grant College System in 1958?**
   (A) Indian Agricultural Research Institute, New Delhi
   (B) G.B. Pant University of Agriculture & technology, Pantnagar
   (C) College of Agriculture, Coimbatore
   (D) College of Agriculture, Kanpur

12. **The most serious economic problem of India are:**
   (A) Poverty and Unemployment
   (B) Unemployment and stagnation
   (C) Underdevelopment and Poverty
   (D) Corruption and Unemployment

13. **By eradicating poverty, who wished to "wipe out every tear from every eye":**
   (A) Jawahar Lal Nehru
   (B) Mahatma Gandhi
   (C) Rabindra Nath Tagore
   (D Vinoba Bhave

14. **The 'Garibhi hatao' slogan was coined by:**
   (A) Jawahar Lal Nehru
   (B) Mahatma Gandhi
   (C) Rabindra Nath Tagore
   (D) Indira Gandhi

15. **The proportion of Indian population living in the villages:**
   (A) One-third
   (B) Half
   (C) Two-third
   (D) Three-fourth

16. **Who said 'If the villages perish, India will perish:**
   (A) Jawahar Lal Nehru
   (B) Mahatma Gandhi
   (C) Rabindra Nath Tagore
   (D Indira Gandhi

17. **Who coined the term Green revolution:**
   (A) Willianm Gadd
   (B) M.S. Swaminathan
   (C) Norman E. Borlaoug
   (D) C. Subramanian

18. **Agricultural Scientists Recruitment Board (ASRB) was set up based on the recommendations of the committee headed by:**
   (A) Mohan Sinha Mehta
   (B) M S Swaminathan
   (C) Gajendra Gadkar
   (D) B P Sinha

19. **The first farm magazine in India is.**
   (A) Kheti
   (B) Phool Phul
   (C) Krishi Chayanika
   (D) Yojana

20. **The general micro-unit of an agro-ecosystem is:**
   (A) District (B) Village
   (C) Farmer (D) Household

Practice Paper 6

1. **The headquarter of Food and Agricultural Organization are at:**
   (A) New York (B) New Delhi
   (C) Geneva (D) Rome

2. **In America which crop is known as yellow gold:**
   (A) Maize (B) Rice
   (C) Soybean (D) Wheat

3. **Scientific name of green varieties of black gram is:**
   (A) *Vigna mungo var viridis*
   (B) *Vigna mungi var niger*
   (C) *Vigna mungo var macropode*
   (D) *Vigna mungi var micropode*

4. **Scientific name of bunch type of ground nut is:**
   (A) *Arachis hypogea* sp *fastigiata*
   (B) *Arachis hypodea* sp *procumbens*
   (C) *Arachis hypogea* sp *viridis*
   (D) *Arachis hypogea* sp *niger*
5. **Scientific name of spreading type of ground nut is:**
   (A) *Arachis hypogea* sp *fastigua*
   (B) *Arachis hypodea* sp *procumbens*
   (C) *Arachis hypogea* sp *viridis*
   (D) *Arachis hypogea* sp *niger*
6. **Scientific name of desi cotton is:**
   (A) *Gossypium herbacium*
   (B) *Gossypium arboreum*
   (C) Both (A) and (B)
   (D) None of these
7. **Scientific name of American cotton is:**
   (A) *Gossypium barbedensii*
   (B) *Gossypium hirsutum*
   (C) Both (A) and (B)
   (D) None of these
8. **Diploid species of cotton are:**
   (A) Desi (B) American
   (C) Both (A) and (B) (D) None of these
9. **Tetraploid species of cotton are:**
   (A) Desi (B) American
   (C) Both (A) and (B) (D) None of these
10. **Poor man's timber or green gold is:**
    (A) Mango wood (B) Bamboo
    (C) Acacia wood (D) Eucalyptus wood
11. **Tendu/Kendu leaves used as wrappers for bidis are got from:**
    (A) *Diospyros ebenum*
    (B) *Dipterocarpus indicus*
    (C) *Dipterocarpus tuberculatus*
    D) *Diospyros melonxylon*
12. **Conservation is:**
    (A) Protection of natural resources
    (B) Management of natural resources
    (C) Proper use of natural resources
    (D) All of these
13. **MAB is:**
    (A) Man and Botany
    (B) Man and Biosphere
    (C) Man and Biotic community
    (D) Man, Antibiotics and Bacteria
14. **Silent valley has been preserved as it has:**
    (A) Recreational value
    (B) Valuable timber plants
    (C) Natural forest
    (D) Rare plant and animals
15. **Silent valley having rare plants and animals is located in:**
    (A) Kerala
    (B) Karnataka
    (C) Jammu and Kashimir
    (D) Andhra Pradesh
16. **Which one is connected with conservation of forests?**
    (A) Gir (B) Silent valley
    (C) Kaziranga (D) Bharatpur
17. **New approach to conservation is establishment of:**
    (A) Reserve forests (B) Santuries
    (C) Biosphere reserves(D) National parks
18. **Niger is a:**
    (A) Pulse crop(B) Fibre crop
    (C) Cereal crop(D) Oilseed crop
19. **Emergence of flag leaf indicates emergence of:**
    (A) Panicle (B) Inflorescence
    (C) Ear (D) Tillers
20. **In wheat crop, crown roots emerge:**
    (A) 2 weeks after sowing
    (B) 3 weeks after sowing
    (C) 4 weeks after sowing
    (D) 5 weeks after sowing

Practice Paper 7

1. **Four major mustard producing states in India are:**
   (A) Rajasthan, U.P., M.P. and Haryana
   (B) Rajasthan, M.P., U.P. and Hrayana
   (C) Rajasthan, Haryana, U.P. and M.P.
   (D) Rajasthan, Punjab U.P. and Haryana
2. **Adulteration of mustard oil with *Argemone mexicana* seed oil caused in the recent past:**
   (A) Tuberculosis (B) Diarrhoea
   (C) Dropsy (D) None of these

3. **Wheat is a:**
(A) Monoecious plant
(B) Diocious plant
(C) Hermaphrodite plant
(D) None of these

4. **First of all coloured cotton developed in India in the state of:**
(A) Madhya Pradesh (B) Maharastra
(C) Gujarat (D) Andhra Pradesh

5. **Botanical name of kodon is:**
(A) *Echinochloa frumentacea* L.
(B) *Eleusine coracana* Gaertn
(C) *Paspalum scrobiculatum*
(D) None of these

6. **First hybrid variety of sorghum was evolved in the year:**
(A) 1946 (B) 1956
(C) 1964 (D) 1974

7. **First hybrid variety of bajra was evolved in the year:**
(A) 1946 (B) 1956
(C) 1965 (D) 1974

8. **According to Manchurian classification soybean has been divided into:**
(A) Two groups (B) Three groups
(C) Four groups (D) Only one group

9. **Pegging of groundnut refers to:**
(A) Flower initiation
(B) Initiation of secondary branches
(C) Entering of fertilized ovary in soil
(D) Grain development

10. **The edible part of cabbage is:**
(A) Root (B) Flower
(C) Leaf (D) Fruit

11. **Evergreen/raindow revolution is related to:**
(A) Potato
B) Food grain production
(C) Milk production
(D) Increasing production in all together

12. **Who says "Everything else can wait, but not agriculture"?**
(A) Shri. J.L. Nehru
(B) Dr. A.P.J. Abul Kalam
(C) Dr. H. Rembold
(D) Mahatma Gandhi

13. **Who says "Agriculture is the culture of the times as/or all culture"?**
(A) Shri. J.L. Nehru
(B) Dr. A.P.J. Abul Kalam
(C) Dr. H. Rembold
(D) Mahatma Gandhi

14. **Who says "We have to have the second green revolution for ever increasing population of the country"?**
(A) Shri. J.L. Nehru
(B) Dr. A.P.J. Abul Kalam
(C) Dr. H. Rembold
(D) Mahatma Gandhi

15. **Who says "India is sitting on gold mines will all its Neem trees"?**
(A) Shri. J.L. Nehru
(B) Dr. A.P.J. Abul Kalam
(C) Dr. H. Rembold
(D) Mahatma Gandhi

16. **Total food grain production in India during 2019- 20 was:**
(A) 275.00 mt (B) 285.21 mt
(C) 296.65 mt (D) 299.00 mt

17. **The Central AGMARK lab is located at:**
(A) Lucknow (B) Nagpur
(C) Bhopal (D) Hyderabad

18. **Cultivation of duram wheat is primarily confined to:**
(A) Central India
(B) Southern India
(C) Central & Southern India
(D) UP hills

19. **Bharat Ratna Dr. C. Subramanian Award is related to:**
(A) Best agriculture teacher
(B) Best technical book
(C) Tribal people upliftment
(D) Best research in agriculture

20. **Father of White Revolution in India is:**
(A) Dr. M.S. Swaminathan
(B) Dr. Vergeese Kurien
(C) Dr. R.S. Paroda
(D) Dr. B.P. Pal

Practice Paper 8

1. **IARI got status of deemed university on:**
   (A) 1958 (B) 1959
   (C) 1960 (D) 1961
2. **IARI established during:**
   (A) Lord Macnaro (B) Lord Curzon
   (C) Lord Carbvalis (D) None of these
3. **Who was first chairman of ICAR:**
   (A) Mohd. Habibulla (B) S.A. Haidery
   (C) B.P.Pal (D) M.S. Swaminathan
4. **When ICAR declared an autonomous body?**
   (A) 1964 (B) 1965
   (C) 1966 (D) 1967
5. **First Director General of ICAR was:**
   (A) Mohd. Habibulla (B) B.P.Pal
   (C) S.A. Haidery (D) M.S. Swaminathan
6. **First five year plan was started on:**
   (A) 1947-1951 (B) 1948-1952
   (C) 1951-1956(D) 1950-1955
7. **ICAR foundation day celebrated on:**
   (A) 16 July (B) 16 August
   (C) 16 September (D) 16 October
8. **Rothamsted Agricultural Research Station started by:**
   (A) Phiphs
   (B) Larves and Gilbert
   (C) Bourlag
   (D) Swaminathan
9. **Maximum agriculture culturable area in which country:**
   (A) America (B) Australia
   (C) Canada (D) India
10. **Queen of food grain is:**
    (A) Wheat (B) Rice
    (C) Maize (D) Potato
11. **Rice was first domesticated approximately 5000 years ago in:**
    (A) South east Asia (B) Souh west Asia
    (C) South Asia (D) Asia Pacific
12. **Weight of a cotton bale is:**
    (A) 150 kg (B) 160 kg
    (C) 170 kg (D) 180 kg
13. **Queen of spices is:**
    (A) Ilaichi (B) Turmeric
    (C) Chilli (D) Coriander
14. **King of fruit is:**
    (A) Banana (B) Papaya
    (C) Guava (D) Mango
15. **Pseudo cereal is:**
    (A) Duram wheat (B) Buck wheat
    (C) Macroni wheat (D) None of these
16. **Which state is most affected by water erosion:**
    (A) West Bengal (B) Uttar Pradesh
    (C) Rajasthan (D) Madhya Pradesh
17. **Which state is most affected by wind erosion:**
    (A) West Bengal (B) Uttar Pradesh
    (C) Rajasthan (D) Madhya Pradesh
18. **Which state has maximum area under alkali soils in India:**
    (A) West Bengal (B) Uttar Pradesh
    (C) Gujarat (D) Bihar
19. **Which state has maximum area under acidic soils in India:**
    (A) West Bengal (B) Uttar Pradesh
    (C) Gujarat (D) Bihar
20. **Maximum fertilizer consumption per hectare is:**
    (A) Punjab (B) Pondicherry
    (C) Haryana (D) Uttar Pradesh

Practice Paper 9

1. **Bowl of rice is known as:**
   (A) West Bengal (B) Kerala
   (C) Tamilnadu (D) Karnataka
2. **Bowl of spices is known as:**
   (A) West Bengal (B) Kerala
   (C) Tamilnadu (D) Karnataka
3. **City of gardens is known as:**
   (A) Bangalore (B) J & K
   (C) New Delhi (D) Hyderabad
4. **Maximum area under forest is:**
   (A) Rajasthan (B) Madhya Pradesh
   (C) West Bengal (D) Uttar Pradesh

5. Which type of cultivation practices mainly responsible for agricultural development in India:

(A) Precision farming
(B) Shifting cultivation
(C) Mutistoryed cultivation
(D) Organic farming

6. Importance of animals and raring of animals was known as era of:

(A) Pastoral age (B) Neolithic age
(C) Bronze age (D) None of these

7. Sorghum is known as camel due to:

(A) Important crop of millet
(B) Drought tolerant
(C) Important feed of camel
(D) Important crop of dry areas

8. Panchgavya is an organic product has the potential to play the role of promoting:

(A) Growth
(B) Productivity
(C) Growth and productivity
(D) None of these

9. Panchgavya consist of:

(A) Cow dung and cow urine
(B) Cow milk and cow curd
(C) Cow ghee
(D) All of these

10. Sorghum is considered as the king of millets also known as:

(A) Camel crop
(B) King of cereals
(C) Queen of cereals
(D) Important crop of kharif

11. IRRI is the abbreviation for:

(A) International Raw Rice Institute
(B) Institute of Rice Research in India
(C) Indian Rice Research Institute
(D) International Rice Research Institute

12. Which crop is called as king of cereal crops?

(A) Rice (B) Wheat
(C) Oat (D) Barley

13. The term 'Tropical oils' refers to:

(A) Palm oil and coconut oil
(B) Olive oil and almond oil
(C) Corn oil and soyabean oil
(D) Sunflower oil and safflower oil

14. 'Touch me not' plant exhibits a reflex action that is due to:

(A) Special knee joint
(B) Sudden temperature control
(C) Exceptional nervous conduction
(D) Powerful water pressure alteration

15. "Tension wood" is the name given to that wood which reacts to:

(A) Drought
(B) Low oxygen
(C) Drooping
(D) Disease

16. The Bengal famine of 1942 was due to failure of paddy crop caused by:

(A) *Helminthosporium oryzae* (Brown spot of rice)
(B) *Xanthomonas compestris pv oryzae*
(C) *Pyricularia oryzae*
(D) *Rhizoctonia solani*

17. Highest number of sugar mills in India is located in:

(A) Maharastra
(B) Tamilnadu
(C) Bihar
(D) Uttar Pradesh

18. Sustainable agriculture can be achieved by:

(A) More resources
(B) More subsidies
(C) More technological support
(D) Better prices of the produce

19. Which of he following crops have been approved by commercial cultivation:

(A) Bt cotton and Bt brinjal
(B) Bt cotton and Golden rice
(C) Bt maize and Bt cotton
(D) Bt cotton only

20. Norman *E. Borlaug* got Nobel Prize in the field of:

(A) Agriculture
(B) Medicine
(C) Peace
(D) Economics

Practice Paper 10

1. The UPOV (Union for the Protection of New Plant Varieties) headquarter is located at:

(A) Geneva (B) Berne
(C) Rome (D) Paris

2. Jojoba yields wax for cosmetics from its:

(A) Stem (B) Leaves
(C) Seeds (D) Fruit pulp

3. IR 36 was developed through breeding of:

(A) Six rice varieties and *Oryza nivara*
(B) 13 rice varieties and *Oryza nivara*
(C) *Oryza sativa* and *O. indica*
(D) *Oryza indica* and *O. nivara*

4. Which one is beverage plant:

(A) *Ilex paraguriensis* (B) *Caltha edulis*
(C) *Poullinia cupana* (D) All of these

5. The three top most major crops of the world in the order of total production are:

(A) Wheat > Rice > Maize
(B) Rice > Wheat > Maize
(C) Wheat > Maize > Rice
(D) Rice > Maize > Wheat

6. Potential food plant is:

(A) Winged bean (Psophocarpus tetragonolobus)
(B) Jajoba (*Simmondsia chinensis*)
(C) Guayule (*Parthenium argentatum)*
(D) *Leucaena*

7. A potential source of rubber is:

(A) Guayule (*Parthenium argentatum*)
(B) *Leucaena leucòcephala*
(C) *Cucurbita foetidissima*
(D) *Citrullus colocynthis*

8. Cola, the common beverage, is obtained from:

(A) Flowers (B) Seeds
(C) Leaves (D) Stems

9. Maximum genetic diversity of crop plants occurs where agriculture is:

(A) Run on commercial lines
(B) Advanced
(C) Rainfed
(D) Primitive

10. Which is more suitable for germplasm collection:

(A) Centre of production
(B) Homelands
(C) Tropical areas
(D) Temperate areas

11. Guarana is:

(A) Fodder plant
(B) Oil plant
(C) Beverage plant
(D) Food plant

12. Basic input of agriculture is:

(A) Irrigation (B) Fertilizer
(C) Soil (D) Seed

13. Who is the present chairman of protection of Plant Varieties and Farmers Right Authority (PPV&FRA):

(A) Dr. K.V. Prabhu
(B) Dr. P.L. Gautam
(C) Dr. R.R. Hanchinal
(D) Dr. S. Nagarajan

14. Which of the following is not an objective of the National Food Security Mission?

(A) Sustainable increase in production of rice, wheat and pulses
(B) Restoring soil fertility and productivity at individual farm level
(C) Promoting use of biopesticides and organic fertilizer
(D) Creation of employment opportunities

15. Agmarket, a portal for the dissemination of agricultural marketing information, is a joint endeavour of:

(A) DMI and NIC
(B) DMI and Ministry of Agriculture
(C) NIC and Ministry of Agriculture
(D) DMI and Directorate of Economics and Statistics

16. According to the approach paper to the 12th Five Year Plan, the basic objective of the 12th plan is:

(A) Inclusive growth
(B) Sustainable growth
(C) Faster, more inclusive and sustainable growth
(D) Inclusive and sustainable growth

17. To address the problems of sustainable and holistic development of rainfed areas including appropriate farming and livelihood system approaches, the Government of India has set up the:

(A) National Rainfed Area Authority

(B) National Watershed Development Project for Rainfed Areas

(C) National Mission on Rainfed Areas

(D) Command Area Development and Water Management Authority

18. Which of the following sub-scheme are not covered under the Rastriya Krishi Vikas Yojna?

(A) Extending the Green Revolution to Eastern India

(B) Development of 60000 pulses and Oilseeds village in identified watersheds

(C) National Mission on Saffron

(D) National Mission on Bamboo

19. Two major races of rice are:

(A) European and Tropical

(B) Indica and japonica

(C) Asiatic and American

(D) Temperate and Tropical

20. Which one of the following is not the goal of sustainable agriculture in India:

(A) Maintaining productive top soil

(B) Mechanical farming

(C) Reduce the use of chemical fertilizer and pesticide

(D) Keep farms economically viable

Answers Key

Practice Paper 1

Q. No.	Ans	Q. No.	Ans	Q. No.	Ans	Q. No.	Ans	Q. No.	Ans	Q. No.	Ans
1.	A	2.	A	3.	B	4.	C	5.	C	6.	B
7.	C	8.	D	9.	D	10.	A	11.	B	12.	A
13.	B	14.	C	15.	D	16.	C	17.	D	18.	A
19.	A	20.	D								

Practice Paper 2

Q. No.	Ans	Q. No.	Ans	Q. No.	Ans	Q. No.	Ans	Q. No.	Ans	Q. No.	Ans
1.	C	2.	B	3.	A	4.	A	5.	B	6.	D
7.	C	8.	A	9.	A	10.	B	11.	A	12.	C
13.	D	14.	C	15.	A	16.	B	17.	D	18.	A
19.	B	20.	D								

Practice Paper 3

Q. No.	Ans	Q. No.	Ans	Q. No.	Ans	Q. No.	Ans	Q. No.	Ans	Q. No.	Ans
1.	A	2.	A	3.	C	4.	B	5.	D	6.	D
7.	B	8.	D	9.	B	10.	D	11.	C	12.	A
13.	C	14.	C	15.	D	16.	D	17.	D	18.	D
19.	C	20.	B								

Practice Paper 4

Q. No.	Ans	Q. No.	Ans	Q. No.	Ans	Q. No.	Ans	Q. No.	Ans	Q. No.	Ans
1.	D	2.	D	3.	A	4.	A	5.	C	6.	D
7.	D	8.	A	9.	A	10.	A	11.	B	12.	C
13.	B	14.	A	15.	B	16.	A	17.	C	18.	B
19.	D	20.	D								

Practice Paper 5

Q. No.	Ans	Q. No.	Ans	Q. No.	Ans	Q. No.	Ans	Q. No.	Ans	Q. No.	Ans
1.	A	2.	B	3.	A	4.	C	5.	B	6.	A
7.	B	8.	C	9.	B	10.	A	11.	A	12.	A
13.	B	14.	D	15.	D	16.	B	17.	A	18.	C
19.	A	20.	B								

■ Practice Paper 6

Q. No.	Ans	Q. No.	Ans	Q. No.	Ans	Q. No.	Ans	Q. No.	Ans	Q. No.	Ans
1.	D	2.	A	3.	A	4.	A	5.	B	6.	C
7.	C	8.	A	9.	B	10.	B	11.	D	12.	D
13.	B	14.	D	15.	A	16.	B	17.	C	18.	D
19.	C	20.	B								

■ Practice Paper 7

Q. No.	Ans	Q. No.	Ans	Q. No.	Ans	Q. No.	Ans	Q. No.	Ans	Q. No.	Ans
1.	A	2.	C	3.	A	4.	B	5.	C	6.	C
7.	C	8.	B	9.	C	10.	C	11.	D	12.	A
13.	A	14.	B	15.	C	16.	C	17.	B	18.	C
19.	B	20.	B								

■ Practice Paper 8

Q. No.	Ans	Q. No.	Ans	Q. No.	Ans	Q. No.	Ans	Q. No.	Ans	Q. No.	Ans
1.	A	2.	B	3.	A	4.	C	5.	B	6.	C
7.	A	8.	B	9.	B	10.	C	11.	A	12.	C
13.	A	14.	D	15.	B	16.	A	17.	C	18.	C
19.	A	20.	B								

■ Practice Paper 9

Q. No.	Ans	Q. No.	Ans	Q. No.	Ans	Q. No.	Ans	Q. No.	Ans	Q. No.	Ans
1.	A	2.	B	3.	A	4.	B	5.	B	6.	A
7.	B	8.	C	9.	D	10.	A	11.	D	12.	B
13.	A	14.	D	15.	C	16.	A	17.	D	18.	C
19.	D	20.	C								

■ Practice Paper 10

Q. No.	Ans	Q. No.	Ans	Q. No.	Ans	Q. No.	Ans	Q. No.	Ans	Q. No.	Ans
1.	A	2.	C	3.	B	4.	D	5.	C	6.	A
7.	A	8.	B	9.	D	10.	B	11.	C	12.	D
13.	A	14.	C	15.	A	16.	C	17.	A	18.	D
19.	B	20.	•								

12. Multiple Choice Questions of Agricultural Climatology

Practice Paper 1

1. **Climate word originated from:**
   (A) Greek (B) Latin
   (C) English (D) Parsian
2. **Climate word derived from:**
   (A) Klima+ Logo (B) Klima+Logos
   (C) Clima + Logo (D) Clima+Logos
3. **The basic meaning of klima is:**
   (A) Incidence of humidity
   (B) Changes in temperature
   (C) Angle incidence of sun light
   (D) All of these
4. **The science that seeks to describe and explain that nature of climate is called:**
   (A) Climatology (B) Meteorology
   (C) Both of these (D) None of these
5. **The study of physical status of climate is called:**
   (A) Applied climatology
   (B) Statistical climatology
   (C) Descriptive climatology
   (D) Physical climatology
6. **To study the physical atmospheric factors viz., temperature, relative humidity, clouds, wind velocity and direction is called:**
   (A) Physical climatology
   (B) Statistical climatology
   (C) Descriptive climatology
   (D) Applied climatology
7. **To study the different factors of climate such as temperature and rainfall, are divided into state level is called:**
   (A) Physical climatology
   (B) Statistical climatology
   (C) Descriptive climatology
   (D) Applied climatology
8. **To study the relationship with other branched of climatology is called:**
   (A) Physical climatology
   (B) Statistical climatology
   (C) Descriptive climatology
   (D) Applied climatology
9. **The weather information collected from observatory are depicted in tables is called:**
   (A) Weather chart (B) Weather map
   (C) Both of these (D) None of these
10. **The weather information are depicted through map is called:**
    (A) Weather chart (B) Weather map
    (C) Both of these (D) None of these
11. **World Meteorological Organisation (WMO) was established at:**
    (A) Geneva (Switzerland)
    (B) Spain
    (C) USA
    (D) Canada
12. **World Meteorological Organisation (WMO) was established in:**
    (A) 1950 (B) 1951
    (C) 1952 (D) 1953
13. **International Meteorological Organisation was established in:**
    (A) 1877 (B) 1878
    (C) 1879 (D) 1880
14. **The science of climatological condition which have direct relation or relevance to agriculture is called:**
    (A) Agro-meteorology (B) Agro-climatology
    (C) Meteorology (D) Climatology
15. **The composite or generalization of the variety of day to day weather conditions is called:**
    (A) Season (B) Weather
    (C) Climate (D) All of these

16. How many years of weather is optimum for climate:

(A) 30 years (B) 31 years
(C) 32 years (D) 33 years

17. The observations recorded from open area of 2 meter height of soil surface is called:

(A) Macro climate (B) Micro climate
(C) Topo climate (D) Local climate

18. When weather data collected from sloppy land is called:

(A) Macro climate (B) Micro climate
(C) Topo climate (D) Local climate

19. When weather data collected form standing crop is called:

(A) Macro climate (B) Micro climate
(C) Topo climate (D) Local climate

20. The data recorded in macro climate is called:

(A) Synoptic Meteorological observation
(B) Asynoptic Meteorological observation
(C) Both of these
(D) None of these

Practice Paper 2

1. Upto 15 km above the earth is called:

(A) Troposphere (B) Mesosphere
(C) Stratosphere (D) None of these

2. 50 km above the earth is called:

(A) Troposphere (B) Mesosphere
(C) Stratosphere (D) None of these

3. In atmosphere mainly which gas is determined to atmospheric activities is:

(A) O_2 (B) Ozone
(C) Both of these (D) None of these

4. In atmosphere gasses found in:

(A) Mixed status
(B) Without mixed status
(C) Alone
(D) None of these

5. Four gasses- nitrogen, oxygen, argon and CO_2 accounts:

(A) 97 % (B) 98 %
(C) 99 % (D) 100 %

6. Nitrogen alone constituents nearly:

(A) One-third by volume
(B) Two-third by volume
(C) Three-fourth by volume
(D) Four-fifth by volume

7. Oxygen alone constituents nearly:

(A) One-third by volume
(B) Two-third by volume
(C) One-fifth by volume
(D) Four-fifth by volume

8. The blanket of gases, suspended liquids and solids the entirely envelops the earth is called:

(A) Atmosphere (B) Climate
(C) Clouds (D) Gasses

9. The lower portion of the atmosphere, extending upto about 8 km at the poles and 16 km at the equator is called:

(A) Troposphere (B) Stratosphere
(C) Ionosphere (D) Exosphere

10. The upper portion of the toposphere, extending upto 50-60 km is called:

(A) Troposphere (B) Stratosphere
(C) Ionosphere (D) Exosphere

11. The upper portion of the stratosphere, extending upto 60-500 km is called:

(A) Troposphere (B) Stratosphere
(C) Ionosphere (D) Exosphere

12. The upper portion of the atmosphere, extending upto 500-1000 km is called:

(A) Troposphere (B) Stratosphere
(C) Ionosphere (D) Exosphere

13. Ozone gas mainly found in which region:

(A) Troposphere (B) Stratosphere
(C) Ionosphere (D) Exosphere

14. High air pressure and its affected area is called:

(A) Anticyclone (B) Cyclone
(C) Depression (D) Rainfall

15. Low air pressure and its affected area is called:

(A) Anticyclone
(B) Cyclone and Depression
(C) Fog
(D) Rainfall

16. Ozonated hexane led to distruption of .... in plants:

(A) Membrane permeability
(B) Protein synthesis
(C) Lipid synthesis
(D) Stem elongation

17. Animals and nongreen plants continuously use oxygen and liberate .... into the air:

(A) Sulfur dioxide
(B) Carbon monoxide
(C) Carbon dioxide
(D) Nitrous oxide

18. The rate of photosynthesis under the conditions of high light intensity may be controlled by:

(A) Light (B) Carbon dioxide
(C) Water (D) None of these

19. North of the equator, surface winds are known as:

(A) Northeast trade winds
(B) Southeast winds
(C) Westerlies
(D) Easterlies

20. In the humid tropics and subtropics the process of weathering (soil formation) is known as:

(A) Podzolization (B) Gleization
(C) Latosolization (D) None of these

Practice Paper 3

1. The diversity pertaining to a plant composition in several plant communities in one range of environment is known as:

(A) Alpha index of diversity
(B) Beta index of diversity
(C) Gamma index of diversity
(D) None of these

2. Among the types of natural succession, the one which begins in water is known as:

(A) Xerarch (xerosere)
(B) Halosere
(C) Hydrarch (hydrosere)
(D) Psammosere

3. In the earth's atmosphere at .... cal/cm²/annum of solar energy is received:

(A) 15×10^{6} (B) 15×10^{8}
(C) 15×10^{10} (D) 15×10^{12}

4. Fixation of each gramme of carbon from carbon dioxide to organic compounds during photosynthesis traps .... of solar energy:

(A) 124 k cal (B) 134 k cal
(C) 144 k cal (D) 154 k cal

5. .... pollutants are very harmful for human health and therefore such polluted water should essentially be purified before their supply for drinking purposes:

(A) Biological (B) Physical
(C) Mechanical (D) Metallic

6. Ionising .... are very important and harmful pollutants:

(A) Reflection (B) Deflection
(C) Radiation (D) Absorptions

7. Fresh water bodies like rivers and streams are known as:

(A) Palustrine
(B) Lotic or riverine
(C) Lentic or lacustrice
(D) None of these

8. On of the basis of temperature variation in the vertical column of water, or thermal stratification, lake ecosystem have been divided into the following zones wherein .... comes in direct contact of sun and wind:

(A) Thermocline (B) Hypolimnion
(C) Epilimnion (D) None of these

9. CFCs methane and nitrogen oxide gases have the capacity to react with ozone to convert into:

(A) Nitrogen (B) Oxygen
(C) Hydrogen (D) None of these

10. The estimates reveal that the maximum NO_2 emission in European countries is in .... in the range of 2Tg/annum (Tetagramme = 10^{12} grammes):

(A) France (B) Germany
(C) England (D) None of these

11. The threshold concentration upto which no visible injury symptoms appear on plants due to NO2 is 1 mg/m2 on:

(A) Daily average (24 hrs)
(B) Weekly average
(C) Monthly average
(D) Annual average

12. Environment Protection Act 1986 came into force in India on 1968:

(A) September 19 (B) November 19
(C) October 19 (D) August 19

13. Of the total fresh water, ground water constitutes:

(A) 90% (B) 22.4%
C) (10% (D) 44.2%

14. The direct effect of strong winds on plants is of which nature:

(A) Chemical (B) Biological
(C) Mechanical (D) None of these

15. Troposphere is the densest part of atmosphere and contains about:

(A) 55% of atmospheric mass
(B) 75% of atmospheric mass
(C) 85% of atmospheric mass
(D) 95% of atmospheric mass

16. The layer separated the stratosphere and mesosphere is called:

(A) Tropopause (B) Stratopause
(C) Mesopause (D) None of these

17. Inasphore is present in:

(A) Troposphere (B) Stratosphere
(C) Mesosphere (D) Thermosphere

18. Heat transfer into the soil is mainly by:

(A) Conduction (B) Convection
(C) Radiation (D) None of these

19. The rate of heat flux into soil depends on:

(A) Temperature gradient
(B) Thermal conductivity
(C) (A) and (B)
(D) None of these

20. Thermal conductivity of soil decrease with:

(A) Increase porosity
(B) Increase organic matter
(C) Decrease soil moisture
(D) All of these

Practice Paper 4

1. What will be the effect of tillage operation on thermal conductivity of soil?

(A) Remain unchanged
(B) Decrease
(C) Increase
(D) First increase then decrease

2. The biological amplication of DDT in the various trophic levels is known as:

(A) Green house effect
(B) Biomagnification
(C) Entraphication
(D) Pollution

3. Environment can be improved by:

(A) Wind erosion (B) Water erosion
(C) Conservation (D) None of these

4. Smog is produced due to:

(A) Air and metal dust
(B) Water and nitrogen oxide
(C) Smoke and fog
(D) None of these

5. The photochemical smog is formed by:

(A) Nitrogen oxides
(B) Hydrocarbons
(C) Nitrogen oxides and hydrocarbons
(D) Solar radiations upon nitrogen oxides and hydrocarbons

6. Water soacked appearance of leaves due to sulphur dioxide pollution is caused by:

(A) Destruction of chlorophyll
(B) Damage to membrane
(C) Both (A) and (B)
(D) Corrosive action over metals

7. Addition of phosphates and nitrate fertilizers into water leads to:

(A) Increased growth decompositors
(B) Reduced algal growth
(C) Increased algal growth
(D) Nutrient enrichment/eutrophication

8. Major reason for soil erosion in India is:

(A) Tilling of land
(B) Deforestation
(C) Overgrazing and deforestation
(D) Over grazing

9. Domestic water is:
(A) Effluent (B) Biodgradable
(C) Nonbiodegradable (D) Air pollutant

10. Inadequate drainage in a soil will lead to:
(A) Floods (B) Soil erosion
(C) Aridity (D) Salination

11. Which is most ecofriendly?
(A) Bacteria (B) Animal species
(C) Plant species (D) Both (B) and (C)

12. Ecological balance of a mining area is restricted through:
(A) Conversion into agricultural area
(B) Revegetation of area
(C) Prevention of soil erosion
(D) Protection against grazing

13. Ultra-violet light causes:
(A) Formation of pyrimidines
(B) Sticky metaphases
(C) Photodynamic action
(D) Destruction of hydrogen bonds between complementary DNA strands

14. Sulphur dioxide affects:
(A) Cell wall
(B) Plasmodesmata
(C) All membrane system
(D) Nucleus

15. Ultra-violet radiation are lethal because of inactivation of:
(A) Proteins, nucleic acids and pigments
(B) Minerals, air and water
(C) Carbohydrates, fats and vitamins
(D) Water, carbon dioxide and oxygen

16. Acid rain is due to increase in atmospheric concentration of:
(A) Ozone and dust (B) CO_2 and CO
(C) SO_3 and CO (D) SO_2 and NO_2

17. Major source of methane in India is:
(A) Rice fields (B) Sugarcane fields
(C) Wheat fields (D) Fruit orchards

18. Phosphate pollution is caused by:
(A) Weathering of phosphate rocks only
(B) Agricultural fertilizers
(C) Phosphate rocks and sewage
(D) Sewage and agricultural fertilizers

19. Increased incidence of floods in the plains of northern India are due to:
(A) Increased deforestation in catchment areas
(B) Increase in incidence of rainfall
(C) Silting of dams
(D) More area under cultivation

20. Adherence to two-children norms shall help in:
(A) Improving food web
(B) Checking pollution
(C) Stabilising ecosystem
(D) Fertility of soil

Practice Paper 5

1. Mist is horizontal visibility of:
(A) 100 m (B) 200 m
(C) 500 m (D) 1 km

2. Fog is formed by:
(A) Smoke
(B) Condensation of water vapours
(C) Condensation of water vapours over dust particles
(D) Both (A) and (C)

3. Succession on sand dune is:
(A) Lithosere (B) Psychrosere
(C) Psammosere (D) Halosere

4. Carbosphere comprises:
(A) Atmosphere (B) Lithosphere
(C) Hydrosphere (D) All of these

5. Adiabatic expansion resulting in cooling is called:
(A) Coriolis effect
(B) Joule-Kelvin effect
(C) Le Chatelier Principle
(D) Principle of Parsimony

6. Cyclic/frontal rainfall generally occurs:
(A) All over the world
(B) Tropical areas
(C) Temperate areas
(D) Boundary of warm tropical air and cold polar air

7. The land of wind mills is:
(A) UK (B) USA
(C) Netherlands (D) HImachal Pradesh

8. **The shining and thunder in clouds during rainy season:**
   (A) Start simulataneosly
   (B) Occur thunder first and then shining starts
   (C) Occur shining first and then thunder starts
   (D) Occur mostly no thunder and no shining

9. **Daily weather forecast is obtained in the country from:**
   (A) Pune (B) New Delhi
   (C) Mumbai (D) Kolkata

10. **The layer of atmosphere having uniform temperature is:**
   (A) Stratosphere (B) Exosphere
   (C) Thermosphere (D) Troposphere

11. **The physical state of the atmosphere at a given place and time is referred to as:**
   (A) Weather (B) Climate
   (C) Both of these (D) None of these

12. **The first meteorological department of India was started at Shimla in:**
   (A) 1863 (B) 1864
   (C) 1865 (D) 1866

13. **Layer of ozone is present in:**
   (A) Stratosphere (B) Thermosphere
   (C) Troposphere (D) None of these

14. **The atmosphere stratum nearest to earth is:**
   (A) Thermosphere (B) Stratosphere
   (C) Troposphere (D) None of these

15. **Part of earth covered by water is about:**
   (A) 73% (B) 50%
   (C) 92% (D) 87%

16. **Floods can be checked by:**
   (A) Making dams (B) Deforestation
   (C) Reforestation (D) None of these

17. **Movement of air occurs in the atmosphere upto height of:**
   (A) 100 km (B) 50 km
   (C) 25 km (D) 10 km

18. **Peak concentration of ozone in the stratosphere (ozonosphere) lies at a height of:**
   (A) 10 km (B) 25 km
   (C) 40 km (D) 50 km

19. **The direct effect of strong winds on plants is of which nature:**
   (A) Chemical (B) Biological
   (C) Mechanical (D) None of these

20. **The unit of atmospheric pressure is:**
   (A) Millibar (B) Bar
   (C) Pascal (D) Gram

Practice Paper 6

1. **Oxygen and ozone at high levels absorb most of which radiation:**
   (A) Ultra violet radiation
   (B) Gama radiation
   (C) Blue radiation
   (D) All of these

2. **The ratio of the amount of radiation reflected to the amount received on a surface is termed as:**
   (A) Albedo (B) Radiation
   (C) Reflection (D) All of these

3. **The ability to admit most of the insolation yet retard losses by reradiation from the earth's surface commonly is known as:**
   (A) Greenhouse effect
   (B) Atmosphere effect
   (C) Both of these
   (D) None of these

4. **The highest per cent of albedo is found:**
   (A) Desert
   (B) Tropical forest
   (C) Deciduous forest
   (D) Coniferous forest

5. **The instrument which measure sunshine and radiation is:**
   (A) Total radiation pyranometer
   (B) Net radiometer
   (C) Foster switch sunshine duration sensor
   (D) All of these

6. **The ratio of the mass of water vapour actually in the air to a unit of mass of air, including the water vapour is called:**
   (A) Specific humidity
   (B) Relative humidity
   (C) Both of these
   (D) None of these

7. The ratio of the amount of water vapour actually in a volume occupied by air to the amount the space could contain at saturation is called:

(A) Specific humidity (B) Relative humidity
(C) Both of these (D) None of these

8. Early morning hours when temperatures are low the relativė humidity will be:

(A) More (B) Less
(C) Not change (D) Unknown

9. The visible aggregates of water droplets, ice particles, or a mixture of both alongwith varying amounts of dust particles is called:

(A) Fog (B) Frost
(C) Clouds (D) All of these

10. The banner clouds above peaks are of which type clouds is called:

(A) Cirrocumulus (B) Cirrostratus
(C) Altocumulus (D) Altostratus

11. The clouds are dense, dome shaped and have flat bases is called:

(A) Cumulus (B) Cirrostratus
(C) Altocumulus (D) Altostratus

12. The heavy showers of rain, snow or hail, often accompanied by lighting and thunder occurs by:

(A) Cirrocumulus (B) Cumulonimbus
(C) Altocumulus (D) Altostratus

13. More than 60 per cent rains occurs by:

(A) South-west monsoon
(B) North-East monsoon
(C) Both of these
(D) None of these

14. Less than 30 per cent rains occurs by:

(A) South-west monsoon
(B) North-East monsoon
(C) Both of these
(D) None of these

15. A counter clock wise rotation of wind in northern hemisphere is:

(A) Cyclone (B) Anticyclone
(C) Rainfall (D) Fog

16. A counter clock wise rotation of wind in southern hemisphere is:

(A) Cyclone (B) Anticyclone
(C) Rainfall (D) Fog

17. Wind speed measured by:

(A) Windvane (B) Anemometer
(C) Both of these (D) None of these

18. The barometric pressure is highest at its centre and decreased out ward is indication of:

(A) Cyclone (B) Anticyclone
(C) Rainfall (D) Fog

19. The moisture index is more in:

(A) Perhumid (B) Humid
(C) Semiarid (D) Arid

20. The moisture index is less in:

(A) Perhumid (B) Humid
(C) Semiarid (D) Arid

Practice Paper 7

1. Human health and comfort suggest another possible approach to defining:

(A) Climatic types
(B) Relative humidity
(C) Temperature
(D) None of these

2. The summation of all atmospheric events an infinite time span does not changes is called:

(A) Weather (B) Climate
(C) Both of these (D) None of these

3. The periods of atmospheric changes on the order of millions of years is:

(A) Paleoclimates (B) Taiga climate
(C) Polar climate (D) Ocean climate

4. The frost occurs on clear nights with a temperature inversion and usually results in formation of ice crystals on bold objects is called:

(A) Advection frost (B) Radiation frost
(C) Black frost (D) None of these

5. Advection frost is also called:

(A) Radiation frost (B) Advection frost
(C) Black frost (D) None of these

6. The frost which results when the temperature at the surface in an air mass is below freezing is called:

(A) Radiation frost (B) Advection frost
(C) Black frost (D) None of these

7. **Smog is produced due to :**
   (A) Air and metal dust
   (B) Water and nitrogen oxide
   (C) Smoke and fog
   (D) None of these
8. **The photochemical smog is formed by:**
   (A) Nitrogen oxides
   (B) Hydrocarbons
   (C) Nitrogen oxides and hydrocarbons
   (D) Solar radiations upon nitrogen oxides and hydrocarbons
9. **Photochemical smog brings about:**
   (A) Defoliation
   (B) Bleaching and blazing of foliage
   (C) Floral bud shedding
   (D) Discolouration and curling of petals
10. **Ecological balance of a mining area is restored through:**
    (A) Conversion into agricultural area
    (B) Revegetation of area
    (C) Prevention of soil erosion
    (D) Protection against grazing
11. **Terrestrial life would become impossible with:**
    (A) Change in orbit of earth
    (B) Change on atmosphere permitting all solar radiations to reach lithosphere
    (C) Disappearance of moon
    (D) Decrease in mean annual temperature by 10 °C
12. **Troposphere of the atmosphere extending upto a height from ground to:**
    (A) 8-18 km (B) 50 km
    (C) 40 km (D) >50 km
13. **Fog, dew, mist, rain etc occur in the layer of:**
    (A) Thermosphere (B) Mesosphere
    (C) Stratosphere (D) Troposphere
14. **Troposphere is the densest part of atmosphere and contains about:**
    (A) 55% of atmospheric mass
    (B) 75% of atmospheric mass
    (C) 85% of atmospheric mass
    (D) 95% of atmospheric mass
15. **The layer separated the stratosphere and mesosphere is called:**
    (A) Tropopause (B) Stratopause
    (C) Mesopause (D) None of these
16. **Inosphore is present in:**
    (A) Troposphere (B) Stratosphere
    (C) Mesosphere (D) Thermosphere
17. **Heat can flow between substance or within substance by:**
    (A) Conduction (B) Convection
    (C) Radiation (D) All of these
18. **Vertical temperature decrease is called lapse rate and the normal lapse rate is:**
    (A) 4.50 C/km (B) 6.50 C/km
    (C) 8.50 C/km (D) None of these
19. **Heat transfer into the soil is mainly by:**
    (A) Conduction (B) Convection
    (C) Radiation (D) None of these
20. **The rate of heat flux into soil depends on:**
    (A) Temperature gradient
    (B) Thermal conductivity
    (C) (A) and (B)
    (D) None of these

Practice Paper 8

1. **Thermal conductivity of soil decrease with:**
   (A) Increase porosity
   (B) Increase organic matter
   (C) Decrease soil moisture
   (D) All of these
2. **What will be the effect of tillage operation on thermal conductivity of soil?**
   (A) Remain unchanged
   (B) Decrease
   (C) Increase
   (D) First increase then decrease
3. **One of the following processes is not a step involved in the water cycle operating in nature:**
   (A) Evaporation (B) Temperature
   (C) Precipitation (D) Photosynthesis
4. **South-west monsoon contributes how much per cent of rainfall?**
   (A) 50 % (B) 75 %
   (C) 80 % (D) 95 %

5. **Average rainfall in India is around:**
(A) 2120 mm (B) 1120 mm
(C) 850 mm (D) 500 mm

6. **Taj may be destroyed by:**
(A) Flood in Yamuna
(B) Temperature mediated spoilage of marble
(C) Air pollutants from Mathura refinery
(D) All of these

7. **Smog occurs in places of:**
(A) Excess SO_2 (B) Low temperature
(C) High temperature (D) Excess NH3

8. **Acid rains is due to increase in atmospheric concentration of:**
(A) Ozone and dust (B) CO_2 and CO
(C) SO_3 and CO (D) SO_2 and NO_2

9. **Which of these statements is not a valid reason for 'the depletion of flora and fauna?**
(A) Agricultural expansion
(B) Grazing and fuel wood collection
(C) Large scale development projects
(D) Rapid industrialization and urbanization

10. **Which factors does not cause depletion of biodiversity?**
(A) Environment pollution
(B) Expansion on animals
(C) Judicious way of collecting products from forests
(D) Excessive hunting

11. **Fog is a colloid of:**
(A) Solid in gas (B) Liquid in gas
(C) Gas in liquid (D) Liquid in gas

12. **The .... zone is absent in ponds:**
(A) Sublittoral (B) Limnetic
(C) Profundal (D) Littoral

13. **Most stable ecosystem is:**
(A) Desert (B) Mountain
(C) Ocean (D) Fores

14. **The gas which causes depletion of ozone layer is the atmosphere is:**
(A) Ethane
(B) Methane
(C) Carbon dioxide
(D) Chloro fluoro carbon

15. **Which one is found in higher concentration in the atmospheric?**
(A) Ozone
(B) Nitrous oxides
(C) Carbon dioxide, carbon monoxide and sulphur dioxide
(D) All of these

16. **The nitrogen molecules present in air can be converted into nitrates and nitrites by:**
(A) A biological process of nitrogen fixing bacteria present in soil
(B) A biological process of carbon fixing factor present in soil
(C) Any of the industries manufacturing nitrogenous compounds
(D) The plants used as cereal crops in field

17. **Depletion of ozone layer is due to:**
(A) Oxides of carbon
(B) Oxides of nitrogen
(C) Oxides of sulphur (D) None of these

18. **Which pigment protects plants from UV damage?**
(A) Chlorophyll (B) Xanthophyll
(C) Carotenoids (D) Phycocyanin

19. **Component of smog is:**
(A) O_3 (B) PAN
(C) PPN and PBN (D) Both (A) and (B)

20. **Ozone hole refers to:**
(A) Hole in ozone layers
(B) Reduction in thickness of ozone layer in stratosphere
(C) Reduction of thickness of ozone in troposphere
(D) Increase concentration of ozone

Practice Paper 9

1. **Of the total water evaporated from oceans, precipitation returns:**
(A) 100% (B) 90%
(C) 75% (D) 50%

2. **Planning Commission advice that the agro-climatic zones in which plan:**
(A) Fifth Five Year Plan
(B) Sixth Five Year Plan
(C) Seventh Five Year Plan
(D) Eight Five Year Plan

3. Ozone is:
(A) Pale blue
(B) Highly reactive
(C) With penetrating odour
(D) All of these

4. An ozone depleting chemical which is used in fire-extinguishers is:
(A) Nitrogen oxide
(B) Methyl bromide
(C) Halon
(D) Chlorofluorocarbon

5. Smoked foods are not favoured now because of the possible toxicity of:
(A) Benzpyrene (B) Nitrogen oxides
(C) Sulphur dioxide (D) All of these

6. Benzpyrene is an ingredient of:
(A) Automobile exhausts
(B) Wood smoke and tobacco smoke
(C) Coal tar
(D) All of these

7. Smoke or dust dispersed in air is called:
(A) Fog (B) Haze
(C) Mist (D) Smog

8. Mist is fog with horizontal visibility of:
(A) 100 m (B) 200 m
(C) 500 m (D) 1000 m

9. Fog is formed by:
(A) Smoke
(B) Condensation of water vapours
(C) Condensation of water vapour over dust particles
(D) Both (A) and (C)

10. The term green house effect was given by:
(A) Fourier (B) Tyndall
(C) Arrhenius (D) Linnaeus

11. Polychlorinated biphenyls are chemicals useful as:
(A) Fire - resistant (B) Fertilizers
(C) Pesticides (D) Germicides

12. Use of polychlorinated biphenyl is restricted because of their:
(A) Extreme toxicity
(B) Persistent toxicity
(C) Biomagnification
(D) Both (B) and (C)

13. Ecofriendly detergents do not contain:
(A) Phosphates and bleaches
(B) Alcohols
(C) Water softners
(D) Perborates

14. Ozone day is observed on:
(A) September 16
(B) June 5
(C) May 1
(D) December 25

15. Nitrogen oxides released in the stratosphere cause ozone depletion due to:
(A) Catalysing conversion of ozone to oxygen
(B) Binding ozone
(C) Reacting with ozone
(D) All of these

16. In foggy and hazy environment, birds are able to locate position of sun and their path by means of:
(A) Infra-red wavelength
(B) Ultra-violet radiations
(C) Magnetic system
(D) Pheromones

17. The amount of biogenic nutrients present in the abiotic environment is called:
(A) Standing state
(B) Standing crop
(C) Detritus
(D) Reserve nutrient

18. Carbon dioxide is being added to the cycling pool by:
(A) Respiration and combustion
(B) Sedimentation of carbonates
(C) Precipitation, skeletons and fossilization
(D) Both (B) and (C)

19. A source of nonbiological carbon dioxide is:
(A) Volcanic eruptions
(B) Hot springs
(C) Weathering
(D) All of these

20. A succession that does not follow the natural course due to human activity is:
(A) Subsere (B) Plagiosere
(C) Prisere (D) Ascre

Practice Paper 10

1. **Which of the following disorder in mango is caused by low temperature?**
   (A) Black heart
   (B) Crotch injury
   (C) Collor injury
   (D) All of these
2. **Use of green plants and their associated microorganisms, soil amendments, and agronomic practices to remove, contain or render harmless environmental contaminants is known as:**
   (A) Phytoremediation
   (B) Phytostabilization
   (C) Both (A) and (B)
   (D) None of these
3. **Use of plants to sequester contaminants is cell wall lignins of root tissues, or to promote its binding to humus is known as:**
   (A) Phytoremediation
   (B) Phytostabilization
   (C) Both (A) and (B)
   (D) None of these
4. **The atmosphere of the earth is heated by radiations which are mainly:**
   (A) Radiated by the sun
   (B) Re-radiated by land
   (C) Re-radiated by water
   (D) Re-radiated by land and water
5. **If there were no atmosphere around the earth, the temperature of the earth will:**
   (A) Increase
   (B) Go on decreasing
   (C) Increase during day and decrease during night
   (D) Be unaffected
6. **What would happen, if all the oxygen present in the environment is converted to ozone :**
   (A) We will be protected more
   (B) It will become poisonous and kill living forms
   (C) Ozone is not stable, hence it will be toxic
   (D) It will help harmful sun radiation to reach earth and damage many life forms
7. **The two forms of oxygen found in the atmosphere are :**
   (A) Water and Ozone
   (B) Water and Oxygen
   (C) Ozone and Oxygen
   (D) Water and Carbon dioxide
8. **Rainfall pattern depend on:**
   (A) The underground watertable
   (B) The density pattern of human population in an area
   (C) The number of water bodies in an area
   (D) The prevailing season in an area
9. **Among the given options, which one is not correct for the use of large amount of fertilizers and pesticides?**
   (A) They are eco-friendly
   (B) They fields barren after some time
   (C) The adversely affect the useful component from the soil
   (D) They destroy the soil fertility
10. **Desert regions receive the rainfall:**
    (A) More than 50 cm (B) Less than 50 cm
    (C) 80-100 cm (D) None of these
11. **High velocity wind whirls are called:**
    (A) Monsoon (B) Cyclones
    (C) Earthquake (D) None of these
12. **Which is the effect of anticyclone?**
    (A) Pleasant weather (B) Heavy rainfall
    (C) Hail (D) Snow fall
13. **The formation of upholds by horizontal compression is called:**
    (A) Rift valley (B) Synclines
    (C) Anticyclones (D) None of these
14. **For converting gifts of nature into a resource, man requires:**
    (A) Skill and technology
    (B) Machines
    (C) Capital and labour
    (D) All of these
15. **Which of the following is not a community resources?**
    (A) Play ground (B) A village pond
    (C) Burial (D) Fruit orchard
16. **Which one of the following is a natural resources:**
    (A) An apartment (B) A river
    (C) A factory (D) A car

17. Rajasthan and Gujarat are the store houses of:

(A) Solar and wind energy
(B) Coal and iron ore
(C) Tidal energy
(D) Hydel energy

18. Which type of resource is solar energy?

(A) Soil resource
(B) Exhaustive resource
(C) Biotic resource
(D) Non-exhaustive resource

19. Part of earth covered by water is about:

(A) 73% (B) 50%
(C) 92% (D) 87%

20. The atmosphere stratum nearest to earth is:

(A) Thermosphere
(B) Stratosphere
(C) Troposphere
(D) None of these

Answers Key

■ Practice Paper 1

Q. No.	Ans	Q. No.	Ans	Q. No.	Ans	Q. No.	Ans	Q. No.	Ans	Q. No.	Ans
1.	A	2.	B	3.	C	4.	A	5.	D	6.	B
7.	C	8.	D	9.	A	10.	B	11.	A	12.	B
13.	B	14.	A	15.	C	16.	B	17.	A	18.	C
19.	B	20.	A								

■ Practice Paper 2

Q. No.	Ans	Q. No.	Ans	Q. No.	Ans	Q. No.	Ans	Q. No.	Ans	Q. No.	Ans
1.	A	2.	C	3.	C	4.	A	5.	C	6.	D
7.	C	8.	A	9.	A	10.	B	11.	C	12.	D
13.	B	14.	A	15.	B	16.	A	17.	C	18.	B
19.	A	20.	C								

■ Practice Paper 3

Q. No.	Ans	Q. No.	Ans	Q. No.	Ans	Q. No.	Ans	Q. No.	Ans	Q. No.	Ans
1.	B	2.	C	3.	B	4.	C	5.	D	6.	C
7.	B	8.	C	9.	B	10.	C	11.	B	12.	A
13.	B	14.	C	15.	C	16.	B	17.	D	18.	A
19.	B	20.	D								

■ Practice Paper 4

Q. No.	Ans	Q. No.	Ans	Q. No.	Ans	Q. No.	Ans	Q. No.	Ans	Q. No.	Ans
1.	B	2.	B	3.	C	4.	C	5.	D	6.	C
7.	D	8.	C	9.	B	10.	D	11.	C	12.	B
13.	D	14.	C	15.	A	16.	D	17.	A	18.	D
19.	A	20.	C								

Practice Paper 5

Q. No.	Ans	Q. No.	Ans	Q. No.	Ans	Q. No.	Ans	Q. No.	Ans	Q. No.	Ans
1.	D	2.	C	3.	C	4.	D	5.	B	6.	D
7.	C	8.	B	9.	A	10.	C	11.	A	12.	B
13.	A	14.	C	15.	A	16.	C	17.	D	18.	B
19.	C	20.	A								

Practice Paper 6

Q. No.	Ans	Q. No.	Ans	Q. No.	Ans	Q. No.	Ans	Q. No.	Ans	Q. No.	Ans
1.	A	2.	A	3.	C	4.	A	5.	D	6.	A
7.	B	8.	A	9.	C	10.	C	11.	A	12.	B
13.	A	14.	B	15.	A	16.	B	17.	C	18.	A
19.	A	20.	D								

Practice Paper 7

Q. No.	Ans	Q. No.	Ans	Q. No.	Ans	Q. No.	Ans	Q. No.	Ans	Q. No.	Ans
1.	A	2.	B	3.	A	4.	B	5.	C	6.	B
7.	C	8.	D	9.	B	10.	B	11.	B	12.	A
13.	D	14.	C	15.	B	16.	D	17.	D	18.	B
19.	A	20.	B								

Practice Paper 8

Q. No.	Ans	Q. No.	Ans	Q. No.	Ans	Q. No.	Ans	Q. No.	Ans	Q. No.	Ans
1.	D	2.	B	3.	D	4.	B	5.	C	6.	C
7.	A	8.	D	9.	B	10.	C	11.	A	12.	C
13.	B	14.	D	15.	D	16.	A	17.	B	18.	C
19.	D	20.	B								

Practice Paper 9

Q. No.	Ans	Q. No.	Ans	Q. No.	Ans	Q. No.	Ans	Q. No.	Ans	Q. No.	Ans
1.	B	2.	C	3.	D	4.	C	5.	A	6.	D
7.	B	8.	D	9.	C	10.	C	11.	A	12.	D
13.	A	14.	A	15.	C	16.	B	17.	A	18.	A
19.	D	20.	B								

Practice Paper 10

Q. No.	Ans	Q. No.	Ans	Q. No.	Ans	Q. No.	Ans	Q. No.	Ans	Q. No.	Ans
1.	D	2.	A	3.	B	4.	D	5.	C	6.	B
7.	C	8.	C	9.	A	10.	B	11.	B	12.	A
13.	C	14.	A	15.	D	16.	B	17.	A	18.	B
19.	A	20.	C								

13. Multiple Choice Questions of Genetics and Plant Breeding

Practice Paper 1

1. **The branch genetics started in**
(A) 1857 (B) 1866
(C) 1900 (D) 1905

2. **Sum total of characters which are transmitted from generation to generation constitute:**
(A) Genetics (B) Phenocopy
(C) Heredity (D) Variation

3. **Rediscovery of Mendel's work was made in:**
(A) 1866 (B) 1900
(C) 1898 (D) None of these

4. **The term Genetics was first used by:**
(A) Johannsen (B) Punnett
(C) Bateson (D) Darlington

5. **Genetics deals with:**
(A) Genes (B) Heredity
(C) Variations (D) Heredity and Variations

6. **The heredity variations are due to:**
(A) Asexual reproduction
(B) Sexual reproduction
(C) Vegetative reproduction
(D) All of these

7. **Concept of 'Pangenes' was given by:**
(A) Wolff (B) Swammerdam
(C) Johannsen (D) Darwin

8. **'Genotype-phenotype concept' was given by:**
(A) Johennsen (B) Charles Darwing
(C) Mendel (D) Bateson

9. **When two genotype produce the same phenotype due to different environments one is called .... of other :**
(A) Phenotype (B) Allelomorph
(C) Phenocopy (D) None of these

10. **Mendel worked on:**
(A) Pigeonpea (B) Garden pea
(C) Sweet pea (D) Chickpea

11. **Mendel originally belonged to:**
(A) Sweden (B) Austria
(C) Australia (D) Czechoslovakia

12. **Mendel published the results of his experiments in:**
(A) 1857 (B) 1866
(C) 1868 (D) 1900

13. **Mendel studies how many pairs of contrasting characters?**
(A) 1 (B) 2
(C) 7 (D) 10

14. **Which of the following characters was not studied by Mendel?**
(A) Plant height
(B) Position of flower
(C) Colour of pod
(D) Length of pod

15. **The term 'Allele' was proposed by:**
(A) Johennsen (B) Bateson
(C) Mendel (D) Punnett

16. **Emasculation is achieved by:**
(A) Removal of stigma
(B) Removal of anthers
(C) Removal of sepals and petals
(D) All of the above

17. **Alleles are:**
(A) Homologous chromosomes
(B) LinKed genes
(C) Dominant genes
(D) Alternate forms of gene

18. **Mendel performed his work on pea plants during:**
(A) 17th century (B) 18th century
(C) 19th century (D) 20th century

19. Who is known as "father of Genetics"?
(A) C. Darwing (B) G. Mendel
(C) Hugo de Vries (D) Weismann

20. One complete set of chromosomes inherited as a unit from one parent is known as:
(A) Genotype (B) Genome
(C) Gene pool (D) Karyotype

Practice Paper 2

1. The geometrical device that helps in visualizing all the possible combinations of male and female gametes is known as:
(A) Bateson square (B) Morgan square
(C) Punnett square (D) Mendel square

2. The condition in which only one allele of a pair is present is called:
(A) Homozygous (B) Hemizygous
(C) Heterozygous (D) Incomplete dominance

3. The term heterosis represents:
(A) Hybrid incompatibility
(B) Hybrid vigour
(C) Hybrid sterility
(D) Structural hybridity

4. Genes which in homozygous condition result in non-viable progeny are:
(A) Duplicate genes (B) Linked genes
(C) Lethal genes (D) Epistatic genes

5. Which is called a 'pleitropic gene'?
(A) Control characters of human beings only
(B) Control character of plants
(C) Control one character in association with other
(D) Control more than one character

6. Exception to Mendel's law of independent assortment was first discovered by:
(A) Morgan
(B) Bateson and Punnett
(C) Sturtevant
(D) None of these

7. Phenomenon of linkage was first discovered in:
(A) Sweet pea (B) Garden pea
(C) Mirabilis (D) Snapdragon

8. Recombinant inbred lines exhibits .... for monogenic trait:
(A) 1:1 segregation ratio
(B) 1:2 segragation ratio
(C) 2:1 segregation ratio
(D) 1:3 segregation ratio

9. Petaloid male sterility is observed in:
(A) Tomato (B) Carrot
(C) Brinjal (D) Cucurbit

10. Okra somatic chromosome number is:
(A) 2n = 2x = 130 (B) 2n = 2x = 140
(C) 2N = 2X =150 (D) 2N = 2X = 160

11. Purelines are:
(A) Homozygous and Homogeneous
(B) Heterozygous and Hetergeneous
(C) Only homozygous
(D) Only homogeneous

12. For the transfer of oligogenic trait .... is mostly followed:
(A) Pureline breeding
(B) Backcross breeding
(C) Pedigree method
(D) Mass selection

13. Highest uniformity is observed in:
(A) Single cross hybrids
(B) Triple cross hybrids
(C) Double cross hybrids
(D) Top cross hybrids

14. Barrier to interspecific hybridization of cucumis due to:
(A) Staminate sterility (B) Stigmatic sterility
(C) Short ovary (D) Long ovary

15. Male sterility source in brinjal is:
(A) *Solanum anguvii*
(B) *Solanum dulcamara*
(C) *Solanum tuberosum*
(D) *Solanum melongena*

16. Mechanism promotes cross pollination are:
(A) Monoecious (B) Dioecious
(C) Male sterility (D) All of these

17. Gene er1 is associated with:
(A) Garden pea (root rot resistant)
(B) Garden pea (Powdery mildew resistant)
(C) Garden pea (Dumping off resistant)
(D) Gram (Powdery mildew resistant)

18. Watermelon is cross comptable with:

(A) Musk melon (B) Round melon
(C) Drum melon (D) Turnip

19. Alleles of F1 hybrid are .... in nature:

(A) Heterozygous (B) Homozygous
(C) Both (D) None

20. Homoeostasis is associated with:

(A) Vertical resistance
(B) Horizontal resistance
(C) Inherent resistance
(D) Innate resistance

Practice Paper 3

1. If a plant heterozygous for tallness is selfed, the F_2 generation has both tall and dwarf plants. This proves the principle of:

(A) Dominance
(B) Segregation
(C) Independent assortment
(D) Incomplete dominance

2. In an experiment on pea, plants with yellow round seeds were crossed with plants with green wrinkled seeds; the phenotypic ratio of F_1 progeny is:

(A) 9 yellow round : 3 round green : 3 wrinkled yellow : 1 green wrinkled
(B) All wrinkled green
(C) All yellow round
(D) 1 Yellow round : 1 Green round : 1 Yellow wrinkled : 1 Green wrinkled

3. Red (RR) flower plant is crossed with white (WW), all the off springs are pink, this indicates that R gene is:

(A) Hybrid
(B) Recessive
(C) Incomplete dominant
(D) Mutant

4. Cross between a heterozygous individual and homozygous recessive parent is called:

(A) Monohybrid cross
(B) Dihybrid cross
(C) Back cross
(D) Test cross

5. An organism with two identical alleles of gene in a cell is:

(A) Dominant (B) Homozygous
(C) Hterozygous (D) Recessive

6. In F_2 progeny of dihybrid cross, the expected genotypic proportions of individuals homozygous for both dominant characters is:

(A) 3/16 (B) 9/16
(C) 12/16 (D) 1/16

7. When dominant epistasis is operative between two gene loci, the classical 9 : 3 : 3 : 1 ratio becomes modifies into:

(A) 9 : 3 : 4 ratio (B) 9 : 6 : 4 ratio
(C) 12 : 3 : 1 ratio (D) 15 : 1 ratio

8. A pure tall pea plant was reared in soil poor in the nutrition and reached the size of a pure dwarf pea plant, if this plant is selfed, then the genotype in the F_1 gen. is most likely to be:

(A) All tall
(B) 50% tall : 50% dwarf
(C) All dwarf
(D) Not possible to predict

9. A cross between a plant heterozygous for two factors and a plant recessive for both the factors, gives a phenotypic ratio of:

(A) 9 : 1 : 1 : 7 (B) 9 : 3 : 3 : 1
(C) 1 : 1 : 1 : 1 (D) 1 : 7 : 7 : 1

10. Which is back cross in a monohybrid cross?

(A) 9 : 3 : 3 : 1 (B) 1 : 1
(C) 1 : 2 : 1 (D) 3 : 1

11. Which term represents a pair of contrasting characters?

(A) Homozygous
(B) Heterozygous
(C) Complementary genes
(D) Allelomorphs

12. In monohybrid cross, the F_2 phenotypic ratio is:

(A) 1 : 2 (B) 2 : 1
(C) 3 : 1 (D) 1 : 3 : 1

13. The genotype of a dominant parent is determined by crossing it with a hybrid; this cross is called:

(A) Back cross (B) Test cross
(C) Long cross (D) Red cross

14. Allosomes are:

(A) Bead like structures in chromosomes
(B) Node like structures on chromosomes
(C) Sex –chromosomes
(D) None of the above

15. 'F_1' generation represents:

(A) First fertile generation
(B) First final generation
(C) First filial generation
(D) First federation generation

16. Independent assortment of genes ours due to the orientation of chromosomes at:

(A) Metaphase of mitosis
(B) Metaphase I of meiosis
(C) Metaphase II of meiosis
(D) Any phase of the cell division

17. The plant which are unable to produce functional pollen and fertilise the female part is known as:

(A) Self incompatible (B) Self pollinated
(C) Male sterile (D) Male fertile

18. Line developed from continuous selfing of cross pollinated species is known as:

(A) Pureline (B) Inbred
(C) Stain (D) Synthetic varieties

19. Seedless watermelon is:

(A) Diploid (B) Tetraploid
(C) Triploid (D) Pentaploid[

20. In apomixes, the progeny closely resembles:

(A) Male parent (B) Female parent
(C) Both of these (D) None of these

Practice Paper 4

1. Incomplete dominance was first reported in 4 o'clock plant by:

(A) Mendel (B) Morgan
(C) Batesen (D) Correns

2. Same genotypic and phenotypic ratios in F_2 generations are obtained in:

(A) Complete dominance
(B) Blending inheritance
(C) Codominance
(D) None of these

3. In which, incomplete dominance is seen?

(A) Wheat (B) Maize
(C) Mouse (D) Snapdragons

4. Lethal factors were reported in plants by:

(A) Cuenot (B) Baur
(C) Batesen (D) Morgan

5. The most common cause of the pleiotropic effect of a gene is due to:

(A) The same product of the given gene being involved in different metabolic pathways
(B) The gene making very different products in different cell types
(C) The DNA sequence of the gene getting changed in cell specific manner
(D) The gene not functioning in some cells

6. Gene interaction in which either of dominant gene is enough to give rise to same phenotypic effect is:

(A) Complementary gene
(B) Duplicate gene
(C) Epistatic gene
(D) Supplementary gene

7. When two pairs of genes interact to give 9 : 7 F_2 ratio, they are:

(A) Supplemetary factors
(B) Complementary factors
(C) Epistatic factors
(D) Duplicate factors

8. 12 : 3 : 1, F_2 ratio is obtained in interaction of .... Factors:

(A) Complementary (B) Supplementary
(C) Epistatic (D) Inhibitory

9. Recombination percentage in a diploid cannot exceed:

(A) 50 (B) 75
(C) 100 (D) 25

10. Two linked genes (A) and (B) show 20% recombination. The individuals of a dihybrid cross between ++/++ x ab/ab shall show gametes:

(A) ++ 80 : ab : 20
(B) ++ 50 : ab : 50
(C) ++ 40 : ab 40 : + a10 : + b : 10
(D) ++ 30 : ab 30 : + a 20 : + b : 20

11. Coupling and repulsion phenomenon are concerned with:

(A) Crossing over (B) Linkage
(C) Mutation (D) None of these

12. Theory of linear arrangement of genes on chromosomes was formulated by:

(A) Sturtevant (B) Morgan
(C) Bateson (D) Mendel

13. When blue flower long pollen plant (BBLL) was crossed with red flowered round pollen plant (bbll) in Lathyrus and their F_1 was crossed with red round parent, then what is the ratio of blue long and red round obtained in this cross?

(A) 1/16 : 1/16 (B) 7/16 : 7/16
(C) 7/16 : 1/6 (D) 9/16 : 3/16

14. In a diploid organism, the genes A, B and C are present on the same chromosome in that order. The AB interval; is 10 map units and B interval is 20 map units. In an ABC/abc heterozygous individual, what will be the proportion of gametes that carry the genotype AbC?

(A) 1 % (B) 10 %
(C) 20 % (D) 30 %

15. If 2n = 20 in a plant, then how many linkage groups are present in this?

(A) 20 (B) 10
(C) 5 (D) 5 or 10

16. Somatic crossing over was discovered in Drosophila by:

(A) Pontecorvo (B) Morgan
(C) Stern (D) Serebrousky

17. Individuals around mean are selected in:

(A) Negative directional selection
(B) Positive directional selection
(C) Disruptive selection
(D) Stabilizing selection

18. Development of embryo from egg cell without fertilization:

(A) Apomixes (B) Portenocarpy
(C) Parthenocarpy (D) Apospory

19. Which of the following method is used for population improvement:

(A) Pureline selection
(B) Mass selection
(C) Recurrent selection
(D) Clonal selection

20. Replacement of purine by pyrimidime or vise versa is known as:

(A) Transition (B) Transversion
(C) Transduction (D) Translocation

Practice Paper 5

1. G. Pontecorvo observed non-germinal cover in:

(A) Maize (B) Drosophila
(C) Aspergillus (D) None of these

2. Most accepted theory of crossing over is:

(A) Copy Choice theory
(B) Precosity theory
(C) Contact first theory
(D) Breakage first theory

3. Phenotypic expression of the heterozygote being greater than that of either homozygote is called:

(A) Dominance (B) Midparent value
(C) Overdominance (D) Pseudodominance

4. Pseudoautosomal gene is present on:

(A) Both X-and Y- chromosomes
(B) Both X- and autosomes
(C) Both Y- and autosomes
(D) Only autosomes

5. Exchange of chromosomal segments between non-homologous chromosomes is:.

(A) Inversion (B) Crossing over
(C) Translocation (D) None of these

6. Which of the following most accurately explains the cause for the abnormal numbers of chromosomes during human reproduction that an result in Down syndrome, Turner's syndrome, or Klinefelter syndrome?

(A) The occurrence of non-disjunction of homologous chromosomes during meiosis
(B) The duplicative production of extra chromosomes during DNA replication
(C) The abnormal pairing of non-homologous chromosomes during prophase of miosis I
(D) The selective loss of particular chromosomes from the sex cells after formation of the mature gamete

7. **The process which involves reverse order of genes in chromosome is:**
   (A) Deficiency
   (B) Aneuploidy
   (C) Inversion
   (D) Translocation

8. **'x' represens:**
   (A) Gametic number
   (B) Basic chromosome number
   (C) Haploid number
   (D) All of the above

9. **Presence of chromosome number, which is different from the multiple of basic chromosome no. is called?**
   (A) Hyperploidy (B) Heteroploidy
   (C) Aneuploidy (D) Euploidy

10. **2n – 2 is a:**
   (A) Monosomic (B) Nullisomic
   (C) Tetrasomic (D) Trisomic

11. **If haploid chromosome no. in an organism is 12, what is the chromosome no. present in its trisomic?**
   (A) 11 (B) 13
   (C) 25 (D) 23

12. **In a three-factor cross among genetic markers A, B and C, the relative frequencies of recombinants were AB (1), AC (0.2) and BC (0.3). Which of the following is the most probable order of these markers on the genome?**
   (A) ABC
   (B) BCA
   (C) CAB
   (D) The data do not reflect the order of genes

13. **Reduction in size of vegetative and floral parts is found in:**
   (A) Autopolyploids
   (B) Allopolyploids
   (C) Haploids
   (D) Diploids

14. **'Gigantism' is mostly found in:**
   (A) Autopolyploids
   (B) Allopolyploids
   (C) Haploids
   (D) None of these

15. **Polyploidy can be artificially induced by:**
   (A) 0.1 M NaCl solution
   (B) 0.25 M sucrose solution
   (C) Zeatin
   (D) Colchicine

16. **In Drosophila Y-chromosome is:**
   (A) Essential for male sex determination
   (B) Required for development of secondary sex-characters
   (C) Required for dosage compensation
   (D) Not required for male sex determination

17. **Character of pureline selected variety is:**
   (A) homozygous – homogeneous
   (B) heterozygous – homogeneous
   (C) homozygous – heterogeneous
   (D) heterozygous – heterogeneous

18. **The reason for GMS is:**
   (A) Single dominant gene
   (B) Single pseudo dominant gene
   (C) Single recessive gene
   (D) All of these

19. **Mechanism which promote self pollination:**
   (A) Cleistogamy
   (B) Dichogamy
   (C) Male sterility
   (D) All of these

20. **Which of the following is responsible for cross pollination in chilli and brinjal?**
   (A) Male sterility
   (B) Protandry
   (C) Dichogamy
   (D) Self incompatibility

Practice Paper 6

1. ***Colchicine* was first obtained from *Colchicines autumnale* by:**
   (A) Dustin (B) Houde
   (C) E.R. Sears (D) Blakeslee

2. **Colchicine is a:**
   (A) Alkaloid (B) Steroid
   (C) Cholesterol (D) None of these

3. **Natural autopolyploidy is found in:**
   (A) Cynodon (B) Grapes
   (C) Watermelon (D) None of these

4. One of the following statements is most appropriate about the Y chromosome in mammals:

(A) Essential for the development of embryo
(B) Non-essential for the development of the embryo but important for male sex determination
(C) Non-essential for sex determination by important for male fertility
(D) Essential for male sex determination and fertility

5. Complete linkage is observed in:

(A) ♂ Drosophila (B) ♀ Drosophila
(C) Birds (D) Maize

6. Percentage of crossing over will be more if:

(A) Linked genes are located far away from each other
(B) Linked genes are located close to each other
(C) Both (A) and (B)
(D) None of the above

7. What are the sexual phenotypes of the following genotypes in Drosophilla : XXY, XO?

(A) Male, female (B) Male, male
(C) Intersex, female (D) Female, male

8. Who is associated with Raphanobrassica?

(A) Muntzing (B) E.R. Sears
(C) Karpechenko (D) None of these

9. 'Triticale' is derived from cross between?

(A) Wheat and aegilops
(B) Wheat and rice
(C) Wheat and maize
(D) Wheat and rye

10. Survival of fittest used by:

(A) Charles Darwin
(B) Herbert Spencer
(C) Jean Baptiste Lamarck
(D) Hugo de Vries

11. Which of the following is called Drosophila of plant kingdom?

(A) *E. coli* (B) *Acetabularia*
(C) *Neurospora* (D) *Saccharomyces*

12. In mice, a mutation results in the production of yellow oat colour, as opposed to the normal gray oat. Crossing yellow mice with normal mice results in equal proportions of normal and yellow mice, whereas crossing yellow mice amongst themselves results in a ratio of 2 yellow : 1 normal progeny. The results suggest that:

(A) More than one gene is involved in conferring the yellow phenotype
(B) The yellow allele is lethal under homozygous condition
(C) The yellow allele exhibits cytoplasmic heredity
(D) The results cannot be explained by any genetic principles

13. A mutation in sheep called 'Ancon sheep' was recorded by:

(A) Hugo de Vries (B) Darwin
(C) Seth Wright (D) None of these

14. Mutagen is:

(A) Organism in which mutation has occurred
(B) Mutation inducing agent
(C) Both (A) and (B)
(D) None of the above

15. Frequency of mutation:

(A) Varies with character and organism
(B) Can be increased by X-rays
(C) Is greatly affected by environmental factors
(D) All of the above

16. A cross between unlike organisms is called:

(A) Test cross (B) Heterosis
(C) Back cross (D) Hybrid

17. Concept of epigenesist was proposed by:

(A) Aristotle (B) Bonnet
(C) Wolf (D) Darwin

18. The chromosomes are most condensed at:

(A) Anaphase (B) Metaphase
(C) Prophase (D) Telophase

19. DNA synthesis takes place in which stage?

(A) G phase (B) G phase
(C) S phase (D) None of these

20. The science of changing and improving the heredity of plants is known as:

(A) Genetics
(B) Plant breeding
(C) Cytogenetics
(D) Plant improvement

Practice Paper 7

1. If a man who is colour-blind marries a woman who is homozygous for normal colour vision, the probability of their son being colour-blind is:

(A) 0 (B) 1
(C) 0.5 (D) 0.75

2. In nature, the rate of mutation is?

(A) 1×10^5 (B) 1×10^8
(C) 1×10^{-5} (D) 1×10^{12}

3. Mutations are

(A) Mostly useful (B) Always useful
(C) Rarely useful (D) Always harmful

4. Who used X-rays to cause mutation?

(A) Morgan (B) Muller
(C) Alterberg (D) Auerbach

5. Consider two linked loci, labeled (A) and (B). Each locus has two alleles, labeled 1 and 2. The frequencies of the alleles in a population are: A1 = 0.6, A2 = 0.4, B1 = 0.7, B2 = 0.3. If there is linkage equilibrium between these two loci in the population, what is the expected frequency of chromosomes carrying a combination of the A1 and B2 alleles?

(A) 0.24 (B) 0.12
(C) 0.18 (D) 1.00

6. Changes due to mutation are:

(A) Small (B) Negligible
(C) Large (D) All of these

7. Pseudo dominance is:

(A) There is incomplete dominance
(B) Dominance of recessive alleles in a single generation
(C) Deficiency in which a recessive allele will behave like a dominant allele
(A) There is no such phenomenon

8. Gene sanctuary refers to:

(A) An area of diversity protected against interference from man
(B) Gene collection
(C) Gene evoalution
(D) Gene mutation

9. In a test cross of an individual heterozygous for each of three linked genes, the most frequent lasses of progeny were ABc/abc and abC/abc, and the least frequent lasses were ABC/abc and abc/abc. What is the order of the genes:

(A) A - C - B (B) B – A - C
(C) A – B - C (D) None of these

10. If an organism is having genotype AaBb, then what are the possible combinations of gametes after meiosis?

(A) AB, ab
(B) Aa, Bb
(C) AB, aB, Ab, ab
(D) Not possible to tell

11. Suppose DNA has base sequence as UAC, GAC, AGC, CGC, ACA, AAA and due to mutation only first base is deleted, what is the effect on coding?

(A) First amino acid will be different
(B) There will be one amino acid less
(C) Complete change in type and sequence of amino acids
(D) No change

12. Why are some autosomal dominant disorders (e.g., Marfan syndrome) seen more commonly in the offspring of older fathers?

(A) Replication errors accumulate as sperm-producing stem cells continue to divide
(B) Rate of nondisjunction increases in older males
(C) Recombination rates increase in older males
(D) None of these

13. Which of the following is non-ionizing radiation?

(A) X- rays (B) β- rays
(C) γ- rays (D) UV-rays

14. Certain types of cancers can be correlated with specific changes in chromosome structure. In patients suffering from myelogenous leukemia, the abnormal chromosome detected was termed Philadelphia chromosome. Which of the following chromosomes is altered in this disease?

(A) Chromosome 10 (B) Chromosome 11
(C) Chromosome 20 (D) Chromosome 22

15. UV-rays as mutagenic agents were shown by:

(A) Auerbach (B) Stadler
(C) Altenberg (D) Muller

16. Chemicals as mutagenic agents were first shown by:

(A) Altenberg (B) Auerbach
(C) Stadler (D) None of these

17. People who carry an allele for normal hemoglobin and an allele for sickle cell hemoglobin are resistant to malaria. This phenomenon is called:

(A) Heterozygote advantage
(B) Polyploidy
(C) Mutation selection balance
(D) Recessive superiority

18. Double cross plant was suggested by:

(A) Beal (B) Shul
(C) Jones (D) Knight

19. Who gave the concept of circular chromosome in *E. coli*?

(A) Cairns (B) Meselson
(C) Watson (D) Stahl

20. Sex in plants was discovered by:

(A) Cameratious (B) Linnaeus
(C) Grew (D) Fairchild

Practice Paper 8

1. Hypophosphatemia is caused by an X-linked dominant gene in humans. If a hypophosphatemia man marries a normal woman, which of their children will have hypophosphatemia?

(A) All daughters
(B) Half of their daughters
(C) All of the sons
(D) Half of their sons

2. Which of the following statements are correct:

P. Down's syndrome/mongolism is due to extra 21st chromosome

Q. Down's syndrome is due to non-disjunction of chromosomes

R. Trisomy has chromosome complement of 2n + 1

S. Monosomics are 2n - 1

(A) P, Q (B) R, S
(C) P, Q, R (D) P, Q, R, S

3. Base substitution caused by 5-bromo uracil and BudR are called:

(A) Inversions (B) Transitions
(C) Transversions (D) Translocations

4. Which of the following statements are correct:

P. Cri-du-chat syndrome is due to chromosomal change involving deletion

Q. Formation of multivalents in meiosis is due to reciprocal translocation

R. Klinefelter's syndrome has 44 + XXY

S. In Down's syndrome of a male child, the sex complement is XY

(A) P, Q (B) R, S
(C) P, Q, R (D) P, Q, R, S

5. Proflavin and Acridine orange induce:

(A) Transitions
(B) Frameshift mutations
(C) Inversions
(D) Transversions

6. Which is not associate with mutations?

(A) Hybrid vigour
(B) Death of organisms
(C) Change in genetic constitution
(D) Better progenies

7. 'Mutations which are recessive are not lost by a gene pool' is called:

(A) Survival of dominance
(B) Survival of recessive
(C) Hardy Weinberg law
(D) None of these

8. **Cytoplasmic inheritance is responsible for:**
   (A) Inheritance of kappa particles
   (B) Iojap inheritance in corn
   (C) Petite character in yeast
   (D) All of these

9. **A decentric chromosome is unstable because:**
   (A) It cannot resynthesize its telomeres during replication
   (B) It pairs with non-homologous chromosomes in meiosis
   (C) It pairs with non-homologous chromosomes in mitosis
   (D) It is often simultaneously drawn to opposing spindle poles in mitosis

10. **Who is credited to have differentiated between heritable variation and environmental variation?**
    (A) Darwin (B) Morgan
    (C) Hugo de Vries (D) Bateson

11. **The resistance developed by some insects against insecticides is due to:**
    (A) Hybrid vigour
    (B) Genetic mutations
    (C) Variations due to sexual rep.
    (D) Genetic recombinations

12. **Who coined the term Heterosis?**
    (A) Mendel (B) Schmidt
    (C) G.H. Shull (D) Schleiden

13. **The force that can change the frequency of an allele in a population are:**
    (A) Forward mutation, gene conversion, neutral evolution and recombination
    (B) Selection, mutation, migration, inbreeding and random genetic drift
    (C) Dominance, family selection, fitness and diversification
    (D) Gene interaction, gene transfer, gene mutation and out breeding

14. **Which of the following crosses would result in 1 : 2 : 1 F_2 ratio?**
    (A) Yg x gg (B) Yg x yg
    (C) Yg x yy (D) YY x gg

15. **'Gene Battery Model' was given by:**
    (A) Jacob and Monod
    (B) McClinktock and Creigheten
    (C) Bawden and Pierre
    (D) Bridden and Davidson

16. **The first artificial 'hybrid' was produced by:**
    (A) Camerarius (B) Fairchild
    (C) Lamarck (D) N. Grew

17. **Sex determination was first of all studied in which plant?**
    (A) Coccinia (B) Melandrium
    (C) Rumex (D) Humulus

18. **As example of a heterozygous but homogeneous population is:**
    (A) Pure line
    (B) Open pollinated variety
    (C) Hybrid variety
    (D) Synthetic variety

19. **The cross of an inbred line with an open pollinated variety is known as:**
    (A) Back cross (B) Top cross
    (C) Three way cross (D) Double

20. **Systematic hybridization on a scientific basis began with the work:**
    (A) Shull – Maize
    (B) Patel – Cotton
    (C) Koelreuter - Nicotiana
    (D) Knight – Feuits

Practice Paper 9

1. **Which of the following shows slowest cell division?**
   (A) Haploid (B) Diploid
   (C) Polyploid (D) Same in all

2. **2n=x=7, represents a:**
   (A) Monoploid (B) Diploid
   (C) Haploid (D) Polyploid

3. **In wheat 2n=6x=42. No. of possible monosomics is:**
   (A) 42 (B) 21
   (C) 7 (D) 6

4. **Translocation was first reported in:**
   (A) Datura (B) Maize
   (C) Oenothera (D) Doob grass

5. **In snapdragon, when yellow leaved var. (Aurea) x Aurea, what is the ratio of yellow: green in progeny?**
(A) 3 : 1 (B) 9 : 3
(C) 1 : 1 (D) 2 : 1

6. **Mutations are mostly:**
(A) Dominant
(B) Incompletely dominant
(C) Codominant
(D) Recesive

7. **Suppose DNA has base sequence AAG GAG GAC CAA CCA, which one represents frameshift mutations?**
(A) AAG GAG GAC CAA CCA
(B) AAG AAG GAC CAA CCA
(C) AAG GAG ACC AAC CAA
(D) AAG GAG GAC CAA CCA

8. **No. of linkage groups is equal to:**
(A) Diploid no. of chromosomes
(B) Haploid no. of chromosomes
(C) Tetraploid no. of chromosomes
(D) None of the above

9. **Who discovered Mendel's principle in 1900 A.D.?**
(A) Avery, McLeod, McCarty
(B) Sutton, Morgan, Bridges
(C) Bateson, Punnett and Bridges
(D) de Vries, Correns and Tschermak

10. **Heterosis is also called:**
(A) Hybrid vigour
(B) Hybrid sterility
(C) Mutation
(D) Variation

11. **Why are haploids superior to diploids in study of mutations?**
(A) They have shorter life time
(B) Smaller no. of chromosomes
(C) They allow expression of recessive mutation immediately
(D) Obtained in large number

12. **The individuals having same maternal and paternal parents are called?**
(A) Epistasis
(B) Hypostasis
(C) Siblings
(D) Gynander

13. **When a pyrimidine is replaced by purine, it is called:**
(A) Frameshift mutation
(B) Transition
(C) Transversion
(D) Reversible mutation

14. **Picking up plants with superior phenotype for further propagation is:**
(A) Mass selection
(B) Pure line selection
(C) Hybrid vigour
(D) Introduction

15. **Picking up only homozygous plants for further cultivation is known as:**
(A) Mass selection
(B) Pure line selection
(C) Clone
(D) Acclimitization

16. **Homozygosity is obtained after inbreeding for:**
(A) Two generations
(B) Four generations
(C) Eight generations
(D) Ten generations

17. **A : T ratio in one strand of DNA duplx is 2, what will be the ratio in the complementary strand?**
(A) 1.0 (B) 0.5
(C) 2.5 (D) 2.0

18. **Number of chromosomes in wheat endosperm:**
(A) 21 (B) 42
(C) 63 (D) 14

19. **Cytoplasmic male sterility is best suited for hybrid seed production in:**
(A) SP crops
(B) CP crops
(C) Vegetatively propagated crops
(D) All of these

20. **Multiple effects of a single gene is known as:**
(A) Pleiotropy
(B) Polyploidy
(C) Heteroploidy
(D) None of these

Practice Paper 10

1. **PV-18, Kalyan sona and S-227 wheat varieties have developed through:**
   (A) Gamma ray irradiation
   (B) Multiple crossing
   (C) Selection
   (D) Genetic engineering

2. **Polyploidy is induced through:**
   (A) Colchicine
   (B) Irradiation
   (C) Mutagenic chemicals
   (D) Ethylene

3. **Best varieties of SugarBeet are:**
   (A) Tetraploids (B) Diploids
   (C) Triploids (D) Allopolyploids

4. **Most varieties of sugarcane are:**
   (A) Tetraploids (B) Triploids
   (C) Hexaploids (D) Aneuploids

5. **Main varieties of potato are:**
   (A) Tetraploids (B) Triploids
   (C) Aneuploids (D) Diploids

6. **Bread Wheat is:**
   (A) Autopolyploids (B) Allopolyploids
   (C) Aneuploids (D) Trisomic

7. **Seedless Water Melon is:**
   (A) Hexaploid (B) Tetraploid
   (C) Triploid (D) Pentaploid

8. **The first domesticated Wheat was:**
   (A) *Triticum boeoticum*
   (B) *T. dicocum*
   (C) *T. dicoccoies*
   (D) *T. monococcum*

9. **Induced mutation have produced a rice variety:**
   (A) Pelita-1 (B) Reimei
   (C) Pelita-2 (D) All the above

10. **Explant is:**
    (A) Plant collected after harvesting
    (B) Exploited part of a plant
    (C) Small part of the plant meant for tissue culture
    (D) Uprooted for transplantation

11. **A technique of micropropagation is:**
    (A) Multiple shoot production
    (B) Multiple shoot production and somatic embryogenesis
    (C) Growth of micro-organisms on culture medium
    (D) Somatic embryogenesis

12. **Which one is hexaploid wheat?**
    (A) *Triticum durum* (B) *T. monococcum*
    (C) *T. tugidum* (D) *T. aestivum*

13. **What is the role of the I gene of phage lambda?**
    (A) It encodes a protein that enables the phage to enter lysis
    (B) Its product helps the prophage to excise from the host DNA
    (C) It encodes the lambda repressor which represses the lytic cycle
    (D) Its gene product helps the prophage to integrate into the host DNA

14. **A clone is:**
    (A) Homozygote produced by sexual reproduction
    (B) Heterozygote produced by sexual method
    (C) Homozygote obtained asexually
    (D) Heterozygote obtained asexually

15. **Hybrid vigour is due to:**
    (A) Heterozygosity (B) Linkage
    (C) Crossing over (D) Chiasma

16. **Crop improvement is possible through:**
    (A) Judicious combination of selection, introduction and hybridization
    (B) Selection
    (C) Scientific improvement of cultivated plants
    (D) Introduction

17. **Most of our crop plants are:**
    (A) Allopolyploid in origin
    (B) Autopolyploid in origin
    (C) Mixed genotypic in origin
    (D) Heterozygous in origin

18. **In Rho-independent transcription termination, stem and loop structure is formed in:**
    (A) The transcript
    (B) The template DNA
    (C) Coding strand
    (D) Both (A) and (B)

19 ***Triticum aestivum*** **is:**
(A) Hexaploid (B) Alloautopolyploid
(C) Heteropolyploid (D) Autopolyploid

20. **Bombay green banana has developed from:**
(A) Mass selection
(B) Pure line selection
(C) Clonal selection
(D) Hybridisation

Practice Paper 11

1. **Removal of anthers of some flowers during plant breeding is:**
(A) Anthesis
(B) Emasculation
(C) Pollination
(D) For collection of pollen

2. **Hybrid potato and tomato is:**
(A) Tomapa
(B) Pomato
(C) Potamato
(D) None of the above

3. **The pair-rule gene proteins activate the transcription of the:**
(A) Gap genes
(B) Pair rule genes
(C) Segment-polarity genes
(D) All of these

4. **The fountain head of evolution is**
(A) Mutations
(B) Acquired characters
(C) Natural selection
(D) Sexual reproduction

5. **C_0 t analysis provides an estimate of the:**
(A) G + C content of the DNA
(B) T_m of the DNA
(C) Complexity of the genome
(D) Hyperchromic shift of the genome

6. **Genetic drift is change of**
(A) Gene frequency in same generation
(B) Appearance of recessive genes
(C) Gene frequency from one generation to next
(D) None of the above

7. **Sum total of genes in a population is**
(A) Genotype
(B) Karyotype
(C) Genetic constitution
(D) Gene pool

8. **Hexaploid Wheat developed through**
(A) Hybridomas
(B) Chromosom doubling
(C) Hybridisation
(D) Hybridisation and chromosome doubling

9. **Lederberg's replica experiment is connected with**
(A) Adaptations
(B) Evolution
(C) Genetic
(D) Artificial selection

10. **Theory of pangenesis was given by**
(A) Darwin
(B) Lamarck
(C) Hugo de Vries
(D) Oparin

11. **Phenomenon of industrial melanism demonstrates**
(A) Geographical isolation
(B) Natural selection
(C) Reproductive isolation
(D) Induced mutation

12. **Genetic drift**
(A) Is random change in gene frequency
(B) Has nothing in common with inbreeding
(C) Is an orderly change in gene frequency
(D) Produces greatest fluctuations in large population

13. **Greater is the distance between the two genes on a chromosome:**
(A) Greater is the linkage strength
(B) Lesser is the linkage strength
(C) Linkage strength remains unchanged
(D) There is no relationship between the two

14. **Crossing over occurs at**
(A) Single-strand stage of chromosomes during prophase
(B) Two-strand stage during zygotene
(C) Four-strand stage of pachytene
(D) Metaphase II of meiosis

15. A colour blind daughter may be born if the:

(A) Father is normal and mother is colour blind

(B) Father is colour blind and mother is normal

(C) Father is normal and mother is carrier

(D) Father is colour blind and mother is a carrier

16. A somatic cell in human male contains:

(A) No gene on sex chromosomes

(B) Only one sex linked gene for each character

(C) Two genes for every sex linked character

(D) Genes only on sex chromosomes

17. Which of the following disease is sex linked?

(A) Colour blindness (B) Malingnancy

(C) Hepatitis (D) Leukemia

18. In transversion:

(A) Purine is replaced by another purine

(B) Purine is replaced by pyrimidine or *vice versa*

(C) Some nitrogen bases are eliminated from a gene

(D) Some nitrogen bases are added in a gene

19. Crossing over may result into:

(A) Gene mutation

(B) Genomatic mutation

(C) Genetic recombination

(D) None of these

20. Test cross ratio in case of duplicate gene action will be:

(A) 1 : 1 (B) 1 : 1 : 1 : 1

(C) 3 : 1 (D) 1 : 3

Answers Key

Practice Paper 1

Q. No.	Ans	Q. No.	Ans	Q. No.	Ans	Q. No.	Ans	Q. No.	Ans	Q. No.	Ans
1.	C	2.	C	3.	B	4.	C	5.	D	6.	B
7.	D	8.	A	9.	C	10.	B	11.	B	12.	B
13.	C	14.	D	15.	B	16.	B	17.	D	18.	C
19.	B	20.	B								

Practice Paper 2

Q. No.	Ans	Q. No.	Ans	Q. No.	Ans	Q. No.	Ans	Q. No.	Ans	Q. No.	Ans
1.	C	2.	B	3.	B	4.	C	5.	C	6.	B
7.	A	8.	A	9.	B	10.	A	11.	A	12.	B
13.	A	14.	B	15.	A	16.	D	17.	B	18.	B
19.	A	20.	B								

Practice Paper 3

Q. No.	Ans	Q. No.	Ans	Q. No.	Ans	Q. No.	Ans	Q. No.	Ans	Q. No.	Ans
1.	B	2.	C	3.	C	4.	D	5.	D	6.	D
7.	C	8.	A	9.	C	10.	B	11.	D	12.	C
13.	B	14.	C	15.	C	16.	B	17.	C	18.	B
19.	C	20.	B								

■ Practice Paper 4

Q. No.	Ans	Q. No.	Ans	Q. No.	Ans	Q. No.	Ans	Q. No.	Ans	Q. No.	Ans
1.	D	2.	B	3.	D	4.	A	5.	A	6.	B
7.	B	8.	C	9.	A	10.	C	11.	B	12.	A
13.	B	14.	A	15.	B	16.	C	17.	D	18.	C
19.	C	20.	B								

■ Practice Paper 5

Q. No.	Ans	Q. No.	Ans	Q. No.	Ans	Q. No.	Ans	Q. No.	Ans	Q. No.	Ans
1.	C	2.	B	3.	C	4.	A	5.	B	6.	A
7.	C	8.	B	9.	C	10.	B	11.	C	12.	D
13.	C	14.	A	15.	D	16.	D	17.	A	18.	C
19.	A	20.	C								

■ Practice Paper 6

Q. No.	Ans	Q. No.	Ans	Q. No.	Ans	Q. No.	Ans	Q. No.	Ans	Q. No.	Ans
1.	B	2.	A	3.	B	4.	D	5.	A	6.	A
7.	D	8.	C	9.	D	10.	B	11.	C	12.	B
13.	C	14.	B	15.	D	16.	D	17.	C	18.	A
19.	C	20.	B								

■ Practice Paper 7

Q. No.	Ans	Q. No.	Ans	Q. No.	Ans	Q. No.	Ans	Q. No.	Ans	Q. No.	Ans
1.	A	2.	C	3.	C	4.	B	5.	C	6.	D
7.	C	8.	A	9.	A	10.	C	11.	C	12.	A
13.	D	14.	D	15.	C	16.	B	17.	A	18.	C
19.	B	20.	A								

■ Practice Paper 8

Q. No.	Ans	Q. No.	Ans	Q. No.	Ans	Q. No.	Ans	Q. No.	Ans	Q. No.	Ans
1.	A	2.	D	3.	B	4.	D	5.	B	6.	A
7.	C	8.	D	9.	D	10.	C	11.	B	12.	C
13.	B	14.	D	15.	D	16.	B	17.	D	18.	C
19.	B	20.	C								

■ Practice Paper 9

Q. No.	Ans	Q. No.	Ans	Q. No.	Ans	Q. No.	Ans	Q. No.	Ans	Q. No.	Ans
1.	C	2.	A	3.	B	4.	C	5.	D	6.	D
7.	C	8.	B	9.	D	10.	A	11.	C	12.	C
13.	C	14.	A	15.	B	16.	D	17.	B	18.	C
19.	C	20.	A								

■ Practice Paper 10

Q. No.	Ans	Q. No.	Ans	Q. No.	Ans	Q. No.	Ans	Q. No.	Ans	Q. No.	Ans
1.	C	2.	A	3.	C	4.	D	5.	A	6.	B
7.	C	8.	D	9.	D	10.	C	11.	B	12.	D
13.	C	14.	C	15.	A	16.	A	17.	B	18.	A
19.	A	20.	C								

■ Practice Paper 11

Q. No.	Ans	Q. No.	Ans	Q. No.	Ans	Q. No.	Ans	Q. No.	Ans	Q. No.	Ans
1.	B	2.	B	3.	C	4.	A	5.	C	6.	C
7.	D	8.	D	9.	A	10.	A	11.	B	12.	A
13.	B	14.	C	15.	D	16.	B	17.	A	18.	B
19.	C	20.	C								

14. Multiple Choice Questions of Agricultural Biotechnology

Practice Paper 1

1. **Biotechnology is:**
   (A) Technology used in biology
   (B) Use of genetic engineering
   (C) Development of technology for medical application
   (D) Use of microorganisms in industrial processes

2. **Who introduced the term Biochemistry?**
   (A) Marshall Nirenberg
   (B) Carl Neuberg
   (C) John H.Northrop & Kunitz
   (D) Van Helmont

3. **Biotechnology is a multidisciplinary field that encompasses inventions that are related to:**
   (A) Basic science (B) Applied science
   (C) Both (A) and (B) (D) None of these

4. **Transgenic plants containing genes for drought and salt tolerance, improved quality for value addition and increase shelf life are produced through:**
   (A) Genetic transformation
   (B) Genetic drift
   (C) Genetic transduction
   (D) Genetic transfection

5. **Agrobacterium - mediated transformati-on an indirect method of transformation that involves:**
   (A) *Agrobacterium tumefaciens*
   (B) *E. coli*
   (C) *S. cerevisiae*
   (D) None of these

6. **Lamda phage infects bacteria *E. coli* and replicates by a:**
   (A) Lysogenic pathway
   (B) Lytic pathway
   (C) Both (A) and (B)
   (D) None of these

7. **The vectors that are hybrids of lambda phages and plasmids, and their DNA can replicates in the cell like that of a plasmid or be packaged like that of a phage is called:**
   (A) Phagemids
   (B) Fosmids
   (C) Plasmids
   (D) Cosmids

8. **Plasmids capable of propagating and transferring (shuuling) genes between two different organism, one of which is typically a prokaryote (*E. coli*) and other a eukaryote (yeast) is called:**
   (A) Cloning vector insertion vector
   (B) Shuttle vector
   (C) Viral vector
   (D) None of these

9. **Cosmid DNA can be packaged into phage particles by using the:**
   (A) *In vitro* system
   (B) *In vivo* system
   (C) Both of these
   (D) None of these

10. **Recombinants can therefore be identifies because the characteristics coded by the inactivated gene is no longer displayed by the host cells is called:**
    (A) Insertional inactivation
    (B) Transduction
    (C) Transfection
    (D) None of these

11. **Virus mediated gene transfer, in which the foreign DNA is packaged inside a virus particle is called:**
    (A) Insertional inactivation
    (B) Transduction
    (C) Transfection
    (D) None of these

12. The use of chemical or physical methods to persuade cells to take up DNA from the culture medium or direct physical introduction of DNA into the cell is called :

(A) Insertional inactivation
(B) Transduction
(C) Transfection
(D) None of these

13. A fundamental tool in molecular genetics which takes advantage of the ability of individual single-stranded nucleic acid molecules to form double stranded molecules is called:

(A) Nucleic acid hybridization
(B) In situ hybridization
(C) Both (A) and (B)
(D) None of these

14. Nucleic acid hybridization can be used to detect homologous DNA or RNA sequences not only in cell extracts, but also in chromosomes or intact cells – a procedure is called:

(A) Nucleic acid hybridization
(B) In situ hybridization
(C) Both (A) and (B)
(D) None of these

15. The technique is used to detect the presence of a known piece of DNA in any given DNA preparation from an organism is known as:

(A) Southern blot (B) Northern blot
(C) Both (A) and (B) (D) None of these

16. A method of estimating the size and abundance of a particular transcript is known as:

(A) Southern blot (B) Northern blot
(C) Both (A) and (B) (D) None of these

17. The determine the 5′-end of transcript, the most reliable method is:

(A) Primer extension
(B) Reporter assays
(C) Expression vector
(D) None of these

18. To findout the expression patterns of a gene within the cell, its promoter can be fused to one of the several reporter genes and transfected inside living cells by:

(A) Primer extension
(B) Reporter assays
(C) Expression vector
(D) None of these

19. To maximize the expression of heterologous genes – these genes that are taken from different organisms and being expressed in bacterial, yeast, animal or plant cells - it is necessary to design a cloning vector, which allows optional transcription of the genes. Such vectors are also called:

(A) Primer extension (B) Reporter assays
(C) Expression vector (D) None of these

20. Recombinants are hybrid DNA molecules consisting of autonomously replicating DNA segment plus inserted elements. Such hybrid molecules are also called:

(A) Chimera (B) Ligation
(C) Transformation (D) Cloning

Practice Paper 2

1. For plasmid purification the bacterial cells extract is deproteinized, the RNA removed, and the DNA concentrated by:

(A) Methane precipitation
(B) Ethanol precipitation
(C) CO_2 precipitation
(D) Ammonia precipitation

2. Autonomously replicating genetic element used to carry a fragment of genomic DNA into a host cell for the purpose of gene cloning:

(A) Cloning vector insertion vector
(B) Shuttle vector
(C) Viral vector
(D) None of these

3. The chromosome of the transgenic animal includes an endogenous coding sequence, preferably the c-myc gene, which is substantially the same as the:

(A) Oncogene sequence
(B) Coding sequence
(C) Viral sequence
(D) Homologous endogenous sequence

4. **The transcription of the oncogene sequence is under the control of a promoter sequence different from the promoter sequence (or a synthetic promoter sequence) controlling transcription of the endogenous:**
   (A) Oncogene sequence
   (B) Coding sequence
   (C) Viral sequence
   (D) Homologous endogenous sequence
5. **The term oncogene encompasses the:**
   (A) Viral sequence
   (B) Homologous endogenous sequence
   (C) Both (A) and (B)
   (D) Coding sequence
6. **The promoter sequence controlling transcription of the oncogene sequence is:**
   (A) Oncogene sequence
   (B) Coding sequence
   (C) Viral sequence
   (D) Inducible
7. **Introduction of the oncogene sequence at the fertilized-oocytes stage ensures that the oncogene sequence will be present in transgenic animals is:**
   (A) Germ cell (B) Somatic cells
   (C) Both (A) and (B) (D) None of these
8. **Introduction of the oncogene sequence at a later embryonic stage might result in the oncogene's absence from some somatic cells of the founder animal, but the descendants of such an animal that inherit the gene will carry the activated oncogene in all of their:**
   (A) Germ cell (B) Somatic cells
   (C) Both (A) and (B) (D) None of these
9. **The process that promotes gene transfer between bacteria by direct cell-to-cell contact, comes closest to sex in higher organisms is called:**
   (A) Conjugation (B) Transduction
   (C) Transformation (D) Transposons
10. **The transfer of bacterial DNA from one cell (the donor) to another (the recipient) via the agency of bacteriophage is called:**
   (A) Conjugation (B) Transduction
   (C) Transformation (D) Transposons
11. **The mechanism of gene transfer involves uptake of free DNA from killed donor cells by a recipient cell is known as:**
   (A) Conjugation (B) Transduction
   (C) Transformation (D) Transposons
12. **The discrete genetic elements with the ability to move from one genomic position to other non homologous positions is known as:**
   (A) Conjugation (B) Transduction
   (C) Transformation (D) Transposons
13. **These single-stranded sequences (GGGCGGCGACCT/AGGTCGCCGCCC) are complementary to each other and are called:**
   (A) COS (Cohesive sites)
   (B) Cis
   (C) CS trans
   (D) Chi
14. **Lambda is representative of a large group of bacteriophages that can propagate either through the standard lytic cycle or through a process called:**
   (A) Lysogeny
   (B) Caspid
   (C) Both (A) and (B)
   (D) None of these
15. **In mature virus particles, these genomes are packaged inside a protein coat called:**
   (A) Lysogeny (B) Caspid
   (C) Both (A) and (B) (D) None of these
16. **The represented phage genome in lysogenic cells is called:**
   (A) Prophage or provirus
   (B) Metaphage
   (C) Transphage
   (D) Post phage
17. **Who describes relative position of various genes on the chromosome(s) of an organism is?**
   (A) Genetic map (B) Synaptic map
   (C) Physical map (D) None of these
18. **For cloning eukaryotic gene in prokaryotes, genes should be isolated from:**
   (A) Genomic library (B) cDNA library
   (C) Eukaryotic host (D) None of these

19. **In order to move from a molecular marker to the linked gene of interest, which of the following approaches can be used?**
    (A) Chromosome jumping
    (B) Chromosome walking
    (C) Colony hybridization
    (D) Chromosome sorting
20. **Essential constituents of a cloning vector are:**
    (A) Replicon, promoter and selectable marker
    (B) Replicon, unique restriction site and selectable marker
    (C) Promotor, operatop and restriction site
    (D) Replicon, unique restriction site, selectable marker and promoter

Practice Paper 3

1. **For patenting a plant variety it should be:**
   (A) DUS (B) DUD
   (C) DDS (D) SUS
2. **Micro propagation can be done by:**
   (A) Through shoot bud proliferation
   (B) Through adventitious buds
   (C) Meristem tip culture
   (D) All of the above
3. **IPR stands for:**
   (A) Intellectual Property Rights
   (B) Indigenous Product Right
   (C) Indian Property Rights
   (D) International Product Rights
4. **Virus free plants can be obtained by:**
   (A) Meristem culture (B) Pollen culture
   (C) Ovary culture (D) Shoot tip culture
5. **Selection of protoplast fusion hybrids can be done by:**
   (A) Visual selection
   (B) Fluorescent labels
   (C) FACS
   (D) Any of these methods
6. **Structural DNA nanotechnology is for:**
   (A) Nanoelectronics (B) Nanorobotics
   (C) Smart material (D) All of these
7. ***In vitro* techniques for germplasm storage are useful for:**
   (A) Clonally propagated crops
   (B) Crops producing recalcitrant seeds
   (C) Crops producing orthodox seeds
   (D) Both (A) and (B)
8. **Which of the following enzyme is used in brewing?**
   (A) β amylase (B) Cellukase
   (C) Dextranase (D) Pectinase
9. **The combination of biology and IT is:**
   (A) Nanotechnology (B) Biotechnology
   (C) Bio-informatics
   (D) Information technology
10. **Technique for sequencing DNA and proteins is:**
    (A) NMR spectroscopy
    (B) X-ray crystallography
    (C) Electron microscopy
    (D) All of these
11. **Cosmid DNA an be packaged into phage particles by using the:**
    (A) *In vitro* system (B) Transduction
    (C) Both (A) and (B) (D) None of these
12. **DNA chip carries a large number of:**
    (A) Oligonucleotides (B) cDNAs
    (C) DNA probes (D) All of the above
13. **Cosmid DNA an be packaged into phage particles by using the:**
    (A) *In vitro* system (B) Transduction
    (C) Both (A) and (B) (D) None of these
14. **DNA chip carries a large number of:**
    (A) Oligonucleotides (B) cDNAs
    (C) DNA probes (D) All of the above
15. **Indian has created a National Biodiversity Authority with headquarters at:**
    (A) New Delhi (B) Kolkata
    (C) Chennai (D) Hyderabad
16. **Transmission of chromosomal DNA from generation to generation is crucial to cell propagation id called:**
    (A) DNA conservation
    (B) DNA replication
    (C) Genetic engineering
    (D) Transgenic plants

17. A nuclei aid sample is resistant to digestion with exonuclease. When heated it does not show typical melting curve of a linear double stranded DNA. On CsCl-ethidium bromide equilibrium density centrifugation it settles at the bottom of the centrifuge tube. The nuclei aid is:

(A) ccc pBR322
(B) Bacteriophage P22 DNA
(C) rRNA
(D) RFIIM13DNA

18. Which technologies and/or raw materials were used to develop Endless Summer tomato by DNA Plant Technology?

(A) Selectable marker *npt*II
(B) CaMV 35S promoter and the ACC synthase gene
(C) Antisense RNA technology and *Agrobacterium tumerfaciens* mediated genetic transformation
(D) All of these

19. Cybrids are produced:

(A) By *in vitro* fusion of gemetes
(B) By fusion of two cells and contain full nuclear genomes of the two parents
(C) By fusion of two cells and contain full nuclear genome of one and partial nuclear genome of the other parent
(D) By fusion of two cells and contain nuclear genome of one parent and cytoplasmic genome of the other parent

20. DNA finger printing is useful for study of:

(A) Forensics
(B) Polymorphism
(C) Identity and relationships
(D) All of these

Practice Paper 4

1. Complete set of cloned DNA fragments of an organism is known as:

(A) Genomic library (B) cDNA library
(C) mRNA library (D) Any of the above

2. During the process of reverse transcription:

(A) mRNA is synthesized from DNA
(B) DNA is synthesized from DNA
(C) Proteins are formed from mRNA
(D) DNA is synthesized from mRNA

3. Cloning is a means to:

(A) Replace original genotype
(B) Preserve genotype
(C) Production of HGH gene in *Escherichia coli*
(D) None of the above

4. The medium is added continuously to the fermentor tank to replace that, which has been fermented. This is true in case of:

(A) Batch technique
(B) Continuous flow technique
(C) Both are true
(D) None is true

5. Eukaryotic DNA replication is carried out by

(A) DNA Pol α (B) DNA Pol β
(C) DNA Pol μ (D) DNA Pol δ

6. The codons causing chain termination are:

(A) AGT, TAG, UGA
(B) UAG, UGA, UAA
(C) TAG, TAA, TGA
(D) GAT, AAT, AGT

7. A restriction endonuclease is a bacterial enzyme that cuts double-stranded DNA by leaving the phosphodiester bonds at specific nucleotide sequences known as:

(A) Restriction site
(B) 200 base-pairs
(C) 256 base-pairs
(D) 4096 base-pairs

8. The molecular scissors used to cut DNA into specific genes of interest are called:

(A) Exonucleases
(B) Restriction endonucleases
(C) Ligases
(D) Polymerases

9. RNA molecules involve in splicing are:

(A) Sn rRNA (B) 5s rRNA
(C) 16s RNA (D) tRNA

10. AIDS lead to destruction of immune system due to depletion of

(A) B-cells (B) T helper cells
(C) Macrophages (D) Plasma cells

11. Virions are ....

(A) DNA viruses
(B) RNA viruses
(C) Naked DNA viruses
(D) Naked RNA viruses

12. Prokaryotic DNA Polymerase comprises five subunits, which are

(A) αβμδε (B) $\alpha_2\beta\beta\sigma$
(C) σεδωω (D) αεδβθ

13. AFLP molecular markers are ..... in nature

(A) Dominant (B) Co-dominant
(C) Recessive (D) Neutral

14. Taq polymerase used in PCR is

(A) Thermotolerant
(B) Highly processive enzyme
(C) Osmotolerant
(D) All of the above

15. Franken food is the 'nick name' of:

(A) Fast food (B) Frozen food
(C) GM food (D) Fortified food

16. Transgenic expression study can be done by:

(A) Transcription profiling
(B) RT-PCR
(C) Western hybridisation
(D) All of these

17. Usually which of the following RNA is used for production of cDNA?

(A) rRNA (B) mRNA
(C) tRNA (D) siRNA

18. Which among the following is not used as vector for gene cloning?

(A) Cosmid (B) Phagemid
(C) PAC (D) MAC

19. GEM stands for:

(A) Genetically Engineered Metazoa
(B) Genetically Engineered Microbes
(C) Genetically Engineered Metabolism
(D) Genetically Engineered Mustard

20. During RNA isolation which chemical is used for breaking of plasma membrane?

(A) Phenol
(B) Isopropanol
(C) CTAB/SDS
(D) Ethanol

Practice Paper 5

1. Bioreactor are used for :

(A) Production of enzymes
(B) Production of secondary metabolites
(C) Production of cell cultures
(D) All of these

2. Minisatellite are also known as:

(A) VNTRs (B) RFLPs
(C) STRs (D) RAPDs

3. Which programmes converts the submitted amino acid sequence into nucleotide sequence and compares it with nucleotide sequence database?

(A) $BLAST_x$ (B) $BLAST_n$
(C) $tBLAST_n$ (D) $tBLAST_x$

4. Which enzyme plays important role in reverse transcription?

(A) DNA polymerase
(B) RNA polymerase
(C) Reverse transcriptase
(D) Ribonuclease

5. GAATIC is restriction sequence of CTTAAG:

(A) Eco RI (B) Bam HI
(C) Sau 3 (D) Nco I

6. Which of the following is a dominant marker?

(A) RAPD (B) RFLP
(C) SSR (D) AFLP

7. High density molecular map can be developed by:

(A) RFLP (B) RAPD
(C) AFLP (D) SNPs

8. Histone octomer comprise of:

(A) HI, H2A, H2B and H3
(B) HI, H2A, H2B, H3 and H4
(C) HI, H2B, H3 and H4
(D) H2A, H2B, H3 and H4

9. Operon comprise of

(A) Operator
(B) Structural gene
(C) Promoter
(D) Repressor
(E) All of the above

10. International Centre for Genetic Engineering and Biotechnology is located at :

(A) Trieste (B) New Delhi
(C) Bangalore (D) Both (A) and (B)

11. Genetic engineering means:

(A) Meiotic division of cells
(B) Nucleotide transfer
(C) Deletion and repair mechanism
(D) None of the above

12. The technique of insertion of a desired gene into DNA of plasmid vector is:

(A) Gene drafting (B) Gene dressing
(C) Gene cloning (D) Gene splicing

13. Which one of the following makes use of RNA template to synthesize DNA?

(A) DNA polymerase
(B) RNA polymerase
(C) Reverse transcriptase
(D) DNA dependent RNA polymerase

14. In the area of biotechnology, which materials can be kept as trade secret?

(A) Cell lines
(B) Strains of microorganism
(C) Production processes
(D) All of these

15. Which of the following portions of a longer duplex DNA segment are likely to be recognition sequences of a restriction enzyme?

(A) 5'-AGTC -3' / 3'-TCAG-5' (B) 5'-ATCG – 3' / 3'-TAGC -5'
(C) 5'-ACCT-3" / 3'-TGGA-5' (D) 5'-ACGT-3' / 3'-TGCA-5'

16. A restriction endonuclease has the recognition sequence G/AATTC, where "/" indicates the cut site. This sequence is found, on average, one every 'x' residues in a chromosome. 'x'=

(A) 146 base-pairs (B) 200 base-pairs
(C) 256 base-pairs (D) 4096 base-pairs

17. A gene or DNA sequence having a known location on a chromosome and associated with a particular trait or gene is used as a:

(A) Genetic marker
(B) Promoter gene
(C) Genetic map
(D) Genetic engineering

18. Recombinants can therefore be identified because the characteristics coded by the inactivated gene is no longer displayed by the host cells is called:

(A) Transfection
(B) Transduction
(C) Insertional inactivation
(D) None of these

19. Recombinants are hybrid DNA molecules consisting of autonomously replicating DNA segment plus inserted elements. Such hybrid molecules are also called:

(A) Chimera (B) Ligation
(C) Transformation (D) Cloning

20. Three terminal codons are:

(A) AUG, GUG, GUC
(B) UAG, UAA, AGA
(C) UAA, UAC, AUG
(D) UAA, UAG, UUA

Practice Paper 6

1. A DNA microarray also known as:

(A) Gene chip (B) DNA chip
(C) Gene array (D) All of these

2. Which sequence in case of eukaryotes is important for mRNA tailing?

(A) GAGAGA (B) GAATTC
(C) UACGAC (D) UACUAAC

3. The technique for correcting defective genes responsible for disease development is called:

(A) Gene therapy
(B) Gene inhibition
(C) Gene cloning
(D) All of these

4. The technique used to distinguish between individual of the same species using only samples of their DNA is called:

(A) DNA fingerprinting
(B) DNA testing
(C) DNA typing/DNA profiling
(D) All of these

5. **Department of biotechnology promotes human resource development through:**
(A) Master's programmes
(B) Post-doctoral programmes
(C) Certificate/Diploma programmes
(D) All of these

6. **RAPD molecular markers are .... in nature**
(A) Dominant
(B) Co-dominant
(C) Recessive
(D) Dominant-Recessive

7. **Which transgenic crops is being grown on commercial scale in India?**
(A) Bt-cotton (B) Bt-brinjal
(C) Bt-rice (D) Bt-tobacco

8. **DNA is fit for making tools in case of nanotechnology. This is due to:**
(A) Branching nature of DNA
(B) Flexibility in DNA conformation
(C) Small size of DNA
(D) All of these

9. **The short tagged tracts of DNA sequence (200 to 500 base pairs) that is operationally unique and has a single occurrence in the genome is called:**
(A) Sequence tagged sites mapping
(B) Expressed sequence tags
(C) Both of these
(D) None of these

10. **DNA replication is:**
(A) Conservation
(B) Semi-conservative
(C) None of these
(D) Both of these

11. **mRNA usually is being extracted using by:**
(A) Poly T resin (B) RNase P
(C) a polyA resin (D) an rRNA trap

12. **Which means that one of the allele is being converted from one type to other, possibly by DNA recombination during meiosis after fusion of the gametes is called:**
(A) Gene conversion
(B) Gene transcription
(C) Gene promoter
(D) Gene disruption

13. **The first step of gene expression when one strand of the double-stranded DNA encompassing the gene is copied to the functional RNA (synthesized from 5′ to 3′ direction) is called:**
(A) Gene conversion
(B) Gene transcription
(C) Gene promoter
(D) Gene disruption

14. **The part of the gene which controls its transcription is called .... that is located upstream of the coding sequence is called:**
(A) Gene conversion
(B) Gene transcription
(C) Gene promoter
(D) Gene disruption

15. **The technique to introduce gene-specific mutations to chromosomes in intact organism is called:**
(A) Gene conversion
(B) Gene transcription
(C) Gene promoter
(D) Gene disruption

16. **The technique of insertion of a desired gene into DNA of plasmid vecter is:**
(A) Gene splicing (B) Gene dressing
(C) Gene cloning (D) Gene drafting

17. **One of the main technique in recombinant DNA technology is:**
(A) DNA cloning (B) DNA molecules
(C) Both of these (D) None of these

18. **DNA fingerprinting is useful for study of:**
(A) Forensics
(B) Polymorphism
(C) Identify and relationships
(D) All of these

19. **A map of the locations of identifiable landmarks on chromosomes is known as:**
(A) Structural gene (B) Genetic marker
(C) Genetic map (D) Eucaryotic gene

20. **The important data bases in bioinformatics are used:**
(A) SWISS-PROT (B) EMBL
(C) GenBank (D) Pfam
(E) All of these

Practice Paper 7

1. **The primary nucleotide sequence databases are:**
   (A) EMBL (B) GenBank
   (C) DDBI (D) All of these
2. **The translated DNA databases are:**
   (A) PROTEIN (NCBI)
   (B) TREMBL
   (C) Both (A) and (B)
   (D) None of these
3. **The sequence derived databases are:**
   (A) PROTEIN (NCBI)
   (B) TREMBL
   (C) Pfam
   (D) SWISS-PROT
4. **The programme is used to compare a nucleotide (translated into all six frames) sequence with a nucleotide (translated into all ix frames0 sequence database is:**
   (A) TBLASTN (B) TBLASTX
   (C) PROSITE (D) PROMOT
5. **A collection of functional sites and sequences patterns found in proteins, it also has search tools for matching patterns/motifs is:**
   (A) TBLASTN (B) TBLASTX
   (C) PROSITE (D) PROMOT
6. **The search tool can be used to match a sequence against the PROSITE database is:**
   (A) TBLASTN (B) TBLASTX
   (C) PROSITE (D) PROMOT
7. **Genome sequencing became possible on a large scale with the discovery of the:**
   (A) Microarray
   (B) Molecular force field
   (C) Shot gun sequencing strategy
   (D) Shot gun approach of sequencing
8. **Which programmes can be used for the identifications/detection of genes from genome sequence of prokaryotes?**
   (A) Glimmer
   (B) GENMARK
   (C) GeneFinder
   (D) Both (A) and (B)
9. **India has not been involved in sequencing the genomes of which organisms as:**
   (A) Pigeon (B) Rat
   (C) Pigeonpea (D) Humans
10. **Which database is used by ENTREZ for bibliographic or citation search?**
    (A) SWISS-PROT (B) PubMed
    (C) TrEMBL (D) PDB
11. **Which is used for comparing the submitted nucleotide sequence with nucleotide database?**
    (A) BLASTn (B) BLASTx
    (C) rBLASTx (D) tBLASTn
12. **Transgenic plants could serve as vectors for transgene dissemination through :**
    (A) Seed
    (B) DNA released from debris of transgenic plants
    (C) Pollen
    (D) All of these
13. **Which approaches can be used for gene identification?**
    (A) Exon trapping (B) cDNA capture
    (C) cDNA selction (D) All of these
14. **DNA sequences/molecules used for preparing spotted DNA arrays may be obtained from:**
    (A) PCR amplification (B) Genomic DNA
    (C) cDNA clones (D) All of these
15. **Shot-gum sequencing was conceptualized by:**
    (A) Craig Venter (B) Michelmore
    (C) Giovannoni (D) Young
16. **Which phenomenon comes under post-translational modifications?**
    (A) Phosphorylation (B) Proteolysis
    (C) Glycosylation (D) All of these
17. **Which PCR strategies monitors the fluorescence emitted during reaction to measure the amount of amplification product generated during the PCR cycle?**
    (A) Anchored PCR
    (B) Touchdown PCR
    (C) Real-Time PCR
    (D) Nested PCR

18. It is a method of breeding crops to increase their nutritional value; it can be achieved through conventional selective breeding or through genetic engineering is called:

(A) Consensus sequence
(B) Conserved sequence
(C) Biofortification
(D) None of these

19. A crop variety expressing a transgene, produced by genetic engineering is called:

(A) Genetically modified crops
(B) Genetically modified food
(C) Genetically modified organism
(D) Genetically modifies plants

20. Food and additives obtained from genetically modifies organisms is called:

(A) Genetically modified crops
(B) Genetically modified food
(C) Genetically modified organism
(D) Genetically modifies plants

Practice Paper 8

1. An organism with a unique gene combination that does not occur naturally, and is produced either by using genetic engineering or protoplast fusion is called:

(A) Genetically modified crops
(B) Genetically modified food
(C) Genetically modified organism
(D) •

2. Transfer of gene from a transgenic variety to non-transgenic varieties, wild species and wild relatives through hybridization is called:

(A) Transgene instability
(B) Transgene silencing
(C) Transgene flow
(D) Gene splicing

3. A transgenic rice that accumulates - carotene in its seeds; expected to combat vitamin A deficiency areas where rice is the staple food is:

(A) Golden rice (B) Hybrid rice
(C) Boiled rice (D) Desi rice

4. A DNA sequence that appears at the end of a transcriptional unit and signals the end of transcription is called:

(A) Termination sequence
(B) Terminal transferase
(C) Supercoiled DNA
(D) Suppressor mutation

5. An enzyme that catalyzes the addition of nucleotide residues of a single kind to the 3' end of DNA chains is called:

(A) Termination sequence
(B) Terminal transferase
(C) Supercoiled DNA
(D) Suppressor mutation

6. Highly repeated, nontranslated segments of DNA in eukaryotic chromosomes; most often associated with the centromeric region is called:

(A) Satellite – DNA
(B) Ribosomal RNA
(C) Small nuclear RNA
(D) Transfer RNA

7. A genetic marker that is detected as differential mobility of a protein/DNA g\fragment or as differential amplification of a DNA sequence present in different individuals is:

(A) Molecular marker
(B) Molecular biology
(C) Molecular breeding
(D) Molecular farming

8. A specific location on a chromosome that is defined by naked eye polymorphism, protein or DNA sequences; the inheritance of a genetic marker can be followed in a a mapping population is known as:

(A) Molecular marker
(B) Genetic marker
(C) Molecular breeding
(D) Molecular farming

9. The plant breeding programme that are supported by the used of DNA-based markers and recombinant DNA technology is called:

(A) Molecular marker
(B) Molecular biology
(C) Molecular breeding
(D) Molecular farming

10. A single protein species produced by T-lymphocytes in response to antigen-activation is:

(A) Interfron α
(B) Interferon β
(C) Interferon γ
(D) Interferon ω

11. A single protein species produced by virus infected cells is:

(A) Interfron α
(B) Interferon β
(C) Interferon γ
(D) Interferon ω

12. Developing the DNA fingerprint of an individual is called:

(A) DNA fingerprint
(B) DNA finger printing
(C) DNA footprinting
(D) DNA sequencing

13. It enables identification of the site on a DNA molecule to which a protein, such as RNA polymerase, binds is called:

(A) DNA fingerprint
(B) DNA finger printing
(C) DNA footprinting
(D) DNA sequencing

14. Determination of the base sequence of a DNA fragment is called:

(A) DNA polymerase
(B) DNA microarray
(C) DNA footprinting
(D) DNA sequencing

15. The enzyme that catalyses synthesis of DNA copies of template DNA, separate enzymes involved in DNA replication and DNA repair is called:

(A) DNA polymerase
(B) DNA microarray
(C) DNA footprinting
(D) DNA sequencing

16. A gene that permits easy and rapid identification of the cell/organism in which it is present and expressed in:

(A) Molecular marker
(B) Genetic marker
(C) Selectable marker
(D) Scorable marker

17. A gene that gives rise to a product involved in the regulation of the expression of another gene; for example, a gene coding for a repressor protein is called:

(A) Structural gene
(B) Regulatory gene
(C) Regulatory sequence
(D) Reverse transcriptase

18. An enzyme that catalyzed the formation of RNA from ribonucleoside 5′-triphosphates, using a strand of DNA or RNA as a template is called:

(A) RNA polymerase
(B) Regulatory gene
(C) Regulatory sequence
(D) Reverse transcriptase

19. Removal of introns and joining of exons in a primary transcript is called:

(A) RNA polymerase
(B) RNA splicing
(C) Regulatory sequence
(D) Reverse transcriptase

20. Which markers systems has been used for developing Pusa 1460?

(A) RFLP
(B) STMS
(C) ISSR
(D) SCAR

Practice Paper 9

1. Spilicing of GU-AG intron:

(A) Involves transesterification reaction
(B) Occurs in prokaryotic mRNA
(C) Form Y-shaped branch structure
(D) Requires maturase activity

2. Which of the following sets of factors are needed for translation termination at an amber codon (UAG) and release of the finished polypeptide chain?

(A) RF2, RF3, EF-G and RRF
(B) RF1, RF3, EF-Tu and RRF
(C) RF1, RF3, EF-G and RRF
(D) RF2, RF3, EF-G and RRF

3. **DNA sequencing by the Sanger method takes advantage of which property of DNA synthesis to generate a sequencing ladder?**
   (A) Addition of nucleotides requires a free OH group on the 3' of the DNA strand
   (B) DNA has a free 5' phosphoryl group
   (C) DNA polymerase has proofreading capabilities
   (D) U is a base only found in RNA

4. **Hydroxyurea blocks DNA replication by:**
   (A) Inhibiting DNA polymerase
   (B) Inhibiting ribonucleotide reductase
   (C) Inhibiting lagging strand DNA synthesis
   (D) Inhibiting replicative helicase

5. **Which of the following is true about RFLP?**
   (A) Acts as genetic marker
   (B) RLFPs have more than two alleles
   (C) Used in case of FISH
   (D) Both (A) and (B)

6. **Khorana et al synthesized RNA with copolymer of UGUGUGUGUG .... It produced a peptide with alternatives cysteine and valine. The codons for the two are:**
   (A) UGU and GUG (B) UUG and GGU
   (C) GUG and UGU (D) UGG and GUU

7. **Electrophoretic separation of RNA molecules based on their molecular weight, requires a denaturant such as formaldehyde to be present in the gel system, because:**
   (A) RNAs are sensitive to nucleases
   (B) RNAs are single stranded
   (C) RNAs are usually present in high amounts
   (D) RNAs are usually small

8. **Khorana et al synthesized RNA with copolymer of UGUGUGUGUG ...... It produced a peptide with alternatives cysteine and valine. The codons for the two are:**
   (A) UGU and GUG (B) UUG and GGU
   (C) GUG and UGU (D) UGG and GUU

9. **Electrophoretic separation of RNA molecules based on their molecular weight, requires a denaturant such as formaldehyde to be present in the gel system, because:**
   (A) RNAs are sensitive to nucleases
   (B) RNAs are single stranded
   (C) RNAs are usually present in high amounts
   (D) RNAs are usually small

10. **RFLP analysis is a technique that:**
    (A) Uses hybridization to detect specific DNA restriction fragments in genomic DNA
    (B) Is used to determine whether a gene is transcribed in specific cells
    (C) Measures the transfer frequency of genes during conjugation
    (D) Is used to detect genetic variation at the protein level

11. **Aminoacyl-tRNA synthetases must be capable of recognizing which of the following:**
    (A) A specific tRNA and a specific amino acid
    (B) A specific rRNA and a specific amino acid
    (C) A specific tRNA and the 40S ribosomal subunit
    (D) A specific amino acid and the 40S ribosomal subunit

12. **In *E. coli*, which of the following codons are reorganized by the release factor RF1?**
    (A) UAG and UGA
    (B) UAA and UGG
    (C) UAG and UAA
    (D) UAG and UUA

13. **An RNA-dependent RNA polymerase is likely to be present in the virion of a:**
    (A) DNA virus that multiplies in the cytoplasm
    (B) DNA virus that multiplies in the nucleus
    (C) Minus-strand RNA virus
    (D) Transforming virus

14. **The saturated molecular linkage maps initially developed in tomato. Potato and maize used which of the following molecular marker systems?**
    (A) AFLP (B) RFLP
    (C) SNP (D) DAF

15. When DNA molecules of same density but different masses are to be separated by:

(A) Velocity sedimentation
(B) Equilibrium sedimentation
(C) Gel Electrophoretic separation
(D) Centrifugal force

16. When DNA molecules of different density but same or different masses are to be separated by:

(A) Velocity sedimentation
(B) Equilibrium sedimentation
(C) Gel Electrophoretic separation
(D) Centrifugal force

17. The genomic genes are often much larger that the:

(A) cDNA (B) mRNA
(C) rRNA (D) cRNA

18. To identify particular locations in the genome by:

(A) Molecular markers
(B) Genomes
(C) Selectable marker
(D) Scorable marker

19. Molecular markers rare:

(A) Based on nucleic acid hybridization (mostly DNA-DNA)
(B) Amplification based (PCR based)
(C) Both (A) and (B)
(D) None of these

20. Molecular markers based on:

(A) Nucleic acid hybridization (mostly DNA-DNA)
(B) Amplification based (PCR based)
(C) Both (A) and (B)
(D) None of these

Practice Paper 10

1. The molecular markers are based on the hybridization of a cloned piece of DNA to total genomic:

(A) Genome (B) Gene
(C) DNA (D) All of these

2. The process of cybrid production; usually, to transfer cytoplasm from one species in to other is called:

(A) Chromosome jumping
(B) Chromosome walking
(C) Cybridization
(D) Hybridization

3. The PAC cloning system combines the features of the vectors derived from the bacteriophage P1 and the :

(A) BAC vectors (B) YAC vectors
(C) PAC vectors (D) None of these

4. Among the given wheat varieties, which one was developed through haploid phase?

(A) Janak (B) Jinghua 1
(C) Arjun (D) Pb 543

5. Somaclone Andro has improvement in which traits?

(A) Rust immunity
(B) Powdery mildew resistance
(C) Wilt resistance
(D) Leaf spot resistance

6. Bio-13 is a somaclone of:

(A) *Mentha arvensis*
(B) *B. juncea*
(C) *Zea mays*
(D) *Cironella java*

7. Sugarcane somaclone Ono was isolated from:

(A) CO 1148 (B) Pindar
(C) BO 54 (D) COS 771

8. Somatic hybrids are also called:

(A) Parasexual hybrids
(B) Cytoplasmic hybrids
(C) Vybrids
(D) Both (A) and (B)

9. Which was the first to be applied to solve a practical problems?

(A) Embryo culture
(B) Anther culture
(C) Somatic hybridization
(D) Embryogenesis

10. Which was the last to be achieved?

(A) Embryo culture
(B) Anther culture
(C) Somatic hybridization
(D) Embryogenesis

11. Virus free plants produced by various in vitro techniques can be used for:
(A) Breeding programmes
(B) For germplasm exchange
(C) Commercial cultivation
(D) Both (A) and (B)

12. 'Genetic erosion' is the consequence of:
(A) Reduction in genetic variability among cultivated varieties
(B) Reduction in genetic variability within cultivated varieties
(C) Extinction of wild relatives of crop species
(D) All of these

13. Which breeding procedures is applicable to only autopolyploid crop species?
(A) Hybrid sorting
(B) Analytical breeding
(C) Somaclonal breeding
(D) Both (A) and (B)

14. Guha and Maheshwari produced haploids from anther culture of:
(A) Nicotiana
(B) Datura
(C) Ginkgo
(D) Hordeum

15. Pusa Jaikisan (Bio 902) is a somaclonal variety of:
(A) Rapeseed
(B) Indian mustard
(C) Groundnut
(D) Sunflower

16. Pusa Jai kisan (Bio 902) somaclone of Indian mustard was isolated from:
(A) Pusa Bold (B) Kranti
(C) Varuna (D) Vaibhav

17. Cryopreservation is based on:
(A) Liquid CO_2 (B) Liquid helium
(C) Liquid nitrogen (D) Both (A) and (C)

18. Somatic hybridization has been the most successful in:
(A) Graminae
(B) Leguminosae
(C) Malvaceae
(D) Solanaceae and Crucifereae

19. Post transcriptional gene silencing takes place at the:
(A) Gene regulation (B) Gene silencing
(C) DNA (D) RNA levels

20. The material could be stored in the form of suspension cultures, meristem, callus cultures and embryoids by:
(A) Dedifferentiation phase
(B) Cybrids
(C) *In vitro* method
(D) *Ex-vitro* method

Practice Paper 11

1. The method for preservation of genetic resources for cultivated as also the wild materials by:
(A) Dedifferentiation phase
(B) Cybrids
(C) *In vitro* method
(D) *Ex-situ* method

2. The method to preserve land races and wild relatives which could be used for future breeding programme by:
(A) *In-situ* method (B) Cybrids
(C) *In vitro* method (D) *Ex-vitro* method

3. A place where stores DNA for future analysis is:
(A) Gene bank (B) DNA bank
(C) Both (A) and (B) (D) None of these

4. IRGSP stand for:
(A) International Rice Genome Sequencing Project
(B) International Rice and Groundnut Sequencing Project
(C) International Research on Groundnut, Sorghum and Pigeonpea
(D) Indian Rice Genetic Resources Programme

5. Who developed pBR322:
(A) Benfy Ris (B) Berg Rosenberg
(C) Boliver Rodriguez (D) Bonner Riggs

6. Which vectors has been constructed for use in two different hosts?
(A) Shuttle vectors (B) Cosmids
(C) Phasmids (D) Phagemids

7. **The sequence (AGGAGG), which is a tibosomal binding site in bacterial and archaoal messenger RNA, generally located around 8 bases upstream of the short codon AUG:**
   (A) Pribnow box
   (B) Hogness box
   (C) Homeo box
   (D) Shine-Dalgarno box

8. **Which vectors has been constructed for use in two different hosts?**
   (A) Shuttle vectors
   (B) Cosmids
   (C) Phasmids
   (D) Phagemids

9. **Nature Gard is a transgenic variety of:**
   (A) Brinjal
   (B) Potato
   (C) Cotton
   (D) Corn

10. **Potato transgenic variety New Leaf is resistant to:**
    (A) Potato virus
    (B) Colorado potato beetle
    (C) Late blight
    (D) Frost

11. **Papaya transgenic Rainberg is resistant to:**
    (A) Papaya Mosaic Virus
    (B) Leaf Curl Virus
    (C) Papaya Ring Spot Virus
    (D) Distorted Ring Spot Virus

12. **Bt cotton contains:**
    (A) Bt gene
    (B) Multi genic
    (C) Cry IAC gene
    (D) Cry genes

13. **The patent Act, 1970 was amended third time in the year of:**
    (A) 1999 (B) 2002
    (C) 2005 (D) 2006

14. **The term of protection shall not end before the expiration of a period of twenty years counted from the date of filing in:**
    (A) Article 33 (B) Article 34
    (C) Article 35 (D) Article 36

15. **Plant varieties may be protected in India under:**
    (A) PVPFRA (B) PVRA
    (C) Both (A) and (B) (D) None of these

16. **Which has enabled successful transformation of monocots using Agrobacterium?**
    (A) Agrobacterium strain
    (B) Exogenous acetosyringone
    (C) Explant tissue
    (D) Plant variety

17. **In case of mice, DNA is microinjected into:**
    (A) Male pronucleus
    (B) Female pronucleus
    (C) Ovum cytoplasm
    (D) Both (A) and (C)

18. **Which marker systems would be preferred for testing the purity of hybrid seeds?**
    (A) RAPD (B) RFLP
    (C) STMS (D) AFLP

19. **RNA acts as the genetic material:**
    (A) Tobacco mosaic virus
    (B) Rhizobium
    (C) Fungi
    (D) Archaebacterium

20. **Which viruses gas acquired a gene from its host genome?**
    (A) Adenoviruses
    (B) Acute transforming retroviruses
    (C) Epstein-Barr Virus
    (D) Replication competent tumour retroviruses

Answers Key

■ Practice Paper 1

Q. No.	Ans	Q. No.	Ans	Q. No.	Ans	Q. No.	Ans	Q. No.	Ans	Q. No.	Ans
1.	D	2.	B	3.	C	4.	A	5.	A	6.	C
7.	D	8.	B	9.	A	10.	A	11.	B	12.	C
13.	A	14.	B	15.	A	16.	B	17.	A	18.	B
19.	C	20.	A								

■ Practice Paper 2

Q. No.	Ans	Q. No.	Ans	Q. No.	Ans	Q. No.	Ans	Q. No.	Ans	Q. No.	Ans
1.	B	2.	A	3.	A	4.	B	5.	C	6.	D
7.	C	8.	C	9.	A	10.	A	11.	C	12.	D
13.	A	14.	A	15.	B	16.	A	17.	A	18.	B
19.	A	20.	B								

■ Practice Paper 3

Q. No.	Ans	Q. No.	Ans	Q. No.	Ans	Q. No.	Ans	Q. No.	Ans	Q. No.	Ans
1.	A	2.	D	3.	A	4.	A	5.	D	6.	D
7.	D	8.	A	9.	C	10.	D	11.	A	12.	D
13.	A	14.	D	15.	C	16.	B	17.	C	18.	D
19.	D	20.	D								

■ Practice Paper 4

Q. No.	Ans	Q. No.	Ans	Q. No.	Ans	Q. No.	Ans	Q. No.	Ans	Q. No.	Ans
1.	A	2.	D	3.	B	4.	B	5.	D	6.	B
7.	A	8.	D	9.	A	10.	B	11.	D	12.	B
13.	B	14.	A	15.	C	16.	D	17.	B	18.	D
19.	B	20.	C								

■ Practice Paper 5

Q. No.	Ans	Q. No.	Ans	Q. No.	Ans	Q. No.	Ans	Q. No.	Ans	Q. No.	Ans
1.	D	2.	A	3.	C	4.	C	5.	A	6.	A
7.	D	8.	D	9.	B	10.	D	11.	C	12.	B
13.	C	14.	D	15.	D	16.	D	17.	A	18.	C
19.	A	20.	B								

■ Practice Paper 6

Q. No.	Ans	Q. No.	Ans	Q. No.	Ans	Q. No.	Ans	Q. No.	Ans	Q. No.	Ans
1.	D	2.	D	3.	A	4.	D	5.	D	6.	A
7.	A	8.	D	9.	A	10.	B	11.	A	12.	A
13.	B	14.	C	15.	D	16.	B	17.	A	18.	D
19.	C	20.	E								

■ Practice Paper 7

Q. No.	Ans	Q. No.	Ans	Q. No.	Ans	Q. No.	Ans	Q. No.	Ans	Q. No.	Ans
1.	D	2.	C	3.	C	4.	B	5.	C	6.	D
7.	C	8.	D	9.	D	10.	B	11.	A	12.	D
13.	D	14.	D	15.	A	16.	D	17.	C	18.	C
19.	A	20.	B								

■ Practice Paper 8

Q. No.	Ans	Q. No.	Ans	Q. No.	Ans	Q. No.	Ans	Q. No.	Ans	Q. No.	Ans
1.	C	2.	C	3.	A	4.	A	5.	B	6.	A
7.	A	8.	B	9.	C	10.	C	11.	D	12.	B
13.	C	14.	D	15.	A	16.	D	17.	B	18.	A
19.	B	20.	B								

■ Practice Paper 9

Q. No.	Ans	Q. No.	Ans	Q. No.	Ans	Q. No.	Ans	Q. No.	Ans	Q. No.	Ans
1.	A	2.	C	3.	A	4.	B	5.	A	6.	A
7.	B	8.	A	9.	B	10.	A	11.	A	12.	C
13.	C	14.	B	15.	B	16.	C	17.	C	18.	D
19.	D	20.	C								

■ Practice Paper 10

Q. No.	Ans	Q. No.	Ans	Q. No.	Ans	Q. No.	Ans	Q. No.	Ans	Q. No.	Ans
1.	C	2.	C	3.	A	4.	B	5.	A	6.	D
7.	B	8.	A	9.	A	10.	C	11.	D	12.	D
13.	B	14.	B	15.	B	16.	C	17.	C	18.	D
19.	D	20.	C								

■ Practice Paper 11

Q. No.	Ans	Q. No.	Ans	Q. No.	Ans	Q. No.	Ans	Q. No.	Ans	Q. No.	Ans
1.	D	2.	A	3.	B	4.	A	5.	C	6.	A
7.	D	8.	A	9.	D	10.	B	11.	C	12.	C
13.	C	14.	A	15.	A	16.	B	17.	A	18.	C
19.	A	20.	B								

15. Multiple Choice Questions of Soil Science

Pedology

Practice Paper 1

1. **Study of soil as a natural resource on the surface of the earth including fertilizer schedule includes: soil formation, classification and mapping; physical, chemical and fertility properties of soils; and these properties in relation to the use and management of soils is called:**
   (A) Soil science (B) Padology
   (C) Climatalogy (D) Microbiology

2. **The different forms of earth on the surface of the rocks. Formed by the breaking down or weathering of rocks is called:**
   (A) Soil science (B) Padology
   (C) Climatalogy (D) Microbiology

3. **To study of soils in their natural environment is called:**
   (A) Soil science (B) Padology
   (C) Climatalogy (D) Microbiology

4. **Padology word derived from:**
   (A) Romen (B) English
   (C) Greek (D) Sanskrit

5. **Pedology is a study of:**
   (A) Locomotion of animals
   (B) Rocks
   (C) Soil
   (D) Crop diseases

6. **The concept of soil as a natural body was first developed by:**
   (A) C.F. Marbut (B) K.D. Glinka
   (C) V.V. Dakuchaev (D) E.W. Hillgard

7. **When is a man a part of biotic community:**
   (A) Always
   (B) Only when he lives in big cities
   (C) Only when he lives in forest
   (D) Never

8. **Soil is originated from word 'Solum'. The word 'Solum' is a:**
   (A) Greek word (B) Latin word
   (C) Spanish word (D) English word

9. **Edaphology refers the study of soil related to:**
   (A) Crop production
   (B) Origin of soil
   (C) Classification of soil
   (D) None of these

10. **Soil is a:**
    (A) Rock (B) Static body
    (C) Dynamic body (D) None of these

11. **Sand particles contain lighter amount of:**
    (A) Quartz (B) Silica
    (C) Iron (D) All of them

12. **Fine particles of soil contain higher amount of:**
    (A) Quartz (B) Silica
    (C) Iron (D) All of them

13. **Different identified soil layers are known as:**
    (A) Horizon (B) Solum
    (C) Profile (D) All of these

14. **A horizon is known as:**
    (A) Organic (B) Elluvial
    (C) Accumulation (D) All of these

15. **O horizon is known as:**
    (A) Organic (B) Elluvial
    (C) Accumulation (D) All of these

16. **Illuvial horizon is a:**
    (A) O horizon (B) A horizon
    (C) B horizon (D) None of these

17. **The colour of different soil layer varies due to presence of:**
    (A) Minerals
    (B) Organic matter
    (C) Minerals and organic matter
    (D) None of these

18. (A) and (B) horizons are collectively known as:

(A) Horizon (B) Solum
(C) Profile (D) All of these

19. Sial is a:

(A) Upper layer of earth's crust
(B) Lower of earth's crust
(C) Upper layer of atmosphere
(D) Lower layer of atmosphere

20. Sial is made of mainly by:

(A) Al silicate (B) Mg silicate
(C) Fe- silicate (D) All of them

Practice Paper 2

1. Sima is situated just:

(A) Upper of Sial (B) Lower of Sial
(C) Middle of soil (D) None of these

2. Sima is made of:

(A) Al silicate (B) Mg silicate
(C) Fe- silicate (D) All of these

3. Eluviation refers to:

(A) Percolation of minerals from upper layers to lower layers
(B) Deposition of percolated materials from upper to lower layers
(C) Erosion of soil
(D) None of these

4. Illuviation refers to:

(A) Percolation of minerals from upper layers to lower layers
(B) Deposition of percolated materials from upper to lower layers
(C) Erosion of soil
(D) None of these

5. Humification helps in:

(A) Formation of Ao layer
(B) Decomposition of organic matter
(C) Synthesis of new organic materials
(D) All of these

6. Podzolization refers to:

(A) Decomposition of humus and sesquioxides in lower soil layers.
(B) Leaching of humus and sesquioxides from upper soil layers
(C) This process occurs in humid temperate regions
(D) All of these

7. Laterization refers to:

(A) Fe and Al remain in upper layers
(B) Leaching of Si to lower layers
(C) This process occurs in humid tropical regions
(D) All of these

8. Classification refers to:

(A) This process in low rainfall areas
(B) It happens due to presence of higher amount of bases in parent materials
(C) $CaCO_3$ is accumulated in soil profile
(D) All of these

9. Solanization refers to:

(A) Accumulation of salts in soil profile
(B) High water table and poor drainage enhance this process
(C) The structure of such soils are very poor
(D) All of these

10. Arrangement of soil particle is called:

(A) Soil texture (B) Soil structure
(C) Bulk density (D) Particle density

11. With increase in organic matter, the density will:

(A) Decrease (B) Increase
(C) Not change (D) Unknown

12. The weight per unit volume of the solid particle is called:

(A) Bulk density (B) Particle density
(C) Soil porosity (D) Soil Consistency

13. Soil particles determine its:

(A) Texture
(B) Field capacity
(C) Water holding capacity
(D) Soil flora

14. A branch of science that deals with the study of origin of the soil, its classification and description is known as:

(A) Entomology (B) Ecology
(C) Pedology (D) Geology

15. Regolith is known as:

(A) Elluvial + Illuvial layers
(B) Solum + Parent material
(C) O+A+B+C layers of soil profile
(D) All of them

16. **The phenomenon slickenside found in:**
(A) Vertisols (B) Alfisols
(C) Aridisols (D) None of these

17. **International Union of Soil Science (IUSS) is situated at:**
(A) Alfisols (B) Vertisols
(C) Both (A) and (B) (D) None of these

18. **International year of soil science observed in:**
(A) 2013 (B) 2014
(C) 2015 (D) 2016

19. **Soil catena is related to:**
(A) Topography
(B) Soil texture
(C) Soil structure
(D) Relief

20. **Formation of soil is the result of:**
(A) Bio-chemical process
(B) Physical process
(C) Chemical process
(D) Mechanical process

Rocks and minerals

Practice Paper 1

1. **Gypsum mainly found in:**
(A) Bihar (B) Rajasthan
(C) Andhra Pradesh (D) Maharashtra

2. **Pyrite mainly found in:**
(A) Bihar (B) Rajasthan
(C) Andhra Pradesh (D) Maharashtra

3. **Iron is an important component of:**
(A) Siroheme (B) Ferrodoxin
(C) Cytochromoes (D) All of these

4. **Clay fraction of Indo-Gangetic alluvium is dominated by the mineral:**
(A) Vermiculite (B) Illite
(C) Kaolinite (D) Chlorite

5. **Which clay mineral is rich in potash?**
(A) Montmorillonite (B) Kaolinite
(C) Illite (D) Chlorite

6. **Potassium content is maximum in:**
(A) Feldspar (B) Pyroxine
(C) Amphibole (D) Olivine

7. **The CEC SC mol (p+) kg^{-1} of black cotton soils of India ranges between:**
(A) 15 to 25 (B) 5 to 20
(C) 35 to 50 (D) None of these

8. **The clay minerals predominant in black cotton soils of India is/are:**
(A) Montmorillonite
(B) Beidellite
(C) Both (A) and (B)
(D) None of these

9. **Argillopedoturbation in black cotton soils of India is the result of:**
(A) Slickensides
(B) Shrinking and swelling of clays
(C) Gilgai
(D) None of these

10. **SiO_2/R_2O_2 of the clay fraction of black cotton soils usually lies between:**
(A) 1.5 to 2.0 (B) 2.0 to 2.5
(C) 2.5 to 3.0 (D) 3.0 to 3.5

11. **The CEC SC mol (p+) kg^{-1} red soils is:**
(A) < 20 (B) 20 to 30
(C) 30 to 40 (D) >40

12. **Red soils have the dominance of the following clay mineral:**
(A) Illite (B) Kaolinite
(C) Smectite (D) Vermiculite

13. **The SiO_2/R_2O_2 ratio of laterite clays varied between:**
(A) 1.00 to 1.25 (B) 1.40 to 2.00
(C) 2.00 to 3.00 (D) > 3.00

14. **Clay and organic matter of the soil have negative charges. They attract positively charged ions like:**
(A) Ca^{2+} (B) Mg^{2+}
(C) K^+ (D) All the above

15. **Non-expanding 1:1 clay mineral is:**
(A) Chlorite
(B) Illite
(C) Vermiculite
(D) Kaolinite

16. Which of the following is a kaolinite?

(A) $Al_4Si_4O_{10}(OH)_8$ (B) $Al_4Si_8O_{20}(OH)_4$
(C) $Al_2Si_4O_{10}(OH)_2K$ (D) None of these

17. Higher silica-sesquioxide ratio of soil indicates that soil has:

(A) Higher CEC (B) Higher WHC
(C) Higher soil fertility (D) All of these

18. Which of the following minerals is important for formation of silicate minerals?

(A) Feldspars (B) Mica
(C) Amphibols (D) All of these

19. Which of the following mineral has the highest anion exchange capacity?

(A) Kaolinite (B) Montmorillonite
(C) Illite (D) Chlorite

20. Which of the following mineral has the largest crystal size?

(A) Kaolinite (B) Montmorillonite
(C) Illite (D) Vermiculite

Practice Paper 2

1. Which of the clay mineral has higher amount of K?

(A) Kaolinite (B) Montmorillonite
(C) Illite (D) Vermiculite

2. Which of the following mineral has the highest CEC?

(A) Kaolinite (B) Montmorillonite
(C) Illite (D) Vermiculite

3. Which of the following is a 2 : 2 type mineral?

(A) Chlorite (B) Illite
(C) Kaolinite (D) None of these

4. Which of the following is a 1 : 1 type clay mineral?

(A) Kaolinite (B) Halloysite
(C) Anauxite (D) All of these

5. Which of the following is a 2 : 1 type expanding mineral?

(A) Vermiculite (B) Montmorillonite
(C) Beidellite (D) All of these

6. Which of the following is a 2 : 1 type non expanding mineral?

(A) Vermiculite (B) Montmorillonite
(C) Illite (D) None of these

7. The soil colloids of tropical and subtropical regions have more:

(A) Ca-H complexes
(B) Ca-Mg complexes
(C) Na-Ca complexes
(D) None of these

8. The soil colloids of arid region have more:

(A) Ca-H complexes
(B) Ca-Mg complexes
(C) Na-Ca complexes
(D) None of these

9. The chemical composition of clay is:

(A) Fixed (B) Dynamic
(C) Loose (D) None of these

10. Which of the following element is found in tetrahedral layer of clay crystals?

(A) Se^{4+} (B) Al^{3+}
(C) Fe^{3+} (D) Zn^{2+}

11. Which of the following element is found in Octahedral layer of clay crystals?

(A) Se^{4+} (B) Al^{3+}
(C) Fe^{3+} (D) Zn^{2+}

12. Which of the following is a montmorillonite?

(A) $Al_4Si_4O_{10}(OH)_8$
(B) $Al_4Si_8O_{20}(OH)_4$
(C) $Al_2Si_4O_{10}(OH)_2K$
(D) None of these

13. List A has soil groups and B main clay minerals as below:

List A	List B
1. Alluvial soil	i. Illite
2. Black soil	ii. Montmorillonite
3. Red soil	iii. Kaolinite and Illite
4. Hilly soils	iv. Illite and chlorite

Which of the pair is correctly matched?

(A) 1. i. 2. ii. 3. iii. 4. iv.
(B) 1. i. 2. iv. 3. iii. 4. ii.
(C) 1. ii. 2. iii. 3. iv. 4. iii.
(D) 1. iv. 2. ii. 3. iii. 4. i.

14. List A has soil orders and B main clay minerals as below:

List A	List B
1. Mollisol	i. Montmorillonite & Illite
2. Vertisol	ii. Montmorillonite
3. Inceptisol	iii. Illite
4. Alfisol	iv. Kaolinite & Montmorillonite

Which of the pair is correctly matched?

(A) 1. i. 2. ii. 3. iii. 4. iv.
(B) 1. i. 2. iv. 3. iii. 4. ii.
(C) 1. ii. 2. iii. 3. iv. 4. iii.
(D) 1. iv. 2. ii. 3. iii. 4. i.

15. Soil colours are generally due to:

(A) Iron oxides
(B) Humus
(C) Clays
(D) Iron oxide and humus

16. Most resistant mineral to weathering is:

(A) Quartz (B) Apatite
(C) Fluospar (D) Corundum

17. More weathering causing mineral is:

(A) Quartz (B) Apatite
(C) Fluospar (D) Corundum

18. Gypsum is a example of:

(A) Igneous rock
(B) Sedimentary rocks
(C) Metamorphic rocks
(D) All of these

19. The list of rocks is given below:

Rocks	Metamorphic rocks
1. Sand stone	i. Quartzite
2. Shale	ii. Slate
3. Limestone	iii Marble
4. Granite	iv Gneiss

Which combinations is correct:

(A) 1. i 2. ii 3. iii 4. iv
(B) 1. i 2. iii 3. ii 4. iv
(C) 1. iv 2. ii 3. iii 4. i
(D) 1. i 2. iv 3. iii 4. ii

20. Celluvial is the result of:

(A) Water (B) Air
(C) Gravity (D) None of these

Practice Paper 3

1. Granite is a:

(A) Acidic (B) Basic
(C) Saline (D) None of these

2. Basalt is a:

(A) Acidic (B) Basic
(C) Saline (D) None of these

3. Sedimetory rocks are:

(A) Layered rocks (B) Layer less rocks
(C) Metamorphic rocks (D) All of these

4. Limonite is the source of:

(A) N (B) P
(C) Fe (D) Zn

5. Dolomite is a source of following elements?

(A) Ca (B) Mg
(C) Ca and Mg (D) K

6. Apatite is a source of following elements?

(A) Ca (B) Mg
(C) P (D) K

7. Which of the following mineral does not belong to primary mineral?

(A) Quartz (B) Orthoclase
(C) Biotite (D) Hematite

8. Size of particle of primary minerals is:

(A) $\geq 2\mu$ (B) $< 2\ \mu$
(C) $\leq 2\ \mu$ (D) All of these

9. According to Mohr's scale, few minerals are arranged in order of its decreasing hardness. Which of the following is a correct order?

(A) Diamond > Corundum>Topaz>Talc
(B) Diamond > Topaz > Corundum>Talc
(C) Talc > Topaz > Corundum > Diamond
(D) None of these

10. Theory suggesting that carbon dioxide produced in respiration helps in mineral absorption is called:

(A) Carbonic acid exchange theory
(B) Contact exchange theory
(C) Active mineral absorption
(D) Donnan equilibrium

11. Passage of minerals from top soil to subsoil through seepage of water is known as:

(A) Leaching (B) Percolation
(C) Conduction (D) Transportation

12. Humus is a:

(A) Amorphous (B) Colloidal
(C) Both (A) and (B) (D) None of these

13. What percent of humic group present in Humus?

(A) 60-80% (B) 50-60%
(C) 30-40% (D) 20-30%

14. Humic group contains:

(A) Polypherals (B) Polyquenones
(C) Both (A) and (B) (D) None of these

15. Non humic group contains:

(A) Polysaccharide (B) Polyquenones
(C) Polyphenols (D) Both (B) and (C)

16. Red soil develops from the parent material of:

(A) Granite (B) Schist
(C) Both of these (D) None of these

17. Quartz, feldspar and mica are:

(A) Primary mineral
(B) Secondary mineral
(C) Tertiary mineral
(D) None of these

18. Dolomite, gypsum, apatite and hematite are:

(A) Primary mineral
(B) Secondary mineral
(C) Tertiary mineral
(D) None of these

19. Geological distribution of clay minerals is of considerable importance, southern are predominantly found:

(A) Illite
(B) Kaolinite
(C) Montmorillonite
(D) None of these

20. The clay minerals having CEC 200 meq/100 g belong to:

(A) Illite
(B) Kaolinite
(C) Vermiculite
(D) Montmorillonite

Practice Paper 4

1. Humus is defined as:

(A) Physical texture of soil
(B) Chemical composition of soil
(C) Decomposed organic matter in soil
(D) None of these

2. Sink is related to:

(A) Phytochrome
(B) Enzyme
(C) Stomata
(D) Transport of minerals

3. Soil contains minerals approximately:

(A) 25 % (B) 45 %
(C) 5 % (D) 10 %

4. Soil contains organic matter approximately:

(A) 2 % (B) 4 %
(C) 6 % (D) 8 %

5. Soil contains porosity approximately:

(A) 25 % (B) 35 %
(C) 50 % (D) 60 %

6. Soil contains water approximately:

(A) 75 % (B) 25 %
(C) 35 % (D) 45 %

7. Chalophilic elements make covalent bonds with:

(A) Sulphide (B) Carbonate
(C) Oxide (D) Hydroxide

8. Igneous rocks are made of:

(A) Lava (B) Lime
(C) Marble (D) All of them

9. Silica percentage in acidic igneous rocks is:

(A) 55-60 % (B) 60-65 %
(C) 65-85 % (D) All of these

10. Silica percentage in basic rocks is:

(A) 55-60 % (B) 60-65 %
(C) 45-55 % (D) All of these

11. Indo-Gangetic plains were developed by:

(A) Alluvial materials
(B) Aeolian materials
(C) Colluvial materials
(D) None of these

12. Sand dunes are the result of:
(A) Alluvial materials
(B) Aeolian materials
(C) Colluvial materials
(D) None of these

13. Peat and muck are the example of:
(A) Cumulose materials
(B) Residual materials
(C) Transported materials
(D) All of these

14. Rock phosphate is used in:
(A) Neutral soil (B) Acidic soil
(C) Alkaline soil (D) Saline soil

15. Micelles or micro cells are:
(A) Micro clay colloids particles
(B) Micro cells
(C) Micro organisms
(D) None of these

16. The soil colloids of humid region have:
(A) Ca^{++} - H^{+} complexes
(B) Ca^{++} - Mg^{++} complexes
(C) Na^{+} - Ca^{++} complexes
(D) None of these

17. Montmorillonite clay has silica and alumina layers respectively in the following proportion:
(A) 2 : 1 (B) 1 : 2
(C) 3 : 2 (D) 2 : 3

18. Most effective phosphatic fertilizer for crops in acidic soil is:
(A) DAP
(B) Rock phosphate
(C) SSP
(D) None of these

19. The general structural formula of montmorrilonite is:
(A) $X_{0.8}(Al_{0.3}Si_{7.7})Al_{2.6}Fe^{3+}{}_{0.9}Mg_{0.5})O_{20}(OH)_4nH_2O$ (B) $K_2Al_2Si_6Al_4O_{20}(OH)_4$
(C) $K_2Al_2Si_6(Fe^{2+},Mg)_6O_{20}(OH)_4$ (D) $Si_4Al_4O_{10}OH_8$

20. The structural formula of muscovite is:
(A) $X_{0.8}(Al_{0.3}Si_{7.7})Al_{2.6}Fe^{3+}{}_{0.9}Mg_{0.5})O_{20}(OH)_4nH_2O$ (B) $K_2Al_2Si_6Al_4O_{20}(OH)_4$
(C) $K_2Al_2Si_6(Fe^{2+},Mg)_6O_{20}(OH)_4$ (D) $Si_4Al_4O_{10}OH_8$

Practice Paper 5

1. The structural formula of biotite is:
(A) $X_{0.8}(Al_{0.3}Si_{7.7})Al_{2.6}Fe^{3+}{}_{0.9}Mg_{0.5})O_{20}(OH)_4nH_2O$
(B) $K_2Al_2Si_6Al_4O_{20}(OH)_4$
(C) $K_2Al_2Si_6(Fe^{2+},Mg)_6O_{20}(OH)_4$
(D) $Si_4Al_4O_{10}OH_8$

2. The structural formula of kaolinite is:
(A) $X_{0.8}(Al_{0.3}Si_{7.7})Al_{2.6}Fe^{3+}{}_{0.9}Mg_{0.5})O_{20}(OH)_4nH_2O$
(B) $K_2Al_2Si_6Al_4O_{20}(OH)_4$
(C) $K_2Al_2Si_6(Fe^{2+},Mg)_6O_{20}(OH)_4$
(D) $Si_4Al_4O_{10}OH_8$

3. The minerals of montmorillonite or smectic isomorphism series are freely expandable layer silicate are:
(A) Beidellite (B) Montmorillonite
(C) Saponite (D) All of these

4. Quartz and feldspar are:
(A) Primary minerals
(B) Secondary minerals
(C) Tertiary minerals
(D) Accessory minerals

5. **Silicate minerls, phyllo-silicates or clay minerals are:**
   (A) Primary minerals
   (B) Secondary minerals
   (C) Tertiary minerals
   (D) Accessory minerals
6. **Apatite, pyroxins, Tourmaline and Magnetite are:**
   (A) Primary minerals
   (B) Secondary minerals
   (C) Tertiary minerals
   (D) Accessory minerals
7. **Feldspar are alluminosilicates of:**
   (A) Potassium (B) Sodium
   (C) Calcium (D) All of these
8. **As the inter-layer potassium in subjected to depletion through chemical equilibrium with the soil solution, the exchangeable ions and inter-layer water is gained and leads to another mineral according to following relation:**
   (A) Micas → Illite → Vermiculite → Montmorillonite
   (B) Micas → Vermiculite → Illite → Montmorillonite
   (C) Montmorillonite → Illite → Vermiculite → Micas
   (D) Vermiculite → Micas → Illite → Montmorillonite
9. **Granite is a .... rock:**
   (A) Metamorphic (B) Igneous
   (C) Sedimentary (D) None of these
10. **Minerals are absorbed by the plants through:**
    (A) Pressure flow (B) Diffusion
    (C) Active transport (D) Osmosis
11. **Colloidal particles have:**
    (A) Positive charge (B) Negative charge
    (C) Both of these (D) None of these
12. **Zeta potential refers to:**
    (A) Magnitude of electric charges of colloidal particles
    (B) Size of colloidal particles
    (C) Movement of colloidal particles
    (D) None of these
13. **Browning movement can be seen in:**
    (A) Colloidal particles
    (B) Crystalloid particles
    (C) Urea
    (D) None of these
14. **Colloid is a Greek word that is made of:**
    (A) Kolla + Eidos
    (B) Colla + Oid
    (C) Co + Lloid
    (D) None of these
15. **Which of the following is not a colloidal material?**
    (A) Starch (B) Zelatin
    (C) Fog (D) Sugar
16. **What is the diameter of colloidal particles:**
    (A) 0-4 mμ (B) 5-100 mμ
    (C) 101-200 mμ (D) None of these
17. **Clay particles are made of:**
    (A) Quartz (B) Feldspar
    (C) Keolinite (D) All of these
18. **Which clay mineral is responsible for cracking in black soil:**
    (A) Montmorillanite
    (B) Vermiculite
    (C) Illite
    (D) None of these
19. **Who is the father of clay mineralogy in India is:**
    (A) N.N. Goswami
    (B) S.K. Mukherji
    (C) J S Kanwar
    (D) D.K. Das
20. **CEC is far minimum in case of:**
    (A) Montmorillanite
    (B) Vermiculite
    (C) Kaolinite
    (D) None of these

Essential plant nutrients

Practice Paper 1

1. Who gave the criteria of essentiality of plant nutrients:
(A) Arnon and Stout (1939)
(B) Arnon and Stout (1941)
(C) Leibig (1949)
(D) None of these

2. The essentiality of nitrogen for higher plants was first established in the year:
(A) 1801 (B) 1802
(C) 1803 (D) 1804

3. The essential nature of nitrogen to higher plants was first established by:
(A) JustusVon Leibig
(B) Sommer and Lipman
(C) Warrington
(D) Theodore de Saussure

4. The essential nature of phosphorus for higher plants was first established experimentally by:
(A) Arnon
(B) E. Gris
(C) Ville
(D) Johonson and Stout

5. The essentiality of phosphorus for higher plants was reported at first in the year:
(A) 1829 (B) 1839
(C) 1849 (D) 1859

6. The phosphorus in plant seeds is mainly stored as:
(A) Phospholipids (B) Nucleic acids
(C) Inorganic –P (D) Phytin

7. The phosphorus in cell membranes is mainly present as:
(A) Phospholipids (B) ATP
(C) NADP (D) Nucleoproteins

8. Phosphorus is a structural component of the cell organelles:
(A) Cell membranes
(B) Cell membranes and chloroplast
(C) Cell membrane, chloroplast and mitochondria
(D) None of these

9. As a constituent of coenzyme NAD and NADP, phosphorus plays an important role in:
(A) Energy transformation
(B) Hydrolytic reactions
(C) Decarboxylation
(D) Oxidation-reduction reactions

10. Potassium was experimentally established as an element essential to higher plants during the year:
(A) 1839 (B) 1876
(C) 1886 (D) 1896

11. The essentiality of potassium to higher plants was at first reported by:
(A) Russel (B) Sprengel
(C) Steinberg (D) Arnon

12. The potassium is almost entirely present as K^+ in the:
(A) Cytoplasm (B) Vacuole
(C) Both (A) and (B) (D) None of these

13. Within the cell, potassium helps to maintain:
(A) Permeability of cellular membranes
(B) Hydration of protoplasm
(C) Both (A) and (B)
(D) None of these

14. Potassium supply to plants has a favourable effect on:
(A) Protein synthesis (B) Respiration
(C) Both (A) and (B) (D) None of these

15. Potassium supply to plants decreases crop lodging through strengthening stalk by:
(A) Promoting cellulose and lignin deposition on walls
(B) Improving turgidity of cells
(C) Increasing parenchymatous tissues
(D) Maintaining electrical charge within the cell

16. Micronutrients mainly function as:
(A) Osmotic constituents of cell sap
(B) Cofactors of enzymes
(C) Constituents of chlorophyll
(D) Components of important biochemicals

17. **At what stage are more nutrients essentially required in cereal crops?**
(A) Growth stages of plants
(B) Formation of leaves
(C) Panicle initiation
(D) Maturity of crop

18. **Which element is now essential for plant growth:**
(A) Na (B) Si
(C) Ni (D) None of these

19. **Phosphorus is called macronutrient because it is:**
(A) Available in soil in relatively large amount
(B) Available in plant in relatively in large amount
(C) Used by plants in relatively large amount
(D) Found in grains in relatively large amount

20. **Which elements are useful in energy storage, transfer and bonding?**
(A) NPK (B) NSP
(C) NKS (D) None of these

Practice Paper 2

1. **Which of the following elements play an important role in nitrogen metabolism?**
(A) Magnesium (B) Manganese
(C) Molybdenum (D) Iron

2. **Most of the plants obtained nitrogen from the soil in the form of:**
(A) Nitrite (B) Nitrate
(C) Free nitrogen gas (D) Nitric acid

3. **Sulphur is needed for which group of the following amino acids:**
(A) Lysine and glutamine
(B) Methionine and cystine
(C) Tryptophane and cystine
(D) None of these

4. **Energy metabolism in plants is performed by the element:**
(A) Potassium (B) Magnesium
(C) Phosphorus (D) Iron

5. **Brayer and others discovered which of the essential elements?**
(A) Copper (B) Boron
(C) Chlorine (D) Molybdenum

6. **Critical elements are:**
(A) Na, K and Ca (B) N, P and Mg
(C) N, P and K (D) Mn, B and Mo

7. **The carbohydrates produced in leaves cannot be translocated to different growing parts of plants in the absence of:**
(A) Manganese (B) Boron
(C) Zinc (D) Iron

8. **Which element causes dispersion of soil particles:**
(A) Na (B) K
(C) Ca (D) Mg

9. **Calcium, magnesium and sulphur plant nutrients are called:**
(A) Primary (B) Secondary
(C) Micro (D) Trace

10. **Non essential but useful plant nutrients are:**
(A) Cobalt & Iodine
(B) Strontium & Flourine
(C) Vanadium & Bromine
(D) All of these

11. **Non essential but useful plant nutrients are:**
(A) Cobalt & Iodine
(B) Strontium & Flourine
(C) Vanadium & Bromine
(D) All of these

12. **Highest P is found in:**
(A) Leaf (B) Stem
(C) Seed (D) Root

13. **Which of the following P ion is available in acidic conditions?**
(A) $H_2PO_4^-$ (B) HPO_4^{--}
(C) PO_4^{---} (D) PO_3^{---}

14. **Which of the following P ion is absorbed by plants?**
(A) $H_2PO_4^-$ (B) HPO_4^{--}
(C) PO_4^{---} (D) PO_3^{---}

15. **Na is essential for:**
(A) Green algae (B) Sugar beet
(C) Diatoms (D) Sugarcane

16. Va is essential for:

(A) Green algae (B) Sugar beet
(C) Rice (D) Wheat

17. Role of Mo is:

(A) Sugar synthesis
(B) NH_3 and NO_3 reduction
(C) N fixation by free living bacteria
(D) All of these

18. Major function of phosphorus is in the formation of:

(A) Cell membranes (B) Cell wall
(C) Enzymes (D) Carbohydrates

19. Magnesium is required for:

(A) DNA replication
(B) RNA synthesis
(C) Stomatal movements
(D) Synthesis of chlorophyll

20. Potassium is employed in:

(A) Chlorophyll synthesis
(B) Protein synthesis
(C) Cementing cells
(D) Opening and closing of stomata

Practice Paper 3

1. Calcium is a component of:

(A) Middle lamella (B) Primary walls
(C) Secondary wall (D) Chlorophyll

2. Both nitrogen and sulphur are required by plants for:

(A) Chlorophyll synthesis
(B) Enzymes
(C) Cell wall
(D) Stomatal movements

3. Role of Cu is:

(A) Synthesis of Indole acetic acid
(B) Control respiration
(C) Synthesis of chlorophyll
(D) All of these

4. Role of Zn is:

(A) Control disintegration of RNA
(B) Water absorption
(C) Sugar translocation
(D) All of these

5. Role of B is:

(A) Translocation of sugars
(B) Pollen formation
(C) N absorption
(D) All of these

6. The uptake of which element is affected least under waterlogged conditions:

(A) N (B) P
(C) K (D) Ca

7. The uptake of which element is affected most under waterlogged conditions:

(A) N (B) P
(C) K (D) Ca

8. Phosphorus and nitrogen ions generally depleted in soil because they usually occur as:

(A) Nautral ions
(B) Negatively charged ions
(C) Positively charged ions
(D) Both positively and negatively charged but disproportionate mixture

9. Potassium is involved in:

(A) Photosynthesis
(B) Promoting many enzymatic activities that regulate plant processes
(C) Providing reddish pigmentation to fruits
(D) Formation of vascular cambium

10. Phosphorus brings about:

(A) Retardation of protein synthesis
(B) Healthy root growth
(C) fruit ripening
(D) Retarded plant growth

11. Zinc is essential for:

(A) Stomatal closing
(B) Stomatal opening
(C) Biosynthesis of IAA
(D) Oxidation of carbohydrates

12. Mineral absorption is mostly:

(A) Physical process (B) Chemical process
(C) Active process (D) Passive process

13. Which is absent in plant ash:

(A) Mineral elements
(B) Nitrogen
(C) Essential elements
(D) Trace elements

14. Bidirectional translocation of mineral takes place through:

(A) Parenchyma (B) Cambium
(C) Xylem (D) Phloem

15. Permeability of protoplasm is accelerated by:

(A) Ca (B) Na
(C) P (D) K

16. Plants require Fe and Mg for:

(A) Synthesis of chlorophyll
(B) Opening and closing of stomata
(C) Energy transfer during photosynthesis and respiration
(D) Translocation of carbohydrate

17. Nitrogen is an important constituent of:

(A) Polyphosphates (B) Carbohydrates
(C) Proteins (D) Lipids

18. Minerals are absorbed in the form of:

(A) Molecules (B) Ions
(C) Compounds (D) Mixtures

19. The absorption of minerals due to difference in the electropotential gradient without use of energy is:

(A) Active absorption
(B) Passive absorption
(C) Osmotic absorption
(D) None of these

20. Passive absorption of mineral salts is not dependent on:

(A) Osmosis
(B) Diffusion
(C) Donnan equilibrium
(D) Ionic exchange

Practice Paper 4

1. Active absorption of minerals salts makes use of:

(A) Ionic exchange
(B) Carriers
(C) Phosphorylation of elements
(D) None of these

2. Which of the following does not operate in the carrier concept of active absorption of minerals:

(A) Electron transport
(B) Contractile protein
(C) Protein synthesis
(D) Ascending water stream

3. Absorbed minerals are transported to other parts of the plants by:

(A) Diffusion (B) Imbibition
(C) Ion exchange (D) Water stream

4. Mineral salts which are absorbed by the root from the soil are in the form of:

(A) Very dilute solution
(B) Dilute solution
(C) Concentrated solution
(D) Very concentrated solution

5. Minerals absorbed by roots move to the leaf through:

(A) Phloem (B) Xylem
(C) Sieve tube (D) None of these

6. Plants absorb minerals salts from the soil solution through:

(A) Semipermeable membrane into the cytoplasm
(B) Perforation at the apex of root hairs only
(C) The cell wall which is semipermeable
(D) None of these

7. Which one plays an independent role in energy metabolisms:

(A) Calcium (B) Sodium
(C) Sulphur (D) Phosphorus

8. Active uptake of minerals requires:

(A) Respiration
(B) Expenditure of energy
(C) Root metabolism
(D) Photophosphorylation

9. A nonessential element is:

(A) Calcium (B) Barium
(C) Potassium (D) Magnesium

10. Nonessential element taking part in plant growth is:

(A) Magnesium (B) Calcium
(C) Cobalt (D) Chlorine

11. An essential element is that which:

(A) Is found in plant ash
(B) Is available in soil
(C) Improves health of plants
(D) Is irreplaceable and indispensable for growth of plants

12. Micronutrients are:
(A) Less important than macronutrients
(B) As important as macronutrients
(C) Having a minor role in plant nutrition
(D) Omitted from culture medium without any detrimental effect

13. A micronutrient is the one which is:
(A) More important than any major element
(B) Less important than major elements
(C) Needed in small quantity but is as important as a major element
(D) Found in small quantities in the soil

14. Framework elements are:
(A) C, H and O (B) Cu, Co and Fe
(C) Mg, Cu and Fe (D) Mn, Ca and N

15. Which can function as carrier in active ion absorption:
(A) Cytochrome (B) Ferredoxin
(C) Lecithin (D) Plastoquinone

16. Ions are absorbed by the plants through:
(A) Molecular diffusion
(B) Carriers and pumps
(C) Difference in water potential
(D) Difference in DPD

17. Mineral absorption occurs:
(A) If soil solution is hypotonic
(B) If soil solution is hypertonic
(C) Independent of water potential
(D) Independent of water absorption

18. Active and passive transports across cell membrane differ in:
(A) Passive transport is nonselective
(B) Passive transport is along the concentration gradient while active transport is due to metabolic energy
(C) Active transport is more rapid
(D) Passive transport is confined to anions while active transport is confined to cations

19. Molybdenum is involved in plant metabolism in:
(A) Translocation of solutes
(B) Nitrate reduction
(C) Tryptophan synthesis
(D) Ascorbic acid synthesis

20. Element responsible for maintaining turgor is:
(A) K (B) Na
(C) Ca (D) Hg

Practice Paper 5

1. Element required is least quantity is:
(A) Zn (B) Mn
(C) Mo (D) Cl

2. Role of inorganic nutrients in plant growth and metabolism was first discovered by:
(A) Knop (B) Woodland
(C) DeSassure (D) Steward

3. Enzyme catalysed reactions can be inhibited by:
(A) Mg^{2+} (B) Cu^{2+}
(C) Hg^{2+} (D) Zn^{2+}

4. Movement of minerals in plant roots through diffusion is:
(A) Endosmosis
(B) Osmosis
(C) Active absorption
(D) Passive absorption

5. Inorganic nutrients are present in the soil as:
(A) Molecules
(B) Atoms
(C) Electrically charged ions
(D) Colloids

6. The entry of inorganic materials into xylem occurs at:
(A) Leaves (B) Stem
(C) Root tips (D) None of these

7. Nitrate assimilation in plant is mainly carried out at:
(A) Toot (B) Stem
(C) Branches (D) Leaves

8. Conversion of organic form to inorganic form is known as:
(A) Mineralization
(B) Immobilization
(C) Volatilization
(D) None of these

9. **Oxy anions are generally:**
   (A) Lithophilic elements
   (B) Chaleophilic elements
   (C) Biophilic elements
   (D) Siderophilic elements

10. **N, O, He and Xe elements are grouped into:**
   (A) Lithophilic elements
   (B) Chaleophilic elements
   (C) Biophilic elements
   (D) Atmosphilic elements

11. **The term functional or metabolic nutrient was proposed by:**
   (A) Stout (B) Nicholus
   (C) Leibig (D) Arnon

12. **The rate of water absorption is decrease by:**
   (A) N (B) P
   (C) B (D) S

13. **P uptake in alkali soil in the form of:**
   (A) PO_4^{3-+} (B) P
   (C) P_2O_5 (D) None of these

14. **Which form of arsenic is more toxic:**
   (A) Al (B) Al^{+}
   (C) Al^{+2} (D) Al^{+3}

15. **Vitamin B12 having element:**
   (A) B (B) Co
   (C) P (D) Cu

16. **Which of he following only nutrient which might be applied completely by interception:**
   (A) P (B) N
   (C) K (D) Ca

17. **Which micronutrient in difficient in most of Indian soil:**
   (A) B (B) Zn
   (C) Cu (D) Na

18. **Fertility of male flower is enhanced by:**
   (A) N (B) Ca
   (C) Cu (D) S

19. **Smithsonite mineral consist which micronutrient:**
   (A) Zn (B) Cu
   (C) Na (D) None of these

20. **Most important metal ions responsible for the binding of phosphates in soils:**
   (A) Al (B) Fe
   (C) Ca (D) All of these

Soil fertility and productivity

Practice Paper 1

1. **Soil fertility refers to:**
   (A) Capacity of soil to provide all essential elements in balanced amounts to plants
   (B) Availability of N, P and K elements
   (C) High crop productivity
   (D) None of these

2. **Soil productivity refers to:**
   (A) Capacity of soil to produce high economic yield
   (B) Capacity of soil to provide essential plant nutrients to plant
   (C) Capacity of soil to produce high soil organisms
   (D) All of these

3. **Lag phase of growth indicates:**
   (A) Slow growth rate
   (B) It is observed at early plant growth stage
   (C) Decreasing growth rate with time
   (D) All of these

4. **Lag plant growth phase refers to:**
   (A) Maximum growth rate
   (B) Increasing growth rate with time
   (C) Maximum dry matter accumulation
   (D) All of these

5. **Who established the relationship between crop yield and growth factor?**
   (A) Wilcox
   (B) Mitscherlich
   (C) Brown
   (D) None of these

6. **The Baule unit refers to:**
   (A) Unit of any growth factor that provides 50% yield of expected yield
   (B) It was proposed by German Scientist, Baule
   (C) It varies with inputs
   (D) All of these

7. **Soil fertility is reduced by:**
   (A) Crop rotation
   (B) Nitrogen fixing bacteria
   (C) Decaying organic matter
   (D) Intensive agriculture

8. **Soil fertility is reduced due to:**
   (A) Over irrigation
   (B) Poor drainage
   (C) Imbalanced use of fertilizers
   (D) Continuous cropping

9. **Chemical formula of pyrite is:**
   (A) FeS (B) CuS
   (C) MnS (D) FeS_2

10. **Sizeable amount of nitrogen could be escaped by volatilization is:**
    (A) Ammonia (B) Nitrate
    (C) Nitrite (D) Amide

11. **Sizeable amount of nitrogen could be escaped by denitrification is:**
    (A) Nitrous oxide or chemical nitrogen
    (B) Nitrate (C) Nitrite
    (D) Amide

12. **Sizeable amount of nitrogen could be escaped by underground water is:**
    (A) Ammonia (B) Nitrate
    (C) Nitrite (D) Amide

13. **Under humid condition potassium fixation more in silicate clay of:**
    (A) Illite (B) Monmorillonite
    (C) Kaolinite (D) None of these

14. **Organic matter in soil is an index of:**
    (A) Available nitrogen
    (B) Available phosphorus
    (C) Available potassium
    (D) Available zinc

15. **For calculating value of K, K_2O is multiplied by:**
    (A) 2.29 (B) 0.437
    (C) 1.20 (D) 3.00

16. **For calculating value of S, SO_4 is multiplied by:**
    (A) 2.29 (B) 0.437
    (C) 1.20 (D) 3.00

17. **For calculating value of Ca, CaO is multiplied by:**
    (A) 1.40 (B) 3.00
    (C) 1.286 (D) 4.43

18. **For calculating value of Ca, $Ca(OH)_2$ is multiplied by:**
    (A) 1.40 (B) 3.00
    (C) 1.286 (D) 4.43

19. **For calculating value of N, NH_4 is multiplied by:**
    (A) 1.40 (B) 3.00
    (C) 1.286 (D) 4.43

20. **For calculating value of N, NO_3 is multiplied by:**
    (A) 1.40 (B) 3.00
    (C) 1.286 (D) 4.43

Practice Paper 2

1. **For calculating value of organic carbon, organic matter is multiplied by:**
   (A) 1.724 (B) 3.00
   (C) 1.286 (D) 4.43

2. **Since last 25 years the Indian soils are experiencing on an average net negative balance of:**
   (A) 8-10 mt/annum (B) 10-12 mt/annum
   (C) 12-14 mt/annum (D) 14-16 mt/annum

3. **Flame photometer is used for the estimation of:**
   (A) Available N (B) Available P
   (C) Available K (D) Available Ca

4. **Walkley and Black method is used for estimation of:**
   (A) N (B) P
   (C) K (D) S

5. **A value technique is used for estimation of:**
   (A) Available K in soil
   (B) Available P in soil
   (C) Available N in soil
   (D) Available Ca in soil

6. **Cunninghmella plague method is used for testing the soil value of:**
   (A) N (B) P
   (C) K (D) S
7. **Inverse nitrogen yield concept was given by:**
   (A) O.W. Wilcox (B) Mitscherlish
   (C) Spillman (D) Tisdale
8. **Mitscherlich's equation is:**
   (A) dy/dx=(A-Y)C (B) Y = K/n
   (C) Y=CX – A (D) None of these
9. **The baule units in lbs/acre for N, P_2O_5 and K_2O, respectively are:**
   (A) 218, 45, 70 (B) 223, 45, 76
   (C) 255, 50, 75 (D) 280, 55, 80
10. **Mineralization of nitrogen means:**
    (A) Conversion of organic N into inorganic N
    (B) Conversion of inorganic N to organic N
    (C) Conversion of inorganic N into N_2
    (D) None of these
11. **Conversion of proteins and organic N into amino acids is known as:**
    (A) Amonification (B) Amminization
    (C) Nitrification (D) None of these
12. **Conversion of amino acids into NH_3 and ammonium compounds are known as:**
    (A) Amonification (B) Amminization
    (C) Nitrification (D) None of these
13. **Ammonification rate will be higher at:**
    (A) 7-8 pH, 30-40 °C and 60-80 % WHC
    (B) 4-5 pH, 30-40 °C and 60-80 % WHC
    (C) 7-8 pH, 20-25 °C and >80 % WHC
    (D) None of these
14. **Maximum number of bacteria are available in:**
    (A) Horizon A
    (B) Solum
    (C) Horizon B
    (D) Horizon C
15. **Nitrification is a:**
    (A) Aerobic process
    (B) Anaerobic process
    (C) Both of these
    (D) None of these
16. **Nitrification rate is equal to:**
    (A) $\xrightarrow{\text{Quantity of } NO_3}$ Time
    (B) $\xrightarrow{\text{Quantity of } NO_2}$ Time
    (C) $\xrightarrow{\text{Quantity of } NO_2 \text{ \& } NO_3}$ Time
    (D) None of these
17. **Optimum soil pH for nitrification is:**
    (A) 5.5-6.0 (B) 6.5-8.0
    (C) 8.5-9.0 (D) >9.5
18. **Which of the following is harmful for nitrification process?**
    (A) Hyposulphite (B) Bacteria
    (C) Presence of Mo (D) None of these
19. **Conversion of nitrate into nitrite and ammonia is called:**
    (A) Dinitrification (B) Assimilation
    (C) Nitrification (D) None of these
20. **Conversion of NO_3 into N_2 and N_2O is called:**
    (A) Denitrification (B) Nitrification
    (C) Mineralization (D) None of these

Acidity and Alkalinity

Practice Paper 1

1. **What is the percentage of total exchangeable cations of CEC in normal fertile soils?**
   (A) 50-60 % (B) 60-70 %
   (C) 70-80 % (D) 80-90 %
2. **What is the percent of exchangeable H^+ in normal fertile soils?**
   (A) <10 % (B) <20 %
   (C) <30 % (D) <40 %
3. **Ca-clay has soil pH:**
   (A) 6.5-7.5 (B) 4.5-5.5
   (C) 8.0-8.5 (D) None of these

4. **In general, the range of soil pH is:**
   (A) 5-7 (B) 2-5
   (C) 4-10 (D) 7-12

5. **Active acidity of soil is due to presence of:**
   (A) H^+ (B) Al^{3+} and H^+
   (C) $Al(OH)^{+2}$ (D) All of these

6. **Residual acidity of soil is due to presence of:**
   (A) H^+ (B) Al^{3+} and H^+
   (C) $Al(OH)^{+2}$ (D) All of these

7. **Which of the acidity is the highest in soils?**
   (A) Active (B) Exchangeable
   (C) Residual (D) All of these

8. **With increasing soil pH, base saturation % is:**
   (A) Decrease
   (B) Increase
   (C) Increases upto pH 7 and then decline
   (D) Decrease upto pH 7 and then decline

9. **Gypsum contains Ca and S:**
   (A) 23 % Ca and 18 % S
   (B) 12 % Ca and 12 % S
   (C) 40 % Ca and 12 % S
   (D) 40 % Ca and 20 % S

10. **Pyrite contains S:**
    (A) 22-24 % (B) 12-18 %
    (C) 28-30 % ($) 30-32 %

11. **1 ds/m EC is equal to:**
    (A) 640 ppm salt concentration
    (B) 64 ppm salt concentration
    (C) 6.4 gm/litre
    (D) None of these

12. **Sodium percentage is equal to:**

    (A) $\xrightarrow[Ca^{++} + Mg^{++} + K^{+} + Na^{+}]{Na^{+} \times 100}$

    (B) $\xrightarrow[Ca^{++} + Mg^{++}]{Na^{+} \times 100}$

    (C) $\xrightarrow[Ca^{++} + Mg^{++}]{\sqrt{Na} + 100}$

    (D) None of these

13. **Commonly used liming factor is:**
    (A) 05-1.0 (B) 1.5-2.0
    (C) 2.5-3.0 (D) 35.-4.0

14. **Lime requirement for acid soils was worked out by:**
    (A) Arnon (1961)
    (B) Kanwar (1972)
    (C) Shoemaker *et al.* (1961)
    (D) Biswas (1965

15. **Generally acid soils have pH range of:**
    (A) <4.0 (B) 4.0-6.8
    (C) 6.5-7.5 (D) >7.5

16. **If the EC of soil is >4 ds m^{-1}, the soil is rated as:**
    (A) Normal (B) Injurious
    (C) Sensitive (D) None of these

17. **SAR is equal to:**
    (A). $Na^+ / \sqrt{Ca^{+2}} + Mg^{+2}$
    (B) $Na^+ / \sqrt{Ca^{+2}} + Mg^{+2}/2$
    (C) Na^+/CEC
    (D) $\sqrt{Ca^{+2}} + Mg^{+2}/2$

18. **How much percentage of soluble salts are present in alkali soils?**
    (A) 1-2 % (B) 0.1-0.2 %
    (C) 1.5-2.5 % (D) 2-3 %

19. **Over liming may cause:**
    (A) P availability reduced
    (B) Fe, Mn, Cu and Zn availability reduced
    (C) Adverse effect on soil particles
    (D) All of these

20. **Total value of pH and POH is equal to:**
    (A) 7 (B) 10
    (C) 14 (D) 21

Practice Paper 2

1. **pH word was first used by:**
   (A) Sorenson (B) Scofield
   (C) Brady (D) Tisdale

2. **Which of the following is more toxic to roots?**
   (A) Nacl (B) Na_2SO_4
   (C) Na_2CO_3 (D) $NaHCO_3$

3. Which of the following method is used for pH determination?

(A) Hydrogen electrode method
(B) Glass electrode method
(C) Calorimetric method
(D) All of these

4. The critical limit of boron in soil is:

(A) 0.7 ppm (B) 1.0 ppm
(C) 1.5 ppm (D) 2.0 ppm

5. The soil pH value of humid region is always:

(A) <7.0 (B) >7.0
(C) 7.0 (D) None of these

6. Mo availability is higher under:

(A) Acidic condition
(B) Neutral
(C) Alkaline
(D) Saline

7. Fe availability is higher at soil pH of:

(A) 5.5 – 7.0
(B) 7.0 – 8.5
(C) 3.5 – 5.5
(D) None of these

8. B availability is highest at soil pH of:

(A) 3.5-5.5 (B) 5.5-8.0
(C) 8.0-8.5 (D) 9.0-9.5

9. At which salt concentration in soil solution, the water absorption by plants is reduced?

(A) 0.5 % (B) 1.0 %
(C) 1.5 % (D) 2.0 %

10. The water absorption by plants is completely checked at salt concentration in soil solution:

(A) 10 % (B) 20 %
(C) 30 % (D) 40 %

11. Ion exchange is a:

(A) Static process
(B) Reversible process
(C) Only cations can be exchanged
(D) Only anions can be exchanged

12. Base saturated soils are those which have fully saturated soil colloids with:

(A) Ca^{2+} (B) Mg^{2+}
(C) H^+ (D) $Al(OH)^+_2$

13. Arid soils have higher soil pH because of:

(A) Higher cations on soil colloids
(B) Lower cations on soil colloids
(C) Higher anions on soil colloids
(D) All of these

14. If the base saturation of a soil is more the pH of soil will be:

(A) More
(B) Low
(C) Normal
(D) None of these

15. Which of the soil has higher CEC?

(A) Alluvial soil
(B) Black soil
(C) Laterite soil
(D) Red soil

16. In general, which of the following cation has higher concentration in soil?

(A) Ca^{2+} (B) Mg^{2+}
(C) H^+ (D) Na^+

17. Which of the following anion has higher concentration in soil?

(A) NO_3^- (B) Cl^-
(C) SH^- (D) $H_2PO_4^-$

18. Base saturation of soil can be calculated by which of the under given formula?

(A) $\frac{100 \times \text{total exchangeable cations}}{\text{CEC}}$

(B) $\frac{100 \times \text{total exchangeable anions}}{\text{CEC}}$

(C) $\frac{100 \times \text{CEC}}{\text{total exchangeable cation}}$

(D) All of these

19. The pH of 0.1 N NaOH solution would theoretically be:

(A) 12 (B) 13
(C) 14 (D) 11

20. Ph meter measures......... activity:

(A) H ions
(B) OH ions
(C) Both (A) and (B)
(D) None of these

Manures and fertilizers

Practice Paper 1

1. **Fertilizer schedule includes:**
   (A) Do se only
   (B) Source only
   (C) Time of application
   (D) Dose, source, time and method of application

2. **The cheaper source of sulphur is:**
   (A) Single super phosphate
   (B) Ammonium sulphate
   (C) Elemental sulphur
   (D) Gypsum

3. **Nitrogen as starter solution is applied in:**
   (A) Paddy (B) Wheat
   (C) Vegetable crops (D) Mustard

4. **FYM is applied to the field:**
   (A) One day before sowing
   (B) 7 days before sowing
   (C) 15 days before sowing
   (D) 30 days before sowing

5. **Complete complex fertilizers contains:**
   (A) Only N (B) Only P
   (C) Only K (D) N, P and K

6. **Complex fertilizers generally contain total nutrients:**
   (A) >20 % (B) <20 %
   (C) 10-15 % (D) 5-10 %

7. **P in nitrophosphate is:**
   (A) Water soluble
   (B) Citrate soluble
   (C) Water and Citrate soluble
   (D) None of these

8. **Which of the following is a nitro phosphate fertilizer?**
   (A) Suphla
   (B) Ammonium chloride
   (C) Super phosphate
   (D) Urea

9. **Which of the following is a complex fertilizer?**
   (A) Urea (B) Muriate of potash
   (C) Mono Ammonium Phosphate
   (D) None of these

10. **Guano contain high P is a:**
    (A) Fish by product (B) Beat of sea birds
    (C) Mixture of cakes (D) Inorganic fertilizer

11. **Trench method of preparing FYM was developed by:**
    (A) Dr. C.N. Acharya (B) Dr. Panjab Singh
    (C) B.N. Ghildyal (D) None of these

12. **Which of the following micronutrient is found highest in FYM:**
    (A) Cu (B) Mn
    (C) Zn (D) B

13. **Average C : N ratio of FYM is:**
    (A) 0.5 : 1 (B) 2 : 1
    (C) 10-12 : 1 (D) None of these

14. **Average N content in FYM nearly on dry weight basis:**
    (A) 0.2 % (B) 0.6 %
    (C) 0.9 % (D) 1.0 %

15. **Organic carbon content in FYM is :**
    (A) 5 % (B) 8 %
    (C) 11 % (D) 15 %

16. **Digested gobar gas slurry contains N % on dry weight basis nearly:**
    (A) 1.8-1.9 % (B) 1.51-1.55 %
    (C) 2.11-2.24 % (D) None of these

17. **Which of the complex fertilizer?**
    (A) Ammonium sulphate nitrate
    (B) Single super phosphate
    (C) Ammonium phosphate
    (D) Calcium ammonium nitrate

18. **Which is the slow release urea?**
    (A) Sulphur coated urea
    (B) Shellack coated urea
    (C) Neem coated urea
    (D) All of these

19. **Low grade fertilizers have total percentage of N, P and K:**
    (A) 14 (B) >14
    (C) <14 (D) None of these

20. **Medium grade fertilizers have total percentage of N, P and K:**
    (A) 15-25 (B) >35
    (C) <15 (D) None of these

Practice Paper 2

1. **High grade fertilizers have total percentage of N, P and K:**
   (A) 15-25 (B) >35
   (C) <15 (D) None of these

2. **Concentrated fertilizers have total percentage of N, P and K:**
   (A) 15-25 (B) >35
   (C) <15 (D) None of these

3. **Indian soils generally low in:**
   (A) Organic matter (B) Phosphorus
   (C) Potassium (D) None of these

4. **Under submerged soils, urea must be placed at:**
   (A) Oxidized zone (B) Reduced zone
   (C) Root zone (D) All of these

5. **Super granulated urea release slow N because of:**
   (A) Higher weight (B) Big size
   (C) More surface area in contact with soil
   (D) Less surface area in contact with soil

6. **Dicyandiamide and S.T. (2-Silphathiazole) are:**
   (A) Nitrification inhibitors
   (B) N slow release fertilizers
   (C) High grade N fertilizer
   (D) None of these

7. **Conditioners are mixed with mixed fertilizers because:**
   (A) To improve physical condition
   (B) Avoid clouding
   (C) To avoid formation of volatile compounds
   (D) All of these

8. **Fillers are mixed in fertilizers just to:**
   (A) Improve physical condition of fertilizers
   (B) Increase weight of fertilizers
   (C) Improve chemical composition of fertilizers
   (D) None of these

9. **Fertilizer control order was announced in year:**
   (A) 1947 (B) 1957
   (C) 1965 (D) 1970

10. **Which phosphatic fertilizer contains phosphoric acid but not soluble in water or citric acid?**
    (A) SSP
    (B) Di calcium phosphate
    (C) Rock phosphate
    (D) DAP

11. **How much urea contains nitrogen:**
    (A) 16% (B) 26%
    (C) 36% (D) 46%

12. **How much calcium ammonium nitrate contains nitrogen:**
    (A) 16% (B) 26%
    (C) 25% (D) 46%

13. **How much ammonium liquid contains nitrogen:**
    (A) 16% (B) 20-25%
    (C) 36% (D) 46%

14. **How much ammonium nitrate contains nitrogen:**
    (A) 16% (B) 26%
    (C) 33-34% (D) 46%

15. **How much Ammonium sulphate nitrate contains nitrogen:**
    (A) 16% (B) 20-21%
    (C) 36% (D) 46%

16. **How much Ammonium anhydrous contains nitrogen:**
    (A) 16% (B) 26%
    (C) 36% (D) 82%

17. **How much oxamide contains nitrogen:**
    (A) 16% (B) 26%
    (C) 31.8% (D) 46%

18. **How much ammonium phosphate contains nitrogen:**
    (A) 16-20% (B) 26%
    (C) 36% (D) 46%

19. **How much ammonium sulphate nitrate contains nitrogen:**
    (A) 16% (B) 26%
    (C) 36% (D) 46%

20. **How much ammonium phosphate contains phosphorus:**
    (A) 16% (B) 20%
    (C) 36% (D) 46%

Practice Paper 3

1. **How much rhenania phosphate contains phosphorus:**
(A) 16% (B) 25%
(C) 36% (D) 46%

2. **How much pelophos contains phosphorus:**
(A) 17% (B) 26%
(C) 36% (D) 46%

3. **How much rock phosphate (imported) contains phosphorus:**
(A) 16% (B) 26%
(C) 30-40% (D) 46%

4. **How much mussorie rock phosphate contains phosphorus:**
(A) 16% (B) 23-24%
(C) 36% (D) 46%

5. **How much Jhabua rock phosphate contains phosphorus:**
(A) 16% (B) 26%
(C) 31-38% (D) 46%

6. **How much bone meal contains phosphorus:**
(A) 16% (B) 20-25%
(C) 36% (D) 46%

7. **How much stream bone meal contains phosphorus:**
(A) 16% (B) 22%
(C) 36% (D) 46%

8. **How much muriate of potash contains potash:**
(A) 16% (B) 26%
(C) 36% (D) 60%

9. **How much sulphate of potash contains potash:**
(A) 16% (B) 26%
(C) 48-52% (D) 46%

10. **How much potassium nitrate contains potash:**
(A) 16% (B) 44%
(C) 36% (D) 46%

11. **How much potassium magnesium sulphate contains potassium:**
(A) 16% (B) 25-30%
(C) 36% (D) 46%

12. **How much bone meal contains nitrogen:**
(A) 1-2% (B) 3-4%
(C) 4-5% (D) 5-6%

13. **How much nitrogen contains in blood meal?**
(A) 10-12% (B) 4-10%
(C) 13% (D) 15.2%

14. **How much nitrogen contains in fish meal?**
(A) 10-12% (B) 4-10%
(C) 13% (D) 15.2%

15. **How much nitrogen contains in horn and hoof meal?**
(A) 10-12% (B) 4-10%
(C) 13% (D) 15.2%

16. **How much nitrogen contains in farm yard manure:**
(A) 0.5-0.6% (B) 0.50%
(C) 1.0-1.5% (D) 1.5%

17. **How much nitrogen contains in sewage and sludge:**
(A) 0.5-0.6% (B) 0.50%
(C) 1.0-1.5% (D) 1.5%

18. **How much nitrogen contains in bird beat:**
(A) 0.60% (B) 1.1-1.6%
(C) 1.50% (D) 3.0%

19. **How much nitrogen contains in vermicompost:**
(A) 0.60% (B) 1.1-1.6%
(C) 1.50% (D) 3.0%

20. **How much potassium contains in bird beat:**
(A) 0.5% (B) 0.50-1.0%
(C) 0.8% (D) 6.1-8.5%

Practice Paper 4

1. **How much potassium contains in vermicompost:**
(A) 0.60% (B) 1.1-1.6%
(C) 1.50% (D) 3.0%

2. **How much sulphur contains in gypsum (Ag grade)?**
(A) 12% (B) 13%
(C) 15% (D) 16%

3. **How much sulphur contains in phosphogypsum?**
(A) 12% (B) 13%
(C) 15% (D) 16%

4. **How much sulphur contains in pyrite (Ag grade)?**
(A) 12% (B) 13%
(C) 15% (D) 22%

5. **How much sulphur contains in zinc sulphate?**
(A) 12%% (B) 11-16%
(C) 15% (D) 13%

6. **How much calcium contains in gypsum (Ag grade)?**
(A) 16-19% (B) 13%
(C) 15% (D) 21%

7. **How much calcium contains in phopshogypsum?**
(A) 21% (B) 13%
(C) 15% (D) 16-19%

8. **How much magnesium contains in magnesium sulphate?**
(A) 12% (B) 13%
(C) 15% (D) 16%

9. **How much boron contains in borax?**
(A) 10.5% (B) 13%
(C) 15% (D) 16%

10. **How much boron contains in solabar?**
(A) 10% (B) 13%
(C) 19% (D) 26%

11. **How much copper contains in copper sulphate?**
(A) 20% (B) 22%
(C) 24% (D) 26%

12. **How much iron contains in ferrus sulphate?**
(A) 12% (B) 14%
(C) 14% (D) 16%

13. **How much mangnese contains in magnesium sulphate?**
(A) 20% (B) 30.5%
(C) 35% (D) 40%

14. **How much zinc contains in zinc sulphate?**
(A) 21 or 33% (B) 22%
(C) 24% (D) 26%

15. **How much molybdenum contains in ammonium molybdate?**
(A) 39% (B) 45%
(C) 54% (D) 60%

16. **Average values (%) of N, P_2O_5 and K_2O in sludge are:**
(A) 3.5, 2.5 and 0.5 (B) 0.5, 0.25 and 0.75
(C) 2.5, 0.5 and 2.5 (D) None of these

17. **NPK removal ratio of crops is:**
(A) 4 : 1 : 4 (B) 1 : 1 : 5
(C) 5 : 1 : 5 (D) 5 : 5 : 1

18. **During 1998-99, consumption of N : P_2O_5 : K_2O ratio was:**
(A) 1.33: 4.10 : 11.32
(B) 4.10: 11.32: 1.33
(C) 11.32 : 4.10 : 11.33
(D) 11.32 : 4.10 : 1.33

19. **The equivalent acidity of $(NH_4)_2SO_4$ fertilizer is:**
(A) 110 (B) 120
(C) 140 (D) 160

20. **One Baule unit of P (lb P2O5/acre) corresponds to:**
(A) 35 (B) 45
(C) 55 (D) 65

Biofertilizers

Practice Paper 1

1. **Biofertilizers are:**
(A) Organic manure
(B) Green manure
(C) Culture of microorganism
(D) None of these

2. **Denitrifying bacteria are:**
(A) Aerobic (B) Anaerobic
(C) Both of these (D) None of these

3. **Denitrifying bacteria belongs to genera:**
(A) Pseudomonas (B) Acromobactor
(C) Bacillus (D) All of these

4. **Which of the following genera of bacteria is found mainly in soil?**
   (A) Pseudomonas and Achromobacter
   (B) Thiobacillus
   (C) Vibrio
   (D) All of these
5. **Denitrifying bacteria uses NO_3 as an electron acceptor in absence of:**
   (A) CO_2 (B) O_2
   (C) H_2O (D) None of these
6. **Optimum temperature for denitrification is:**
   (A) $<25^0C$ (B) $>25^0C$
   (C) 25-30 ^0C (D) $>30^0C$
7. **The name of rhizobia for lucerne group is:**
   (A) *Rhizobium meliloti*
   (B) *Rhizobium japonicum*
   (C) *Rhizobium trifoli*
   (D) *Rhizobium phaseoli*
8. **Symbiotic N fixation governed by the group of bacteria is:**
   (A) Azotobacter
   (B) Rhizobia
   (C) Azospirillum
   (D) All of these
9. **Nitrogen fixation is a:**
   (A) Endothermic process
   (B) Exothermic process
   (C) Hypothermic process
   (D) None of these
10. **The main enzyme of non symbiotic N fixation is:**
    (A) Nitrogenase (B) Hydrogenase
    (C) Isomerase (D) All of these
11. **Azotobacter are active at soil pH of:**
    (A) >6.5-7.0 (B) 7.0-7.5
    (C) 7.5-8.0 (D) 5.5-6.0
12. **Average quantum of N fixed by non-symbiotic process is:**
    (A) 10-15 kg/ha (B) 25-30 kg/ha
    (C) 50-60 kg/ha (D) None of these
13. **Bio fertilizer supplied nitrogen to the extent of:**
    (A) 10 kg/ha (B) 20 kg/ha
    (C) 40 kg/ha (D) 60 kg/ha
14. **Cross inoculation group of clover is:**
    (A) *Rhizobium meliloti*
    (B) *Rhizobium trifoli*
    (C) *Rhizobium lupini*
    (D) *Rhizobium phaseoli*
15. **Cross inoculation group of alfalfa is:**
    (A) *Rhizobium lupini*
    (B) *Rhizobium trifoli*
    (C) *Rhizobium meliloti*
    (D) *Rhizobium phaseoli*
16. **The species of rhizobim present in the root nodules of soybean:**
    (A) *Rhizobium japonicum*
    (B) *R. leguminosorum*
    (C) *R. trifoli*
    (D) *R. phaseoli*
17. **Blue green algae fix nitrogen mainly in:**
    (A) Cereals field
    (B) Rice field
    (C) Sugarcane field
    (D) Legumes field
18. **S reduction is mainly caused by:**
    (A) Desulphovibrio (B) Azotobacter
    (C) Azospirillum (D) None of these
19. **Which of the following statement is true?**
    (A) Nitrosomonas produces nitrite and nitrobacter produces nitrate
    (B) Conversion rate of nirite to nitrate is faster than ammonium to nitrite
    (C) Nitrosomonas oxidizes 35-40 mole NH_4, while nitrobacter 70-100 mole nitrite
    (D) All of these
20. **Mo is required for the activity of:**
    (A) Azotobacter (B) Legume bacteria
    (C) Vibrio (D) None of these

Practice Paper 2

1. **B is required mainly by:**
   (A) Azotobacter (B) Azospirillum
   (C) Legume bacteria (D) None of these
2. **Facultative free living 'N' fixing bacterium:**
   (A) Rhizobium (B) Klebsiella
   (C) Azotobacter (D) None of these

3. **Cyst formation is characteristics feature of:**
(A) *Acetobacter* (B) *Arthrobacter*
(C) *Azotobacter* (D) None of these

4. **Phosphate absorber bio fertilizer is:**
(A) VAM (B) BGA
(C) FYM (D) Azolla

5. **Which of the following is the phosphate mobilizing bio-fertilizers?**
(A) *Bacillus polymyza*
(B) *Pseudomonas strata*
(C) *Aspergillus niger*
(D) All of these

6. **Bacteria having especially high rate of respiration:**
(A) *Rhizobium* (B) *Azotobacter*
(C) *E. coli* (D) *Acetobacter*

7. **"Ray Fungi" is the name given to:**
(A) Rhizopus (B) Yeast
(C) Actinomycetes (D) Mycoplasma

8. **Actinomycetes are:**
(A) Gram +ve, aerobic
(B) Gram +ve, anaerobic
(C) Gram –ve, aerobic
(D) Gram –ve, anaerobic

9. **Example for microaerophillic N_2 fixer:**
(A) *Rhizobium* (B) *Azotobacter*
(C) *E. coli* (D) *Frankia*

10. **Phosphorus solublizing microorganism:**
(A) VAM (B) Mucor
(C) Rhizopus (D) *E. coli*

11. **Low pH of media inhibits the growth of:**
(A) Bacteria (B) Moulds
(C) Both of the above (D) None of these

12. **Which of the following genera is most likely to contain organisms capable of surviving high temperature?**
(A) Clostridium (B) Penicillium
(C) Torula (D) Pseudomonas

13. **Which of the following is/are Bio-insecticides?**
(A) *Bacillus thuringiensis*
(B) *Bacillus popilliae*
(C) Both (A) and (B)
(D) None of these

14. **Funaric acid is produced by:**
(A) *Aspergillus terrus*
(B) *Rhizopus nigricans*
(C) *Aspergillus niger*
(D) *Asergillus fumigatus*

15. **ATP needed for N_2 fixation:**
(A) 12 (B) 16
(C) 28 (D) 32

16. **Factor which does not affect legume root nodulation?**
(A) Temperature
(B) Combined N_2
(C) Light
(D) H^+ ion conc.

17. **Microaerophillic N_2 fixing bacteria:**
(A) *Azotobacter* (B) *Azospirillium*
(C) *Clostridium* (D) *Enterobacter*

18. **Bacteria which can ferment sugar faster than yeast is:**
(A) *Clostridium* (B) *Zygomonas*
(C) *Thiobacillus* (D) *Bacillus*

19. **In leghaemoglobin the heme portion is specified by:**
(A) Plantgenes (B) Bacterial genes
(C) Both of the above (D) None of these

20. **Leghaemoglobin is structurally and functionally similar to:**
(A) +Haemoglobin (B) Actin
(C) Myosin (D) Myogolobin

Practice Paper 3

1. **Anaerobic free living 'N' fixing bacterium:**
(A) Azotobacter (B) Acetobacter
(C) Kelbseilla (D) Clostridium

2. **Microorganism which are associated with non legumes:**
(A) Anabaena (B) Klebseilla
(C) Both of them (D) None of these

3. **Bacteria are commonly seen in soils such as:**
(A) Acidic (B) Neutral
(C) Alkaline (D) Saline

4. **Nitrat reductase enzyme present in:**
   (A) Cytoplasm (B) Chloroplast
   (C) Mitochondria (D) Nucleus
5. **Microbes involved in fibre retting:**
   (A) *Micrococcus* sp.
   (B) *Bacillus subtilis*
   (C) *Clostridium butrycium*
   (D) *Lactobacillus sp.*
6. **Microbe involved in tobacco curing:**
   (A) *Micrococcus* sp. (B) *Bacillus subtilis*
   (C) *Lactobacillus* sp. (D) *Aspergillus niger*
7. **Urea degrading bacteria:**
   (A) *Bacillus pasturi*
   (B) *Clostridium pasteurianum*
   (C) *A. niger*
   (D) *Micrococcus* sp.
8. **'P' solubilizing bacteria:**
   (A) *Bacillus megatherium*
   (B) *Clostridium pasteurianum*
   (C) *A. niger*
   (D) *Micrococcus* sp.
9. **The size of microbe as:**
   (A) Viriod < Virus < Bacteria < Fungi
   (B) Virus < Viriod < Bacteria < Fungi
   (C) Viriod < Virus < Fungi < Bacteria
   (D) Virus < Viriod < Fungi < Bacteria
10. **Type of relationship between *Acetobacter diazotrophicus* and sugarcane is:**
    (A) Symbiotic (B) Associative
    (C) Endophytic (D) Free living
11. **Example of VAM fungi:**
    (A) Pisoletecis (B) Sclerotenia
    (C) Glomus (D) Trichoderma
12. **How much nitrogen is fixed in the soil by *Crotolaria juncea*?**
    (A) 75-85 kg/ha (B) 70-78 kg/ha
    (C) 55-65 kg/ha (D) 45-52 kg/ha
13. **How much nitrogen is fixed in the soil by *Sesbania aculeata*?**
    (A) 75-85 kg/ha (B) 70-78 kg/ha
    (C) 55-65 kg/ha (D) 45-52 kg/ha
14. **How much nitrogen is fixed in the soil by *Phaseolus tribolus*?**
    (A) 75-85 kg/ha (B) 70-78 kg/ha
    (C) 55-65 kg/ha (D) 45-52 kg/ha
15. **How much nitrogen is fixed in the soil by *Vigna radiata*?**
    (A) 75-85 kg/ha (B) 70-78 kg/ha
    (C) 55-65 kg/ha (D) 45-52 kg/ha
16. **How much nitrogen is fixed in the soil by *Vigna mungo*?**
    (A) 42-53 kg/ha (B) 55-58 kg/ha
    (C) 60-67 kg/ha (D) 120-125 kg/ha
17. **How much nitrogen is fixed in the soil by *Vigna sinensis*?**
    (A) 42-53 kg/ha (B) 55-58 kg/ha
    (C) 60-67 kg/ha (D) 120-125 kg/ha
18. **How much nitrogen is fixed in the soil by *Cyamopsis tetragonoloba*?**
    (A) 42-53 kg/ha (B) 55-58 kg/ha
    (C) 60-67 kg/ha (D) 120-125 kg/ha
19. **How much nitrogen is fixed in the soil by *Melilotus alba*?**
    (A) 42-53 kg/ha (B) 55-58 kg/ha
    (C) 60-67 kg/ha (D) 120-125 kg/ha
20. **How much nitrogen is fixed in the soil by *Trifolium alexandrinum*?**
    (A) 42-53 kg/ha (B) 58-64 kg/ha
    (C) 62-75 kg/ha (D) 120-125 kg/ha

Practice Paper 4

1. **How much nitrogen is fixed in the soil by *Lathyrus sativus*?**
   (A) 42-53 kg/ha
   (B) 58-64 kg/ha
   (C) 62-75 kg/ha
   (D) 120-125 kg/ha
2. **Which rhizobium species is suitable for pea and lentil?**
   (A) *Rhizobium leguminosorum*
   (B) *Rhizobium phaseoli*
   (C) *Rhizobium meliloti*
   (D) *Rhizobium japonicum*
3. **Which rhizobium species is suitable for arhar, mung, urd and lobia?**
   (A) *Rhizobium leguminosorum*
   (B) *Rhizobium phaseoli*
   (C) *Rhizobium meliloti*
   (D) *Rhizobium japonicum*

4. **Which rhizobium species is suitable for lucern?**
   (A) *Rhizobium leguminosorum*
   (B) *Rhizobium phaseoli*
   (C) *Rhizobium meliloti*
   (D) *Rhizobium japonicum*

5. **Which rhizobium species is suitable for soyabean?**
   (A) *Rhizobium leguminosorum*
   (B) *Rhizobium phaseoli*
   (C) *Rhizobium meliloti*
   (D) *Rhizobium japonicum*

6. **Which rhizobium species is suitable for lupin?**
   (A) *Rhizobium lupine*
   (B) *Rhizobium trifoli*
   (C) *Rhizobium meliloti*
   (D) *Rhizobium japonicum*

7. **Which rhizobium species is suitable for berseem?**
   (A) *Rhizobium lupine*
   (B) *Rhizobium trifoli*
   (C) *Rhizobium meliloti*
   (D) *Rhizobium japonicum*

8. **Phosphate Solubilizing Bacteria (PSB) improves soil fertility by:**
   (A) Fixing atmospheric nitrogen
   (B) Fixing atmospheric phosphorus
   (C) Releasing fixed phosphorus in the soil
   (D) All of these

9. **Seed inoculation in pulses is done with:**
   (A) Azotobacter (B) Rhizobium
   (C) Azospirillum (D) Phosphotica

10. **What culture should be given priority in groundnut cultivation?**
   (A) Myccorhiza (B) Rhizobium
   (C) Phosphobacteria (D) Azospirillum

11. **Azotobacter is useful in fixing of nitrogen form air in:**
   (A) Non-leguminnous crops like rice and wheat
   (B) Leguminous crops
   (C) Soybean
   (B) None of these

12. **Mycorrhizae refers to:**
   (A) Root environment
   (B) Soil-root interaction
   (C) Fungus – root association
   (D) Mycology

13. **Rhizobium is mainly associated with:**
   (A) Pulses (B) Spices
   (C) Cereals (D) Oilseeds

14. **Azolla is nitrogen fixing:**
   (A) Bacteria (B) Fern
   (C) Fungi (D) Algae

15. **Iron containing enzyme responsible for the biological nitrogen fixation is:**
   (A) Nitrate reductase
   (B) Nitrite reductase
   (C) Nitrogenase
   (D) Peroxidase

16. **Decomposition of organic matter in submerged soil is carried out by:**
   (A) Bacteria
   (B) Actinomycetes
   (C) Fungi
   (D) Earthworm

17. **Decomposition of organic matter in submerged soil is carried out by:**
   (A) Bacteria
   (B) Actinomycetes
   (C) Fungi
   (D) Earthworm

18. **Bactria responsible for converting nitrite into nitrate is called:**
   (A) Pseudomonas
   (B) Nitrobacter
   (C) Heterosomonas
   (D) Rhizobium

19. **Azotobacter fixes atmospheric nitrogen:**
   (A) Symbiotically
   (B) Non-symbiotically
   (C) Both symbiotically and non-symbiotically
   (D) None of these

20. **Which organism is responsible for phosphorus availability to plants?**
   (A) Azotobactor
   (B) Nitrosomonas
   (C) Mycorrhiza
   (D) Earthworms

Practice Paper 5

1. **Ammonia is converted into nitrite with the help of:**
   (A) Nitrosomonas (B) Nitrobacter
   (C) Desulfovibrio (D) Bacillus
2. **Which of the following bacteria helps in conversion of nitrite to nitrate:**
   (A) Nitrosomonas (B) Nitrobacter
   (C) Desulfovibrio (D) Bacillus
3. **Optimum soil pH for nitrogen fixing bacteria is:**
   (A) 4.5-5.5 (B) 5.5-6.5
   (C) 6.5-8.0 (D) >8.0
4. **Mycorrhiza is generally found in:**
   (A) N and P deficient soils
   (B) P and K deficient soils
   (C) Ca deficient soils
   (D) All of these
5. **Highest activity of actinomycetes is found at soil pH:**
   (A) 4.5-5.5 (B) 5.5-6.5
   (C) 6.5-8.0 (D) >8.0
6. **Which of the following is not a fungi?**
   (A) Nocardia (B) Mucor
   (C) Rhizopus (D) Yeast
7. **Which of the following is not a actinomycetes?**
   (A) Nocardia (B) Mucor
   (C) Micromonospora (D) Streptomycetes
8. **Among soil organisms, earthworms account nearly:**
   (A) 50-75 % weight of total soil organisms
   (B) 75-85 % weight of total soil micro organisms
   (C) > 85 % weight of total soil micro organisms
   (D) None of these
9. **How much dry soil is produced by earthworms annually per acre?**
   (A) 10 tonnes (B) 15 tonnes
   (C) 20 tonnes (D) None of these
10. **What will be the effect of earthworms on C/N ratio of soil?**
    (A) Increase (B) Decrease
    (C) Same (D) All of these
11. **Earthworms increase the highest availability of:**
    (A) K (B) Mg
    (C) N (D) P
12. **Number of earthworms may be present per m^2 area of arable land is:**
    (A) 30-300 (B) 500-800
    (C) 1000-12000 (D) None of these
13. **At which soil pH, the activity of earthworms is highest?**
    (A) < 4.5 (B) 4.5-5.5
    (C) 5.5-6.6 (D) 7.0-8.5
14. **Which of the following reduces the population of earthworms:**
    (A) Drought (B) Frost
    (C) Cold (D) All of these
15. **Pasture lands and non-ploughed soils have earthworms population in comparison to ploughed soils:**
    (A) 6.5-11 % higher
    (B) 12-15 % higher
    (C) 16-18 % higher
    (D) > 20 % higher
16. **Which of the following is not a soil microorganisms?**
    (A) Protozoans (B) Nematodes
    (C) Rotifers (D) Sludge and snails
17. **Mycorrhiza is a**
    (A) Association of fungi with roots of higher plants
    (B) Association of fungi and bacteria
    (C) Association of bacteria and higher plants
    (D) None of these
18. **Rotifers are:**
    (A) Saprophytes (B) Parasite
    (C) Predators (D) None of these
19. **Higher population of rotifers are available in:**
    (A) Dry region (B) Temperate region
    (C) Humid region (D) None of these
20. **The .... oxidation of ammonia has been termed as nitrification:**
    (A) Physical (B) Chemical
    (C) Biological (D) Physico-chemical

Soils and its properties

Practice Paper 1

1. **Component of soils are:**
(A) Solid phase (B) Liquid phase
(C) Gasseous phase (D) All of these

2. **The organic matter of soil has the value:**
(A) 25% (B) 50%
(C) 40% (D) 10%

3. **The size of clay particles is:**
(A) Less than 0.002 mm
(B) More than 0.002 mm
(C) 0.2 mm to 2 mm
(D) None of these

4. **Which of the following has smallest of soil particles:**
(A) Loam soil (B) Sand
(C) Clay (D) Silt

5. **Major constituent of soil is:**
(A) Organic matter (B) Inorganic matter
(C) Soil air (D) Soil water

6. **Formation of organic matter takes place by:**
(A) Chemical weathering
(B) Soil micro-organisms
(C) Herbivores
(D) Soil water

7. **Formation of soil takes place by:**
(A) Weathering (B) Pedogenesis
(C) Melting (D) Both (A) and (B)

8. **White colour of soil is due to the process of?**
(A) Silica (B) Iron
(C) Manganese oxide (D) All of these

9. **According to modern classification, soils have been divided into:**
(A) Order-Sub order-Great soil group-Sub group
(B) Order-soil group-family-series
(C) Order-family-series-types
(D) All of these

10. **Soil orders have been divided into:**
(A) Zonal – Interzonal – Azonal soils
(B) Zonal – Azonal – Series
(C) Zonal – Interzonal – Series – Azonal
(D) None of these

11. **Which of the following soil type is not a interzonal soil?**
(A) Hydromorphic soil
(B) Halomorphic
(C) Calcimorphic soil
(D) Alluvial soil

12. **Calcimorphic soils have:**
(A) Higher lime content
(B) Higher Na content
(C) Low lime content
(D) None of these

13. **Azonal soils have three following great soil groups:**
(A) Lithosol – Regosol – Alluvial
(B) Podzol – Planosol – Chernozems
(C) Latosol – Podzol – Lithosol
(D) None of these

14. **Which is correct for Regosols?**
(A) Soil profiles are not well developed
(B) Soils are made of marls, sand dunes and ash
(C) Parent materials affect the characters of soil
(A) All of these

15. **Chernozems are like:**
(A) Black cotton soils (B) Red soils
(C) Brown soils (D) Alluvial soils

16. **Soil classification 7th approximation was developed in year:**
(A) 1960 in America (B) 1950 in America
(C) 1960 in UK (D) 1960 in Russia

17. **Black cotton soils of India are poor in:**
(A) Nitrogen (B) Phosphorus
(C) Both (A) and (B) (D) None of these

18. **Hematite is changed into limonite by the process of:**
(A) Hydrolysis (B) Hydration
(C) Reduction (D) Carbonation

19. **The colour of hematite is:**
(A) Blue (B) Red
(C) Black (D) Yellow

20. **The % pore space can be calculated by which of the under given formula:**

(A) $\frac{\text{Bulk density}}{\text{Solid space}}$ (B) $\frac{\text{100–Bulk density}}{\text{Particle density}}$

(C) 110-Bulk density (D) All of these

Practice Paper 2

1. **The porosity of soil will decrease with increase of:**
(A) Soil depth
(B) Soil organic matter
(C) Soil texture
(D) All of these

2. **Which of the under given soil texture has highest porosity?**
(A) Sandy soils (B) Loam soils
(C) Loamy clay soils (D) Clay loam soils

3. **Which of the following soil structure has the highest porosity?**
(A) Columnar (B) Compact
(C) Oblique (D) Granular

4. **Red soil of India occupy an area of nearly:**
(A) 2000 km^2 (B) 3000 km^2
(C) 4000 km^2 (D) 5000 km^2

5. **Red soils of India are:**
(A) Calcareous
(B) Rich in soluble salts
(C) Both (A) and (B)
(D) None of these

6. **The colour of the red soils is due to:**
(A) Wide diffusion of iron
(B) High content of iron oxide
(C) Both of these
(D) None of these

7 **The laterites of India occupy nearly an area of:**
(A) 50000 km^2 (B) 12500 km^2
(C) 250000 km^2 (D) 500000 km^2

8. **Laterite soils have developed as:**
(A) in situ formation
(B) Sedimentary formations
(C) Both (A) and (B)
(D) None of these

9. **Laterite soils are characterized by:**
(A) Compact, honey combed/vesicular sub-soil mass
(B) Pallid zone/Plinthite above the water table
(C) Both (A) and (B)
(D) None of these

10. **Laterites have a pH of:**
(A) 4.5 to 6.0 (B) 6.0 to 7.0
(C) 7.0 to 8.0 (D) > 8.0

11. **Laterite soils in India are commonly found in:**
(A) Uttar Pradesh (B) Assam
(C) Both of these (D) None of these

12. **The north-east alluvium of Uttar Pradesh:**
(A) Light textured (B) Heavy textured
(C) High calcareous (D) Non calcareous

13. **In alluvial soils of India, deficiency of the following nutrient is most wide spread:**
(A) Iron (B) Zinc
(C) Copper (D) Manganese

14. **Black cotton soils of India have originated from the rocks:**
(A) Sand stones (B) Shales
(C) Lime stones (D) Deccan trap

15. **Laterite soils is rich in:**
(A) Calcium (B) Magnesium
(C) Iron (D) Nitrogen

16. **Arable land constitutes:**
(A) 36.6% (B) 30%
(C) 22% (D) 11%

17. **Soil forms the surface layer in percentage of land:**
(A) 80% (B) 60%
(C) 50% (D) 30%

18. **Soil transported by air is:**
(A) Alluvial (B) Colluvial
(C) Glacial (D) Eolian

19. **River water deposits:**
(A) Loamy soil (B) Alluvial soil
(C) Laterite soil (D) Sandy soil

20. **Alluvial soils are found in:**
(A) Northern India
(B) Southern India
(C) Gangetic Jamuna plains
(D) Eastern India

Practice Paper 3

1. **Ao layer of the soil is made of:**
   (A) Decaying litter
   (B) Rocky matter
   (C) Soil mixed with organic matter
   (D) Soil rich in inorganic matter
2. **Development of soil from rocky matter is:**
   (A) Pedology
   (B) Pedogenesis
   (C) Edaphic factors
   (D) Edaphic climax
3. **Particle density is equal to:**
   (A) $\frac{\text{Mass of soil solids}}{\text{Volume of soil solids}}$
   (B) $\frac{\text{Mass of soil solids}}{\text{Total soil volume}} \times 100$
   (C) $\frac{\text{Mass of soil solids}}{\text{Volume of solid water}}$
   (D) $\frac{\text{Mass of soil solids}}{\text{Volume of soil air}} \times 100$
4. **Effective saturation is equal to:**
   (A) $\frac{\text{Volume of soil water}}{\text{Volume of soil air}}$
   (B) $\frac{\text{Mass of soil water}}{\text{Volume of soil water}} \times 100$
   (C) $\frac{\text{Volume of soil}}{\text{Volume of soil water}}$
   (D) None of these
5. **Void ratio is equal to:**
   (A) $\frac{\text{Volume of soil pores}}{\text{Volume of soil air}}$
   (B) $\frac{\text{Volume of soil}}{\text{Volume of soil air}} \times 100$
   (C) $\frac{\text{Volume of soil pores}}{\text{Volume of soil solids}}$
   (D) None of these
6. **Epipedons are:**
   (A) Diagnostic surface horizons
   (B) Diagnostic subsurface horizons
   (C) Bed rocks
   (D) None of these
7. **Number of soil epipedons are:**
   (A) 6 (B) 10
   (C) 16 (D) 21
8. **As per new soil classification - 7th approximation, soil has been classified into following 6 categories**
   (A) Order – Suborder - Great soil group – Family - Series
   (B) Order – Soil group – Family – Series – Type - Phase
   (C) Order –Soil group – Family – Type – Series - Phase
   (D) None of these
9. **Number of soil orders are:**
   (A) 10 (B) 11
   (C) 12 (D) 13
10. **Which of the following soil order covers the highest area in the world?**
    (A) Vertisol (B) Entisol
    (C) Aridisol (D) Alfisol
11. **Usar was derived from 'Ustra word. Ustra is a word of:**
    (A) Latin (B) Greek
    (C) Sanskrit (D) Bangla
12. **Which of the following state has highest area under saline and alkaline soils:**
    (A) UP (B) Punjab
    (C) Gujarat (D) Rajasthan
13. **First stage in the evaluation of saline soils is:**
    (A) Solonetz (B) Solonchek
    (C) Solodi (D) None of these
14. **Number of soil orders recognized in the taxonomy are:**
    (A) 10 (B) 11
    (C) 12 (D) 13
15. **Organic matter content of most Indian soils is:**
    (A) Less than 1.0%
    (B) Between 1.0 to 5.0%
    (C) Between 6.0 to 10.0%
    (D) More than 10.0 %

16. **Least porous soil is:**
(A) Sandy soil (B) Clay soil
(C) Loam soil (D) Gravelly soil

17. **Deep black soil is productive due to high proportion of:**
(A) Sand and Zinc
(B) Gravel and Calcium
(C) Clay and Humus
(D) Silt and Earthworm

18. **Type of soil in Gangetic basin is:**
(A) Eolian (B) Colluvial
(C) Residual (D) Alluvial

19. **The soil without a well developed soil profile characteristics are known as azonal soil and include:**
(A) Planosols (B) Latosols
(C) Lithosols (D) None of these

20. **Which of the following word is used for saline and alkaline soils?**
(A) Rakar (B) Usar
(C) Kaller (D) All of these

Practice Paper 4

1. **Saline soil is also known as:**
(A) White alkali soil (B) Black alkali soil
(C) Solonetz soil (D) All of these

2. **Alkali soil is also known as:**
(A) White alkali soil (B) Black alkali soil
(C) Usar soil (D) All of these

3. **Which of the following is the characteristics of saline soils?**
(A) EC > 4 ds/m, ESP < 15, pH < 8.5
(B) EC > 4 ds/m, ESP > 15, pH < 8.5
(C) EC < 4 ds/m, ESP > 15, pH > 8.5
(D) None of these

4. **Which of the following is the characteristics of alkali soils?**
(A) EC > 4 ds/m, ESP < 15, pH < 8.5
(B) EC > 4 ds/m, ESP > 15, pH < 8.5
(C) EC > 4 ds/m, ESP > 15, pH > 8.5
(D) None of these

5. **Which of the following is the characteristics of saline and alkali soils?**
(A) EC > 4 ds/m, ESP < 15, pH < 8.5
(B) EC > 4 ds/m, ESP > 15, pH < 8.5
(C) EC < 4 ds/m, ESP < 15, pH > 8.5
(D) None of these

6. **Which of the anion is not found in saline soils?**
(A) CO_3^- (B) Cl^-
(C) SO_4^- (D) HCO_3^-

7. **Saline and alkaline soils are mainly found in:**
(A) Arid and semiarid region
(B) Humid region
(C) Temperate region
(D) All of these

8. **Which of the following chemical is used for reclamation of saline and alkaline soils?**
(A) Zypsum (B) Phosphozypsum
(C) El.Sulphur (D) All of these

9. **Saline and alkaline soils were classified by:**
(A) D. Sigmoid (B) Arnon
(C) Wilcox (D) None of these

10. **Which of the following process takes place in the development of saline and alkaline soils?**
(A) Salinazation (B) Alkalization
(C) Salodization (D) All of these

11. **Which of the following type of soil has the largest area?**
(A) Alluvial soil
(B) Black cotton soil
(C) Red soil
(D) Laterite and lateritic soil

12. **Old alluvial plains are known as:**
(A) Bangar (B) Khadar
(C) Regur (D) None of these

13. **New alluvial plains are known as:**
(A) Bangar (B) Khadar
(C) Regur (D) None of these

14. **Alluvial soils are generally deficient in:**
(A) N (B) Humus
(C) P (D) All of these

15. **Black cotton soils are black due to presence of:**
(A) Titaniferrous magnetite
(B) $MgCO_3$ (C) $CaCO_3$
(D) All of these

16. Karail is a:

(A) Type of black soils
(B) Black soils of lower lands of both banks of Ganga in UP
(C) These soils have high lime and Mg
(D) All of these

17. Which of the following is correct for black cotton soils?

(A) Excess in lime and K_2O
(B) Deficient in N, P_2O_5 and organic matter
(C) Excess in Fe, $CaCO_3$ and $MgCO_3$
(D) All of these

18. Which of the following is correct for red soils?

(A) Made of gneiss, schist and granite rocks
(B) Deficient in N, humus, P and lime
(D) Excess of Fe and Al oxides
(D) All of these

19. Which of the following is correct for laterite soils?

(A) Lower layers are made of Al and Fe oxides
(B) Deficient in P, K and lime
(C) Generally rich in organic matter
(D) All of these

20. Rakar is a:

(A) Red soil (B) Black soil
(C) Alluvial soil (D) None of these

Practice Paper 5

1. Which of the following is correct for Tarai soils?

(A) Deficient in Zn
(B) High organic matter
(C) High soil fertility
(D) All of these

2. Arid and desert soils are characterized by:

(A) High evaporation
(B) Low organic matter
(C) High $CaCO_3$
(D) All of these

3. Efflorescent deposits are found in:

(A) Saline and alkaline soils
(B) Alluvial soils
(C) Black soils
(D) None of these

4. Which of the following is correct for peat and marshy soils?

(A) Low pH
(B) High organic matter
(C) Blue colour due to Fe^{++}
(D) All of these

5. Dhakar soils are characterized as:

(A) Good for rice soils
(B) Found in eastern UP
(C) Light acidic soils
(D) All of these

6. Saline and alkali soils are characterized as:

(A) Excess of Na_2CO_3 and NaCl
(B) Absence of $CaSO_4$
(C) Hard rock at subsoil
(D) All of these

7. In general, Indian soils have been classified into:

(A) 4 types (B) 6 type
(C) 8 types (D) None of these

8. Which of the following order of Indian soil has highest area:

(A) Alfisols (B) Entisols
(C) Inceptisols (D) Vertisols

9. Who classified Indian soils into 25 great soil groups?

(A) Govindarajan (1965) and Dutta Vishwash (1968)
(B) FAO (1965)
(C) ICAR (1937)
(D) Brady (1972)

10. Black soils are classified in order:

(A) Vertisols (B) Alfisols
(C) Spodosol (D) Histosols

11. Red soils are classified in order:

(A) Vertisols (B) Alfisols
(C) Spodosol (D) Histosols

12. Which of the following epipedon is present in mollisols?

(A) Mollic (B) Umbric
(C) Ochric (D) Histic

13. Which of the following is correct for histosols?

(A) They are bog and half bog soils
(B) They are developed under waterlogged conditions
(C) They contain 20-30% organic carbon
(D) All of these

14. Soil monoliths are:

(A) Permanent soil profile records
(B) Temporary soil profile records
(C) Types of soil classification
(D) None of these

15. Soil is divided on the basis of land capability into:

(A) 4 classes (B) 6 classes
(C) 8 classes (D) 10 classes

16. Land capability classification is based on:

(A) Utility of land
(B) Soil erosion
(C) Techniques to protect soil
(D) All of these

17. Which of the class is unfit for agricultural uses:

(A) Class I (B) Class IV
(C) Class VIII (D) Class VI

18. The categories of soil classification and its number are given in 'A' and 'B' list below:

A	B
1. Orders	i. 54
2. Suborders	ii. 11
3. Greatgroups	iii. 254
4. Subgroups	iv. 1200

Which of the pair is matched correctly:

(A) 1. i 2. ii 3. iii 4. iv
(B) 1. iii 2. ii 3. iv 4. i
(C) 1. ii 2. i 3. iii 4. iv
(D) 1. iv 2. ii 3. i 4. iii

19. List 'A' and 'B' represent the land capability classes and colour on soil maps, respectively below:

A	B
1. Class I	i. Light green
2. Class II	ii. Yellow
3. Class III	iii. Red
4. Class IV	iv. Blue

Which of the pair is matched correctly:

(A) 1. i 2. ii 3. iii 4. iv
(B) 1. iii 2. ii 3. iv 4. i
(C) 1. ii 2. i 3. iii 4. iv
(D) 1. iv 2. ii 3. i 4. iii

20. Peat soils may contains organic matter:

(A) 20-95% (B) 0.1-5 %
(C) 5-10 % (D) > 10 %

Soil water

Practice Paper 1

1. Amount of water available to the plants in the soil is called as:

(A) Holard (E+C) (B) Chresard
(C) Echard (D) None of these

2. Soil water potential (w) can be expressed as:

(A) $\psi s = \psi g + \psi m + \psi o$
(B) $\psi s = \psi g - \psi m - \psi o$
(C) $\psi s = \psi g + \psi m - \psi o$
(D) $\psi s = \psi g + \psi m$

3. Which of the soil has the highest infiltration rate?

(A) Clay soil
(B) Loam soil
(C) Silt loam
(D) Clay loam

4. Very slow infiltration rate will be called if the infiltration rate is:

(A) < 0.1 inch/hr
(B) 0.1-0.5 inch/hr
(C) 0.5-1.0 inch/hr
(D) > 1.0 inch/hr

5. **The water is absorbed by soil due to the force of:**
   (A) Suction force (B) Cohesion force
   (C) Adhesion force (D) None of these
6. **The water particles are attracted to other water particle due to force of:**
   (A) Adhesion force (B) Cohesion force
   (C) Suction force (D) All of these
7. **Soil water available to root is:**
   (A) Surface water
   (B) Hygroscopic water
   (C) Gravitational water
   (D) Capillary water
8. **Water held tightly by soil particles as thin film over their surface is:**
   (A) Hygroscopic water
   (B) Capillary water
   (C) Chemical water
   (D) Gravitational water
9. **Which governs the solubility and availability of nutrients in the soil:**
   (A) Temperature (B) pH
   (C) Aeration (D) Porosity
10. **Water-logging commonly occurs in:**
    (A) Sandy soil (B) Gravelly soil
    (C) Clayey soil (D) Loam soil
11. **pF is the unit of:**
    (A) Force (D) Distance
    (C) Velocity (D) Weight
12. **Hygroscopic coefficient is equal to:**
    (A) 1/3 bar (B) 5 bar
    (C) 31 bar (D) None of these
13. **Soil metrix is also known as:**
    (A) Soil solids (B) Soil water
    (C) Soil air (D) Soil pores
14. **Soil showing permanently wilted plants has:**
    (A) Gravitational water
    (B) Capillary water
    (C) Hygroscopic water
    (D) Ground water
15. **Water is a resource:**
    (A) Nondegradable nonmaintainable
    (B) Degradable maintainable
    (C) Renewable
    (D) Non renewable
16. **Type of water vapour movement occurs in soils is:**
    (A) Internal (B) External
    (C) Both of these (D) None of these
17. **Capillary movement of water is complemented by:**
    (A) Stem elongation (B) Root extension
    (C) Leaf orientation (D) None of these
18. **When a plant is under soil moisture stress, the roots:**
    (A) Shrink in size
    (B) Expand in size
    (C) Remain unaffected
    (D) Both a and b
19. **The positive and high value of redox potential indicates:**
    (A) Reducing conditions
    (B) Strong oxidizing conditions
    (C) Equally reducing and oxidizing conditions
    (D) None of these
20. **Soil showing permanently wilted plants has:**
    (A) Gravitational water
    (B) Capillary water
    (C) Hygroscopic water
    (D) Ground water

Practice Paper 2

1. **Water is a resource:**
   (A) Nondegradable nonmaintainable
   (B) Degradable maintainable
   (C) Renewable
   (D) Non renewable
2. **The relationship between the soil water content on weight and volume basis can be expressed as:**
   (A) $Pv = \frac{Pw}{Bd}$ (B) $Pv = Pw \times Bd$
   (C) $Pv = \frac{Bd}{Pw}$ (D) $Pw = Pv \cdot Bd$
3. **1 bar is equal to:**
   (A) pF of 3 (B) pF of 2
   (C) pF of 4 (D) None of these

4. **Osmotic potential is also known as:**
(A) Solute potential (B) Pressure potential
(C) Metrix potential (D) None of these

5. **Pascal is a unit of:**
(A) Work (B) Water potential
(C) Distance (D) None of these

6. **The curve indicating available water against soil water tension is called:**
(A) Soil water curve
(B) Soil water tension curve
(C) Soil tension curve
(D) Water tension curve

7. **Which of the following water is available water is available to plant communities?**
(A) Hygroscopic water
(B) Gravitational water
(C) Capillary water
(D) All of these

8. **Moisture equivalent is equal to:**
(A) 1/3 bar (B 15 bar
(C) 0.3 - 1 bar (D) 1/3 – 31 bar

9. **Which of the following is not responsible for soil water loss?**
(A) Evaporation
(B) Transpiration
(C) Percolation
(D) Hydraulic conductivity

10. **When all soil macrospores are water filled, the water movement in soil is called:**
(A) Unsaturated flow (B) Saturated flow
(C) Run-off (D) None of these

11. **How much heat is required for vaporization of 1 g water at 20 °C?**
(A) 500 calories (B) 540 calories
(C) 560 calories (D) None of these

12. **Saturated flow can be expressed by following lay?**
(A) Darcy's law (B) Newton's law
(C) William's law (D) None of these

13. **Under Darcy law, water flow is inversely related to:**
(A) Hydraulic conductivity
(B) Hydraulic heat
(C) Flow length
(D) Cross section area

14. **Which of the following is not used for measuring soil water:**
(A) Tensiometer (B) Zypsum block
(C) Pressure bomb (D) Neutron meter

15. **Tensiometer can measure soil water up to soil moisture tension of:**
(A) 0.0-0.85 bar (B) 0.0-0.50 bar
(C) -0.33-15.0 bar (D) All of these

16. **Gypsum block can measure the soil water potential up to:**
(A) 1-15 bar (B) 15-31 bar
(C) 31-1000 bar (D) All of these

17. **Which of the following is the source of neutrons in neutron moisture meter?**
(A) Thorium (B) Radium
(C) P^{32} (D) H^3

18. **If the soil moisture increases, the heat conductivity of soil will?**
(A) Increase (B) Decrease
(C) Remain same (D) None of these

19. **Which of the following can not diffuse across the semi permeable membrane?**
(A) Colloids (B) Crystalloid
(C) Solid particles (D) None of these

20. **The pF value at PWP is equal to:**
(A) 4.2 (B) 5.2
(C) 3.2 (D) None of these

Practice Paper 3

1. **Which of the soil type has the lowest PWP?**
(A) Sandy loam (B) Loam
(C) Clay loam (D) Clay

2. **Water movement through soil columns is called:**
(A) Permeability (B) Percolation
(C) Surface runoff (D) None of these

3. **Water available in between hygroscopic coefficient and field capacity is known as:**
(A) Capillary water
(B) Gravitational water
(C) Hygroscopic water
(D) None of these

4. **Which of the soil type has the highest field capacity?**
(A) Loam (B) Silt loam
(C) Clay loam (D) Clay

5. **Sticky point refers to:**
(A) Level of soil moisture at which water is completely absorbed by soil colloids
(B) Water status of soil
(C) Water level at which irrigation is applied
(D) None of these

6. **Critical moisture point is equal to:**
(A) Field capacity
(B) Permanent wilting point
(C) Sticky point
(D) None of these

7. **What is the water tension of saturated flow of soil water?**
(A) 1/3 bar (B) 15 bar
(C) 0 bar (D) None of these

8. **Soil water flow is governed by:**
(A) Conductivity of the medium
(B) Driving force of flow
(C) Proportionality factor
(D) All of these

9. **Which of the following sail has the highest rate of saturated soil water flow?**
(A) Sandy soil
(B) Loamy soil
(C) Clay
(D) None of these

10. **Saturated soil water flow is inversely related to:**
(A) Pressure difference
(B) Viscosity of liquid
(C) Radius of tube
(D) All of these

11. **According to Darcy's law, flow velocity of any liquid is directly proportional to:**
(A) Hydraulic conductivity
(B) Length of porous medium
(C) Viscosity of liquid
(D) All of these

12. **Unsaturated water flow can be understood by:**
(A) Darcy's law
(B) Poiseuille law
(C) Capillary principle
(D) All of these

13. **Soil water is lost by:**
(A) Evaporation (B) Transpiration
(C) Runoff (D) All of these

14. **Which of the following elements become toxic to plants under waterlogged conditions?**
(A) Fe and Mn (B) B and Mo
(C) N and P (D) N and K

15. **At ultimate wilting point of pF value is:**
(A) 2.5 (B) 3.0
(C) 4.77 (D) 5.12

16. **Which of the flowing clay mineral absorbs less water?**
(A) Montmorillonite
(B) Kaolinite
(C) Vermiculite
(D) None of these

17. **Which of the following ion may cause more water absorption on soil colloids?**
(A) Na^+ (B) Ca^{++}
(C) H^+ (D) All of these

18. **Attraction force between two water molecules is known as:**
(A) Cohesion
(B) Adhesion
(C) Electromagnetic force
(D) None of these

19. **Attraction force between two different molecules is known as:**
(A) Cohesion
(B) Adhesion
(C) Electromagnetic force
(D) None of these

20. **Height of rise of capillary tube is directly proportional to:**
(A) Surface tension
(B) Mass flow
(C) Both (A) and (B)
(D) None of these

Deficiency symptoms

Practice Paper 1

1. **Phosphorus deficiency in plants appears at first in:**
(A) Older leaves (B) Younger leaves
(C) Both (A) and (B) (D) None of these

2. **The critical value of P in phosphorus deficient leaves of maize plants before silking stage is:**
(A) 0.25 per cent (B) 0.35 per cent
(C) 0.45 per cent (D) 0.55 per cent

3. **Indicator plant for P deficiency indication is:**
(A) Maize
(B) Tomato
(C) Maize and tomato both
(D) None of these

4. **Good 'Indicator plant' for N deficiency is:**
(A) Rice (B) Wheat
(C) Maize (D) Sunflower

5. **Deficiency symptoms of nitrogen in plants appear at first in:**
(A) Lower older leaves
(B) Middle leaves
(C) Top young leaves
(D) All of these

6. **Excessive N supply to plants promotes:**
(A) Mildew and rusts (B) Root-rot
(C) Bacterial blight (D) Leaf mosaic

7. **Potassium deficiency leads to increase the severity of:**
(A) Cotton wilt (B) Potato rot
(C) Both (A) and (B) (D) None of these

8. **Cauliflower crop is highly sensitive to:**
(A) Mo deficiency (B) Fe deficiency
(C) Zn deficiency (D) B deficiency

9. **In general, potassium deficiency influences crop maturity by:**
(A) Delaying maturity period
(B) Preceding maturity period
(C) Not changing maturity period
(D) None of these

10. **Potassium deficiency symptoms in plants appear at first from:**
(A) Lower most leaves
(B) Upper most leaves
(C) Middle leaves
(D) Scattered throughout the entire leaves

11. **Deficiency of K in plants leads to an accumulation of:**
(A) Methionine (B) Putrescine
(C) Ornithine (D) Isoleucine

12. **In general, the shoot/root ratio of K-deficient crops is around:**
(A) 1.0 (B) 2.0
(C) 3.0 (D) 4.0

13. **The critical value of K percentage in petioles of deficient potato plants is:**
(A) 4.0 (B) 5.0
(C) 6.0 (D) 7.0

14. **The critical value for K deficiency in maize is:**
(A) 0.5 per cent (B) 1.0 per cent
(C) 1.5 per cent (D) 2.0 per cent

15. **Indicator plant for monitoring K deficiency is:**
(A) Mustard (B) Pea
(C) Sunflower (D) Potato

16. **Little leaf/leaf resetting is deficiency symptoms of:**
(A) Zn (B) Mn
(C) Fe (D) B

17. **Chlorosis results from deficiency of:**
(A) Chlorine (B) Sodium
(C) Calcium (D) Magnesium

18. **Internal cork of apple/browning of cauliflower is deficiency symptoms of:**
(A) Mo (B) B
(C) Zn (D) Cu

19. **Grey speck of oats are caused by dwficiency of:**
(A) Cu (B) Mn
(C) Zn (D) Cl

20. **Whiptail of crucifers is due to deficiency of:**
(A) Mo (B) Cu
(C) Zn (D) P

Practice Paper 2

1. **Deficiency of Molybdenum causes:**
(A) Poor development of vasculature
(B) Bending of leaf tips
(C) Yellowing of leaves
(D) Mottling and marginal necrosis of leaves

2. **On the basis of symptoms of chlorosis in leaves a student inferred that this Ws due to deficiency of nitrogen. This inference could be correct only if we assume that yellowing of leaves appeared first in:**
(A) Young leaves
(B) Old leaves
(C) Young leaves followed by mature leaves
(D) Mature leaves followed by young leaves

3. **Chlorosis in plants occurs due to:**
(A) High sunlight intensity
(B) Low sunlight intensity
(C) Absorption of yellow pigments from the soil
(D) Deficiency of Mg and Fe in the soil

4. **In plants a common symptom caused by deficiency of P, K, Ca and Mg is:**
(A) Bending of leaf tips
(B) Formation of anthocynin
(C) Poor development of vasculature
(D) Appearance of dead necrotic cells

5. **Yellowing of leaves or chlorosis is the first deficiency symptoms of which element:**
(A) Calcium (B) Nitrogen
(C) Phosphorus (D) Chlorine

6. **Premature leaf fall is due to deficiency of:**
(A) Phosphorus (B) Nitrogen
(C) Iron (D) Calcium

7. **Fall of immature leaves is due to deficiency of:**
(A) Potassium (B) Sodium
(C) Phosphorus (D) Zinc

8. **Chlorosis is caused by deficiency of:**
(A) Iron (B) Calcium
(C) Copper (D) Boron

9. **Brown heart disease is due to deficiency of:**
(A) Iron (B) Boron
(C) Potassium (D) Molybdenum

10. **Deficiency of Magnesium causes:**
(A) Necrosis (B) Plasmolysis
(C) Hydrolysis (D) Chlorosis

11. **Iron deficiency results in:**
(A) Leaf tip necrosis
(B) Small leaves
(C) Decreased protein synthesis
(D) Intervenial chlorosis appearing first in young leaves

12. **Finger and Toe disease in crucifer is caused in:**
(A) Acidic soil (B) Neutral soil
(C) Alkaline soil (D) Saline soil

13. **Deficiency of nitrogen in maize causes:**
(A) Yellowing of leaves from tip to base in V shaped pattern
(B) Chlorotic spots at base of leaves near the margins
(C) Yellow streaking of leaf edges
(D) Whitish-yellow spots on leaves

14. **Browning in cauliflower can be corrected by:**
(A) Ammonium molybdate
(B) Boric acid
(C) Zinc sulphate
(D) Urea

15. **Exothema is a result of deficiency of:**
(A) Cu (B) Ca
(C) P (D) Na

16. **Yellow white mottling of cotton is caused by:**
(A) K deficiency (D) N deficiency
(C) S deficiency (D) Zn deficiency

17. **Necrosis on lower leaf margin is caused by:**
(A) K deficiency (D) N deficiency
(C) S deficiency (D) Zn deficiency

18. **Potato scab disease is caused by:**
(A) Bacteria (B) Fungi
(C) Actinomycetes (D) Algae

19. **Sweet potato pox is caused by:**
(A) Bacteria (B) Fungi
(C) Actinomycetes (D) Algae

20. **Tea yellow is caused by:**
(A) K deficiency (D) N deficiency
(C) S deficiency (D) P deficiency

Practice Paper 3

1. **Interveinal chlorosis on lower leaves is caused by:**
   (A) Mg (B) Ca
   (C) P (D) Cl
2. **Khaira disease of rice is caused by:**
   (A) K deficiency (B) N deficiency
   (C) Zn deficiency (D) P deficiency
3. **Little leaf or mottle leaf in citrus is caused by:**
   (A) K deficiency (B) P deficiency
   (C) S deficiency (D) Zn deficiency
4. **White bud of maize is the deficiency symptoms of:**
   (A) Zn deficiency (B) N deficiency
   (C) S deficiency (D) Mg deficiency
5. **Yellow spot of citrus is caused by:**
   (A) K deficiency (B) N deficiency
   (C) Mo deficiency (D) P deficiency
6. **Which of the following diseases are caused by B deficiency:**
   (A) Heart rot of sugar beet
   (B) Yellow tip of alfalfa
   (C) Brown heart of turnip
   (D) All of these
7. **Soil sickness is caused due to:**
   (A) Accumulation of autotoxins
   (B) N-deficiency
   (C) Accumulation of phosphates
   (D) Zinc deficiency
8. **Deficiency of which of the following element causes general yellowing and poor growth of the plant?**
   (A) Phosphorus (B) Nitrogen
   (C) Calcium (D) boron
9. **Scorching or burning on margins of bottom leaves manifests the deficiency of:**
   (A) Nitrogen (B) Zinc
   (C) Copper (D) Potassium
10. **Browning in cauliflower can be corrected by use of:**
    (A) Ammonium sulphate
    (B) Zinc sulphate
    (C) Boric acid
    (D) Urea
11. **White bud of maize is caused by the deficiency of:**
    (A) Fe (B) Mn
    (C) Zn (D) B
12. **Iron deficiency is prominent in:**
    (A) Neutral soil
    (B) Calcareous soil
    (C) Acidic soil
    (D) Waterlogged soil
13. **Cu deficiency may be corrected by application of:**
    (A) Copper sulphate
    (B) Cupperous oxide
    (C) Cupric oxide
    (D) All of these
14. **Blossem-end rot in tomato has been found to be caused by the deficiency of:**
    (A) Nitrogen
    (B) Phosphorus
    (C) Calcium
    (D) Sulphur
15. **Plants growing in urea sprayed but Mg deficient soil will show:**
    (A) Deep green foliage
    (B) Yellowing of leaves
    (C) Early flowering
    (D) Loss of pigmentation in petals
16. **Stem and root tips die in deficiency of:**
    (A) Nitrogen (B) Phosphorus
    (C) Calcium (D) Sulphur
17. **Pale green to yellow in venial in maize due o the deficiency of:**
    (A) Mn (B) Cl
    (C) Zn (D) Mo
18. **Sand drown disease of tobacco is due to deficiency of:**
    (A) Ca (B) Mg
    (C) S (D) N
19. **Blossom End Rot disease is caused due to deficiency of:**
    (A) N (B) Mg
    (C) Ca (D) K
20. **Bronzing disorder in rice is associated with Fe level at:**
    (A) 100 ppm (B) 150 ppm
    (C) 200 ppm (D) 300 ppm

Answers Key

Pedology

■ Practice Paper 1

Q. No.	Ans	Q. No.	Ans	Q. No.	Ans	Q. No.	Ans	Q. No.	Ans	Q. No.	Ans
1.	A	2.	A	3.	B	4.	A	5.	C	6.	C
7.	A	8.	B	9.	A	10.	C	11.	A	12.	B
13.	A	14.	B	15.	A	16.	C	17.	C	18.	B
19.	A	20.	A								

■ Practice Paper 2

Q. No.	Ans	Q. No.	Ans	Q. No.	Ans	Q. No.	Ans	Q. No.	Ans	Q. No.	Ans
1.	B	2.	B	3.	A	4.	B	5.	D	6.	D
7.	D	8.	D	9.	D	10.	B	11.	A	12.	B
13.	A	14.	C	15.	C	16.	A	17.	B	18.	C
19.	D	20.	A								

Rocks and Minerals

■ Practice Paper 1

Q. No.	Ans	Q. No.	Ans	Q. No.	Ans	Q. No.	Ans	Q. No.	Ans	Q. No.	Ans
1.	B	2.	C	3.	D	4.	B	5.	C	6.	A
7.	C	8.	C	9.	B	10.	D	11.	A	12.	B
13.	A	14.	D	15.	D	16.	A	17.	D	18.	D
19.	A	20.	A								

■ Practice Paper 2

Q. No.	Ans	Q. No.	Ans	Q. No.	Ans	Q. No.	Ans	Q. No.	Ans	Q. No.	Ans
1.	C	2.	D	3.	A	4.	D	5.	D	6.	C
7.	B	8.	C	9.	B	10.	A	11.	B	12.	B
13.	A	14.	A	15.	D	16.	A	17.	C	18.	B
19.	A	20.	C								

■ Practice Paper 3

Q. No.	Ans	Q. No.	Ans	Q. No.	Ans	Q. No.	Ans	Q. No.	Ans	Q. No.	Ans
1.	A	2.	B	3.	A	4.	C	5.	C	6.	C
7.	D	8.	A	9.	A	10.	A	11.	A	12.	C
13.	A	14.	C	15.	A	16.	C	17.	A	18.	B
19.	B	20.	A								

■ **Practice Paper 4**

Q. No.	Ans	Q. No.	Ans	Q. No.	Ans	Q. No.	Ans	Q. No.	Ans	Q. No.	Ans
1.	C	2.	D	3.	B	4.	B	5.	C	6.	B
7.	A	8.	A	9.	C	10.	C	11.	A	12.	B
13.	A	14.	B	15.	A	16.	A	17.	A	18.	B
19.	A	20.	B								

■ **Practice Paper 5**

Q. No.	Ans	Q. No.	Ans	Q. No.	Ans	Q. No.	Ans	Q. No.	Ans	Q. No.	Ans
1.	C	2.	D	3.	D	4.	A	5.	B	6.	D
7.	D	8.	A	9.	C	10.	C	11.	B	12.	A
13.	A	14.	A	15.	D	16.	B	17.	C	18.	A
19.	B	20.	C								

Essential Plant Nutrients

■ **Practice Paper 1**

Q. No.	Ans	Q. No.	Ans	Q. No.	Ans	Q. No.	Ans	Q. No.	Ans	Q. No.	Ans
1.	A	2.	D	3.	D	4.	C	5.	B	6.	D
7.	A	8.	C	9.	D	10.	A	11.	B	12.	C
13.	C	14.	A	15.	A	16.	B	17.	A	18.	A
19.	C	20.	C								

■ **Practice Paper 2**

Q. No.	Ans	Q. No.	Ans	Q. No.	Ans	Q. No.	Ans	Q. No.	Ans	Q. No.	Ans
1.	C	2.	B	3.	C	4.	C	5.	C	6.	B
7.	A	8.	B	9.	B	10.	D	11.	D	12.	C
13.	B	14.	A	15.	B	16.	A	17.	D	18.	A
19.	D	20.	D								

■ **Practice Paper 3**

Q. No.	Ans	Q. No.	Ans	Q. No.	Ans	Q. No.	Ans	Q. No.	Ans	Q. No.	Ans
1.	A	2.	B	3.	D	4.	D	5.	D	6.	B
7.	C	8.	B	9.	B	10.	B	11.	C	12.	C
13.	B	14.	D	15.	D	16.	A	17.	C	18.	B
19.	B	20.	A								

■ **Practice Paper 4**

Q. No.	Ans	Q. No.	Ans	Q. No.	Ans	Q. No.	Ans	Q. No.	Ans	Q. No.	Ans
1.	B	2.	D	3.	D	4.	A	5.	B	6.	A
7.	D	8.	B	9.	B	10.	C	11.	D	12.	B

Q. No.	Ans	Q. No.	Ans	Q. No.	Ans	Q. No.	Ans	Q. No.	Ans	Q. No.	Ans
13.	C	14.	A	15.	C	16.	B	17.	D	18.	B
19.	B	20.	A								

■ **Practice Paper 5**

Q. No.	Ans	Q. No.	Ans	Q. No.	Ans	Q. No.	Ans	Q. No.	Ans	Q. No.	Ans
1.	C	2.	A	3.	C	4.	D	5.	C	6.	C
7.	D	8.	A	9.	A	10.	D	11.	B	12.	C
13.	A	14.	D	15.	B	16.	A	17.	B	18.	C
19.	A	20.	D								

Soil Fertility and Productivity

■ **Practice Paper 1**

Q. No.	Ans	Q. No.	Ans	Q. No.	Ans	Q. No.	Ans	Q. No.	Ans	Q. No.	Ans
1.	A	2.	D	3.	D	4.	D	5.	B	6.	D
7.	D	8.	C	9.	D	10.	A	11.	A	12.	B
13.	A	14.	A	15.	C	16.	D	17.	A	18.	B
19.	C	20.	D								

■ **Practice Paper 2**

Q. No.	Ans	Q. No.	Ans	Q. No.	Ans	Q. No.	Ans	Q. No.	Ans	Q. No.	Ans
1.	A	2.	A	3.	C	4.	A	5.	B	6.	B
7.	A	8.	A	9.	B	10.	A	11.	B	12.	A
13.	A	14.	B	15.	A	16.	A	17.	B	18.	A
19.	B	20.	A								

Acidity and Alkalinity

■ **Practice Paper 1**

Q. No.	Ans	Q. No.	Ans	Q. No.	Ans	Q. No.	Ans	Q. No.	Ans	Q. No.	Ans
1.	D	2.	B	3.	A	4.	C	5.	A	6.	C
7.	C	8.	B	9.	A	10.	A	11.	A	12.	A
13.	B	14.	C	15.	B	16.	A	17.	B	18.	B
19.	D	20.	C								

■ **Practice Paper 2**

Q. No.	Ans	Q. No.	Ans	Q. No.	Ans	Q. No.	Ans	Q. No.	Ans	Q. No.	Ans
1.	A	2.	C	3.	D	4.	D	5.	A	6.	C
7.	A	8.	B	9.	A	10.	C	11.	B	12.	A
13.	A	14.	A	15.	B	16.	A	17.	C	18.	A
19.	B	20.	A								

Manures and Fertilizers

■ Practice Paper 1

Q. No.	Ans	Q. No.	Ans	Q. No.	Ans	Q. No.	Ans	Q. No.	Ans	Q. No.	Ans
1.	D	2.	D	3.	C	4.	D	5.	D	6.	A
7.	C	8.	A	9.	C	10.	B	11.	A	12.	B
13.	C	14.	C	15.	C	16.	A	17.	C	18.	D
19.	C	20.	A								

■ Practice Paper 2

Q. No.	Ans	Q. No.	Ans	Q. No.	Ans	Q. No.	Ans	Q. No.	Ans	Q. No.	Ans
1.	B	2.	B	3.	A	4.	B	5.	D	6.	A
7.	D	8.	B	9.	B	10.	C	11.	D	12.	C
13.	B	14.	C	15.	B	16.	D	17.	C	18.	A
19.	B	20.	B								

■ Practice Paper 3

Q. No.	Ans	Q. No.	Ans	Q. No.	Ans	Q. No.	Ans	Q. No.	Ans	Q. No.	Ans
1.	B	2.	A	3.	C	4.	B	5.	C	6.	B
7.	B	8.	D	9.	C	10.	B	11.	B	12.	B
13.	A	14.	B	15.	C	16.	A	17.	D	18.	B
19.	C	20.	D								

■ Practice Paper 4

Q. No.	Ans	Q. No.	Ans	Q. No.	Ans	Q. No.	Ans	Q. No.	Ans	Q. No.	Ans
1.	C	2.	B	3.	D	4.	D	5.	B	6.	A
7.	A	8.	D	9.	A	10.	C	11.	C	12.	A
13.	B	14.	A	15.	C	16.	A	17.	C	18.	D
19.	A	20.	B								

Biofertilizers

■ Practice Paper 1

Q. No.	Ans	Q. No.	Ans	Q. No.	Ans	Q. No.	Ans	Q. No.	Ans	Q. No.	Ans
1.	C	2.	B	3.	D	4.	A	5.	B	6.	C
7.	A	8.	B	9.	A	10.	A	11.	B	12.	B
13.	B	14.	B	15.	C	16.	A	17.	B	18.	A
19.	D	20.	A								

■ Practice Paper 2

Q. No.	Ans	Q. No.	Ans	Q. No.	Ans	Q. No.	Ans	Q. No.	Ans	Q. No.	Ans
1.	C	2.	B	3.	C	4.	A	5.	D	6.	B
7.	C	8.	A	9.	D	10.	A	11.	A	12.	A
13.	C	14.	B	15.	B	16.	C	17.	B	18.	B
19.	B	20.	A								

■ Practice Paper 3

Q. No.	Ans	Q. No.	Ans	Q. No.	Ans	Q. No.	Ans	Q. No.	Ans	Q. No.	Ans
1.	D	2.	C	3.	B	4.	A	5.	C	6.	A
7.	A	8.	A	9.	A	10.	C	11.	C	12.	A
13.	B	14.	C	15.	D	16.	A	17.	B	18.	C
19.	D	20.	C								

■ Practice Paper 4

Q. No.	Ans	Q. No.	Ans	Q. No.	Ans	Q. No.	Ans	Q. No.	Ans	Q. No.	Ans
1.	B	2.	A	3.	B	4.	C	5.	D	6.	A
7.	B	8.	C	9.	B	10.	C	11.	A	12.	C
13.	A	14.	B	15.	C	16.	A	17.	B	18.	B
19.	B	20.	C								

■ Practice Paper 5

Q. No.	Ans	Q. No.	Ans	Q. No.	Ans	Q. No.	Ans	Q. No.	Ans	Q. No.	Ans
1.	A	2.	B	3.	C	4.	A	5.	C	6.	A
7.	B	8.	A	9.	B	10.	B	11.	A	12.	A
13.	C	14.	D	15.	A	16.	B	17.	A	18.	B
19.	C	20.	C								

Soils and its Properties

■ Practice Paper 1

Q. No.	Ans	Q. No.	Ans	Q. No.	Ans	Q. No.	Ans	Q. No.	Ans	Q. No.	Ans
1.	D	2.	D	3.	A	4.	C	5.	B	6.	B
7.	D	8.	C	9.	A	10.	A	11.	D	12.	A
13.	A	14.	D	15.	A	16.	A	17.	D	18.	B
19.	B	20.	B								

■ Practice Paper 2

Q. No.	Ans	Q. No.	Ans	Q. No.	Ans	Q. No.	Ans	Q. No.	Ans	Q. No.	Ans
1.	A	2.	D	3.	D	4.	A	5.	D	6.	A
7.	B	8.	C	9.	C	10.	A	11.	B	12.	D

Q. No.	Ans	Q. No.	Ans	Q. No.	Ans	Q. No.	Ans	Q. No.	Ans	Q. No.	Ans
13.	B	14.	D	15.	C	16.	D	17.	A	18.	D
19.	B	20.	C								

■ Practice Paper 3

Q. No.	Ans	Q. No.	Ans	Q. No.	Ans	Q. No.	Ans	Q. No.	Ans	Q. No.	Ans
1.	A	2.	B	3.	A	4.	A	5.	C	6.	A
7.	A	8.	A	9.	B	10.	C	11.	C	12.	A
13.	B	14.	A	15.	A	16.	B	17.	C	18.	D
19.	C	20.	D								

■ Practice Paper 4

Q. No.	Ans	Q. No.	Ans	Q. No.	Ans	Q. No.	Ans	Q. No.	Ans	Q. No.	Ans
1.	A	2.	B	3.	A	4.	C	5.	B	6.	A
7.	A	8.	D	9.	A	10.	D	11.	A	12.	A
13.	B	14.	D	15.	A	16.	D	17.	D	18.	D
19.	D	20.	A								

■ Practice Paper 5

Q. No.	Ans	Q. No.	Ans	Q. No.	Ans	Q. No.	Ans	Q. No.	Ans	Q. No.	Ans
1.	D	2.	D	3.	A	4.	D	5.	D	6.	D
7.	C	8.	C	9.	A	10.	A	11.	B	12.	A
13.	D	14.	A	15.	C	16.	D	17.	C	18.	C
19.	A	20.	A								

Soil Water

■ Practice Paper 1

Q. No.	Ans	Q. No.	Ans	Q. No.	Ans	Q. No.	Ans	Q. No.	Ans	Q. No.	Ans
1.	B	2.	A	3.	B	4.	A	5.	C	6.	B
7.	D	8.	A	9.	B	10.	C	11.	A	12.	C
13.	A	14.	C	15.	B	16.	C	17.	B	18.	A
19.	B	20.	C								

■ Practice Paper 2

Q. No.	Ans	Q. No.	Ans	Q. No.	Ans	Q. No.	Ans	Q. No.	Ans	Q. No.	Ans
1.	B	2.	B	3.	A	4.	A	5.	B	6.	B
7.	C	8.	C	9.	D	10.	B	11.	B	12.	A
13.	C	14.	C	15.	A	16.	A	17.	B	18.	A
19.	B	20.	A								

■ Practice Paper 3

Q. No.	Ans	Q. No.	Ans	Q. No.	Ans	Q. No.	Ans	Q. No.	Ans	Q. No.	Ans
1.	A	2.	B	3.	D	4.	D	5.	A	6.	B
7.	C	8.	D	9.	A	10.	B	11.	A	12.	C
13.	D	14.	A	15.	C	16.	B	17.	A	18.	A
19.	B	20.	A								

Deficiency Symptoms

■ Practice Paper 1

Q. No.	Ans	Q. No.	Ans	Q. No.	Ans	Q. No.	Ans	Q. No.	Ans	Q. No.	Ans
1.	A	2.	A	3.	C	4.	C	5.	A	6.	A
7.	C	8.	A	9.	B	10.	A	11.	B	12.	A
13.	D	14.	B	15.	D	16.	A	17.	D	18.	C
19.	B	20.	A								

■ Practice Paper 2

Q. No.	Ans	Q. No.	Ans	Q. No.	Ans	Q. No.	Ans	Q. No.	Ans	Q. No.	Ans
1.	D	2.	B	3.	D	4.	D	5.	B	6.	B
7.	C	8.	A	9.	B	10.	D	11.	D	12.	A
13.	A	14.	B	15.	A	16.	A	17.	A	18.	C
19.	C	20.	C								

■ Practice Paper 3

Q. No.	Ans	Q. No.	Ans	Q. No.	Ans	Q. No.	Ans	Q. No.	Ans	Q. No.	Ans
1.	A	2.	C	3.	D	4.	A	5.	C	6.	D
7.	A	8.	B	9.	D	10.	C	11.	B	12.	B
13.	D	14.	C	15.	B	16.	C	17.	A	18.	B
19.	C	20.	D								

16. Multiple Choice Questions of Horticulture

Floriculture

Practice Paper 1

1. **Growing of flowers, fruits, ornamental plants and vegetables in small grdens or farm is called:**
(A) Horticulture (B) Agriculture
(C) Sericulture (D) Apiculture

2. **NHB stands for:**
(A) National Horticulture Board
(B) Non-Horticulture Board
(C) Nutritional Health Board
(D) None of these

3. **Growing of flowers and ornamental plants is called:**
(A) Pomology (B) Olericulture
(C) Floriculture (D) None of these

4. **First Red King is generally referred to in:**
(A) Merigold (B) Rose
(C) Vinta (D) Larkspur

5. **Which type of garden are of geometric design and symmetrical shape are called:**
(A) Formal garden (B) Informal garden
(C) Free style garden (D) Natural garden

6. **The style reflects naturalistic effect of total view and represents natural beauty is called:**
(A) Formal garden (B) Informal garden
(C) Free style garden (D) Natural garden

7. **The combination of both formal and informal garden is called:**
(A) Formal garden (B) Informal garden
(C) Free style garden (D) Natural garden

8. **Rose garden of Ludhiana is an example of:**
(A) Formal garden (B) Informal garden
(C) Free style gardem (D) Natural garden

9. **Persian and mughal garden is an example of:**
(A) Formal garden (B) Informal garden
(C) Free style garden (D) Natural garden

10. **Japanese garden is an example of:**
(A) Formal garden (B) Informal garden
(C) Free style garden (D) Natural garden

11. **Decarnation of a tract of land with plants and other garden materials so as to produce a picturesque and naturalistic effect in a limited space is called:**
(A) land scape gardening
(B) Informal gardening
(C) Free style gardening
(D) Natural gardening

12. **Plants have special structures to climbs and support is called:**
(A) Climbers (B) Creepers
(C) Both (A) and (B) (D) None of these

13. **Important annual climbers are:**
(A) Monrning glory
(B) *Thunbergia gradiflora*
(C) *Clerodendrom splendens*
(D) All of these

14. **Ideal climbers for pergola are:**
(A) *Quisqualis*
(B) *Poiveria densiflorum*
(C) *Banisyeria taurifolia*
(D) All of these

15. **Mughal garden is famous for:**
(A) Rose (B) Chrysanthemum
(C) Dahelia (D) Merigold

16. **Brandavan garden is situated in:**
(A) Mysore (B) Mathura
(C) New Delhi (D) Agra

17. **Queen of winter is also called:**
(A) Rose (B) Merigold
(C) Chrysanthemum (D) Dahalia

18. **Cymbidium is a kind of:**
(A) Orchid (B) Rose
(C) Vegetables (D) Fig

19. **Growing of plants in a small containers is called:**
(A) Herbaceous border
(B) Shrubary border
(C) Nursery
(D) Bonsai

20. **Valley of flowers is found in:**
(A) Nilgiri Hills
(B) North Eastern Himalayas
(C) North Western Himalayas
(D) Satpura Hills

Practice Paper 2

1. **The Moribana meaning "Piled flowers" denoted the arrangement of flowers in :**
(A) Shallow containers
(B) Deep containers
(C) Paper wrappings
(D) Paper cones

2. **Rose cut flower must be cut at tight bud stage at least .... cm stem length for export:**
(A) 20 (B) 40
(C) 60 (D) 80

3. **The Mandor Garden, Jodhpur (Rajasthan) was established by Raja .... (1724-1749)**
(A) Abhai Singh (B) Ajay Singh
(C) Hari Singh (D) Man Singh

4. **Lloyd Botanical Garden, Darjeeling (West Bengal) was laid out by .... on a piece of land donated by William Lloyd:**
(A) Sir Robert Llyod
(B) Sir George King
(C) Sir Peter Hunt
(D) Sir Holdgate Baird

5. **Rose is mainly propagated by:**
(A) Budding (B) Grafting
(C) Seed (D) Simple layering

6. **Rose oil is extracted from:**
(A) *Rosa damascene* (B) *Rosa borboniana*
(C) *Rosa moschata* (D) *Rosa chinensis*

7. **Gulkand is prepared from:**
(A) *Rosa damascene* (B) *Rosa borboniana*
(C) *Rosa moschata* (D) *Rosa chinensis*

8. **Zinnia attains seed maturity at:**
(A) 20 DAA (B) 30 DAA
(C) 40 DAA (D) 50 DAA

9. **Lotus can be pollinated by:**
(A) Birds (B) Insects
(C) Bees (D) Wind

10. **Marigold attains seed maturity at:**
(A) 20-25 DAA (B) 30-40 DAA
(C) 40-50 DAA (D) 50-60 DAA

11. **Rose attains seed maturity at:**
(A) 100 DAA (B) 125 DAA
(C) 150 DAA (D) 175 DAA

12. **Chrysanthemum is a:**
(A) Self pollinated crop
(B) Often cross pollinated crop
(C) Cross pollinated crop
(D) Self incompatible crop

13. **The highest per capita consumption of flowers in the world is in:**
(A) The USA (B) India
(C) Switzerland (D) The Netherlands

14. **Kemin yellow oil from Kemin Industries, USA is the product of:**
(A) Rose (B) Carnation
(C) Merigold (D) Calendula

15. **A popular lawn grass species Zoysia japonica is commonly known as:**
(A) Blue green grass
(B) Doob grass
(C) Japanese or Korean grass
(D) Nut grass

16. **Which of the following cacti is almost spineless:**
(A) Zygocactus (B) Opuntia
(C) Hawarthia (D) Echinopsis

17. **Zen garden are also known as:**
(A) Mughal garden
(B) English garden
(C) Japanese garden
(D) Italian garden

18. **The earliest example of vertical garden is in:**
(A) Ryoan-ji garden in Japan
(B) Hanging garden of Babylon
(C) Villa Medici at Fiesola in Itely
(D) Vaux-le-Vicomte in France

19. Which of the following Indian rose varieties is believed to have contributed chocolate brown colour to the present day roses?

(A) Pusa Muskan (B) Mohini
(C) Mrinalini (D) Manatha

20. Major constituent of rose oil is:

(A) Geraniol (B) Citrobellol
(C) Citral (D) Rosal

Practice Paper 3

1. Fruit of rose is known as:

(A) Hips (B) Bulbs
(C) Caropsis (D) Berry

2. Series of arches is known as:

(A) Pergola (B) Arch
(C) Terrace (D) Topiary

3. To avoid continuous shade of certain portion of a multispan green house, the most ideal orientation is:

(A) East – West (B) North – South
(C) South – East (D) North – East

4. Terrarium is an art of growing plants:

(A) In terraces (B) In water
(C) In rocks (D) In glass container

5. International Registration Authority of Bougainvillea is located at:

(A) The Netherland (B) New Delhi
(C) China (D) Thailand

6. Dried rose petals are known as:

(A) Gul Roghan (B) Gulkand
(C) Pankhuri (D) Hips

7. Wazid Ali Shah is a variety of :

(A) Barleria (B) Rose
(C) Bougainvillea (D) Chrysanthemum

8. The quickest method of laying a lawn is:

(A) Seedling (B) Turfing
(C) Dibbling (D) Dung plastering

9. The Rashtrapati garden is example of which type of garden?

(A) Mughal (B) Japanese
(C) English (D) Italian

10. Canna is also known as:

(A) Spanish day (B) African lily
(C) Indian shot (D) Century plant

11. Art of training plants into shape of different statues, birds or animals, umbrella etc. known as:

(A) Edge (B) Topiary
(C) Hedge (D) Pergola

12. Gladiolus in hills is planting during:

(A) February – March
(B) March – April
(C) May – June
(D) September – October

13. Stenting is practiced in:

(A) Dahlia (B) Ficus
(C) Rose (D) Ranunculus

14. Which is national flower of south Africa?

(A) Chrysanthemum (B) Bird of Paradise
(C) Protea (D) Gladiolus

15. Gladiolus fruit is a:

(A) Berry (B) Capsule
(C) Pod (D) Drupe

16. Epinasty in poinsettia is due to:

(A) Ethylene
(B) Gibbrellin
(C) Maleic hydrazide
(D) Aluminium sulphate

17. The first botanical garden in the world was made during 16th century at:

(A) Pauda (Itely) (B) Kyato (Japan)
(C) Singapore (D) San Francisco

18. Lipstick is a variety of:

(A) Carnation
(B) Red Ginger
(C) Gerbera
(D) *Heliconia psittacorum*

19. Which is common method of propagation in Dahlia for taking large sized flowers?

(A) Cuttings (B) Seeds
(C) Budding (D) Grafting

20. Hanging garden, a seven wonder located at:

(A) Batican city (B) New Delhi
(C) Babylonia (D) Moscow

Practice Paper 4

1. **Work on Hibiscus Improvement and Development of Varieties was extensively carried at:**
(A) IIHR, Bangluru
(B) NBRI, Lucknow
(C) IARI, New Delhi
(D) YSRUHF, Solan

2. **Rajat Rekha is a mutant of:**
(A) Rose (B) Carnation
(C) Tube rose (D) Chrysanthemum

3. **Star Gazar is a popular variety of:**
(A) Tiger lily (B) Orietal lily
(C) Asiatic lily (D) Torch lily

4. **Deepika is a variety of:**
(A) Zinnia (B) Balsam
(C) Hollyhock (D) Rose

5. ***Rosa wichuriana* is a:**
(A) Tetraploid (B) Diploid
(C) Triploid (D) Aneuploid

6. **Repetition of same object at equidistance in landscape is called:**
(A) Design (B) Rhythm
(C) Focal point (D) Layout

7. **First Islamic garden in India was developed by:**
(A) Ferozshah (B) Jahgir
(C) Akbar (D) Babar

8. **Contour system of orchard planting is generally followed in:**
(A) Punjab (B) Hills
(C) Saline soils (D) UP

9. **"T" buding is also referred to as:**
(A) Shield budding
(B) Patch budding
(C) Annular budding
(D) Ring budding

10. **To check the variability in rootstock one should opt for:**
(A) Clonal rootstock
(B) Seeding rootstock
(C) Seasonal rootstock
(D) Any of the above

11. **"Hillocks are made to create a natural mountains scenery" are the characteristics feature of (informal style):**
(A) Persian garden
(B) Mughal garden
(C) Japanese garden
(D) Rose garden of Ludhiana

12. **Garden lanterns, garden pergola, garden brides, tea garden and flat garden are the unique feature of:**
(A) English garden
(B) Mughal garden
(C) Japanese garden
(D) Rose garden

13. **Running water, high protecting wall, terminal budding and Baradari are the important features of:**
(A) Persian garden (B) Mughal garden
(C) English garden (D) Rose garden

14. **Which of the following garden is known as nature in miniatures?**
(A) Mughal (B) Japanese
(C) English (D) Persian

15. **Charbagh is developed in:**
(A) Persian style (B) Japanese style
(C) French style (D) None of these

16. **Most ancient type of garden is:**
(A) Truck garden (B) Kitchen garden
(C) Market garden (D) None of these

17. **Chrysanthemum is a:**
(A) Short day plant (B) Long day plant
(C) Day neutral plant (D) None of these

18. **Ornamental value of bougainvillea lies in:**
(A) Green foliage (B) Seeds
(C) Colourful bracts (D) Fruits

19. **Aurocaria is a tree suited to following region:**
(A) Cold desert (B) Marshy land
(C) Sub-tropical (D) None of these

20. **Pinching is generally done in quality production of:**
(A) Gladiolus (B) Carnation
(C) Dahlia (D) Aster

Pomology

Practice Paper 1

1. **Growing of fruits in garden and farm is called:**
(A) Pomology (B) Olericulture
(C) Floriculture (D) Vegetable Science

2. **The citrus diseases namely 'Reclamation Disease' and 'Die back' are caused by the deficiency of:**
(A) Zinc (B) Iron
(C) Copper (D) Molybdenum

3. **Fastening of vines to a upright support of a tree called panagara (*Erithrina indica*) is known as:**
(A) Kniffen system
(B) Head system
(C) Single stake system
(D) Cordon system

4. **A fruit that develops from a hollow pear shaped fleshy receptable (inflorscence: hypanthodium) which encloses a number of minute male and female flowers is known as:**
(A) Aggregate fruit (B) Hesperidium
(C) Pome fruit (D) Syconus fruit

5. **Guava is the native of:**
(A) Tropical South-east Asia
(B) Tropical Africa
(C) Tropical America
(D) Sub-tropical India

6. **Allabadi safeda is the variety of:**
(A) Mango (B) Guava
(C) Loquat (D) Grapes

7. **............................ is grownins a shade tree in coffee, areca and cardamom plantations:**
(A) Jack fruit (B) Coconut
(C) Mango (D) Guava

8. **Fruit setting is most adversely affected by:**
(A) Moist air
(B) Moist and cold air
(C) Dry and cold air
(D) Dry and hot air

9. **Which of the following is a natural colouring agent of chilli?**
(A) Oleoresin (B) Turmeric
(C) Resinocine (D) None of these

10. **The cutting are planted by:**
(A) Putting 2/3 part in soil and 1/3 part above the soil surface
(B) Half in soil and half above the soil
(C) 2/3 part in air and 1/3 part in soil
(D) All of these

11. **Scion is a:**
(A) Fruit bearing part of the grafted plant
(B) Root bearing part of the grafted plant
(C) Both (A) and (B)
(D) None of these

12. **Pear is mainly propagated by:**
(A) Air layering
(B) Stool layering
(C) Surpentine layering
(D) Simple layering

13. **Litchi is mainly propagated by:**
(A) Air layering
(B) Grafting
(C) Gootee
(D) Simple layering

14. **Mango is mainly propagated by:**
(A) Grafting
(B) Gootee
(C) Surpentine layering
(D) Seed

15. **Guava is mainly propagated by:**
(A) Seed
(B) Root cutting
(C) Surpentine layering
(D) Grafting

16. **Which of the fruit plant is propagated by root cuttings?**
(A) Apple (B) Litchi
(C) Guava (D) All of these

17. **The soft cutting are generally planted during:**
(A) Spring season (B) Summer season
(C) Winter season (D) Rainy season

18. Suitable season for planting of root cutting is:
(A) Summer season (B) Spring season
(C) Rainy season (D) Winter season

19. Cuttings of ever green plants are planted in:
(A) December-January (B) February-March
(C) May –June (D) June-July

20. Equilateral triangular system of orchard layout is also known as:
(A) Square (B) Hexagonal
(C) Rectangular (D) Quincunx

Practice Paper 2

1. The suitable method of orchard layout near the cities is:
(A) Square (B) Rectangular
(C) Quincunx (D) Hexagonal

2. In hills, the suitable method of orchard layout is:
(A) Contour (B) Square
(C) Hexagonal (D) Rectangular

3. Which of the very common method of orchard layout in India is:
(A) Square (B) Rectangular
(C) Hexagonal (D) Quincunx

4. In hexagonal system how much plants can be accommodated than square system?
(A) 10% more (B) 15% more
(C) 20% more (D) 25% more

5. Most suitable planting timings for deciduous fruit plants is:
(A) Winter season (B) Spring season
(C) Rainy season (D) Summer season

6. Dehorning refers to:
(A) Removal of branching from fruit trees
(B) Removal of terminal portion of branches
(C) Removal of flowers
(D) All of these

7. Pinching refers to:
(A) To remove top portion of plant with few leaves
(B) It is like a head back pruning
(C) It is done in ornamental plants
(D) All of these

8. Stripping refers to:
(A) Removal of barks in stripes of 1.25–2.5 cm length
(B Removal of extra flowers
(C) Removal of terminal portion of branches
(D) None of these

9. Pollinizers are:
(A) Fruit plants that provide pollens to other plants
(B) Helps in fertilization and fruiting
(C) They are planted middle of fruiting plants
(D) All of these

10. Fruiting can be enhanced by:
(A) Root pruning (B) Ringing
(C) Notching (D) All of these

11. Fruiting can be induced by:
(A) NAA (B) 2,4-D
(C) IAA (D) All of these

12. Spliting of fruits can be checked by spraying of:
(A) 1 ppm NAA (B) Orange rind oil
(C) Bordeaux mixture (D) All of these

13. Periclinal chimeras in fruit plants can be produced by:
(A) Colchicine (B) X-rays
(C) High temperature (D) All of these

14. Flowering in mangoes occurs on:
(A) One year old terminal buds
(B) Two years old terminal buds
(C) Whole plants
(D) Old branches

15. Polygamous plants are those plants that produce:
(A) Male, female and hermaphrodite flowers
(B) Male flowers only
(C) Female flowers only
(D) Hermaphrodite flowers

16. Which of the root stock has dwarfing effect on Dashhari?
(A) Rumani (B) Olour
(C) Dogridge (D) Rangur lime

17. Olor is a poly embryonic variety of:
(A) Guava (B) Citrus
(C) Mango (D) Strawberry

18. Which of the following dwarfing root stock is used for mango?
(A) Olour
(B) Totapari Red small
(C) Dashhari
(D) Alfanso

19. Which of the following is a early maturing variety of mango?
(A) Bombay green (B) Alfanso
(C) Gopal Bhog (D) All of these

20. Which of the following is a medium maturing variety of mango?
(A) Langra (B) Dashhari
(C) Safeda (D) All of these

Practice Paper 3

1. Which of the following is a late maturing variety of mango?
(A) Fazli (B) Neelam
(C) Kanchan (D) All of these

2. Which of the following is a polyembryonic variety?
(A) Mazagaon (B) Chandra karan
(C) Neleswar (D) All of these

3. Which of the following is a variety of sweet orange?
(A) Wheetaker (B) Blood Red
(C) Pine apple (D) All of these

4. Which of the following variety of sweet orange is cultivated in south India?
(A) Sathgudi (B) Batavian
(C) Mosambi (D) All of these

5. Which of the following is a early maturing variety of litchi?
(A) Early large red (B) Early seed less
(C) Pickling (D) All of these

6. Which of the following is a medium maturing variety of litchi?
(A) Dehradun (B) Rose scented
(C) Pyazi (D) All of these

7. Which of the following is a late maturing variety of litchi?
(A) Kalkattia (B) Late seed less
(C) Large seed red (D) All of these

8. Which of the following is a early maturing variety of loquat?
(A) Golden yellow
(B) Improved golden yellow
(C) Thames pride
(D) All of these

9. Which of the following is a medium maturing variety of loquat?
(A) Pale yellow (B) Large Agra
(C) Fire ball (D) All of these

10. Which of the following variety of grape is suitable for North India?
(A) Parlett (B) Black prince
(C) Bedana (D) All of these

11. Which of the following variety of grape is suitable for kismish preparation is?
(A) Sultana
(B) Thomson seed less
(C) Munakka
(D) All of these

12. Which of the following variety of grape is suitable for vine preparation?
(A) Muscat (B) Black prince
(C) Bangalare blue (D) All of these

13. Which of the following is a imported grape variety?
(A) Muscat hamburg (B) Foster's seedless
(C) Black Prince (D) All of these

14. Diploid variety of apple are?
(A) Self fruitful (B) Self unfruitful
(C) Both (A) and (B) (D) All of these

15. Triploid varieties of apple are?
(A) Self fruitful (B) Self unfruitful
(C) Both (A) and (B) (D) All of these

16. Which of the following is a diploid variety of apple?
(A) Jonathan (B) Rome beauty
(C) Newton wonder (D) All of these

17. Which of the following is a triploid variety of apple?
(A) Baldvin
(B) Blenhenn orange
(C) Cox's orange pippin
(D) All of these

18. Which of the following is a early maturing variety of apple?

(A) Summer golden (B) Pippin
(C) James grieve (D) All of these

19. Which of the following is a mid season apple variety is?

(A) Orange pippin (B) Jonathan
(C) King of pippin (D) All of these

20. Which of the following is a variety of apple is grown in Uttrakhand hills?

(A) Blenhein orange pippin
(B) Delicious
(C) Rome Beuty
(D) All of these

Practice Paper 4

1. Which of the following is a variety of apple mainly grown in Kashmir valleys?

(A) Ambari (B) Blood red
(C) Baldwin (D) All of these

2. Which of the following is a variety of apple mainly grown in kullu valleys?

(A) Red astrachan (B) Red delicious
(C) Golden delicious (D) All of these

3. Which of the following apple variety should be planted with rymer variety for better fruiting?

(A) Golden delicious (B) Buckinghum
(C) Early shan burry (D) Red delicious

4. Peach variety J.H. Hale is a:

(A) Self fruitful (B) Self unfruitful
(C) Both (A) and (B) (D) None of these

5. Which of the following variety of peach can be grown in plains?

(A) Florida Sun (B) Sharvati
(C) Sun Red (D) All of these

6. Which of the following variety of peach is grown at high hills?

(A) Alexander (B) Elberta
(C) Noblesse (D) All of these

7. Which of the following variety of straw berry?

(A) Kavaliar (B) Elberton
(C) Red Ford (D) All of these

8. Which of the following variety of straw berry is grown in Uttaranchal hills?

(A) Chaubatia Arun
(B) Chaubatia Manjul
(C) Chaubatia Abhichal
(D) All of these

9. Which of the following variety of pine apple?

(A) Kew (B) Queen
(C) Mauritius (D) All of these

10. Ratna variety of mango is a cross of:

(A) Neelam x Dashhari
(B) Dashhari x Neelam
(C) Neelam x Alphonso
(D) None of these

11. Training of fruit trees means:

(A) Remove few unwanted branches to make plant more beautiful and strong
(B) Remove few flowers
(C) Remove and cut few roots
(D) None of these

12. Temperate fruit plants can be trained by:

(A) Leader method
(B) Open centered method
(C) Modified leader system
(D) All of these

13. Which of the training is considered more suitable?

(A) Closed centered system
(B) Open centered system
(C) Modified leader system
(D) All of these

14. In quincunx system, the total plants are nearly double than:

(A) Square method
(B) Rectangular method
(C) Contour method
(D) Hexagonal method

15. In quincunx system, the filler plants are generally:

(A) Fast growing and early fruiting
(B) Slow growing but early fruiting
(C) Late fruiting
(D) All of these

16. Which direction is most sensitive to sun burn effects in plants:

(A) East – South (B) South – West
(C) West – North (D) North – East

17. Frame working refers to:

(A) Change into old plant into high grade plants
(B) All side branches are grafted with high quality scions
(C) Only top part of plant is removed
(D) All of these

18. Pruning is an essential cultural activity in fruit plants because:

(A) To provide proper shape
(B) To provide strong structure
(C) To beautify the plant
(D) All of these

19. Notching of fruit plants refers to:

(A) To make a V shaped notch just below N above the bud in fruit plants
(B) It helps in vegetative growth
(C) It is common in Fig
(D) None of these

20. Which of the following plant hormone is used to control pre harvest fruit drop?

(A) 2, 4-D (B) NAA
(C) 2, 4-D and NAA (D) None of these

Practice Paper 5

1. Propping refers to:

(A) Staking of banana stem at the time of fruiting to avoid bending or breaking of stem due to heavy weight of fruits
(B) To remove extra suckers
(C) To provide shade to the fruits
(D) None of these

2. In spur system of pruning of grape is:

(A) The branches are removed from base leaving only 2-3 buds
(B) Removal of new branches from base
(C) One year old branches are removed from base
(D) All of these

3. In cane system of pruning in grape:

(A) Half portion of one year old branches is removed
(B) Only upper portion of last season fruiting spurs are removed
(C) Both (A) and (B)
(D) None of these

4. Which of the following part has heavy fruiting in grape?

(A) Foundation spur (B) Fruiting spur
(C) Spur (D) All of these

5. Which of the following is a variety of Ber?

(A) Banarasi Karaka (B) Umran
(C) Narma Thornless (D) All of these

6. Flowering occurs in Aonla in the month of:

(A) January-February
(B) March-April
(C) July-August
(D) September-October

7. Cultar is a name of:

(A) Paclobutrazol (B) Methyl oxide
(C) Calcium carbonate (D) None of these

8. The coconut water is present in:

(A) Exocarp (B) Endocarp
(C) Endosperm (D) None of these

9. Copra is obtained from:

(A) Ripe coconut (B) Unripe coconut
(C) Coconut leaves (D) None of these

10. The coconut sap is also known as:

(A) Peera (B) Jeera
(C) Neera (D) Beera

11. Bangkok golden apple is a variety of:

(A) Guava (B) Pineapple
(C) Avacado (D) Pome granate

12. Stooling method of propagation is generally used in:

(A) Grapes (B) Walnut
(C) Guava (D) Apple

13. In Nigiries the main crop of mandarin orange is harvested in:

(A) July
(B) August – October
(C) December – January
(D) March

14. Kniffin system of training is use in:

(A) Litchi (B) Papaya
(C) Mango (D) Grapes

15. Rains during ripening of grapes causes:

(A) Low TSS
(B) Fruit drop
(C) Berry cracking
(D) Higher marketable yield

16. Spongy tissue of mango is a:

(A) Bacterial disease
(B) Fungal disease
(C) Viral disease
(D) Physiological disorder

17. Semi hardwood cutting is mostly used to propagate the following:

(A) Evergreen fruit plants
(B) Apple
(C) Vegetable crops
(D) Peach

18. Most of the varieties of mango grown in south India are:

(A) Disease free (B) Seedless
(C) Monoebryonic (D) Polyembryonic

19. McIntosh is a variety of:

(A) Apple (B) Plum
(C) Peach (D) Pear

20. Nectarine is fuzzyness mutant of:

(A) Plum (B) Almond
(C) Peach (D) Apricot

Practice Paper 6

1. Best time of pruning peach in:

(A) Spring (B) Mid winter
(C) Autumn (D) Mid summer

2. Sweet orange is generally trained to:

(A) Single branches (B) Two branches
(C) Multiple branches (D) None of these

3. Frontier is a mid season variety of:

(A) Apricot (B) Grape
(C) Plum (D) Peach

4. Most peach varieties bear flowers of following colour:

(A) Blue (B) Pink
(C) Yellow (D) White

5. Quincunx system of planting is used in:

(A) Fruit trees (B) Flowers
(C) Potato (D) Vegetable crops

6. Mango and banana are:

(A) Non-climacteric fruits
(B) Climacteric fruits
(C) Nature of ripening depends upon varieties
(D) None of these

7. Increase in ethylene production with ripening is characteristic of:

(A) Citrus fruits
(B) Grapes
(C) Non-climacteric fruits
(B) Climacteric fruits

8. The variety of mango seriously affected by spongy tissue is:

(A) Neelam (B) Dashahari
(C) Amrapali (D) Alphanso

9. The plant bioregulator that acts as antigibberellin is:

(A) Paclobutazole
(B) Indole acetonitrite
(C) Zeatin
(D) Morphactin

10. The pear fruit tree is trained in:

(A) Open center system
(B) Head system
(C) Modified central leader system
(D) None of these

11. Following is the dwarfing rootstock of apple:

(A) M25 (B) M15
(C) M 9 (D) M1

12. Which of the following is rich source of fat:

(A) Aonla (B) Avocado
(C) Bael (D) Pomegranate

13. The origin place of the fruit guava is:

(A) China (B) Australia
(C) India (D) Tropical America

14. Mango tree bears fruits on:

(A) Last year growth
(B) Current season growth
(C) Both (A) and (B)
(D) None of these

15. Seedlessness in most grape varieties is due to:

(A) Stenospermocarpy
(B) Parthenocarpy
(C) Polyembryony
(D) Apomixis

16. The ultra dwarf rootstock of apple is:

(A) M27 (B) MM106
(C) M9 (D) MM104

17. **The dwarfin rootstock of guava is:**
(A) Brazilian guava (B) Cattleya guava
(C) Pineapple guava (D) Chinese guava

18. **In peach the traditional system of training is:**
(A) Open centre
(B) Multiple stem
(C) Centre leader
(D) Modified central leader

19. **Which one of the medicinal plant is called Indian Gooseberry?**
(A) Arjuna (B) Aonla
(C) Bohera (D) Harar

20. **Most of the banana species are:**
(A) Triploid (B) Diploid
(C) Haploid (D) Hexaploid

Practice Paper 7

1. **Avenue term is mostly related to:**
(A) Shrubs only (B) Annuals only
(C) Trees only (D) None of these

2. **Sweetest sugar in fruit is:**
(A) Glucose (B) Fructose
(C) Sucrose (D) Galactose

3. **Lye peeling is done by:**
(A) 10% hot acid solution
(B) 10% cool acid solution
(C) 1% cool alkali solution
(D) 1% hot alkali solution

4. **Mango has bearing habit which is:**
(A) Axillary (B) Terminal
(C) Alternate (D) None of these

5. **Dwarf mutant od papaya is:**
(A) Pusa nanha (B) Pusa giant
(C) Pusa dwarf (D) None of these

6. **Transgenic are mostly made these days for which of the following fruit crops?**
(A) Mango (B) Grapes
(C) Guava (D) Banana

7. **Which of the following variety of apple has comparatively longer self life?**
(A) Ambari (B) Red delicious
(C) Golden delicious (D) Ambrich

8. **Yellow coloured fruits are rich in:**
(A) Vitamin A (B) Vitamin B
(C) Vitamin C (D) Vitamin D

9. **Which fruit is known as "Miracle fruit of China"?**
(A) Kiwi fruit (B) Mango
(C) Litchi (D) Loquat

10. **Pomegranate, fig, olive and grape have been placed in .... salt tolerant group of fruit trees:**
(A) High (B) Low
(C) Medium (D) Very high

11. **Kinnow cultivator of citrus is a hybrid between:**
(A) King × willow leaf (B) King × Owari
(C) King × cleopatra (D) King × Dancy

12. **Which of the following is nown as "goldmine of cold desert":**
(A) Apricot (B) Sea buckthorn
(C) Walnut (D) Apple

13. **Vegetative parthenocarly is common in:**
(A) Banana (B) Pineapple
(C) Fig (D) All of these

14. **Leaf bronzing in guava is due to:**
(A) Al toxicity (B) Fe toxicity
(C) Zn deficiency (D) K deficiency

15. **Stone fruit formation is a disorder of:**
(A) Apticot (B) Annona
(C) Apple (D) Pear

16. **Banana variety suitable for chips making:**
(A) Rasthali (B) Nendran
(C) Poovan (D) Monthan

17. **Which of the following is known as "Alligator Pear":**
(A) Annona (B) Pear
(C) Loquat (D) Avocado

18. **Pineapple widely practiced planting method in India:**
(A) Single row (B) Double row
(C) Hedge system (D) Trench method

19. **Winter cherry commonly known as:**
(A) Aloe (B) Babchi
(C) Safed nusli (D) Ashgand

20. **Soft nose in mango caused by due to deficiency of:**
(A) N (B) B
(C) K (D) Mn

Vegetable Science

Practice Paper 1

1. **Growing of vegetables in farm is called:**
(A) Pomology
(B) Olericulture
(C) Floriculture
(D) Vegetable Science

2. **What is blanching?**
(A) It is a method of fruit ripening
(B) It is a method of bunding
(C) It is a method of protecting fruits from light and hot
(D) It is a method of plant protecting from insects

3. **Blindness in cauliflower means:**
(A) Full development of flower/fruit
(B) Full development of plant without flower
(C) Started plant growth
(D) All of these

4. **Which of the following is a early varieties of cabbage?**
(A) Pride of India (B) Red cabbage
(C) Pusa drum head (D) All of these

5. **Which of the following is a late varieties of cabbage?**
(A) Pride of India
(B) Coperhegan market
(C) Pusa drum head
(D) All of these

6. **Which group belongs to chilli varieties?**
(A) NP 46 A, Pant C-1, Pusa Jwala, Andhra Jyoti
(B) NP 46, MSFH-8, Pant Kanchan, Pusa Jwala
(C) Pant C-1, Pusa long, Nantes, Pusa Mukti
(D) All of these

7. **Which of the following is a variety of brinjal?**
(A) Pant Smart (B) Pant Rituraj
(C) Pant Bagain – 4 (D) All of these

8. **The bitterness in colocasia corms is due to the presence of:**
(A) Calcium oxalate (B) Calcium carbonate
(C) Calcium chloride (D) Potassium oxalate

9. **Which of the following country is largest producer of cut flowers:**
(A) Netherlands (B) USA
(C) Germany (D) India

10. **How much seed of cauliflower is required for sowing of one hectare area?**
(A) 500-600 g (B) 800-1000 g
(C) 1000-1250 g (D) 1500-2000 g

11. **What is optimum sowing time of knoll khol in hills:**
(A) May-June
(B) July-August
(C) November-December
(D) January-February

12. **The main gradient of garlic oil is:**
(A) Alicin
(B) Alile propel disulphide
(C) Alicin and Alicin propel disulphide
(D) Acetelin

13. **Brocoli is a :**
(A) Cole crop (B) Root crop
(C) Fruit crop (D) Legume crop

14. **Among vegetable crops, leaving aside potatoes the maximum production in the world:**
(A) Cucumber (B) Brinjal
(C) Onion (D) Tomato

15. **Seed production of early cauliflower can be done:**
(A) In Europe only (B) In plains
(C) In hills only (D) None of these

16. **Pusa mukta variety of cabbage is resistant to:**
(A) Virus
(B) Bolting
(C) Cabbage caterpillar
(D) Black rot

17. **Which kind of varieties of tomato needs stalking?**
(A) Dwarf (B) Determinaate
(C) Indeterminate (D) None of these

18. **Which of the following variety of radish is round in shape:**
(A) Pusa Himani (B) Japanese white
(C) RRWT (D) Chinese pink

19. Large numbers of hybrids have been developed for which of the following vegetables:
(A) Tomato (B) Onion
(C) Brinjal (D) Potato

20. Cauliflower is:
(A) Self pollinated
(B) Cross pollinated
(C) Often cross pollinated
(D) None of these

Practice Paper 2

1. How much seed of garlic is required for sowing of one hectare area?
(A) 500-800 kg cloves
(B) 800-1000 kg cloves
(C) 1000-1200 kg cloves
(D) 1200-1500 kg cloves

2. Optimum sowing time of muskmelon is:
(A) December-January
(B) January-February
(C) February-March
(D) June-July

3. Optimum seedrate of muskmelon is:
(A) 2-3 kg/ha (B) 3-4 kg/ha
(C) 4-5 kg/ha (D) 5-6 kg/ha

4. The pungent constituents of ginger are:
(A) Oxymethyl phenols
(B) Terpinyl acetate
(C) Ethyl alcohol (D) None of these

5. Turmeric is an:
(A) antidepressant (B) Antipyretic
(C) Antioxidant (D) All of these

6. Which is the cultivar of bottle guard:
(A) Pusa Meghdoot (B) CO-1
(C) Autralian Green (D) Straight Eight

7. Garden pea was an important crop in eleventh century in:
(A) Holland (B) Poland
(C) England (D) New Zealand

8. Garden pea was probably distinguished from field pea in 1536, but their common use began after:
(A) 1586 (B) 1600
(C) 1675 (D) 1700

9. Browning of cauliflower is caused by the deficiency of:
(A) Zinc (B) Manganese
(C) Iron (D) Boron

10. Fruits of tomato known as:
(A) Bulbs (B) Hips
(C) Caropsis (D) Berry

11. Which of the following is dioecious ?
(A) Bitter gourd (B) Pointed gourd
(C) Snake gourd (D) Bottle gourd

12. Red colour of carrot is due to presence of:
(A) Anthrocyanin (B) Carotene
(C) Xanthophyll (D) None of these

13. Which is known as queen of spices:
(A) Cardamon (B) Black pepper
(C) Turmeric (D) Ginger

14. Which vegetable crop is known as vegetable of Twentieth century?
(A) Winged bean (B) Cluster bean
(C) Chekkurmanis (D) Amaranthus

15. Large cardamon is commercially propagated through:
(A) Suckers (B) Seeds
(C) Layers (D) Cuttings

16. The part used as vegetable in gerusalem artichoke is:
(A) Flower bud (B) Root tuber
(C) Stem (D) Leaf stalk

17. Bitter principle in bitter gourd is:
(A) Capsorubin
(B) Allyl propyl disulphide
(C) Cucurbitacin
(D) Phenol

18. Which is recommended for maintaining freshness in cut leafy vegetables:
(A) Cytokinin (B) ABA
(C) Auxin (D) Gibberellin

19. Generally seed less watermelon is:
(A) Diploid (B) Triploid
(C) Tetraploid (D) Hexaploid

20 Early cau;iflower seed production is not possible in plains due to:
(A) High wind speed
(B) High temperature
(C) High relative humidity
(D) Both (B) and (C)

Practice Paper 3

1. **Black rot resistant gene in cauliflower is:**
   (A) xca1bd (B) Xca1bc
   (C) Xca1cb (D) X1cbac
2. **........... is a male sterile gene controlled by single recessive gene in muskmelon:**
   (A) ms-5 (B) ms-4
   (C) Ar-2 (D) None
3. **Karamsag is mostly cultivated in:**
   (A) Andhra Pradesh (B) Himachal Pradesh
   (C) Uttar Pradesh (D) Uttarakhand
4. **Molybdenum deficiency causes whiptail in cauliflower in:**
   (A) Alkaline soil (B) Acidic soil
   (B) Netral soil (D) Both (A) and (B)
5. **Micro propagation is mostly popularly followed in:**
   (A) Okra (B) Chilli
   (C) Potato (D) Cabbage
6. **In tomato female to male ratio maintained in hybrid seed production is:**
   (A) 10 : 1 (B) 15 : 1
   (C) 12 : 1 (D) 8 : 1
7. **Processing cultivars of tomato should contain pH less rgan:**
   (A) 6.5 (B) 5.5
   (C) 3.5 (D) 4.5
8. **Heterostyly is found in :**
   (A) Potato (B) Okra
   (C) Brinjal (D) Radish
9. **Vegetable highly tolerant to soil salinity is:**
   (A) Beetroot (B) Potato
   (C) Cabbage (D) Onion
10. **Vegetable crop recently cultivated in International Space Station by NASA was:**
    (A) Cabbage (B) Lettuce
    (C) Potato (D) Palak
11. **The inheritance of Scerotinia stalk rot in cauliflower is controlled by:**
    (A) Polygenic recessive genes
    (B) Polygenic dominant genes
    (C) Dominant gene
    (D) Both (A) and (B)
12. **Tomato for seed extraction fruits are harvested at:**
    (A) Red ripe stage (B) Firm green stage
    (C) Pink stage (D) Green stage
13. **................... is known as king of vegetable:**
    (A) Potato (B) Radish
    (C) Cabbage (D) Brinjal
14. **Waxing is mostly followed in:**
    (A) Brinjal (B) Banana
    (C) Potato (D) Cabbage
15. **Onion for breeder seed production .... is commonly followed:**
    (A) Seed to bulb method
    (B) Bulb to seed method
    (C) Both (A) and (B)
    (D) None of these
16. **Arka Kaushal s a variety of:**
    (A) Tomato (B) Brinjal
    (C) Cabbage (D) Okra
17. **Pusa Pragati is a variety of:**
    (A) Radish (B) Garden pea
    (C) Cabbage (D) Okra
18. **Anti nutrition compound found in sweet potato is;**
    (A) Trypsin inhibitor
    (B) Pectin inhibitor
    (C) Hormone synthesis inhibitor
    (D) None of these
19. **Pusa bedana is a variety of:**
    (A) Watermelon (B) Bitter gourd
    (C) Muskmelon (D) Pumpkin
20. **Pusa rasdar is a variety of:**
    (A) Watermelon (B) Bitter gourd
    (C) Muskmelon (D) Pumpkin

Practice Paper 4

1. **White flowered cucurbit is:**
   (A) Bitter gourd (B) Muskmelon
   (C) Snake gourd (D) Water melon
2. **Shot hole symptoms in cauliflower is caused by:**
   (A) Shoot borer
   (B) Diamond black moth
   (C) Aphid
   (D) Thrips

3. **Seed plot technique is used to produce:**
   (A) Virus free seed tuber in north Indian plains
   (B) Virus free seed tuber in central Indian plains
   (C) Virus free seed tuber in south
   (D) Virus free seed tuber in North-East Indian plains
4. **Blanching in leaf petiole is followed in:**
   (A) Pumplin (B) Celery
   (C) Carrot (D) Radish
5. **.................. is an Asiatic variety of turnip:**
   (A) Pusa rudhira (B) Pusa sweti
   (C) Pusa kranti (D) Pusa ageti
6. **Garden pea shows .... germination:**
   (A) Epigeal (B) Hypogeal
   (C) Both (D) None
7. **Splitting in turnip is caused by:**
   (A) Heavy irrigation after prolonged drought
   (B) Light irrigation
   (C) Drought
   (D) None
8. **Low chilling temperature is most essential for bolting in:**
   (A) Cauliflower (B) Cabbage
   (C) Radish (D) Turnip
9. **Double bean is the acronym of:**
   (A) Rajma (B) Lima bean
   (C) Cowpea (D) Febabean
10. **Black spek is a disorder of:**
    (A) Potato (B) Sweet potato
    (C) Coriander (D) Cassawa
11. **Favism is an allergic problem associated with:**
    (A) Broad bean (B) Lima bean
    (C) Cowpea (D) Feba bean
12. **Ideal storage temperature for pea is:**
    (A) 5 °C (B) 0 °C
    (C) 10 °C (D) 8 °C
13. **Degree days is the maturity index of:**
    (A) Okra (B) Cabbage
    (C) Radish (D) Potato
14. **Grain amaranthus species is**
    (A) *Amaranthus acanthochiton*
    (B) *Amaranthus viridis*
    (C) *Amaranthus anderssonii*
    (D) *Amaranthus hypochondriachus*
15. **Wind pollinated crop is:**
    (A) Brinjal (B) Amaranth
    (C) Okra (D) Turnip
16. **Wild progenitor species of cole crops naturally found in:**
    (A) Europe
    (B) America
    (C) Mediterranean region
    (D) India
17. **Proginator species for cauliflower is:**
    (A) *Brassica cretica* (B) *Brassica olerasia*
    (C) *Brassica napus* (D) *Brassica cerenua*
18. **Chow chow (Sechium edule) belongs to family:**
    (A) Solanaceae (B) Cucurbitaceae
    (C) Brassicaceae (D) Cypraceae
19. ***Benincasa hispida* is the scientific name of:**
    (A) Bottle gourd (B) Wax gourd
    (C) Bitter gourd (D) Spongy gourd
20. **Gajendra is a variety of:**
    (A) Sweet potato
    (B) Elephant foot yam
    (C) Potato
    (D) Pea

Practice Paper 5

1. **Non heading type of Chinese cabbage is:**
   (A) Pak choi (B) Dak choi
   (C) Sak choi (D) Mak choi
2. **Pithiness is observed in:**
   (A) Onion (B) Brinjal
   (C) Tomato (D) Radish
3. **Curing of bulbs followed for:**
   (A) Removal of excess moisture from the skin
   (B) Removal of excess moisture from fruit
   (C) Removal of excess moisture from the fruit flesh
   (D) Removal of the skin
4. **Fragrant species of Allium is:**
   (A) *Allium cepa* (B) *Allium tuberosum*
   (C) *Allium aaseae* (D) *Allium abramsii*

5. Bhot jholokia which is the hottest cultivar of pepper belongs to:

(A) *Capsicum chinense*
(B) *Capsicum annum*
(C) *Capsicum fruitescence*
(D) None of these

6. Anthrocyanin rich red or purled coloured fruit in tomato is controlled by:

(A) Zm gene (B) Arn gene
(C) Aft gene (D) At gene

7. Tomato was introduced to India by Royal Agri-Horticultural Society, Calcutta in:

(A) 1808 (B) 1828
(C) 1838 (D) 1812

8. Ideal seed production method of cabbage is:

(A) Head cracking method
(B) Head intact method
(C) STF
(D) Seed plot technique

9. Multiplier onion is cultivated through out the year in:

(A) Tamilnadu (B) Maharastra
(C) Andhra Pradesh (D) Uttar Pradesh

10. Drying ratio of Pusa white round is:

(A) 1 : 8 (B) 8 : 1
(C) 7 : 1 (D) 9 : 1

11. Pusa barkha is the variety of cucumber in:

(A) Rainy season (B) Rabi season
(C) Summer season (D) All season

12. Bitterness in brinjal due to:

(A) Solasodine (B) Melanine
(C) Tepharine (D) Bittacine

13. Riceyness in cauliflower is due to the elongation of:

(A) Pedicel (B) Carpel
(C) Petal (D) Stamen

14. Greening is the physiological disorder of:

(A) Sweet potato (B) Potato
(C) Radish (D) Cauliflower

15. Pusa Safed baigan is oval fruited variety of:

(A) Turnip (B) Brinjal
(C) Cauliflower (D) Tomato

16. Pusa hara baigan 1 is antioxidant richest variety of:

(A) Brinjal (B) Turnip
(C) Cauliflower (D) Tomato

17. Asparagus belongs to the family:

(A) Poaceae (B) Liliaceae
(C) Cyperaceae (D) Tiliaceae

18. Triploid variety of Cassava is:

(A) Sree Naresh (B) Sree mukesh
(C) Sree harsha (D) Danish

19. Toxic substance found in the cassava is:

(A) HCN (B) ABA
(C) NAA (D) IAA

20. 3n-butyl pthalamide is found in:

(A) Radish (B) Celery
(C) Brinjal (D) Tomato

Preservation of Fruits and Vegetables

Practice Paper 1

1. IICPT stands for:

(A) Indian Institute of Crop Processing Technology
(B) Indian Institution of Crop Processing Technology
(C) Indian Institute of Crop Production Technology
(D) None of these

2 The transparency of apple juice is measured in:

(A) NTU (B) BTU
(C) CTU (D) GTU

3. Jam is produced from:

(A) Fruit juice (B) Pulp
(C) Clear extracts (D) Fruit slices

4. Jelly is produced from:

(A) Fruit juice (B) Pulp
(C) Clear extracts (D) Fruit slices

5. Jelly is a:

(A) Transparent (B) Opaque
(C) Semi-transparent (D) All of these

6. **Which of the following is not a fruit preservation method?**
(A) Drying (B) Canning
(C) Freezing (D) Grafting

7. **Which of the following fruit is used for Jam preparation?**
(A) Apple (B) Grape
(C) Papaya (D) All of these

8. **Which of the following method is used for temporary preservation of vegetables?**
(A) Drying
(B) Use of low concentration of antiseptic
(C) Pasteurization (D) All of these

9. **Which of the following method is used for permanent preservation of vegetables?**
(A) Cold storage (B) Bottling
(C) Drying (D) All of these

10. **Marmalade is produced from:**
(A) Fruit juice (B) Pulp
(C) Clear extracts (D) Both (A) and (B)

11. **Jams are solid gel which involves:**
(A) Pulp (B) Sugar
(C) Pectin (D) All of these

12. **The pectin mainly present in the:**
(A) Pulp (B) Seeds
(C) Skin (D) Juice

13. **The acid added to jam to attain desired acidity is:**
(A) Citric (B) Malic
(C) Tartaric (D) All of these

14. **Pressure canners used in canning is known as:**
(A) Steam blanchers (B) Retorts
(C) Plate exchangers (D) Sublimators

15. **Fruits and vegetables discolor when cut due to:**
(A) Caramelization (B) Sulfating
(C) Dehydration (D) Enzymatic browning

16. **Jam should have minimum per cent of soluble solids:**
(A) 28 (B) 48
(C) 68 (D) 88

17. **Marmalade is obtained by ...... of fruit juice**
(A) Boiling (B) Fermentation
(C) Homogenization (D) Irradiation

18. **Which of the following is a high pectin fruit:**
(A) Gooseberries
(B) Peaches
(C) Pears
(D) Figs

19. **Pectin is a polymer of:**
(A) Polygalacturonic acid
(B) Maltose
(C) Glucose
(D) None of these

20. **Fruit Product Order came in:**
(A) 1954 (B) 1955
(C) 1956 (D) 1957

Practice Paper 2

1. **Self decomposition of fruits and vegetables products can be prevented by:**
(A) Inactivating food enzymes
(B) By keeping them at room temperature
(C) By ripening
(D) None of these

2. **Which of the following is not as method of preservation:**
(A) Pickle making
(B) Dehydration
(C) Modified atmospheric packaging
(D) None of these

3. **Canning involves:**
(A) Low heating
(B) Low temperature processing
(C) Sterilization
(D) Pasteurization

4. **Curing, pickling and chutney preparation are examples of:**
(A) Preservation
(B) Deterioration
(C) Fermentation
(D) None of these

5. **Commonly added preservative in fruit juice is:**
(A) Potassium metasulphite
(B) Sodium benzoate
(C) Both (A) and (B)
(D) None of these

6. The presence of salt for preservation causes:
(A) Precipitation of enzyme protein
(B) Reduction in sulfur linkage in enzyme protein
(C) Reaction with aldehyde
(D) None of these

7. The tenderization of vegetables is done by:
(A) Cellulase
(B) Protease
(C) Glucosidase
(D) Pectinase

8. Soft pickles is due to:
(A) *Penicillium*
(B) *Fusarium*
(C) Both (A) and (B)
(D) None of these

9. Black pickles is due to:
(A) Yeast
(B) Mold
(C) Hydrogen sulfide produced by bacteria
(D) All of these

10. Slippery pickles results by action of:
(A) Yeast
(B) Mold
(C) Bacteria
(D) All of these

11. Papain is commercially used for:
(A) Tenderization of meat
(B) Fixation of red colour of meat
(C) Removal of fat from meat
(D) All of these

12. The enzyme is responsible for cut fruits and vegetables browning:
(A) PPO
(B) Lipase
(C) Phosphorylase
(D) Catalase

13. Enzyme which causes the loss of eating and nutritional quality in fruits and vegetables:
(A) Lipoxygenase
(B) PPO
(C) Polygalacturonase
(D) All of these

14. Function of blanching is:
(A) Reduce number of contaminating micro organisms
(B) Inactive enzymes
(C) Softens vegetable tissue
(D) All of these

15. Coating of fruits and vegetables:
(A) Improve appearances
(B) Enhance flavours
(C) Modify texture
(D) All of these

16. Dehydrofreezing of fruits and vegetables involves:
(A) Only freezing
(B) Dehydration before freezing
(C) Dehydration after freezing
(D) Dehydration

17. Preservative is used in jam, jellies and pickles:
(A) Benzoates
(B) Sorbates
(C) Acetate
(D) Both (A) and (B)

18. Scalding of vegetables is a:
(A) Freezing treatment
(B) Heat treatment
(C) Irradiation
(D) None of these

19. Which of the following preservative is used in the preparation of squash?
(A) Formaldehyde
(B) NaCl
(C) Potassium-meta bi-sulphide
(D) Na benzoate

20. Sodium benzoate is added for the preservation of fruit juice at the rate of:
(A) 0.01-0.5%
(B) 0.06-0.10%
(C) 1.10-1.5%
(D) 1.6-2.0%

Answers Key

Floriculture

■ Practice Paper 1

Q. No.	Ans	Q. No.	Ans	Q. No.	Ans	Q. No.	Ans	Q. No.	Ans	Q. No.	Ans
1.	A	2.	A	3.	C	4.	B	5.	A	6.	B
7.	C	8.	C	9.	A	10.	B	11.	A	12.	A
13.	D	14.	D	15.	A	16.	A	17.	C	18.	A
19.	D	20.	C								

■ Practice Paper 2

Q. No.	Ans	Q. No.	Ans	Q. No.	Ans	Q. No.	Ans	Q. No.	Ans	Q. No.	Ans
1.	A	2.	C	3.	A	4.	B	5.	A	6.	A
7.	B	8.	C	9.	A	10.	B	11.	C	12.	C
13.	D	14.	A	15.	C	16.	A	17.	C	18.	B
19.	B	20.	B								

■ Practice Paper 3

Q. No.	Ans	Q. No.	Ans	Q. No.	Ans	Q. No.	Ans	Q. No.	Ans	Q. No.	Ans
1.	A	2.	B	3.	B	4.	D	5.	B	6.	C
7.	C	8.	B	9.	A	10.	C	11.	B	12.	B
13.	C	14.	C	15.	B	16.	A	17.	A	18.	D
19.	A	20.	C								

■ Practice Paper 4

Q. No.	Ans	Q. No.	Ans	Q. No.	Ans	Q. No.	Ans	Q. No.	Ans	Q. No.	Ans
1.	A	2.	C	3.	B	4.	C	5.	A	6.	B
7.	D	8.	B	9.	C	10.	A	11.	A	12.	C
13.	B	14.	B	15.	A	16.	B	17.	A	18.	C
19.	C	20.	C								

Pomology

■ Practice Paper 1

Q. No.	Ans	Q. No.	Ans	Q. No.	Ans	Q. No.	Ans	Q. No.	Ans	Q. No.	Ans
1.	A	2.	C	3.	C	4.	D	5.	C	6.	B
7.	A	8.	D	9.	A	10.	A	11.	A	12.	B
13.	A	14.	A	15.	B	16.	D	17.	D	18.	B
19.	D	20.	B								

■ Practice Paper 2

Q. No.	Ans	Q. No.	Ans	Q. No.	Ans	Q. No.	Ans	Q. No.	Ans	Q. No.	Ans
1.	C	2.	A	3.	A	4.	B	5.	B	6.	A
7.	D	8.	A	9.	D	10.	D	11.	D	12.	D
13.	D	14.	A	15.	A	16.	A	17.	C	18.	A
19.	D	20.	D								

■ Practice Paper 3

Q. No.	Ans	Q. No.	Ans	Q. No.	Ans	Q. No.	Ans	Q. No.	Ans	Q. No.	Ans
1.	D	2.	D	3.	D	4.	D	5.	D	6.	D
7.	D	8.	D	9.	D	10.	D	11.	D	12.	D
13.	D	14.	A	15.	B	16.	D	17.	D	18.	D
19.		20.									

■ Practice Paper 4

Q. No.	Ans	Q. No.	Ans	Q. No.	Ans	Q. No.	Ans	Q. No.	Ans	Q. No.	Ans
1.	D	2.	D	3.	B	4.	B	5.	D	6.	D
7.	D	8.	D	9.	D	10.	C	11.	A	12.	D
13.	C	14.	A	15.	A	16.	A	17.	D	18.	D
19.	A	20.	C								

■ Practice Paper 5

Q. No.	Ans	Q. No.	Ans	Q. No.	Ans	Q. No.	Ans	Q. No.	Ans	Q. No.	Ans
1.	A	2.	A	3.	C	4.	B	5.	D	6.	B
7.	A	8.	B	9.	A	10.	C	11.	A	12.	C
13.	B	14.	D	15.	C	16.	D	17.	A	18.	D
19.	A	20.	C								

■ Practice Paper 6

Q. No.	Ans	Q. No.	Ans	Q. No.	Ans	Q. No.	Ans	Q. No.	Ans	Q. No.	Ans
1.	B	2.	A	3.	C	4.	B	5.	A	6.	B
7.	A	8.	D	9.	A	10.	C	11.	C	12.	B
13.	D	14.	B	15.	A	16.	A	17.	B	18.	C
19.	D	20.	A								

Practice Paper 7

Q. No.	Ans	Q. No.	Ans	Q. No.	Ans	Q. No.	Ans	Q. No.	Ans	Q. No.	Ans
1.	C	2.	B	3.	D	4.	B	5.	A	6.	D
7.	A	8.	C	9.	A	10.	C	11.	A	12.	B
13.	D	14.	C	15.	B	16.	B	17.	D	18.	D
19.	D	20.	A								

Vegetable Science

Practice Paper 1

Q. No.	Ans	Q. No.	Ans	Q. No.	Ans	Q. No.	Ans	Q. No.	Ans	Q. No.	Ans
1.	D	2.	C	3.	B	4.	A	5.	C	6.	A
7.	D	8.	A	9.	A	10.	C	11.	A	12.	C
13.	A	14.	D	15.	B	16.	D	17.	C	18.	C
19.	A	20.	C								

Practice Paper 2

Q. No.	Ans	Q. No.	Ans	Q. No.	Ans	Q. No.	Ans	Q. No.	Ans	Q. No.	Ans
1.	A	2.	A	3.	B	4.	A	5.	B	6.	A
7.	A	8.	D	9.	D	10.	D	11.	B	12.	A
13.	A	14.	A	15.	A	16.	B	17.	C	18.	A
19.	B	20.	D								

Practice Paper 3

Q. No.	Ans	Q. No.	Ans	Q. No.	Ans	Q. No.	Ans	Q. No.	Ans	Q. No.	Ans
1.	B	2.	A	3.	B	4.	B	5.	C	6.	C
7.	D	8.	C	9.	A	10.	B	11.	A	12.	A
13.	D	14.	A	15.	C	16.	B	17.	B	18.	A
19.	A	20.	B								

Practice Paper 4

Q. No.	Ans	Q. No.	Ans	Q. No.	Ans	Q. No.	Ans	Q. No.	Ans	Q. No.	Ans
1.	C	2.	B	3.	A	4.	B	5.	B	6.	B
7.	A	8.	B	9.	B	10.	B	11.	A	12.	B
13.	A	14.	D	15.	B	16.	C	17.	A	18.	B
19.	B	20.	B								

■ Practice Paper 5

Q. No.	Ans	Q. No.	Ans	Q. No.	Ans	Q. No.	Ans	Q. No.	Ans	Q. No.	Ans
1.	A	2.	D	3.	A	4.	B	5.	A	6.	C
7.	B	8.	B	9.	A	10.	B	11.	D	12.	A
13.	A	14.	C	15.	B	16.	A	17.	B	18.	C
19.	A	20.	B								

Preservation of Fruits and Vegetables

■ Practice Paper 1

Q. No.	Ans	Q. No.	Ans	Q. No.	Ans	Q. No.	Ans	Q. No.	Ans	Q. No.	Ans
1.	A	2.	A	3.	B	4.	C	5.	A	6.	D
7.	D	8.	D	9.	D	10.	D	11.	D	12.	C
13.	D	14.	B	15.	D	16.	C	17.	B	18.	A
19.	A	20.	B								

■ Practice Paper 2

Q. No.	Ans	Q. No.	Ans	Q. No.	Ans	Q. No.	Ans	Q. No.	Ans	Q. No.	Ans
1.	A	2.	D	3.	C	4.	A	5.	C	6.	A
7.	A	8.	C	9.	C	10.	A	11.	A	12.	A
13.	D	14.	D	15.	D	16.	B	17.	D	18.	B
19.	C	20.	B								

17. Multiple Choice Questions of Entomology

Practice Paper 1

1. **The branch of zoology concerned with the study of insects is called:**
   (A) Zoology (B) Entomolgy
   (C) Both of these (D) None of these
2. **A small arthropod animals that has six legs and generally one or two pairs of wings is called:**
   (A) Insect (B) Spider
   (C) Mammals (D) Pests
3. **To control of insects by means of chemicals is called:**
   (A) Pesticides (B) Miticide
   (C) Insecticide (D) Rodenticide
4. **DDT and 2,4-D were introduced:**
   (A) Towards end of World War II
   (B) Towards beginning of World War II
   (C) During World War I
   (D) Just after the World War I
5. **Which is organochlorine:**
   (A) Malathion (B) Parathion
   (C) Aldicarb (D) DDT
6. **Which one is an organophosphate:**
   (A) BHC (B) Aldrin
   (C) Fenitrothion (D) Endosulphan
7. **A carbamate pesticide is:**
   (A) Propoxur
   (B) Simazine
   (C) Atrazine
   (D) Tetraethyl pyrophosphate
8. **Urea derivative pesticides belong to the category of:**
   (A) Organochlorines
   (B) Organophosphates
   (C) Carbamates
   (D) Triazines
9. **Carbamates employed as insecticide are:**
   (A) Phenylcarbamates (B) Thiocabamates
   (C) Propoxur (D) Dithiocarbamates
10. **Carbamates used as fungicides are:**
    (A) Thiocarbamates
    (B) Dithiocarbamates
    (C) Carbofuron
    (D) Carbofuron and Aldicarb
11. **The pesticides having very low biodegradation and strong affinity for fatty tissues are:**
    (A) Organochlorines
    (B) Organophosphates
    (C) Triazines
    (D) Pyrethroids
12. **Pesticide ingredient methyl isocyanate which killed thousands in Bhopal gas tragedy is intermediate of:**
    (A) Mustard gas
    (B) Nitrous acid
    (C) Carbaryl
    (D) Carbon tetrachloride
13. **Pesticides comprise:**
    (A) Herbicides and insecticides
    (B) Insecticides only
    (C) Insecticides, nematicides and rodenticides
    (D) Fungicides, herbicides, insecticides, nematicides and rotendicides
14. **Household pesticides consist of:**
    (A) Finit, Baygon, Mosquitorepellent cakes and coils
    (B) Finit and Baygon
    (C) Finit/Flit
    (D) DDT
15. **First commercial pesticides was:**
    (A) Burgandy mixture
    (B) Bordeaux mixture
    (C) DDT
    (D) 2,4-D
16. **Mode of pesticides action is similar in:**
    (A) DDT and parathion
    (B) Organochlorine and organophosphates
    (C) Organochlorine and carbamates
    (D) Organophosphates and carbamates

17. The wheat seeds are treated with chlorpyriphos 20 EC against termite @:

(A) 3 to 4 ml/kg (B) 15-20 ml/kg
(C) 25 to 30 ml/kg (D) 35 to 40 ml/kg

18. Eggs of *Atherigona naqvi* are laid in the:

(A) Buds
(B) Flowers
(C) Inside the leaves
(D) At the base of stem and under side of leaves

19. The field rates are controlled by:

(A) Zinc chloride (B) Zinc phosphide
(C) Zinc oxide (D) Copper oxide

20. Which of the following is not attractant?

(A) Pheromones (B) Sound
(C) Light (D) Chemosterilants

Practice Paper 2

1. Pest is:

(A) Organism feeding on and damaging others
(B) Fungal pathogen attacking plants and animals
(C) Bacterial pathogen attacking plants and animals
(D) Viral pathogen attacking organisms

2. Which one of the following is 'June Beetle':

(A) *Holotrichia consanguinea*
(B) *Xanthopimia* sp.
(C) *Rodolia* sp.
(D) *Coccinella* sp.

3. Sugarcane shoot borer damages:

(A) Roots (B) Stem
(C) Leaves (D) All of these

4. 'Bunchy top' in sugarcane is caused by:

(A) Root borer (B) Pyrilla
(C) White fly (D) Scale insects

5. 'Anal-tufts' are found in the:

(A) Adult pyrilla
(B) Nymphs of pyrilla
(C) Nymphs and adults both
(D) None of these

6. *Beauvaria brassiana* is effective against control of:

(A) Termite (B) Pod borer
(C) Grasshopper (D) Grubs

7. There is a pupal stage in:

(A) Pyrilla (B) Termite
(C) White fly (D) Jassids

8. The female of which insect has a tuft of hairs at its anal end?

(A) Sugarcane top borer
(B) Sugarcane leaf hopper
(C) Sugarcane root borer
(D) Sugarcane white grubs

9. White grubs damage:

(A) Roots (B) Stems
(C) Leaves (D) Flowers

10. *Amrasca biguttula* is:

(A) Jassid (B) Leaf roller
(C) Whitefly (D) Stem weevil

11. 'Hibernacula' is found in:

(A) Flower beetle (B) Stem weevil
(C) Spotted boll worm (D) Pink boll worm

12. Yellow stem borer of rice is:

(A) *Scirpophaga innotata*
(B) *Scirpophaga incertulas*
(C) *Orseolia oryzae*
(D) *Dicladispa armigera*

13. Spotted boll worm of cotton pupaes is:

(A) Inside the boll
(B) Out of the boll but naked
(C) Out of the boll in a coccon
(D) Inside the seeds

14. Which one of the following is cotton leaf roller?

(A) *Oxycarenus laetus*
(B) *Sylepta derogata*
(C) *Dysdercus cingulatus*
(D) *Pempherulus affinis*

15. The larvae of *Sylepta derogata* are:

(A) Greenish bluish (B) Pinkish
(C) Blackish (D) Yelowish

16. Nymphs of red cotton bug also damage:

(A) Wheat (B) Sugracane
(C) Gram (D) Lady's finger

17. Which one of the following insects come from store to infest crop in field?

(A) Red cotton bug
(B) Pink boll worm
(C) Leaf roller
(D) Spotted boll worm

18. In which of the following insects a greenish strip is found in forewings?
(A) *Earia vitella*
(B) *Earias insulana*
(C) *Oxycarenus leatus*
(D) *Pectinophora gossypii*

19. Which one of the following lays stalked eggs?
(A) *Bemisia tabaci*
(B) *Aphis gossypii*
(C) *Myllocerms* sp
(D) *Pempherulus affinis*

20. Larvae of *Helicoverpa armigera* damage:
(A) Pods (B) Roots
(C) Stem (D) Buds

Practice Paper 3

1. Termite can be controlled by use of:
(A) BHC (B) Aldrin
(C) Chlordane (D) All of these

2. Japanese beetle is controlled by bacteria:
(A) *Bacillus popilliae*
(B) *Bacillis lentimarbus*
(C) Both (A) and (B)
(D) None of these

3. Acaricides are used against:
(A) Beetles (B) Mites
(C) Moths (D) Grubs

4. Which of the following is a chemosterilants?
(A) Tepa (B) Matepa
(C) Colchison (D) All of these

5. Which of the following is a systemic poison?
(A) Lindane (B) Thiodan
(C) Aldrin (D) All of these

6. Dimethoate is very effective against:
(A) Housefly (B) Fruit fly
(C) Aphids (D) All of these

7. Which of the following is a respiratory poison?
(A) Hydrogen sulphide
(B) Carbon monoxide
(C) Phosphin
(D) All of these

8. Which one of the following is better for the control of gram pod borer?
(A) Parathion
(B) Phosphomidon
(C) Chlorpyriphos
(D) Malathion

9. Which of the following is a physical repellent?
(A) Water barriers
(B) Tar
(C) Oil bands around tree trunks
(D) All of these

10. Cabamates include:
(A) Fungicides, herbicides and insecticides
(B) Insecticides only
(C) Herbicides, insecticides and nematicides
(D) Insecticides, nematicides and rotendicides

11. The first organic pesticide to be used commercially was:
(A) Bordeaux mixture
(B) Burgandy mixture
(C) DDT
(D) 2,4-D

12. Bordeaux mixture is:
(A) Organochlorine
(B) Antibiotic
(C) Organophosphate
(D) Inorganic fungicide

13. The chemical present in Temik is:
(A) Malathion (B) Propoxur
(C) Carbofuran (D) Aldicarb

14. The chemical present in flit (finit) is:
(A) Malathion (B) DDT
(C) BHC (D) Aldicarb

15. The mode of action of insecticides is through the attack on:
(A) Respiratory system
(B) Nervous system
(C) Muscular system
(D) Both (A) and (B)

16. One of earliest pesticides employed by humans was:
(A) Nicotine
(B) Margosa (Neem) leaves
(C) Pyrethrum
(D) DDT

17. Most effective pesticides are:
(A) Carbamates (B) Organochlorine
(C) Organophosphates (D) All the above

18. DDT content in Indian population has touched:
(A) 1-12 ppm (B) 11-30 ppm
(C) 25-50 ppm (D) 48-98 ppm

19. Which one is not matched:
(A) Organochlorine – DDT
(B) Pyrethroid – Atrazine
(C) Organophosphate – Malathion
(D) Carbamate – Carbofuran

20. Pests resistant to insecticides have:
(A) Enzymes for metabolizing toxins
(B) Ability to store toxin in fat
(C) Less permeable cuticles
(D) All the above

Practice Paper 4

1. 'Dead heart' is caused by:
(A) Paddy stem borer
(B) Termite
(C) Rice gandhi bug
(D) Paddy stem weevil

2. 'Silver shoot' is caused by:
(A) Grass hopper (B) Stem borer
(C) White fly (D) Gall midge

3. 'Chaffy appearance' in the paddy is caused on which part of plant:
(A) Roots (B) Tillers
(C) Flowers (D) Ears

4. *Odontotermis obesus* is controlled by the application of:
(A) Phosphamidon
(B) Aluminium chloride
(C) Chlorpyriphos
(D) Brestanol

5. Nuptial flight is found in:
(A) Termite (B) Grasshopper
(C) White fly (D) Gujhia weevil

6. Which one of the following is called as 'Army worm'?
(A) Leptocorisa (B) Mythimna
(C) Spodoptera (D) Tanymecus

7. Wheat shoot fly attacks the plants at:
(A) Seedling stage (B) Flowering stage
(C) Milky stage (D) Maturity stage

8. *Atherigona naqvii* belongs to family:
(A) Curculionidae (B) Aphididae
(C) Anthonomidae (D) Pyralidae

9. Which one of the following is Gujhia weevil?
(A) *Tanymcuns*
(B) *Mylabris pustulatus*
(C) *Proceras indicus*
(D) *Tryporyza incertulas*

10. BHC is:
(A) Benzene hexachloride
(B) Benzine heptachloride
(C) Benzene hydrochloride
(D) Bromide herbicide

11. *Drasicha mangiferae* is known as:
(A) Mango mealy bug
(B) Pink boll worm
(C) Cutworm
(D) None of these

12. Herbicides prevent:
(A) Transpiration
(B) Water absorption
(C) Translocation of salts
(D) Photolysis of water

13. Pesticides generally attacks:
(A) Muscular system
(B) Respiratory system
(C) Nervous system
(D) Circulatory system

14. Pesticides with very low biodegradation but strong affinity for fatty tissues are:
(A) Organochlorines
(B) Organophosphates
(C) Pyrethrines
(D) Triazines

15. Baygon contains:
(A) Malathion (B) Carbofuran
(C) Propoxur (D) Aldicarb

16. Which ingredient killed hundreds of people in Bhopal gas tragedy:
(A) Nitrous acid
(B) Mustard gas
(C) Carbon tetrachloride
(D) Methyl isocyanate

17. **DDT is:**
    (A) Biodegradable
    (B) Non-biodegradable
    (C) Antibiotic
    (D) Growth regulator

18. **Zinc phosphide is a pesticide used for:**
    (A) Spraying (B) Fumigation
    (C) Killing of weevils (D) Killing rodents

19. **Rotenone is:**
    (A) A bioherbicide
    (B) A natural hormone
    (C) An insect hormone
    (D) A natural insecticide

20. **Organophosphate inhibit:**
    (A) Cholinesterases (B) Enterokinase
    (C) Urease (D) Carbohydrases

Practice Paper 5

1. **Fungicide Bordeaux mixture is:**
   (A) Magnesium hydroxide
   (B) Copper sulphate + Sodium hydroxide
   (C) Copper sulphate + Calcium hydroxide
   (D) Magnesium sulphate + Calcium hydroxide

2. **Which of the following is cholinesterase inhibitor:**
   (A) Endosulphan (B) Aldrin
   (C) BHC (D) Malathion

3. **Which is major component of Bordeaux mixture:**
   (A) Copper sulphate
   (B) Calcium chloride
   (C) Sodium chloride
   (D) Magnesium sulphate

4. **Malathion, parathion and fumitrothion belong to group of pesticides:**
   (A) Triazines (B) Carbamates
   (C) Pyrethroids (D) Organophosphates

5. **Ethylene dibromide is:**
   (A) Stomach poison (B) Contact poison
   (C) Fumigant (D) Biological control

6. **A widely but periodically occupying plant disease is:**
   (A) Andemic (B) Epidemic
   (C) Sporodic (D) Epiphytotic

7. **Insecticides generally attack:**
   (A) Respiratory system
   (B) Nervous system
   (C) Muscular system
   (D) Circulatory system

8. **Carbamates used as fungicides are:**
   (A) Thiocarbamates
   (B) Dithiocarbamates
   (C) Carbofuran
   (D) Carbofuran and Aldicarb

9. **A major pest eradicated through release of sterile males is:**
   (A) Screw worm (B) Aphids
   (C) Ladybug (D) Praying mantis

10. **Rotenone is a:**
    (A) Natural herbicide (B) Bioherbicides
    (C) Natural insecticide (D) Insect hormone

11. **Nerve poisons are:**
    (A) Chlorinated hydrocarbons
    (B) Organo phosphates
    (C) Plant origin insecticides
    (D) All of these

12. **Which of the following is a contact poison?**
    (A) Chlorinated hydrocarbons
    (B) Organo phosphate
    (C) Carbamates
    (D) All of these

13. **Pyrethrin is a common ingredient of:**
    (A) Mosquito coils (B) Fly sprays
    (C) Mosquito mats (D) All of these

14. **Thurioside is:**
    (A) Insecticide (B) Fungicide
    (C) Antibiotic (D) Weedicide

15. **Antifeedant property occurs in:**
    (A) Nicotine (B) Azadirachtin
    (C) Rotenone (D) Cinerin

16. **Fruit fly is pest of bottle guard and damaged fruits so, fruit fly can be cotrolled by:**
    (A) By staking
    (B) BY seed treatment
    (C) Bt spray of Aldrin dust
    (D) By spray of 0.2 % Sevin solution at flowering stage

17. Which of the following poison is used to control of rats?

(A) Zinc phosphide (B) Barium carbonate
(C) Arsenium oxide (D) All of these

18. Which of the fumigants can be used against control of rats?

(A) Sulphos (B) Cynogas dust
(C) Hydrocynic acid (D) All of these

19. Which of the following is not a ingredient of universal antidote for insecticide poisoning?

(A) Charcoal (B) Tannic acid
(C) Milk of magnesia (D) All of these

20. Army worm is a serious pest of:

(A) Wheat (B) Chickpea
(C) Rice (D) Groundnut

Practice Paper 6

1. Which one is an effective plant insecticide:

(A) Pyrethrin (B) Cinerin
(C) Nicotine (D) All of these

2. The earliest pesticide was:

(A) Margosa (B) Pyrethrum
(C) Nicotine (D) DDT

3. Pod borer of pulses can be controlled by:

(A) Spray of Endosulphan 35 EC @ 1.5 l/ha
(B) Spray of Monocrotophos @ 0.75 l/ha
(C) Spray of Quinolphos 25 EC @ 1.25 l/ha
(D) All of these

4. Blue triable on insecticide container indicates:

(A) Highly poisonous (B) Poisonous
(C) Medium poisonous (D) Less poisonous

5. The dark yellow triangle on insecticide container indicates:

(A) Highly poisonous (B) Poisonous
(C) Medium poisonous (D) Less poisonous

6. Dark green triable on insecticide container indicates:

(A) Highly poisonous
(B) Poisonous
(C) Medium poisonous
(D) Less poisonous

7. Which of the following poison is used in killing bottle?

(A) Potassium cynide
(B) Hydrocynic acid
(C) H_2S
(D) None of these

8. A third generation pesticide is:

(A) Pheromone
(B) Weedicide
(C) Juvenile hormone analogue
(D) Sterilogastrone

9. Genus *Radopholus* of nematode was first described in 1893 on the basis of a male specimen obtained from necrotic roots of .... growing in Fiji Islands:

(A) Sugarbeet (B) Sugarcane
(C) Banana (D) Coconut

10. The nematode *Radopholus similes* has shown the maximum numbers at a soil depth of:

(A) 5-15 m (B) 20-30 m
(C) 32-42 m (D) 1-4 m

11. Stem borer can be controlled by:

(A) Putting mixture of chloroform and carbon bisulphide into the holes
(B) Spray of monocrotophos
(C) Insect may be killed with the help of wire
(D) All of these

12. Red spider can be controlled by:

(A) Spray of sulphur (B) Spray of zinc
(C) Spray of iron (D) Spray of 2,4-D

13. Ultra-Low Volume (ULV) or Low Volume Concentration (LVC) refers to the use of spray volume (liquid) less than :

(A) 5 litres (B) 7 litres
(C) 9 litres (D) 11 litres

14. Malathion is a insecticide:

(A) Slow fumigant
(B) Contact plus systemic
(C) Contact plus stomach
(D) Systemic

15. Metasystox - R is a .... insecticide:

(A) Sytemic
(B) Fumigant
(C) Contact
(D) Contact plus stomach

16. Phorate is a .... insecticide:
(A) Systemic
(B) Contact plus systemic
(C) Contact
(D) Contact plus stomach

17. Phosdrin is a .... insecticides:
(A) Slow fumigant
(B) Systemic
(C) Systemic plus contact
(D) Contact

18. Pyrilla can be controlled by spraying of:
(A) Endosulphan (B) Rogor
(C) Phosphemidon (D) All of these

19. Which of the preventive measure should be adopted for safe storage?
(A) Fumigation of godowns with E.D.C.T.
(B) Bags should be treated with 0.1 % malathion solution
(C) Put E.D.B. ampules 3.0 ml/q grain
(D) All of these

20. Which of the following is the antidote of insecticide poisoning?
(A) Atropine sulphate
(B) Diacetyl monohexane
(C) Pyridine aldoxymet
(D) All of these

Practice Paper 7

1. One kilometer of locust swarm weights as much as .... tones of locusts:
(A) 100 (B) 200
(C) 300 (D) 400

2. The duration of locust plague period lasts for:
(A) 2-3 years (B) 3-4 years
(C) 5-10 years (D) 11-16 years

3. An average locust swarm spreads over:
(A) 5 sq km (B) 10 sq km
(C) 15 sq km (D) 20 sq km

4. Dead heart in rice is the symptom of insect damage:
(A) Stem borer (B) Leaf roller
(C) Grass hopper (D) Gandhi bug

5. *Agrotis flammatra* larvae cut the:
(A) Leaves (B) Buds
(C) Flowers (D) Stems

6. A dark smoky spot is present on the centre of fore wings is:
(A) *Helicoverpa armigera*
(B) *Agrotis ipsilon*
(C) *Etiella zinckenella*
(D) *Exelastris atomosa*

7. Which of the stage of fruit fly is more harmful:
(A) Adult (B) Nymph
(C) Maggot (D) Adult and Nymph

8. Grasshoppers lay eggs:
(A) On soil surface
(B) Below soil surface
(C) In nest
(D) All of these

9. Suprannal plate is also called as:
(A) 11th tergite (B) Podical plate
(C) Hook (D) All of these

10. Holopneustic respiration system refers to:
(A) All spiracles open and work effectively
(B) Only first one pair spiracles open and work
(C) Only last one pair spiracles open and work
(D) Only first and last pair of spiracles open and work

11. Propneustic respiration system refers to:
(A) All spiracles open and work effectively
(B) Only first one pair spiracles open and work
(C) Only last one pair spiracles open and work
(D) Only first and last pair of spiracles open and work

12. Metapneustic respiration system refers to:
(A) All spiracles open and work effectively
(B) Only first one pair spiracles open and work
(C) Only last one pair spiracles open and work
(D) Only first and last pair of spiracles open and work

13. Johnson's organ is a:
(A) Same organ (B) Respiratory organ
(C) Circulatory organ (D) All of these

14. Johnson's organ is found at:
(A) 2nd segment of antennal
(B) Tympanum
(C) Mouth cavity
(D) Overyduct

15. Stadium is a:
(A) Stage of insects during moulting
(B) Time between two moulting
(C) Play ground for insects
(D) None of these

16. Apneustic respiration means:
(A) All spiracles remain closed
(B) Only few spiracles closed
(C) All spiracles open
(D) Only few spiracles open

17. Sound producing and receiving organs are called :
(A) Scolophorous organs
(B) Olfactory organs
(C) Gustatory organs
(D) None of these

18. A metomorphosis or Ameta bola is a process in which:
(A) No moulting occurs
(B) No metamorphosis occurs
(C) Only growth in size and colour may occur
(D) All of these

19. Levelling of insect should indicate the information:
(A) Name of insect and its host plant
(B) Date and place of collection
(C) Name of collector
(D) All of these

20. At which stage, Gandhi bug damages the rice crop:
(A) Tillering stage
(B) Panicle initiation stage
(C) Milking stage
(D) Maturity stage

Practice Paper 8

1. Yellowing of flour is caused by:
(A) Rice weevil
(B) Red rust flour beetle
(C) Khapra beetle
(D) All of these

2. Khapra beetle is also known as:
(A) Pai (B) Grain moth
(C) Red surahi (D) All of these

3. Bihar hairy caterpillar can be controlled by:
(A) Chlorpyriphos (B) Endosulphan
(C) Nuvan (D) All of these

4. Which one of the following caste of termite is sterile:
(A) King (B) Queen
(C) Soldiers (D) Winged termite

5. Red pumpkin beetle is a harmful insects of:
(A) Tomato (B) Cucurbits
(C) Rice (D) Potato

6. Ovoviviparous insects can:
(A) Adult can yield nymph either directly or from eggs
(B) Adult produces only larvae
(C) Adult can produce nymphs for eggs only
(D) All of these

7. Ovoviviparous insect is:
(A) Grasshopper (B) Mustard aphid
(C) Fruit fly (D) Shoot borer

8. Which of the following is a natural enemy of mustard aphid as:
(A) *Cocciniella septumpunctata*
(B) *Chrysopa* spp.
(C) *Syrphid maggot*
(D) All of these

9. Scientific name of gram caterpiller is:
(A) *Heliothis armigera*
(B) *Athalia spp*
(C) *Nezara viridula*
(D) None of these

10. Scientific name of mustard saw fly is:
(A) *Athalia proxima*
(B) *Bagrada cruciferaum*
(C) *Trogogerma granarium*
(D) None of these

11. Which of the following stage of mealy bug is harmful?
(A) Larvae only
(B) Female adult only
(C) Male adult only
(D) Larvae and female adult

12. **Most harmful stage of khapra beetle is:**
(A) Grub (B) Adult
(C) Pupa (D) All of these

13. **Which of the following stage of lesser grain moth is harmful to grains?**
(A) Grub only (B) Adult only
(C) Grub and adult (D) Nymph only

14. **Red rust flour remain most active during:**
(A) Winter season (B) Summer season
(C) Rainy season (D) All season

15. **Which of the life stage of grain moth is harmful?**
(A) Egg and grub
(B) Caterpillar
(C) Nymph and adult
(D) Adult

16. **Net like structures are formed in flour due to presence of:**
(A) Khapra beetle
(B) Red rust flour beetle
(C) Rice weevil
(D) None of these

17. **Following insecticides may be used to control of mealy bug is:**
(A) Phosphomedon 100 EC
(B) Malathion 50 EC
(C) Thiodan 35 EC
(D) All of these

18. **Mango leaf hopper is most active in the month of:**
(A) February – November
(B) December – January
(C) September – October
(D) All of these

19. **Which of the following part of grains is damaged by khapra beetle?**
(A) Seed
(B) Endosperm
(C) Embryo
(D) Cotyledons

20. **Azospirillum biofertilizer induce resistance in sorghum against pest of:**
(A) Shoot fly
(B) Head borer
(C) Cut worm
(D) None of these

Practice Paper 9

1. **To insects ccdysone takes part in:**
(A) Growth and development
(B) Maturation into adult
(C) Moulting till pupa is formed
(D) Secretion of cuticle

2. **Which of the storage pest may survive at ≤ 10 % moisture in stored grains?**
(A) Rice weevil (B) Khapra beetle
(C) Pulse beetle (D) All of these

3. **Cut worm generally attack on plants at:**
(A) Day only (B) Day and night
(C) Night only (D) None of these

4. **Which of the following is a parasite insect of Gurdaspur borer?**
(A) *Trichogramme* spp
(B) *Apanteles flavipes*
(C) *Sturmiopsis interens*
(D) All of these

5. **Dead heart in sugarcane is caused by:**
(A) Top borer (B) Pyrilla
(C) Gandhi bug (D) Shoot borer

6. **Sugarcane crop is damaged severely by shoot borer in the month of:**
(A) April – June
(B) July – September
(C) October – December
(D) January –March

7. **The virus of Bunchy top disease of banana is transmitted by:**
(A) Fruit fly (B) Pentalonia
(C) Thrips (D) None of these

8. **Antibiosis refers to:**
(A) Ability of plant to suppress the reproduction potential of the insect
(B) Production of antibodies by insect
(C) Insecticide to kill insects
(D) None of these

9. **Phytoecdysone is a:**
(A) Plant origin hormone similar to moulting hormone of insect
(B) Insect moulting hormone
(C) Plant steroids to kill insects
(D) None of these

10. Baygon acts as antifedant against:

(A) Cotton ball weevil (B) Moth worm
(C) Cut worm (D) None of these

11. Which of the following is a protoplasmic poison?

(A) Nitrophenols (B) Lead arsinate
(C) Sodium tetra borate (D) All of these

12. Which of the following is a fumigants?

(A) Methyl bromide (B) HCN
(C) EDCT (D) All of these

13. Which of the following is a plant origin?

(A) Nicotine sulphate (B) Pyrethrins
(C) Ryania (D) All of these

14. Which of the following is a stomach poison?

(A) Arsenic salt
(B) Flourine compounds
(C) Chlorinated hydrocarbons
(D) All of these

15. Sterility method as a synergist if it is mixed with?

(A) Pyrethrum (B) Lindane
(C) Endosulphan (D) None of these

16. The chemicals that make insects sterile are known as:

(A) Systemic poison (B) Chemosterilants
(C) Contact poison (D) Pheromones

17. DDVP is known as:

(A) Nuvan (B) Malathoate
(C) Thiodan (D) Sulfec

18. Cannibalism is a process in which insects:

(A) Fight and kill each other
(B) Protect each other
(C) Attract for copulation
(D) None of these

19. The grasshopper always move towards:

(A) South-east direction
(B) South-west direction
(C) North-east direction
(D) North-west direction

20. Which of the following is the optimum temperature for insect growth and development?

(A) 5-10 °C (B) 13-40 °C
(C) 42-55 °C (D) 15-20 °C

Practice Paper 10

1. *Corpora allata* is removed from a nymph. It will:

(A) Remain nymph for life
(B) Become adult
(C) Change to next nymph and remain in that state for ever
(D) Die after some time

2. Typical or primitive type of mouth parts are found in:

(A) Grubs (B) Honey bee
(C) Bugs (D) Mosquitoes

3. Trachea and tracheoles are found in:

(A) RBC (B) WBC
(C) Blood (D) All of these

4. Which is the most important pest of cauliflower is?

(A) Sawfly (B) Aphids
(C) Cabbage borer (D) Flea beetle

5. Which is the most important pest of onion is?

(A) Thrips (B) Maggot
(C) Tobacco caterpiller (D) Flea beetle

6. Which is the most important pest of bitter guard is:

(A) Red Pumpkin beetle
(B) Aphid
(C) Cutworm
(D) Fruitfly

7. Which one of the following is a cut worm:

(A) Plume moth (B) Pod borer
(C) *Agrotis* sp (D) *Heliothis* sp.

8. Which is the most important pest of carrot is:

(A) Cut worm (B) Wire worm
(C) White grub (D) All of these

9. Sweet potato aphinx attacks mainly on:

(A) Leaves (B) Stem
(C) Roots (D) All of these

10. Which of the following rat species is found in agricultural farms only:

(A) *Bandicota bengalensis*
(B) *Nesokia indica*
(C) *Tetera indica*
(D) All of these

11. Red triable on insecticide container indicates:
(A) Highly poisonous
(B) Poisonous
(C) Medium poisonous
(D) Less poisonous

12. Relaxing insects of insect means:
(A) To collect dead insects in a special bottle for establishing its bodies
(B) To kill the insects
(C) To mount the insects
(B) None of these

13. Which of the following insects are insect parasites?
(A) *Trichogramma* spp.
(B) *Microbacon* spp.
(C) *Sarcophaga* spp.
(D) All of these

14. Which of the following is a nocturnal insect?
(A) Butterfly (B) Beetle
(C) Moth (D) Honey bee

15. Which of the following is known as incomplete metamorphosis?
(A) A metamorphosis
(B) Complete metamorphosis
(C) Simple metamorphosis
(D) None of these

16. Mciller's organ is a:
(A) Sense organ
(B) It is a group of small many sense organs
(C) It is found in Tympanum
(D) All of these

17. Complete metamorphosis is found in:
(A) Exopterygota (B) Endopterygota
(C) Apterygota (D) None of these

18. Mustard aphid has a mouth organs:
(A) Sucking type
(B) Piercing and sucking type
(C) Siphoning type
(D) Chewing type

19. Which of the following stage of mustard aphid is harmful to crop plants?
(A) Nymph (B) Egg
(C) Adult (D) Nymph and Adult

20. Which stage of mustard saw fly is:
(A) Egg (B) Grubs
(C) Adult (D) Grub and Adult

Practice Paper 11

1. Harmful stage of painted mustard bug is:
(A) Nymph (B) Egg
(C) Adult (D) Nymph and Adult

2. Harmful stage of red pumpkin beetle is:
(A) Nymph
(B) grubs
(C) Adult
(D) Grubs and Adult

3. Which of the following is a natural enemy of mustard aphid is:
(A) *Cocciniella septumpunctata*
(B) *Chrysopa* spp.
(C) *Syrphid maggot*
(D) All of these

4. Harmful stage of mango leaf hopper is:
(A) Nymph (B) Egg
(C) Adult (D) Nymph and Adult

5. Mango leaf roller commonly known as:
(A) Mango chanpa (B) Lassi
(C) Tella (D) All of these

6. Mango leaf hopper has mouth parts:
(A) Piercing type
(B) Sucking type
(C) Piercing and sucking type
(D) Chewing type

7. Scientific name of mango hopper is:
(A) *Amritodus atkinsoni*
(B) *Athalia* spp
(C) *Heliothis* spp
(D) None of these

8. Polymorphism is found in:
(A) Mustard aphid
(B) Lemon butterfly
(C) Cotton leaf roller
(D) None of these

9. Mustard aphid is a serious pest of:
(A) Rape seed and mustard
(B) Chickpea
(C) Pigeonpea
(D) Sesame

10. Scientific name of mustard aphid is:

(A) *Lipaphis erysimi*
(B) *Lipophis cruciferaum*
(C) *Athalia proxima*
(D) None of these

11. Which of the following is a natural enemies of honey bees?

(A) Wax moth (B) *Acherontia styx*
(C) *Vespa cincta* (D) All of these

12. The colour of the lac insect is:

(A) Pink and Red (B) Green – red
(C) Blue – red (D) Black

13. Brood lac is a:

(A) The branch of a plant that contains lac cells
(B) Caterpiller of lac insect
(C) Pupa of lac insect
(D) None of these

14. Which of the following is a rat species?

(A) *Rattus rattus* (B) *Rattus norvegicus*
(C) *Mus musculus* (D) All of these

15. Brown or Norvey rat is:

(A) *Rattus rattus* (B) *Rattus norvegicus*
(C) *Mus musculus* (D) None of these

16. The mouth organ of mites is:

(A) Cutting, piercing and sucking type
(B) Chewing and lapping type
(C) Sponging type
(D) None of these

17. The middle stage between egg and adult of mites is called as:

(A) Nymph (B) Larvae
(C) Maggot (D) None of these

18. 'Hopper burn' is caused by:

(A) *Leptocorisa varicornis*
(B) *Bemisia tabaci*
(C) *Spodoptera litura*
(D) *Nilaparvata lugens*

19. Cutworms pupate:

(A) Inside the pods
(B) Inside the stem
(C) On upper surface of soil
(D) In an earthen shell inside the soil

20. The plume moth belongs to the family:

(A) Noctudae (B) Hespiridae
(C) Pterophoridae (D) Gelehiidae

Practice Paper 12

1. The phenomenon of using a predator for controlling a pest is:

(A) Biological control
(B) Genetic engineering
(C) Artificial control
(D) Confusion technique

2. Which of the following is a predator insects?

(A) Trichogramma (B) Coccinellids
(C) Apanteles (D) None of these

3. Trycho card is very effective against control of:

(A) Termite (B) Pod borer
(C) Grasshopper (D) Cutworm

4. First bioinsecticide developed on commercial scale was:

(A) Quinone (B) DDT
(C) Organophosphate (D) Sporeine

5. Biopesticides include:

(A) Bioherbicides
(B) Bioinsecticides
(C) Bioherbicides and bioinsecticides
(D) Bioherbicides, bioinsecticdes and biofertilizers

6. Pheromone are chemicals that are used in animals for:

(A) Sex attractants
(B) Sending alarm signals
(C) Marking trails
(D) All the above

7. Pheromone trap is used for capturing:

(A) Larvae (B) Pupae
(C) Adult males (D) Adult females

8. A natural insecticide obtained from plants is:

(A) Azadirachtin
(B) Rotenone
(C) Nicotine, pyrethrum and cinerin
(D) All the above

9. Seed treatment of *Beauvania brassiana* @ 10 g/kg is effective against control of:

(A) Termite (B) Pod borer
(C) Grasshopper (D) Cutworm

10. Against *Helicoverpa armigera,* NPV is applied @:

(A) 250 LE/ha (B) 100 LE/ha
(C) 50 LE/ha (D) 10 LE/ha

11. *Epiricanea melanoleuca* is a parasite of pyrilla:

(A) Nymphs only
(B) Adults only
(C) Nymphs and adult both
(D) Eggs

12. Pheromone traps have:

(A) Sterile males
(B) Male pheromones
(C) Female pheromones
(D) Sterile females

13. In confusion technique, the pheromone coated paper strips are thrown over an area in order to:

(A) Attract insects and kill them
(B) Repel insects from an area
(C) Confuse males so that they are unable to locate females
(D) Confuse females so that they are unable to locate males

14. Juvenile hormone causes:

(A) Early sexual maturity
(B) Late sexual maturity
(C) Completion of larval stages
(D) Development of giant larva without ability to undergo maturation

15. Pyrethrin is extracted from:

(A) *Azadiracta indica*
(B) *Helianthus annus*
(C) *Poa indica*
(D) *Chrysanthemum cinerarifolium*

16. Ladybug is a predator of:

(A) Moths (B) Beetles
(C) Bacteria (D) Aphids

17. Which is a microbial insecticide:

(A) Bacillus thuringensis
(B) *B. subtilis*
(C) *B. polymixa*
(D) *B. brexis*

18. Thurioside is proteinaceous toxin obtained from:

(A) Biofertilier (B) Green manure
(C) Bacterial origin (D) Farmyard manure

19. Which of the following is a predator insects?

(A) Trichogramma (B) Coccinellids
(C) Apanteles (D) None of these

20. *Cocobacillis acridiorum* bacteria can be used against:

(A) Locust (B) Housefly
(C) Grasshopper (D) Mites

Practice Paper 13

1. Which of the polyphagous insect?

(A) Termite
(B) Bihar hairy caterpillar
(C) Locust
(D) All of these

2. Six segmented head is found in:

(A) Myriapods (B) Arthopoda
(C) Annelida (D) None of these

3. Homeostasis refers to:

(A) Stability of population in a ecosystem
(B) Change in insect population
(C) Increase of insect population
(D) None of these

4. Gyplure is a:

(A) Synthetic compound related to natural compounds of gypsy moth
(B) Natural compound of gypsy moth
(C) Natural sex hormone of fruit fly
(D) None of these

5. The only plant that remained untouched by *Schistocerca gregarta* (Desert locust):

(A) Mango (B) Ber
(C) Neem (D) None

6. Which of the stage of pyrilla is more harmful:

(A) Adult
(B) Nymph
(C) Pupa
(D) Adult and Nymph

7. Monophagy refers to:

(A) Insect feeding on simple plant species
(B) Insect feeding on large variety of plants
(C) Insect feeding on few plants
(D) None of these

8. **Oligophagy refers to:**
(A) Insect feeding on simple plant spp.
(B) Insect feeding on large number of plants
(C) Insect feeding on few plants
(D) None of these

9. **Insect requires relative humidity for its better growth and development?**
(A) 30-40 % (B) 40-50%
(C) 50-60 % (D) 60-80 %

10. **Gandhi bud is active in south India during:**
(A) June – October
(B) November – December
(C) February – March
(D) May - June

11. **Gandhi bud is active in north India during:**
(A) June – October
(B) November – December
(C) February – March
(D) May – June

12. **Order of red pumpkin beetle is a:**
(A) Orthoptera (B) Coleoptera
(C) Isoptera (D) Hemiptera

13. **Pyrilla is a severe insect of:**
(A) Rice (B) Sugarcane
(C) Maize (D) Wheat

14. **Gurdaspur borer is a pest of:**
(A) Rice (B) Sugarcane
(C) Maize (D) Wheat

15. **Fruit fly lays eggs:**
(A) On leaves (B) Inside fruits
(C) In soil (D) All of these

16. **Pheromone is released by:**
(A) Male insects
(B) Female insects
(C) Male and female insects
(D) None of these

17. **Coccinellids are predator on:**
(A) *Coccinellids* (B) *Chrysopa* spp
(C) *Mantis* (D) All of these

18. ***Rodolia cardinalis* (lady bird beetle) is a predator insect of:**
(A) *Icerya purchasi*
(B) *Pseudococcus gahani*
(C) *Pseudococcus nipoe*
(D) None of these

19. ***Pseudoapliyeus utilis* is a parasite insect of:**
(A) Mealy bug
(B) Sugarcane weevil
(C) Cottony cushion scale
(D) None of these

20. ***Cytrohinus mundulus* is a predator insect of:**
(A) Mealy bug
(B) Sugarcane weevil
(C) Cottony cushion scale
(D) None of these

Answers Key

■ Practice Paper 1

Q. No.	Ans	Q. No.	Ans	Q. No.	Ans	Q. No.	Ans	Q. No.	Ans	Q. No.	Ans
1.	B	2.	A	3.	C	4.	A	5.	D	6.	C
7.	A	8.	D	9.	C	10.	B	11.	A	12.	C
13.	D	14.	A	15.	B	16.	D	17.	A	18.	D
19.	B	20.	D								

■ Practice Paper 2

Q. No.	Ans	Q. No.	Ans	Q. No.	Ans	Q. No.	Ans	Q. No.	Ans	Q. No.	Ans
1.	A	2.	A	3.	B	4.	A	5.	B	6.	A
7.	C	8.	A	9.	A	10.	A	11.	D	12.	B
13.	C	14.	B	15.	A	16.	D	17.	B	18.	B
19.	A	20.	A								

■ Practice Paper 3

Q. No.	Ans	Q. No.	Ans	Q. No.	Ans	Q. No.	Ans	Q. No.	Ans	Q. No.	Ans
1.	D	2.	C	3.	D	4.	D	5.	D	6.	D
7.	D	8.	D	9.	D	10.	A	11.	C	12.	D
13.	D	14.	A	15.	D	16.	B	17.	C	18.	B
19.	B	20.	D								

■ Practice Paper 4

Q. No.	Ans	Q. No.	Ans	Q. No.	Ans	Q. No.	Ans	Q. No.	Ans	Q. No.	Ans
1.	A	2.	D	3.	D	4.	C	5.	A	6.	B
7.	A	8.	C	9.	A	10.	A	11.	A	12.	D
13.	C	14.	A	15.	C	16.	D	17.	B	18.	D
19.	D	20.	A								

■ Practice Paper 5

Q. No.	Ans	Q. No.	Ans	Q. No.	Ans	Q. No.	Ans	Q. No.	Ans	Q. No.	Ans
1.	C	2.	D	3.	A	4.	D	5.	C	6.	D
7.	A	8.	B	9.	A	10.	C	11.	D	12.	D
13.	D	14.	A	15.	B	16.	D	17.	D	18.	D
19.	D	20.	C								

■ Practice Paper 6

Q. No.	Ans	Q. No.	Ans	Q. No.	Ans	Q. No.	Ans	Q. No.	Ans	Q. No.	Ans
1.	D	2.	A	3.	A	4.	C	5.	B	6.	D
7.	A	8.	C	9.	C	10.	B	11.	D	12.	A
13.	A	14.	C	15.	C	16.	B	17.	A	18.	D
19.	D	20.	D								

■ Practice Paper 7

Q. No.	Ans	Q. No.	Ans	Q. No.	Ans	Q. No.	Ans	Q. No.	Ans	Q. No.	Ans
1.	C	2.	C	3.	B	4.	A	5.	A	6.	A
7.	C	8.	B	9.	D	10.	A	11.	B	12.	C
13.	A	14.	A	15.	A	16.	A	17.	A	18.	D
19.	D	20.	C								

■ Practice Paper 8

Q. No.	Ans	Q. No.	Ans	Q. No.	Ans	Q. No.	Ans	Q. No.	Ans	Q. No.	Ans
1.	B	2.	A	3.	D	4.	C	5.	B	6.	A
7.	B	8.	D	9.	A	10.	A	11.	D	12.	A
13.	C	14.	C	15.	B	16.	B	17.	D	18.	A
19.	B	20.	A								

■ Practice Paper 9

Q. No.	Ans	Q. No.	Ans	Q. No.	Ans	Q. No.	Ans	Q. No.	Ans	Q. No.	Ans
1.	C	2.	B	3.	C	4.	D	5.	A	6.	A
7.	B	8.	A	9.	A	10.	A	11.	D	12.	D
13.	D	14.	D	15.	A	16.	B	17.	A	18.	A
19.	A	20.	B								

■ Practice Paper 10

Q. No.	Ans	Q. No.	Ans	Q. No.	Ans	Q. No.	Ans	Q. No.	Ans	Q. No.	Ans
1.	A	2.	A	3.	C	4.	B	5.	A	6.	A
7.	B	8.	A	9.	A	10.	D	11.	A	12.	A
13.	D	14.	C	15.	C	16.	D	17.	B	18.	B
19.	D	20.	B								

■ Practice Paper 11

Q. No.	Ans	Q. No.	Ans	Q. No.	Ans	Q. No.	Ans	Q. No.	Ans	Q. No.	Ans
1.	D	2.	D	3.	D	4.	D	5.	D	6.	C
7.	A	8.	A	9.	A	10.	A	11.	D	12.	A
13.	A	14.	D	15.	B	16.	A	17.	A	18.	D
19.	D	20.	C								

■ Practice Paper 12

Q. No.	Ans	Q. No.	Ans	Q. No.	Ans	Q. No.	Ans	Q. No.	Ans	Q. No.	Ans
1.	A	2.	A	3.	B	4.	D	5.	C	6.	D
7.	C	8.	D	9.	A	10.	A	11.	C	12.	C
13.	C	14.	D	15.	D	16.	D	17.	A	18.	C
19.	A	20.	A								

■ Practice Paper 13

Q. No.	Ans	Q. No.	Ans	Q. No.	Ans	Q. No.	Ans	Q. No.	Ans	Q. No.	Ans
1.	D	2.	A	3.	A	4.	A	5.	C	6.	B
7.	A	8.	C	9.	D	10.	A	11.	B	12.	B
13.	B	14.	B	15.	B	16.	B	17.	D	18.	A
19.	A	20.	B								

18. Multiple Choice Questions of Plant Pathology

Practice Paper 1

1. **Studies of plant diseases is called:**
   (A) Virology (B) Plant Pathology
   (C) Mycology (D) None of these

2. **Horizontal resistance is:**
   (A) Resistance against one race of pathogen
   (B) Resistance against all the races of pathogen
   (C) Resistance against lodging
   (D) None of these

3. **Plant Pathology word derived from three Greek words. Which of the following is not related to it?**
   (A) Phyton (B) Pathos
   (C) Logus (D) Logy

4. **Who is known as father of plant pathology ?**
   (A) Anton de Bery (B) Luis Pasteur
   (C) R.L. Tulasne (D) Julious Coon

5. **Who laid the foundation of experimental plant pathology?**
   (A) Anton de Bery (B) Luis Pasteur
   (C) Morren (D) Tillet

6. **Downey mildew fungus was discovered by:**
   (A) Luis Pasteur (B) Anton de Bery
   (C) Morren (D) Tillet

7. **Virus is found in cells in form of:**
   (A) Crystalline groups
   (B) Liquid from
   (C) Suspension form
   (D) None of these

8. **Indian Phytopathological Society was established by:**
   (A) E.T. Butler (B) S.N. Dasgupta
   (C) R. Verma (D) B.B. Mundkur

9. **Who is the father of Indian plant pathology?**
   (A) E.T. Butler (B) K.C. Mehta
   (C) R.S. Singh (D) R. Verma

10. **Who told, 'Virus is a contagium vivum fluidum'?**
    (A) Ivanowski (B) Beijerinck
    (C) Meyor (D) R. Kotch

11. **Dr. K.C. Mehta was associated with:**
    (A) Cereal rust in India
    (B) Smut research
    (C) Bacterial research
    (D) Virus research

12. **Who is associated with the study of wheat rust in India:**
    (A) B.B. Mundkur (B) E.J. Butler
    (C) K.C. Mehta (D) J.F. Dastur

13. **Directorate of Plant Protection Quarantine and Storage was started in:**
    (A) 1907 (B) 1914
    (C) 1921 (D) 1946

14. **Destructive Insects and Pest Act was launched in India in:**
    (A) 1907 (B) 1914
    (C) 1921 (D) 1946

15. **England famine in 1846 was caused mainly by:**
    (A) Late blight of potato
    (B) Helminthosporium leaf spot of rice
    (C) Rust of wheat
    (D) Leaf curl virus in tomato

16. **One of the most important reason of Bengal famine in 1943 was:**
    (A) Late blight of potato
    (B) Helminthosporium leaf spot of rice
    (C) Rust of wheat
    (D) Leaf curl virus in tomato

17. **On germination, the Teliospore produces:**
    (A) Germ tube
    (B) Germ sporangium
    (C) Promycelium
    (D) None of these

18. Which one is considered as the perfect spore:

(A) Pycniospore (B) Aeciospore
(C) Teliospore (D) Uredospore

19. Heteroecious rust completes its life cycle on:

(A) One host
(B) Two unrelated hosts
(C) Three hosts
(D) None of these

20. Pustules are often arranged in rows in:

(A) Black rust (B) Brown rust
(C) Yellow rust (D) None of these

Practice Paper 2

1. *Puccinia recondita* is the causal organism of:

(A) Black rust (B) Brown rust
(C) Yellow rust (D) White rust

2. Appearance of wheat rusts in India takes place through the dissemination of:

(A) Pycniospores (B) Aeciospores
(C) Uredospores (D) Teliospores

3. *Erysiphe graminis tritici* is:

(A) Obligate saprophyte
(B) Obligate parasite
(C) Facultative parasite
(D) Facultative saprophyte

4. Ear cockle of wheat is a :

(A) Fungal disease (B) Bacterial disease
(C) Viral disease (D) Nematode disease

5. The pathogen of loose smut of wheat survives:

(A) With the seeds (B) On the seeds
(C) Within the seeds (D) None of these

6. Which one is the abiotic disease of rice:

(A) Blast (B) False smut
(C) Khaira (D) Brown spot

7. Only the naked rachis remains in the last in case of:

(A) Covered smut of barley
(B) Loose smut of wheat
(C) Grain smut of Jowar
(D) Bajra smut

8. Solar energy treatment against loose smut of wheat was evolved by:

(A) Mehta (B) Mundkur
(C) Butler (D) Luthra and Sattar

9. The causal organism of ear cockle of wheat is:

(A) *Ustilago tritici*
(B) *Puccinia triticina*
(C) *Anguina tritici*
(D) *Alternaria triticina*

10. False smut is the name because its causal organism does not belong to:

(A) Masticomycotina (B) Zygomycotina
(C) Ascomycotina (D) Basidiomyotina

11. Disease development is rapid in Karnal bunt due to:

(A) Frequent light rains
(B) Cloudy weather
(C) Low temperature
(D) Combination of above three factors

12. Which disease has caused the famous 'Bengal famine':

(A) Rust of wheat
(B) Red rot of sugarcane
(C) Tikka disease of groundnut
(D) Brown spot of paddy

13. The toxin 'Cochliobolin' is produced by the causative fungus of:

(A) Chickpea grey mould
(B) Sugarcane wilt
(C) Rice brown spot
(D) Rice blast

14. In blast of rice, the symptoms appear on:

(A) Leaves only
(B) Leaf-sheath only
(C) Panicle only
(D) On all above parts

15. A disease which occurs widely but periodically is:

(A) Epidemic disease (B) Endemic disease
(C) Sporadic disease (D) Pandemic disease

16. A disease which occurs at a short interval and its severity is more called:

(A) Epidemic disease
(B) Endemic disease
(C) Sporadic disease
(D) Pandemic disease

17. Smuts, in general are considered as:
(A) Obligate parasites
(B) Obligate saprophytes
(C) Facultative parasites
(D) Facultative saprophytes

18. The conidia of *Pyricularia oryzae* are:
(A) One celled (B) Two celled
(C) Three celled (D) Many celled

19. Mycellium of powdery mildew pathogen is:
(A) Ectophytic (B) Endophytic
(C) Intercellular (D) Intracellular

20. In red rot of sugarcane, the reddening is found in:
(A) Parenchyma (B) Pith
(C) Phloem (D) Xylem

Practice Paper 3

1. In *Pyricularia oryzae*, the spots on leaves are mostly:
(A) Reniform (B) Pyriform
(C) Doliform (D) Spherical

2. In Ascochyta blight, the spots on leaves are mostly:
(A) Angular (B) Circular
(C) Oblong (D) Irregular

3. The perfect size of *Colletotrichum falcatum* is:
(A) *Pyrenophora graminae*
(B) *Mycosphaerella arachidicola*
(C) *Glumerellatu cumanensis*
(D) *Cochliobolus miyabeanus*

4. In red rot of sugarcane, symptoms also appear on:
(A) Tip of the leaves
(B) Margin of the leaf
(C) Midrib of the leaves
(D) None of these

5. In *Colletrotrichum falcatum*, the conidia are produced in:
(A) Pycnidia (B) Acervuli
(C) Synnemata (D) Sporodochia

6. Bacterial blight of cotton is also called as:
(A) Tar leaf spot (B) Angular leaf spot
(C) Zonate leaf spot (D) Red leaf spot

7. Bacterial blight of cotton does not produce the symptoms on:
(A) Seedlings (B) Stems
(C) Bolls (D) Roots

8. The disease is a typical vascular wilt in rice:
(A) Blast
(B) Brown leaf spot
(C) Bacterial leaf streak
(D) Bacterial leaf blight

9. The main source of primary inoculum of bacterial blight of cotton is:
(A) Seed (B) Soil
(C) Air (D) Water

10. Kresek phase is found in the rice disease:
(A) Bacterial leaf streak
(B) Bacterial blight
(C) Blast
(D) Sheath blight

11. Which factor does not favour the rapid spread of bacterial blight of rice:
(A) Rainy weather
(B) Strong wind
(C) High humidity
(D) Temperature below 20 °C

12. The symptoms only appear on leaf in case of:
(A) Angular leaf spot of cotton
(B) Bacterial leaf streak of rice
(C) Blast of rice
(D) Brown spot of rice

13. In the genus *Fusarium*, conidia are produced in:
(A) Pycnidia (B) Acervuli
(C) Sporodocia (D) Synnemata

14. Wilt pathogen is mainly:
(A) Soil borne (B) Seed borne
(C) Air borne (D) None of these

15. The size of pesticide molecules, if all other conditions are constant, greatly influences it is:
(A) Absorption (B) Solubilization
(C) Adsorption (D) Volatilization

16. Some pesticides with positively charged groups are also absorbed by:
(A) Illite clays
(B) Kaolinite clays
(C) Monmorillonite clays
(D) Silicate clays

17. In general, herbicides are .... than either fungicides or insecticides:
(A) More mobile (B) Less mobile
(C) Equally mobile (D) More immobile

18. Some degradation products of pesticides are even .... than the original pesticides.
(A) Less toxic (B) More toxic
(C) Non-toxic (D) Equally toxic

19. The total number of bacteria frequently .... after fumigation:
(A) Increases (B) Decreases
(C) Remains constant (D) None of these

20. The process of ammonification is often .... by the use of pesticides:
(A) Harmful (B) Benefitted
(C) Not affected (D) None of these

Practice Paper 4

1. The causal organism of black rust in wheat is:
(A) *Pucinia recondite tritici*
(B) *Puccinia streiformis*
(C) *Puccinia graminis tritici*
(D) *Ustilago nuda tritici*

2. Karnal bunt or partial bunt was first observed in experimental seed plots at:
(A) Kernal (B) Pantnagar
(C) New Delhi (D) Hisar

3. The causal agent of anthracnose diseases in sorghum is:
(A) *Puccinia purpurea*
(B) *Helminthosporium turcicum*
(C) *Colletotrichum graminicola*
(D) *Gleourcospora sorghi*

4. The chickpea wilt is caused by:
(A) *Fusarium oxysporum*
(B) *Ascochyta rabi*
(C) *Bortrylis cineraria*
(D) *Scelerotinia sclerotiorum*

5. The most destructive disease of sugar cane is:
(A) Red rot (B) Wilt
(C) Smut (D) Albino

6. In bacterial brown rot and wilt of potato the amount of .... produced by pathogens are proportional to severity of symptoms:
(A) Gibberellins
(B) Lipids
(C) Polysaccharides
(D) Proteins

7. Some .... varieties of apple exude toxic wax on leaf surface to prevent germination of conidia of *Podosphaera leucotricha*:
(A) Scab resistant
(B) Leaf blight resistant
(C) Powdery mildew resistant
(D) Rust resistant

8. Red scales of .... (red varieties) contain protocatechuic acid and catechol which may exude in drops to resist the attack of *Colletrotrichum circinans*:
(A) Garlic (B) Onion
(C) Leek (D) Celery

9. Some varieties of .... are resistant to wilt due to the presence of hydrocyanides (HCN) in root exudates:
(A) Sunflower (B) Safflower
(C) Castor (D) Linseed (flax)

10. Leaf curl is a serious disease of:
(A) Moong (B) Pea
(C) Gram (D) Lentil

11. Ear cockle disease is found in:
(A) Sugarcane (B) Wheat
(C) Cotton (D) Rice

12. Khaira disease is caused due to deficiency of:
(A) Iron (B) Zinc
(C) Sulphur (D) Nitrogen

13. Wilt is an important disease of:
(A) Pea (B) Arhar
(C) Mustard (D) Wheat

14. Which one the important disease of groundnut is:
(A) Phyllody
(B) Wilt
(C) Tikka
(D) Cercospora leaf spot

15. Which one the important disease of sesame is:
(A) Phyllody
(B) Wilt
(C) Tikka
(D) Cercospora leaf spot

16. Sterility mosaic is a disease related to:
(A) Fungal (B) Bacterial
(C) Viral (D) Physiological

17. Black tip is a common disease of:
(A) Banana (B) Grape
(C) Orange (D) Mango

18. Which wheat rust normally appears earlier in plains of Uttar Pradesh:
(A) Black rust (B) Yellow rust
(C) Brown rust (D) None of these

19. All plant pathogenic bacteria are:
(A) Gram positive
(B) Gram negative
(C) Gram positive and gram negative
(D) None of these

20. Multiplication of bacteria is caused by:
(A) Mitosis (B) Meiosis
(C) Binary fussion (D) All of these

Practice Paper 5

1. Infectious disease is caused by:
(A) Parasites
(B) Non parasites
(C) Virus
(D) All of these

2. Non infectious disease is caused by:
(A) Nutrient deficiency
(B) Mineral deficiency
(C) Pollutants
(D) All of these

3. Saprophytic pathogens feed on:
(A) Only dead matter
(B) Only living organisms
(C) Both (A) and (B)
(D) None of these

4. Parasites feed on:
(A) Only dead matter
(B) Only living organisms
(C) Both (A) and (B)
(D) None of these

5. Which of the following has its own metabolic activity?
(A) Fungus (B) Virus
(C) Viroid (D) Plasmids

6. Litchen is a example of:
(A) Mutalistic association
(B) Symbiotic association
(C) Parasitism
(D) All of these

7. A plant disease spreads after irregular intervals at different locations in big way, is called:
(A) Endemic disease (B) Epidemic disease
(C) Sporadic disease (D) All of these

8. An obligate pathogen can:
(A) Grow only on living organism
(B) Grow on dead matters
(C) Grow on both living and non living matters
(D) None of these

9. Yellow mosaic in soybean spread by:
(A) Seeds (B) Soils
(C) Air (D) White fly

10. The fungi grown on dung of herbivorous animals are called as:
(A) Coprophilous fungi
(B) Keratinaceous fungi
(C) Amphibious fungi
(D) All of these

11. The fungi grown on dead bodies are called as:
(A) Coprophilous fungi
(B) Keratinaceous fungi
(C) Amphibious fungi
(D) All of these

12. The root like structure of fungi is called as:

(A) Rhizoid
(B) Stroma
(C) Rhizomorph
(D) Sclerotium

13. The pigeonpea wilt is caused by:

(A) *Fusarium oxysporum*
(B) *Fusarium udum*
(C) *Ascochyta rabi*
(D) *Bortrytis cineraria*

14. Bakanae disease is caused by:

(A) Fungus
(B) Alga
(C) Bacterium
(D) Virus

15. Aflatoxin is produced by:

(A) *Aspergillus flavus*
(B) *Phytophthora infestans*
(C) *Albugo candida*
(D) *Puccinia graminis tritici*

16. Which of the following is a facultative parasite?

(A) Fusarium
(B) Rhizoctonia
(C) Alternaria
(D) All of these

17. Which of the following is a facultative saprophyte?

(A) Smut
(B) Helminthosporium
(C) Leaf curl fungus
(D) All of these

18. The important symptoms of black rot in cauliflower is:

(A) Chlorosis
(B) Drying of leaves
(C) Burning of leaves
(D) Black spots on petioles and veins

19. Purple black of onion is caused by:

(A) Fungus
(B) Bacteria
(C) Virus
(D) None of these

20. Carrot yellow is caused by:

(A) Fungus (B) Algae
(C) Bacteria (D) Virus

Practice Paper 6

1. Bacterial soft rot is the result of:

(A) High temperature
(B) Very low temperature
(C) Waterlogging
(D) None of these

2. Leaf curl of chilli is mainly spread by:

(A) Air (B) White fly
(C) Water (D) Nematode

3. Which of the following disease is the most harmful to mango orchard?

(A) Mango wither tip (B) Anthracnose
(C) Bunchy top (D) Black tip

4. Which of the following disease is not found in papaya orchard?

(A) Anthracnose (B) Collar rot
(C) Papaya mosaic (D) Black tip

5. Panama disease of banana is caused by:

(A) *Fusarium oxysporum*
(B) *Bacillus* spp.
(C) Actinomycetes
(D) Virus

6. Bunchy top of banana is caused by:

(A) Bacteria (B) Fungi
(C) Virus (D) Actinomycetes

7. Which of the following disease is most harmful to citrus plants?

(A) Citrus canker (B) Gummosis
(C) Anthracnose (D) All of these

8. Citrus canker is caused by:

(A) *Phytophthora infestans*
(B) *Xanthomonas citri*
(C) *Collectotrichum* spp.
(D) *Bacillus moniliformis*
(D) All of these

9. Rust fungi mainly a:

(A) Saprophyte
(B) Parasite
(C) Facultative parasite
(D) All of these

10. Which of the alkaloid is produced by ergot fungi?

(A) Ergotin (B) Penicillin
(C) Cutrinin (D) All of these

11. Smut fungi are:

(A) Obligate parasite
(B) Obligate saprophyte
(C) Facultative saprophyte
(D) Facultative parasite

12. Rust fungi are:

(A) Polymorphic (B) Homothallic
(C) Monoecious (D) All of these

13. Autocious fungi complete their life cycle on:

(A) Single host (B) Two host
(C) Dead matter (D) None of these

14. Heterocious fungi complete their life cycle on:

(A) Single host
(B) More than one host
(C) Dead matter
(D) None of these

15. Covered smut of barley is a disease of:

(A) Externally seed borne
(B) Internally seed borne
(C) Air borne
(D) None of these

16. Which of the following sequence of rust occurrence in UP:

(A) Leaf rust > Stripe rust > Stem rust
(B) Stripe rust > Leaf rust > Black rust
(C) Black rust > Brown rust > Yellow rust
(D) Yellow rust > Leaf rust > Black rust

17. The pathogen that can grow only under controlled environment is called:

(A) Facultative parasite
(B) Bio-chemically obligate
(C) Ecologically obligate
(D) All of these

18. Bajra smut is a:

(A) Air borne
(B) Seed borne
(C) Soil borne
(A) Soil and seed borne

19. The sorus of bajra smut is formed usually on:

(A) Leaf (B) Stem
(C) Ovary (D) All aerial parts

20. The main symptoms of early blight of potato is:

(A) Formation of concentric ringes on infected spots
(B) Formation of cottony growth
(C) Formation of chlorosis
(D) All of these

Practice Paper 7

1. The main symptoms of false smut of rice is:

(A) Apearance of round bluish spots on leaves
(B) Conversion of few seeds into scleratial bodies covering by green colour layer
(C) Formation of black powder in whole panicle
(D) All of these

2. The main symptoms of rice blast is:

(A) Round black spots on leaves
(B) Oval shaped round spots on leaves
(C) Boat shaped spots with grey red borders and grey ash colour centre
(D) None of these

3. Rice blast is also known as neck rot because of:

(A) Neck of panicle becomes shriveled due to severity of disease
(B) Neck becomes stronger as fungus produces pyricularin toxin
(C) It produces hyaline spores
(D) None of these

4. Which type of bacteria can be killed by Penicillin?

(A) Gram +ve (B) Gram – ve
(C) Both (A) and (B) (D) None of these

5. The pathogenic bacteria are generally:

(A) Gram +ve (B) Gram –ve
(C) Both (A) and (B) (D) None of these

6. The main function of mesosomes in bacterial cell is:

(A) To help in cell division
(B) To help in respiration
(C) To assist in photosynthesis
(D) All of these

7. **Bacteria having flagella at its whole body is known as:**
(A) Monotrichous (B) Amphitrichous
(C) Petritrichous (D) None of these

8. **Bacteria having a bunch of flagella at one end is known as:**
(A) Monotrichous (B) Lophotrichous
(C) Petritrichous (D) None of these

9. **Which of the following is a autotrophic bacteria?**
(A) *Chlorobium* (B) *Thiospirillum*
(C) *Nitrobacter* (D) All of these

10. **Which of the following is a heterotrophic bacteria?**
(A) *Chlorobium* (B) *Pseudomonas*
(C) *Xantthomonas* (D) All of these

11. **The carrier of leaf curl tomato virus is:**
(A) White fly (B) Beetle
(C) Bacteria (D) Fungus

12. **Cyst formation is very common in which of the following bacteria?**
(A) Azotobacter (B) Myxococcus
(C) Chondromyces (D) All of these

13. **Which of the following feature of gram positive bacteria?**
(A) Homogenous cell wall that contains >85% mucopeptides and its polymers
(B) Teichoic acids, ribitol, glutamic acid and alanine are found in more quantity
(C) Violet or blue colour after staining with solution of crystal violet and iodine
(D) All of these

14. **Which of the following feature of gram negative bacteria?**
(A) Cell wall is thin and made of mainly lipoprotein and its polymers
(B) They react with counter stain safranine and become red to pink in colour
(C) Teichoic acid is not formed
(D) All of these

15. **Generally plant virus posses a:**
(A) Single stranded RNA
(B) Double stranded RNA
(C) Doubled stranded DNA
(D) None of these

16. **Viroid is a:**
(A) Naked RNA particles
(B) Naked DNA particles
(C) Naked RNA and DNA particles
(D) None of these

17. **Prions are:**
(A) Infectious DNA particles
(B) Infectious protein particles
(C) Both (A) and (B)
(D) None of these

18. **The virus enzymes are collectively known as:**
(A) Peplomers (B) Polymerase
(C) Transcriptase (D) None of these

19. **Virus is a:**
(A) Obligate parasite
(B) Obligate saprophyte
(C) Facultative parasite
(D) Facultative saprophyte

20. **Chemically virus is a:**
(A) Proteir (B) Nucleoprotein
(C) Nucleic acid (D) All of these

Practice Paper 8

1. **Ultra microscopic particle capable of infecting live cells is called:**
(A) Plasmic (B) Nucleiod
(C) Virion (D) None of these

2. **Virus can be seen only with:**
(A) Naked eyes
(B) Microscope
(C) Electron microscope
(D) All of these

3. **Nematodes are:**
(A) Saprophyte (B) Predator
(C) Pathogenic (D) All of these

4. **Nematodes may cause following symptoms?**
(A) Cellular hypertrophy/hyperplasia
(B) Suppression of mitosis
(C) Cell necrosis
(D) All of these

5. **Plant nematodes always have:**
(A) Stylet
(B) Respiratory organs
(C) Circulatory system
(D) None of these

6. **Which of the following system is not present in plant nematodes?**
(A) Respiratory system
(B) Circulatory system
(C) Reproductive system
(D) None of these

7. **Which of the following nematicides are used to control of nematodes?**
(A) Nemagon (B) Nemaphos
(C) Hexanema (D) All of these

8. **Which stage of root knot nematode attacks to plant?**
(A) First larvae (B) Second larvae
(C) Third larvae (D) Fourth Larvae

9. **Reverse transcription is found only in:**
(A) Bacteria (B) Virus
(C) Protozoa (D) Fungus

10. **Which of the following disease is caused by MLO's?**
(A) Papaya bunchy top
(B) Sugarcane grassy shoot
(C) Witch's broom of groundnut
(D) All of these

11. **Damping off of tobacco is caused by:**
(A) *Pythium aphanidermatum*
(B) *Pythium infestans*
(C) *Phytophthora infestans*
(D) None of these

12. **Late blight of potato is caused by:**
(A) *Phytophthora infestans*
(B) *Albugo candida*
(C) *Peronospora pisi*
(D) None of these

13. **Downy mildew of pea disease is caused by:**
(A) *Phytophthora pisi* (B) *Peronospora pisi*
(C) *Albigo candida* (D) None of these

14. **Green ear disease is caused by:**
(A) *Albugo candida*
(B) *Sclerospora graminicola*
(C) *Erysiphe graminis*
(D) None of these

15. **White blister of crucifers disease is caused by:**
(A) *Albugo candida*
(B) *Erysiphe graminis*
(C) *Peronospora pisi*
(D) None of these

16. **Downy mildew of brassica is caused by:**
(A) *Albugo candida*
(B) *Peronospora brassicae*
(C) *Phytophthora infestans*
(D) None of these

17. **Downey mildew of maize is caused by:**
(A) *Sclerospora sacchari*
(B) *Sclerospora rayssiae var zeae*
(C) *Sclerospora philippinensis*
(D) All of these

18. **Powdery mildew of pea is caused by:**
(A) *Erysiphe graminis*
(B) *Uncinula necator*
(C) *Erysiphi polygoni*
(D) All of these

19. **Powdery mildew of cereals is caused by:**
(A) *Erysiphe graminis*
(B) *Erysiphi polygoni*
(C) *Taphrina deformans*
(D) All of these

20. **Powdery mildew of apple is caused by:**
(A) *Uncinula necator*
(B) *Podosphaera leuxotricha*
(C) *Sclerotium oryzae*
(D) All of these

Practice Paper 9

1. **Stem rot of paddy is caused by:**
(A) *Sclerotium oryzae*
(B) *Claviceps purpurea*
(C) *Ustilaginoidea virens*
(D) All of these

2. **False smut of paddy is caused by:**
(A) *Claviceps purpurea*
(B) *Ustilaginoidea virens*
(C) *Erysiphe graminis*
(D) None of these

3. **Ergot of bajra is caused by:**
   (A) *Sclerotium oryzae*
   (B) *Claviceps purpurea*
   (C) *Claviceps microcephala*
   (D) None of these

4. **Black rust of wheat is caused by:**
   (A) *Puccinia graminis var tritici*
   (B) *Puccinia recondita*
   (C) *Puccinia striformis*
   (D) *Melampsora lini*

5. **Leaf rust of wheat is caused by:**
   (A) *Puccinia graminis*
   (B) *Puccinia recondita*
   (C) *Puccinia striformis*
   (D) *Melampsora lini*

6. **Stripe rust of wheat is caused by:**
   (A) *Puccinia graminis*
   (B) *Puccinia recondita*
   (C) *Puccinia striformis*
   (D) *Melampsora lini*

7. **Stem rust of wheat is also known as:**
   (A) Black rust (B) Brown rust
   (C) Yellow rust (D) White rust

8. **Leaf rust of wheat is also known as:**
   (A) Black rust (B) Brown rust
   (C) Yellow rust (D) White rust

9. **Stripe rust of wheat is also known as:**
   (A) Black rust (B) Brown rust
   (C) Yellow rust (D) White rust

10. **Causal organism of pea rust is:**
    (A) *Uromyces pisi*
    (B) *Uromyces aestivum*
    (C) *Puccinia recondita*
    (D) None of these

11. **Causal organism of linseed rust is:**
    (A) *Uromycese pisi*
    (B) *Uromyces fabae*
    (C) *Melampsora lini*
    (D) None of these

12. **Causal organism of loose smut of wheat is:**
    (A) *Ustilago nuda*
    (B) *Ustilago nuda var tritici*
    (C) *Ustilago hordei*
    (D) *Tolyposporium penicillariae*

13. **Causal organism of grain smut of bajra is:**
    (A) *Sphacelotheca sorghi*
    (B) *Sphacelotheca hordi*
    (C) *Tolyposporium penicillariae*
    (D) None of these

14. **Causal organism of smut of bajra is:**
    (A) *Fusarium oxysporum*
    (B) *Tolyposorium penicillariae*
    (C) *Ustilago nuda*
    (D) None of these

15. **Causal organism of early blight of potato is:**
    (A) *Alternaria solani*
    (B) *Alternaria infestans*
    (A) *Pyricularia oryzae*
    (D) *Cercospora personata*

16. **Causal organism of Alternaria leaf blight of wheat is:**
    (A) *Alternaria tritici*
    (B) *Alternaria aestivum*
    (C) *Pyricularia triticina*
    (D) None of these

17. **Causal organism of tikka disease of groundnut is:**
    (A) *Cercospora personata*
    (B) *Cercospora arachidicola*
    (C) Both (A) and (B)
    (D) None of these

18. **Which of the following causal organism of tikka disease of groundnut is more fatal?**
    (A) *Cercospora personata*
    (B) *Cercospora arachidicola*
    (C) Both (A) and (B)
    (D) None of these

19. **Which of the following disease is caused by mycoplasma like organisms (MLO)?**
    (A) Rice yellow dwarf
    (B) Sugarcane grassy shoot
    (C) Potato witch's broom
    (D) All of these

20. **Which of the following disease is caused by Rickettsia like microorganisms (RLO's)?**
    (A) Ratoon stunting
    (B) Aster yellow
    (C) Khaira disease
    (D) All of these

Practice Paper 10

1. **Which of the following characteristics is not associated with MLO's?**
   (A) Unicellular
   (B) Usually immobile
   (C) Cell wall absent
   (D) Cell wall present

2. **Khaira disease is a example of:**
   (A) Endemic disease
   (B) Epidemic disease
   (C) Non infectious disease
   (D) Infectious disease

3. **Green ear disease is a example of:**
   (A) Endemic disease
   (B) Epidemic disease
   (C) Sporodic disease
   (D) Infectious disease

4. **Late blight of potato disease is a example of:**
   (A) Endemic disease
   (B) Epidemic disease
   (C) Sporodic disease
   (D) Infectious disease

5. **Necrotrophs are pathogens that:**
   (A) Kill host plants
   (B) Do not kill host plant
   (C) Provide nutrition to host plants
   (D) None of these

6. **Facultative parasites are:**
   (A) Primarily saprophytes but can attack on living organisms under favourable conditions
   (B) Primarily saprophytes and do not attack on living organisms
   (C) Primarily obligate parasites
   (D) All of these

7. **Main symptoms of karnal bunt is:**
   (A) Formation of black powder below the pericarp of seed
   (B) Produces rotten fish odour
   (C) Bad odour is due to Trimethylamine
   (D) All of these

8. **Which of the following culture is a natural medium for isolation of pathogen?**
   (A) Potato dextrose agar medium
   (B) Milk
   (C) Egg
   (D) All of these

9. **Root knot disease of tomato is caused by:**
   (A) *Meloidogyne javanica*
   (B) *Heterodera avenaei*
   (C) *Anguina tritici*
   (D) None of these

10. **Irish blight is also known as:**
    (A) Late blight of potato
    (B) White blight of mustard
    (C) Yellow rust
    (D) Loose smut of wheat

11. **The fruit like apple, pear and grapes are spoiled by:**
    (A) *Penicillium expensum*
    (B) Penicillium notatum
    (C) *Penicillium flavum*
    (D) *Penicillium citrinum*

12. **Which of the *penicillium spp* is harmful for food products?**
    (A) *Penicillium expensum*
    (B) Penicillium glaccusm
    (C) *Penicillium digitatum*
    (D) All of these

13. **Aspergillin antibiotic was prepared by which fungus?**
    (A) *Aspergillus niger*
    (B) Aspergillus fumigatus
    (C) *Aspergillus flavus*
    (D) All of these

14. **Most important symptoms of ergot disease is:**
    (A) Transformation of kernels into black structure
    (B) Dispersion of spores
    (C) Both (A) and (B)
    (D) None of these

15. **Virus can be precipitated with the help of:**
    (A) Acetone
    (B) Safranin
    (C) Ethyl alcohol
    (D) All of these

16. If the virus multiplies in host bacterium which can undergoes complete lysis to release the daughter phages, this type of virus is called:

(A) Lytic bacterio phages
(B) Lysogenic bacteriophages
(C) Transformation
(D) None of these

17. Which of the following disease is caused by MLO's?

(A) Little leaf of egg plant
(B) Citrus greening
(C) Purple top roll of potato
(D) All of these

18. Silica content in leaf helps rice plant to resist against:

(A) Blast
(B) False smut
(C) Bacterial blight
(D) None of these

19. Conidia are produced in basidiomycetes on:

(A) Primary mycelium
(B) Secondary mycelium
(C) Both (A) and (B)
(D) None of these

20. Which of the following is not the collateral host of yellow rust?

(A) *Agropyron semicostatum*
(B) *Bromes cathartics*
(C) *Puccinia striiformis*
(D) All of these

Practice Paper 11

1. Which of the following plant is the natural alternate host of *Puccinia recondita*?

(A) *Berberis vulgaris*
(B) *Isopyron fumarioides*
(C) *Avena fatua*
(D) *Triticum aestivum*

2. The infected seed of bajra smut are converted into:

(A) Sori (B) Nod
(C) Mycelium (D) None of these

3. Grain smut of jowar may cause following disease in humen beings:

(A) Cold (B) Cholera
(C) Hey fever (D) None of these

4. Loose smut is a:

(A) Internally seed borne disease
(B) Air borne disease
(C) Soil borne disease
(D) Externally seed borne disease

5. The perfect stage of tikka disease of groundnut is:

(A) *Mycosphaerella berkeleyii*
(B) *Mycosphaerella arachidicola*
(C) Both (A) and (B)
(D) None of these

6. Rotten neck disease is known as:

(A) Blast of rice
(B) Red rot of sugarcane
(C) Leaf rust of wheat
(D) None of these

7. Antibiotics are prepared mainly from:

(A) Actinomycetes (B) Rhizobium
(C) Pseudomonas (D) Azotobacter

8. Grain smut disease of jowar is a disease of:

(A) Internally seed borne
(B) Externally seed borne
(C) Air borne
(D) Soil borne

9. Ear cockle of wheat or Sehun disease is caused by:

(A) *Anguina tritici*
(B) Heterodera avenae
(C) *Meloidogyne javanica*
(D) All of these

10. Molya disease of barley is caused by:

(A) Bacteria
(B) Virus
(C) Nematodes
(D) None of these

11. The causal organism of molya diseases is:

(A) *Heterodera avenae*
(B) *Melodogne spp.*
(C) *Anguina spp.*
(D) None of these

12. **Ear cockle of wheat is often associated with:**
(A) Loose smut
(B) Yellow ear rot or Tundu disease
(C) Rust
(B) None of these

13. **Mycoses disease is caused by:**
(A) Fungi (B) Bacteria
(C) Virus (D) Fly

14. **Browning of cauliflower is caused by the deficiency of:**
(A) Zinc (B) Manganese
(C) Iron (D) Boron

15. **The citrus diseases namely 'Reclamation Disease' and 'Die back' are caused by the deficiency of:**
(A) Zinc (B) Iron
(C) Copper (D) Molybdenum

16. **Mechanism of wilting can be understood by:**
(A) Plugging of vessels by fungus hyphae
(B) Toxin theory-production of toxins like Fusaric acids
(C) Enzyme theory-Production of enzymes like Pectolytic enzyme
(D) All of these

17. **The pathogen that can grow only under controlled environment is called:**
(A) Ecologically obligate
(B) Bio-chemically obligate
(C) Facultative parasite
(D) All of these

18. **Fungi have been originated from:**
(A) Polygeneric line
(B) Phytogeneric line
(C) Polygendric line
(D) None of these

19. **The carrier of leaf curl tomato virus is:**
(A) White fly
(B) Beetle
(C) Bacteria
(D) Fungus

20. **Leaf curl of tomato is caused by:**
(A) Tobacco virus 16
(B) Tomato virus 16
(C) Virus No. 10
(D) Virus No.16

Practice Paper 12

1. **Bacterial nucleus is characterized by:**
(A) Undeveloped nucleus
(B) Not covered by nucleus membrane
(C) Multiplied by binary fission
(D) All of these

2. **Loose smut of wheat can be controlled by:**
(A) Seed treatment by vitavex before sowing
(B) Seed treatment with hot water
(C) Use of disease resistant varieties
(D) All of these

3. **The important symptoms of black rot in cauliflower is:**
(A) Chlorosis
(B) Drying of leaves
(C) Burning of leaves
(D) Black spots on petioles and veins

4. **Purple black of onion is caused by:**
(A) Fungus (B) Bacteria
(C) Virus (D) None of these

5. **Which of the following disease is the most harmful to mango orchard?**
(A) Mango wither tip
(B) Anthracnose
(C) Bunchy top
(D) Black tip

6. **Citrus canker is caused by:**
(A) *Phytophthora infestans*
(B) *Xanthomonas citri*
(C) *Collectotrichum spp.*
(D) Bacillus moniliformis

7. **RSD is caused by:**
(A) Gram (-)fastidious bacteria
(B) Gram (+) fastidious bacteria
(C) Clavibacter xyli
(D) Both (B) and (C)

8. **MLO and phyllody disease of sesame spread by:**
(A) Leaf hopper (B) White fly
(C) Aphids (D) Jassids

9. **Sterility mosaic disease of pigeonpea spread by:**
(A) Virus (B) Mites
(C) Aphids (D) White fly

10. Alternate host of black rust is:

(A) Barberi (B) Bajra
(C) Jowar (D) Wheat

11. In north India and central India, the black rust inoculums cause from:

(A) South (B) Hilly areas
(C) USA (D) Nepal

12. Viruses contain:

(A) RNA
(B) DNA
(C) Both RNA and DNA
(D) Either RNA or DNA, never both

13. The mature particle of a plant virus in generally known as:

(A) Capsid (B) Viroid
(C) Virion (D) Nucleo capsid

14. The sexual, thick walled and resting spores of the rust fungi is known as:

(A) Basidiospores (B) Urediospores
(C) Teliospores (D) Aeciospores

15. For quick and accurate detection of viruses can be done by:

(A) ELISA (B) IEM
(C) HADAS (D) All of these

16. Which one of the following cannot be detected by ELISA technique?

(A) Virus (B) Bacteria
(C) Viroid (D) Fungus

17. When plant showed the partial resistance against the races of pathogen then it is a type of?

(A) Horizontal resistance
(B) Vertical resistance
(C) Induced resistance
(D) No-host resistance

18. The major storage fungi that affects the food grain is:

(A) Rhizobium (B) Mucor
(C) Cercospora (D) Aspergillus

19. A bacteria which have a large number of flagella all over the cell is categorized in:

(A) Atrichous
(B) Petritrichous
(C) Lophotrichous
(D) Amphitrichous

20. In bacteria, variability is caused by:

(A) Conjugation (B) Transformation
(C) Transduction (D) All of these

Practice Paper 13

1. Bordeaux mixture is a:

(A) Systemic fungicide
(B) Fungicide and bactericide
(C) Tropical inorganic fungicide
(D) Tropical organic fungicide

2. Nimbin and Nimbidin are extracted from:

(A) Neem (B) Ashwagandha
(C) Amla (D) Corainder

3. Onion smut can be avoided by:

(A) Spray of Dithane M-45
(B) Seed treatment by captan/Thiram
(C) Weed control
(D) Irrigation

4. Citrus canker can be controlled by :

(A) Blitox-50 @ 0.25% solution
(B) Endosulphan @ 0.2% solution
(C) 2,4-D @ 1.5 kg a.i./ha
(D) None of these

5. Powdery mildew of apple can be controlled by:

(A) Spray of Krathane @ 0.1% solution
(B) Spray of Endosulphan @ 0.2% solution
(C) Spray of Diathane -78 @ 0.1% solution
(D) All of these

6. Safflower rust can be controlled by:

(A) Spray of Zineb
(B) Seed treatment with thiram
(C) Destroying the infected material
(D) Following the 3 to 4 years crop rotation

7. Brown stripe downy mildew of maize can be controlled by use of:

(A) Bordeaux mixture (4:450)
(B) Bavistin
(C) Karathane-45
(D) Dithane M-45

8. Karathane is basically used for the control of:

(A) Downey mildew (B) Powdery mildew
(C) Wilt (D) Blast

9. **Bordeaux mixture is a:**
(A) Copper fungicide
(B) Sulphur fungicide
(C) Dithiocarbamate
(D) Systemic fungicide

10. **Treatment of cotton seed with organo-mercurial compound(s) helps to minimize the disease:**
(A) Soil borne
(B) Air borne
(C) Water borne
(D) Seed borne

11. **To avoid rotting in early planting of potato, seeds should be treated with:**
(A) 0.5 per cent Dithane M-45
(B) 2.0 per cent Dithane M-45
(C) 1.5 per cent Aretan
(D) 2.0 per cent Tafasan

12. **Treat the sugarcane seed setts with .... against fungaldise:**
(A) Diathane M-45 (B) Zineb
(C) Thiram (D) Aretan

13. **Downey mildew in pulses can be controlled by spraying of:**
(A) Mencozeb @ 0.2 %
(B) Thiofinate methyl 70% @ 0.15 %
(C) Grow resistant variety
(D) All of these

14. **Powdery mildew in pea can be controlled by spraying of:**
(A) Dusting of Pendimethalin @ 2.0-3.0 l/ha
(B) Karathan @ 0.06% spray
(C) Stomp @ 2.0 l/ha
(D) None of these

15. **Which of the following is used for tree surgery?**
(A) Bordeaux mixture
(B) Chaubatia paste
(C) Santar A
(D) All of these

16. **Which of the following is used for tree surgery?**
(A) Bordeaux mixture
(B) Chaubatia paste
(C) Santar A
(D) All of these

17. **Citrus canker can be controlled by:**
(A) Blitox-50 @ 0.25% solution
(B) Endosulphan @ 0.2% solution
(C) 2,4-D @ 1.5 kg a.i./ha
(D) None of these

18. **Loose smut of wheat can be controlled by:**
(A) Seed treatment by vitavex before sowing
(B) Seed treatment with hot water
(C) Use of disease resistant varieties
(D) All of these

19. **Bacterial disease are controlled by:**
(A) Antibiotics
(B) Viricide
(C) Fungicide
(D) Kethane

20. **Onion smut can be avoided by:**
(A) Spray of Dithane M-45
(B) Seed treatment by captan/Thiram
(C) Weed control
(D) Irrigation

Answers Key

■ Practice Paper 1

Q. No.	Ans	Q. No.	Ans	Q. No.	Ans	Q. No.	Ans	Q. No.	Ans	Q. No.	Ans
1.	B	2.	B	3.	D	4.	A	5.	A	6.	B
7.	A	8.	D	9.	A	10.	B	11.	A	12.	C
13.	D	14.	B	15.	A	16.	B	17.	C	18.	C
19.	B	20.									

■ Practice Paper 2

Q. No.	Ans	Q. No.	Ans	Q. No.	Ans	Q. No.	Ans	Q. No.	Ans	Q. No.	Ans
1.	B	2.	C	3.	B	4.	D	5.	C	6.	C
7.	B	8.	D	9.	C	10.	D	11.	D	12.	D
13.	C	14.	D	15.	A	16.	B	17.	D	18.	C
19.	A	20.	B								

■ Practice Paper 3

Q. No.	Ans	Q. No.	Ans	Q. No.	Ans	Q. No.	Ans	Q. No.	Ans	Q. No.	Ans
1.	B	2.	B	3.	C	4.	C	5.	B	6.	B
7.	D	8.	D	9.	A	10.	B	11.	D	12.	B
13.	C	14.	A	15.	C	16.	D	17.	A	18.	B
19.	A	20.	B								

■ Practice Paper 4

Q. No.	Ans	Q. No.	Ans	Q. No.	Ans	Q. No.	Ans	Q. No.	Ans	Q. No.	Ans
1.	A	2.	A	3.	C	4.	A	5.	A	6.	C
7.	C	8.	B	9.	D	10.	A	11.	B	12.	B
13.	B	14.	C	15.	A	16.	C	17.	D	18.	B
19.	C	20.	C								

■ Practice Paper 5

Q. No.	Ans	Q. No.	Ans	Q. No.	Ans	Q. No.	Ans	Q. No.	Ans	Q. No.	Ans
1.	A	2.	D	3.	A	4.	B	5.	A	6.	A
7.	C	8.	A	9.	D	10.	A	11.	B	12.	A
13.	B	14.	A	15.	A	16.	D	17.	D	18.	D
19.	D	20.	D								

Practice Paper 6

Q. No.	Ans	Q. No.	Ans	Q. No.	Ans	Q. No.	Ans	Q. No.	Ans	Q. No.	Ans
1.	C	2.	B	3.	C	4.	D	5.	A	6.	C
7.	A	8.	B	9.	B	10.	A	11.	C	12.	A
13.	A	14.	B	15.	A	16.	A	17.	C	18.	A
19.	C	20.	A								

Practice Paper 7

Q. No.	Ans	Q. No.	Ans	Q. No.	Ans	Q. No.	Ans	Q. No.	Ans	Q. No.	Ans
1.	B	2.	C	3.	D	4.	A	5.	B	6.	A
7.	C	8.	B	9.	D	10.	D	11.	A	12.	D
13.	D	14.	D	15.	A	16.	A	17.	B	18.	C
19.	A	20.	B								

Practice Paper 8

Q. No.	Ans	Q. No.	Ans	Q. No.	Ans	Q. No.	Ans	Q. No.	Ans	Q. No.	Ans
1.	C	2.	C	3.	D	4.	D	5.	A	6.	A
7.	D	8.	B	9.	B	10.	D	11.	A	12.	A
13.	B	14.	B	15.	A	16.	B	17.	B	18.	C
19.	A	20.	B								

Practice Paper 9

Q. No.	Ans	Q. No.	Ans	Q. No.	Ans	Q. No.	Ans	Q. No.	Ans	Q. No.	Ans
1.	A	2.	B	3.	C	4.	A	5.	B	6.	C
7.	A	8.	B	9.	C	10.	A	11.	C	12.	B
13.	A	14.	B	15.	A	16.	A	17.	C	18.	A
19.	D	20.	A								

Practice Paper 10

Q. No.	Ans	Q. No.	Ans	Q. No.	Ans	Q. No.	Ans	Q. No.	Ans	Q. No.	Ans
1.	D	2.	C	3.	C	4.	B	5.	A	6.	A
7.	D	8.	D	9.	A	10.	A	11.	A	12.	D
13.	A	14.	A	15.	D	16.	A	17.	D	18.	A
19.	C	20.	C								

■ Practice Paper 11

Q. No.	Ans	Q. No.	Ans	Q. No.	Ans	Q. No.	Ans	Q. No.	Ans	Q. No.	Ans
1.	B	**2.**	A	**3.**	C	**4.**	A	**5.**	C	**6.**	A
7.	A	**8.**	B	**9.**	A	**10.**	C	**11.**	A	**12.**	B
13.	A	**14.**	D	**15.**	C	**16.**	D	**17.**	A	**18.**	B
19.	A	**20.**	A								

■ Practice Paper 12

Q. No.	Ans	Q. No.	Ans	Q. No.	Ans	Q. No.	Ans	Q. No.	Ans	Q. No.	Ans
1.	D	**2.**	D	**3.**	D	**4.**	A	**5.**	C	**6.**	B
7.	D	**8.**	A	**9.**	B	**10.**	A	**11.**	A	**12.**	D
13.	C	**14.**	C	**15.**	D	**16.**	C	**17.**	A	**18.**	D
19.	B	**20.**	D								

■ Practice Paper 13

Q. No.	Ans	Q. No.	Ans	Q. No.	Ans	Q. No.	Ans	Q. No.	Ans	Q. No.	Ans
1.	C	**2.**	A	**3.**	B	**4.**	A	**5.**	A	**6.**	B
7.	A	**8.**	B	**9.**	A	**10.**	D	**11.**	A	**12.**	D
13.	D	**14.**	B	**15.**	D	**16.**	D	**17.**	A	**18.**	D
19.	A	**20.**	B								

19. Multiple Choice Questions of Agricultural Economics

Priciples of Agricultural Economics

Practice Paper 1

1. **Science which deals the principles and methods of economics are applied to the special conditions of agricultural industry" is called:**
   (A) Agricultural economics
   (B) Economics
   (C) Industrial economics
   (D) Pure economics

2. **The scope of agricultural economics is exactly covered by the subject of agricultural economics. It covers mainly three things:**
   (A) Subject matter of agricultural economics
   (B) Nature of agricultural economics
   (C) Limitations
   (D) All of these

3. **Agricultural economics puts facts before us.**
   (A) Positive science
   (B) Normative science
   (C) Applied science
   (D) None of these

4. **Agricultural economics establishes certain norms or ideals.**
   (A) Positive science
   (B) Normative science
   (C) Applied science
   (D) None of these

5. **Agricultural economics is also an applied science because the principles and practices of economics solves the different agricultural problems of farmers.**
   (A) Positive science
   (B) Normative sciencest pa
   (C) Applied science
   (D) None of these

6. **The cost of production can be minimized by using the principle of:**
   (A) Law of return
   (B) Law of opportunity cost
   (C) Law of least-cost combination
   (D) Cost principle

7. **Who defined Economics is a wealth of science:**
   (A) Smith (B) Marshall
   (C) Robins (D) Pigue

8. **Which of the following is a subject matter of economics?**
   (A) Consumption
   (B) Production and distribution
   (C) Public finance
   (D) All of these

9. **Which is not a factor of production?**
   (A) Land (B) Crops
   (C) Capital (D) Labour

10. **Which of the following function is related to an economic system?**
   (A) What goods and services are to be produced?
   (B) How to best organize and combine recourse of production?
   (C) Allocation or distribution of income to owners of resources
   (D) All of these

11. **The fourth Five Year Plan gave first priority to:**
   (A) Agriculture
   (B) Transport
   (C) Health and Education
   (D) Industries

12. **The Agricultural Prices are recommended by:**
   (A) I.C.A.R. (B) C.S.I.R.
   (C) C.A.C.P. (D) R.B.I.

13. The Agricultural Prices are recommended on the basis of:
(A) Average cost of production
(B) Minimum cost of production
(C) Bulk line cost
(D) Maximum cost of production

14. Which of the following organization refinance agriculture:
(A) I.D.B.I.
(B) I.C.I.C.
(C) N.A.B.A.R.D.
(D) S.B.I.

15. Regional Rural Banks provide credit to:
(A) Big farmers
(B) Weaker section of rural population
(C) Weaker section of urban population
(D) None of these

16. Loan for purchase of tractor is known as:
(A) Short term loan
(B) Medium term loan
(C) Long term loan
(D) Cop plan

17. Agricultural Produce and Marketing Act was passed in:
(A) 1937 (B) 1935
(C) 1947 (D) 1949

18. The growth rate of food production in India after 1980 has been:
(A) Increasing
(B) Decreasing
(C) Constant
(D) Negative

19. The export of Agricultural commodities since Nineties is:
(A) Increasing
(B) Decreasing
(C) Stand still
(D) Negative

20. I.R.D.P. Programme is meant for:
(A) Poverty alleviation
(B) Employment generation
(C) Disposal of agricultural produce
(D) Agro-processing

Practice Paper 2

1. The Zamidari Abolition Act was passed in UP in:
(A) 1950 (B) 1952
(C) 1954 (D) 1956

2. In India land reforms have been equated with total:
(A) Urban development
(B) Industrial development
(C) Agrarian development
(D) Educational development

3. CAPART expands as:
(A) Council fro Action and people Advancement in Rural Technology
(B) Council for Advancement of People Action and Rural Technology
(C) Commission for Action and People Advancement in Rural Technology
(D) Committee for Advancement of People Action and Rural Technology

4. CAPART is an autonomous wing of:
(A) Ministry of Agriculture and Cooperation
(B) Ministry of Rural Development
(C) ICAR
(D) Ministry of HRD

5. A monthly newsletter published by Ministry of Rural Development:
(A) Grameen Bharat
(B) Grama Sabha
(C) Panchayat Sabdh
(D) Rural Panchayat

6. The Right to Information Act passed in the year:
(A) 2005 (B) 2006
(C) 2007 (D) 2008

7. Zamidari system was introduced by:
(A) British (B) Mughals
(C) Marathas (D) All of these

8. The land revenue under Zamidari system was charged:
(A) Through village panchayat
(B) Through landlords
(C) Directly from the farmers
(D) All of these

9. 'Zamidari' system was abolished in India on:

(A) 1952 (B) 1954
(C) 1968 (D) 1960

10. In Mahalwari system land revenue was charged:

(A) Through landlords
(B) Directly from the farmers
(C) Through village panchayat
(D) All of these

11. The land revenue in Ryotwari system was charged:

(A) Through landlords
(B) Directly from the farmers
(C) Through village panchayat
(D) All of these

12. What is globalisation? Select the most appropriate answer:

(A) It is the process of rapid integration or interconnection between countries
(B) It is increase in foreign trade
(C) It is the process under which developing countries are using modern technology
(D) It is the process of movement of people from one region to another

13. Which of the following is not a feature of foreign trade?

(i) It connects different markets of the world
(ii) Choice of goods in the market rises
(iii) Prices of similar goods in the two markets tend to become equal
(iv) Developing countries gain more that the developed countries

(A) Only (i) and (ii)
(B) Only (iv)
(C) Only (iii) and (iv)
(D) All of these

14. .... is the process of rapid integration or interconnection between countries:

(A) Privatisation (B) Liberalisation
(C) Globalisation (D) None of these

15. Which of the following has been the major factor that has stimulated the globalization process:

(A) Technology (B) Ship industry
(C) Democracy (D) Socialism

16. Which of the following is a trade barrier?

(i) Technology
(ii) Tax on imports
(iii) Tax on exports

(A) Only (i) (B) Only (i) and (ii)
(C) Only (ii) and (iii) (D) Only (iii)

17. Why the countries use trade barriers?

(i) Countries put up trade barriers to increase or decrease foreign trade
(ii) Countries put up trade barriers to decide what kind of goods and how much of each, should come into the country
(iii) To create social and political equality

(A) Only (i) (B) Only (i) and (ii)
(C) Only (iii) (D) Only (i) and (ii)

18. The Indian government, after independence has put trade barriers to foreign trade and foreign investments:

(A) To protect the producers within the country from foreign competition
(B) To promote Swadeshi Movement
(C) Because foreign goods were very expensive
(D) Because foreign goods were not liked by the Indian people

19. When was India's foreign trade liberalized?

(A) 1960s (B) 1970s
(C) 1980s (D) 1990s

20. Removing barriers or restrictions setup by the government is known as:

(A) Privatisation (B) Globalisation
(C) Liberalisation (D) Industrialisation

Practice Paper 3

1. Which of the following is/are result of liberalization?

(i) Businesses are allowed to make decisions freely
(ii) Government removes unwanted restrictions from trade
(iii) Government provides political and fundamental rights to all
(iv) Poverty is removed

(A) Only (i) & (ii) (B) Only (i), (ii) & (iii)
(C) Only (iii) & (iv) (D) Only (iv)

2. **Indian government wants to check the free flow of cheap Chinese electronic goods into Indian market. Which of the following method can government use?**
   (A) Trade barrier
   (B) Complaint to WTO
   (C) Government can increase export tax
   (D) Government can complaint to World Bank

3. **What is WTO?**
   (A) World Tower Organisation
   (B) World Trade Organisation
   (C) World Tax Organisation
   (D) World traders Organisation

4. **What is the basic aim of WTO?**
   (A) To check the free flow of goods and services across the world
   (B) To bridge gap between rich and poor countries
   (C) To liberalise international trade
   (D) To support MNC

5. **How many countries are the members of WTO?**
   (A) 95 (B) 120
   (C) 149 (D) 200

6. **Which of the following organization/s liberalises international trade?**
   (i) WTO
   (ii) MNC
   (iii) Commercial bank
   (A) Only (i)
   (B) Only (i) and(ii)
   (C) Only (ii) and (iii)
   (D) Only (iii)

7. **Which of the following has benefitted the most from the globalization?**
   (A) Small producers (B) Farmers
   (C) Poor (D) Consumers

8. **The planning commission was setup in:**
   (A) 1949 (B) 1950
   (C) 1951 (D) 1952

9. **The planning commission of India is a:**
   (A) Constitutional body
   (B) Statutory body
   (C) Non-statutory body
   (D) Autonomous body

10. **Who was the first deputy chairman of planning commission of India:**
   (A) Jawaharlal Nehru
   (B) Rajendra Prasad
   (C) V T Krishnamachari
   (D) C Rajagopalachari

11. **Harrod-Domar model was adopted in:**
   (A) First five year plan
   (B) Second five year plan
   (C) Fifth five year plan
   (D) Sixth five year plan

12. **Micro level planning is an example for:**
   (A) Imperative planning
   (B) Five year planning
   (C) Regional planning
   (D) Target group planning

13. **Planning in India as it is at present can be termed as:**
   (A) State planning
   (B) Centralized planning
   (C) Decentralized planning
   (D) Indicative planning

14. **Five year plan is a:**
   (A) Perspective plan
   (B) Medium term plan
   (C) Adhoc plan
   (D) Contingent plan

15. **A rolling plan is a plan for:**
   (A) One year
   (B) Two year
   (C) Three year
   (D) Year to year basis

16. **Which of the following is/are Indian MNC's?**
   (i) Samsung (ii) Nokia
   (iii) Tata motors (iv) Asian Paints
   (A) Only (i) and (ii)
   (B) Only (ii) ad (iii)
   (C) Only (iii) and (iv)
   (D) Only (i) and (iv)

17. **Which of the following has benefitted the most from the process of globalisation?**
   (A) IT
   (B) Agro based industries
   (C) Textile industries
   (D) Iron and steel industry

18. Which of the following have suffered the most because of the globalisation?

(A) IT
(B) Small manufectures of vegetable oil, capacitors, plastic etc.
(C) MNC's
(D) Automobile industries

19. Which of the following has benefitted the most from the globalization?

(A) People with education, skill and wealth
(B) People living below poverty line
(C) farmers
(D) Local traders

20. Which of the following method can be used by the government for a fair globalization?

(i) It can impose trade barriers
(ii) It can negotiate at the WTO for fairer rules
(iii) It can align with other developing countries
(iv) It can close its market for foreign trade

(A) Only (i) and (ii)
(B) Only (ii) and (iv)
(C) Only (i), (ii) and (iii)
(D) All of these

Practice Paper 4

1. Which one is the most effective method is identifying real rural needs:

(A) Election method
(B) Selection method
(C) Workshop method
(D) Socio-metric method

2. Which one of the following is not an objective of India's economic planning:

(A) Population growth
(B) Industrial growth
(C) Self reliance
(D) Poverty alleviation

3. A tool for optimum utilization of the available resources in the maximization of the attainable income of the farmers is known as:

(A) Farm planning
(B) Home planning
(C) Programme planning
(D) Extension planning

4. For which of following a cooperative society is an example:

(A) Formal group (B) Informal group
(C) Primary group (D) Territory group

5. For which of following logical framework technique is greatly used:

(A) Project planning
(B) Input utilization
(C) Work scheduling
(D) Policy formulation

6. Land Development Banks in India are owned by whom:

(A) Govt. of India
(B) State Governments
(C) Commercial banks
(D) Cooperative banks

7. Mandi ordinance of agricultural production was enforced in:

(A) 1949 (B) 1950
(C) 1955 (D) 1959

8. Mention the type of correlation that operates between rurality and population density:

(A) Positive (B) Negative
(C) Accidental (D) Scattered

9. Planning at districts and block levels is generally considered as:

(A) Micro planning (B) Meso planning
(C) Macro planning (D) All of these

10. Planning in India as it is at present can be termed as which of the following:

(A) State planning
(B) Centralized planning
(C) Decentralized planning
(D) Indicative planning

11. The family in an agrarian society is:

(A) Patriarchal (B) Matriarchal
(C) Matrilocal (D) Nuclear

12. The Panchayati Raj institutions are intended to strengthen by which:

(A) Structural planning
(B) Indicative planning
(C) Multilevel planning
(D) Functional planning

13. The Planning Commission was set in India in accordance with the directive principles in:

(A) Article 38 (B) Article 39
(C) Article 42 (D) Article 51

14. Profits are defined as π = R-(FC + VC), where, FC = total fixed cost and VC = total variable cost and ....:

(A) R = Revenue
(B) R = Reduction
(C) R = Regression
(D) R = Co-efficient

15. Balwantrai Mehta Team stimulated an active consideration of .... through democratic bodies:

(A) Mobilization (B) Centralization
(C) Stabilization (D) Decentralization

16. Micro economics is concerned with:

(A) Analyzing the economic behaviour of the individual farm or business
(B) Analyzing the total economy of nation
(C) Only deals with scarcity of resources
(D) None of these

17. Macro economics deals with:

(A) Analyzing the economic behaviour of the individual farm or business
(B) Analyzing the total economy of nation
(C) Scarcity only
(D) Net profit only

18. The law of diminishing returns states that:

(A) The output increases with increasing input levels continuously
(B) The output increases with input levels irregularily
(C) The output increases with increasing input levels upto certain level and than declines
(D) All of these

19. Which of the following function is related to production function analysis?

(A) Total physical product
(B) Average physical product
(C) Marginal physical product
(D) All of these

20. Principle of least-cost combination explains:

(A) Output is the function of two or more than two variable inputs
(B) Use of two variable inputs in such a way to get maximum profit
(C) Use of two variable inputs in such a way to get same level of output at comparatively lowest cost
(D) All of these

Practice Paper 5

1. Law of equi-marginal return states that:

(A) Each unit of input is invested in different enterprises in such a way so that a farmer can get maximum marginal return
(B) Each unit of input is invested to get maximum return
(C) Each unit of input invested to get maximum average return
(D) None of these

2. Principal of equi-marginal return states that:

(A) Maximum satisfaction is obtained from a given level of expenditure on two goods when the last penny spent on either good gives the same amount of satisfaction
(B) Equal marginal return is obtained by substitution of one good by other
(C) Both (A) and (B)
(D) None of these

3. Factor-product relationship is also known as:

(A) Input-output relationship
(B) Output-output relationship
(C) Demand-supply relationship
(D) Output-Input relationship

4. Law of equi-marginal return also known as:

(A) Law of opportunity cost
(B) Alternative cost principle
(C) Both (A) and (B)
(D) Law of equi-marginal return

5. Law of variable proportions is also known as:

(A) Law of returns
(B) Law of factor-product relationship
(C) Law of input-output relationship
(D) All of these

6. Law of variable proportions deals with:

(A) Input-output relationship where the output is the function of input
(B) Cost of output and input
(C) Supply of output
(D) All of these

7. Maximum net return from one variable input will occur when:

(A) Additional return from an additional unit of output from employing an additional unit of input is equal to the additional cost of that unit of input
(B) Value of marginal return is equal to cost of the input
(C) Marginal return × selling price of product = marginal input cost
(D) All of these

8. The production possibilities curve shows:

(A) Alternative combination of two outputs that may be produced with a given amount of resource during a period of time
(B) Combination of inputs that can produce equal level of production
(C) Combination of outputs that can be produced with same amount of input
(D) All of these

9. Loan for marriage known as:

(A) Direct productive loan
(B) Unproductive loan
(C) Indirect productive loan
(D) None of these

10. Economy of a developing country is marked by:

(A) Predominance of agriculture
(B) Predominance of industry
(C) Predominance of transportation
(D) None of these

11. The Regional Rural Bank was established through:

(A) RRB Act 1974 (B) RRB Act 1975
(C) RRB Act 1976 (D) None of these

12. At point of inflection in production function curve, MR is:

(A) > AR (B) < AR
(C) MR = AR (D) All of these

13. Marginal cost is equal to:

(A) $\frac{\text{Change in total cost}}{\text{Additional output}}$

(B) $\frac{\text{Change in total revenue}}{\text{Additional output}}$

(C) $\frac{\text{Change in total value product}}{\text{Additional input}}$

(D) All of these

14. Marginal rate of substitution is equal to:

(A) $\frac{\text{Change in number of y units}}{\text{Change in number of x units}}$

(B) $\frac{\text{Change in price of x commodity}}{\text{Change in price of y commodity}}$

(C) $\frac{\text{Price of x commodity}}{\text{Price of y commodity}}$

(D) None of these

15. The price which necessarily covers the variable cost and also the fixed cost is:

(A) The long run price
(B) The short run price
(C) The market price
(D) The equilibrium price

16. Read the following statements and mark your answer according to code given below:

I. For variable input; APP = MPP in the Ist stage
II. For variable input; APP = MPP at the end of Ist stage
III. For variable input; APP = MPP at the start of Ist stage
IV. For variable input; APP = MPP in the IInd stage

Answer code

(A) I and III are correct
(B) II and III are correct
(C) I and IV are correct
(D) II and IV are correct

17. MPP (Marginal Physic Product) of a variable input is negative:

(A) Only in I stage (B) Only in II stage
(C) Only in III stage (D) In II and III stage

18. Match list I with list II and select the correct answer using the code give below:

List I (Farm Income)	List II (Measure)
a. Family labour income	1. G.I. – cost A1
b. Farm business income (owner farmer)	2. G.I. – cost A2
c. Net income	3. G.I. – cost B
d. Farm business income (tenant farmer)	4. G.I. – cost C

Code:

	a	b	c	d
(A)	2	1	3	4
(B)	1	2	4	3
(C)	1	2	3	4
(D)	3	1	4	2

19. Marketing surplus is:

(A) Total production – Total family requirement
(B) Total consumption – Marginal consumption
(C) Total cost – marginal cost
(D) Total produce – Total cost

20. Line passing through the least cost points in the isoquants map is called:

(A) Isoline (B) Ridge line
(C) Expansion path (D) Isoquant curve

Practice Paper 6

1. The demand of the commodity is different mainly by its:

(A) Quality (B) Price
(C) Storage (D) Supply

2. Formula which indicates the least-cost combination point is:

(A) MR = MC
(B) ΔY/Δx1
(C) MRSY1, Y2
(D) MRSx1 x2 = PX2/PX1

3. Level of optimum production is available in:

(A) I stage of production
(B) II stage of production
(C) III stage of production
(D) None of these

4. Which one of the following kind of budgeting would you advise to a new farm planner:

(A) Enterprise budgeting
(B) Complete budgeting
(C) Partial budgeting
(D) None of these

5. Which of the following is the most effective way to overcome the defects of agricultural marketing?

(A) Cooperative marketing
(B) Forward marketing
(C) Regulated marketing
(D) Monopoly

6. A farm plan can be prepared by using:

(A) Partial budgeting
(B) Complete budgeting
(C) Input-output model
(D) Break even point

7. When marginal product in increasing at increasing rate then:

(A) Total product increases at a decreasing rate
(B) Total product increases at a increasing rate
(C) Average product decreases
(D) None of these

8. If the increase or decrease one product increases or decreases the other product, the relationship between both products is known as:

(A) Complementary (B) Supplementary
(C) Competitive (D) Symbiotic

9. If the increase or decrease in the production level of one product does not affect the production level of the other product, the relationship is called as:

(A) Complementary
(B) Supplementary
(C) Competitive
(D) Symbiotic

10. If the increase in the production level of one product necessarily decreases the production level of the other product, the relationship is called as:

(A) Complementary
(B) Supplementary
(C) Competitive
(D) Symbiotic

11. Production possibility curve is also called as:

(A) Iso-revenue curve
(B) Opportunity curve
(C) Iso-return curve
(D) All of these

12. Law of demand indicates:

(A) Inverse relationship between demand and price
(B) Linear relationship between demand and price
(C) Inverse relationship between market demand and marketable surplus
(D) Linear relationship between supply and demand

13. Price line or budget line indicates:

(A) A capacity of a consumer to purchase within a given budget
(B) A capacity of a consumer to provide marketable surplus
(C) The graph between demand and satisfaction
(D) None of these

14. Price elasticity of demand is always:

(A) Positive (B) Zero
(C) Negative (D) None of these

15. The supply elasticity of bulky products is normally:

(A) Less elastic (B) High elastic
(C) Complete elastic (D) All of these

16. Supply of a commodity is governed by:

(A) Price of the commodity
(B) Price of production factors
(C) State of technology
(D) All of these

17. The price elasticity of demand for agricultural commodities is always:

(A) < 1 (B) > 1
(C) Zero (D) Infinite

18. The price elasticity of luxury items is always:

(A) < 1 (B) > 1
(C) Zero (D) Infinite

19. If the price elasticity is >1, the total revenue will be increased at:

(A) Decrease in price
(B) Increase in price
(C) No change in price
(D) Unknown

20. If the price elasticity is <1, the total revenue will be increased at:

(A) Decrease in price
(B) Increase in price
(C) No change in price
(D) Unknown

Practice Paper 7

1. Marginal return is equal to:

(A) $\frac{\Delta Y}{\Delta X}$ (B) $\frac{Y}{X}$
(C) $\frac{Px}{Px}$ (D) $\frac{Px}{Py}$

2. In rational stage of production function, both marginal and average returns are:

(A) At increasing rate
(B) At decreasing rate
(C) Both are equal
(D) None of these

3. An isoquant line shows:

(A) Combination of two inputs that may produce a given level of output
(B) Combination of two inputs that may produce maximum level of output
(C) Different marginal returns
(D) None of these

4. An isocost line shows:

(A) Combination of two variable inputs that cost different amount
(B) Combination of two variable inputs that cost the same amount
(C) Combination of two variable inputs that produce the same level of production
(D) All of these

5. **Which of the following principle is employed in production when resources are scare?**
   (A) Equi-marginal principle
   (B) Comparative advantage
   (C) Law of returns
   (D) None of these

6. **Production possibility curve indicates:**
   (A) Various combinations of the products y_1 and y_2 obtained with fixed level of resources combinations
   (B) Maximum production with unlimited sources
   (C) Marginal return with limited resources
   (D) None of these

7. **Operating ratio is equal to:**
   (A) $\frac{\text{Total expenses}}{\text{Gross income}}$ (B) $\frac{\text{Fixed expenses}}{\text{Gross income}}$
   (C) $\frac{\text{Operating expenses}}{\text{Gross income}}$
   (D) None of these

8. **Gross ratio is equal to:**
   (A) $\frac{\text{Total expenses}}{\text{Gross income}}$ (B) $\frac{\text{Fixed expenses}}{\text{Gross income}}$
   (C) $\frac{\text{Operating expenses}}{\text{Gross income}}$
   (D) None of these

9. **Fixed ratio is equal to:**
   (A) $\frac{\text{Total expenses}}{\text{Gross income}}$ (B) $\frac{\text{Fixed expenses}}{\text{Gross income}}$
   (C) $\frac{\text{Operating expenses}}{\text{Gross income}}$
   (D) None of these

10. **Fixed cost is also known as:**
    (A) Over head cost (B) Prime cost
    (C) Alternate cost (D) All of these

11. **Variable cost is also known as:**
    (A) Overhead cost (B) Prime cost
    (C) Alternate cost (D) All of these

12. **Opportunity cost is also known as:**
    (A) Overhead cost (B) Prime cost
    (C) Alternate cost (D) All of these

13. **Indifference curve shows a:**
    (A) Relative willingness to consume one good rather than another
    (B) Relationship of demand and supply
    (C) Relationship between demand and price
    (D) Relationship between supply and price

14. **If price elasticity is -1, it indicates that:**
    (A) An equal change in demand and price
    (B) Percentage change on quantity in smaller than the percentage price change
    (C) Percentage quantity change is greater than the percentage price change
    (D) None of these

15. **If the price elasticity of demand is more than one, the elasticity is known as:**
    (A) Elastic
    (B) Inelastic
    (C) Infinity
    (D) None of these

16. **If the price elasticity of demand is less than one, the elasticity is known as:**
    (A) Elastic
    (B) Inelastic
    (C) Infinity
    (D) None of these

17. **Changes in supply elasticity occurs mainly during:**
    (A) Short period
    (B) Long period
    (C) Winter season
    (D) Rainy or summer season

18. **The elasticity of the agricultural commodity in very long run will be always:**
    (A) Elastic
    (B) Inelastic
    (C) Both (A) and (B)
    (D) None of these

19. **The supply slope is normally:**
    (A) Negative
    (B) Positive
    (C) Both (A) and (B)
    (D) None of these

20. **The luxury good has income elasticity:**
    (A) 1 (B) < 1
    (C) > 1 (D) Between o to 1

Practice Paper 8

1. **The rural development programmme launched on the basis of area development approach:**
   (A) DPAP (B) MWP
   (C) RWP (D) IRDP
2. **The Ganga Kalyan Yojana is for:**
   (A) Consumption purpose of farmers
   (B) Irrigation for small and marginal farmers
   (C) Credit support scheme for rural women
   (D) •
3. **Bhoodan movement, 1951was started by:**
   (A) Acharya Vinoba Bhave
   (B) Mahatma Gandhi
   (C) Jai Prakash Narayan
   (D) Jawaharlal Nehru
4. **Bhodan movement was started at:**
   (A) Santhinietan
   (B) Wardha
   (C) Panchampalli
   (D) Pavnar
5. **The land collected in bhoodan is distributed among:**
   (A) Marginal farmers
   (B) Landlless peasants
   (C) Medium farmers
   (D) Krishi pandits
6. **Which of the following network connect agricultural produce, wholesale markets and rural markets:**
   (A) AFPINET
   (B) AGMARKNET
   (C) HORTNET
   (D) VISTARNET
7. **Role of different agencies for village development is included in:**
   (A) Resource mapping
   (B) Time line
   (C) Transect walk
   (D) Chapati diagram
8. **Rural development depends on:**
   (A) Research
   (B) Research – Extension
   (C) Research – Teaching – Extension
   (D) None of these
9. **The basic unit of development under IRD program is:**
   (A) A village
   (B) A community development block
   (C) A family
   (D) A district
10. **From among the following identify one which is three - tier system of the local self - government:**
    (A) Gram panchayat – Khand samiti – Zila parishad
    (B) Gram panchayat –Village school– Village cooperative
    (C) IRD programme – TRYSEM – NREP
    (D) Lab to land – KVK – KGK
11. **The motto of land reform in India is:**
    (A) Land to tiller
    (B) Land to landlord
    (C) Land to Talukedar
    (D) All of these
12. **The land revenue was charged in cash first time during the reign of:**
    (A) Ashoka
    (B) Chandra Gupta I
    (C) Tamur
    (D) Akbar
13. **The land revenue during Kautilya era was:**
    (A) $1/6^{th}$ of total produce
    (B) $1/5^{th}$ of total produce
    (C) $1/4^{th}$ of total produce
    (D) $1/3^{rd}$ of total produce
14. **The land revenue during Sher Shah Suri reign was:**
    (A) $1/6^{th}$ of total produce
    (B) $1/5^{th}$ of total produce
    (C) $1/4^{th}$ of total produce
    (D) $1/3^{rd}$ of total produce
15. **The land revenue during Akbar regime was:**
    (A) $1/6^{th}$ of total produce
    (B) $1/5^{th}$ of total produce
    (C) $1/4^{th}$ of total produce
    (D) $1/3^{rd}$ of total produce

16. Agricultural holding refers to:
(A) Total area of land which is held for cultivation as a single unit
(B) Total area of land used as a single unit
(C) Total area owned by a cultivator
(D) All of these

17. Which of the following step(s) are involved in farm survey?
(A) Planning of survey and collection of data
(B) Analysis of data
(C) Presentation and interpretation of the findings
(D) All of these

18. Which type of sampling method is used for collection of data if the population heterogeneous?
(A) Random sampling
(B) Stratified sampling
(C) Purposive sampling
(D Sequential sampling

19. Sequential sampling method is widely used in industries for:
(A) Quantity control
(B) Efficiency measures
(C) Quality control
(D) None of these

20. Cluster sampling method is used for:
(A) Quality control
(B) Geological surveys
(C) Efficiency measures
(D) All of these

Practice Paper 9

1. Operational holding refers to:
(A) Total area of land which is held for cultivation as a single unit
(B) Total area including fallow and barren used as a single unit
(C) Total area owned by a owner
(D) All of these

2. Fragmentation of holding refers to:
(A) It is a breaking up of a single holdings into a number of small pieces situated at a distance from each other
(B) Consolidating the small holdings
(C) Breaking of holdings into economic holdings
(D) None of these

3. An optimum holding refers to the size that:
(A) Should not be more than 3 times of economic holding
(B) Should be > 3 times of economic holding
(C) Should be equal to operational holding
(D) Should be <2 ha land

4. Which of the non-institutional agency of that provides agricultural loan?
(A) Traders and money lenders
(B) Commercial banks
(C) Land Development Bank
(D) Regional Rural Banks

5. Which of the following credit agency that provides institutional loan to farmers?
(A) Cooperative Societies
(B) Commercial Banks
(C) Land Development Bank
(D) All of these

6. Marketing involves the activities from:
(A) Production to end consumer
(B) Production only
(C) Production to middlemen
(D) Production to storage

7. Monopoly market refers to:
(A) Single producer or seller in market
(B) Single buyer in market
(C) More number of producers and single buyers
(D) None of these

8. Duopsony refers to:
(A) Market having only two sellers
(B) Market having only two buyers
(C) Market having only two buyers and sellers each
(D) None of these

9. Oligopsony refers to:
(A) Few buyers in a market
(B) Only one buyer in a market
(C) Only one seller in a market
(D) None of these

10. Marketable surplus refers to:
(A) Actual amount that is marketed
(B) Excess amount than requirement for consumption
(C) Both (A) and (B)
(D) None of these

11. Marketed surplus refers to:
(A) Actual amount that is marketed
(B) Excess amount than requirement for consumption
(C) Both (A) and (B)
(D) None of these

12. Market structure refers to:
(A) Organizational structure that affects the competition and price in market
(B) Behavioral pattern of buyers and sellers through which prices are affected
(C) It indicates about the role of commission agents in market only
(D) All of these

13. Market conduct refers to:
(A) Organizational structure that affects the competition and price of a commodity
(B) Behavioral pattern of market that affect the commodity price
(C) Involvement of brokers in marketing
(D) All of these

14. Which of the marketing system was prevalent in the ancient time?
(A) Barter system
(B) Modern system
(C) European system
(D) None of these

15. Computer based market refers:
(A) Prices are decided on interest or phone
(B) Exchange of commodities are based on computer information
(C) Payment can also be · made through computer
(D) All of these

16. Market clearing refers to:
(A) Equilibrium state of market price
(B) Stage of marketing at which brokers mediates in deciding the price of a commodity
(C) Change in price with increase supply
(D) None of these

17. Price discovery refers to:
(A) Process through which buyer and sellers decide a price of a commodity at a given place
(B) Stage of deciding equilibrium market price
(C) Search of market price
(D) All of these

18. Price spread is also known as:
(A) Farm price
(B) Marketing margin
(C) Marketing cost
(D) None of these

19. Which of the statement is correct for monopoly market?
(A) MC = MR (B) MC > AC
(C) MC < AC (D) None of these

20. The price which necessarily covers the variable cost and also the fixed price is:
(A) The long run price
(B) The short run price
(C) The market price
(D) The equilibrium price

Practice Paper 10

1. Marginal Physical Product of a variable input is negative:
(A) Only in I stage
(B) Only in II stage
(C) Only in III stage
(D) In I and II stage

2. Average cost is equal to:
(A) Total cost + opportunity cost
(B) $\frac{\text{Total cost}}{\text{Output}}$
(C) $\frac{\text{Average variable cost}}{\text{Output}}$
(D) Cost B + Cost C

3. Fixed cost is equal to:
(A) $\frac{\text{Total fixed cost}}{\text{Output}}$ (B) $\frac{\text{Total cost}}{\text{Output}}$
(C) $\frac{\text{Total variable cost}}{\text{Output}}$ (D) None of these

4. Variable cost is equal to:

(A) $\frac{\text{Total cost}}{\text{Output}}$

(B) $\frac{\text{Total variable cost}}{\text{Unit of output}}$

(C) Cost A + Cost B + Cost C

(D) All of these

5. Marginal cost is equal to:

(A) $\frac{\text{Total cost}}{\text{Output}}$

(B) $\frac{\text{Increase in variable cost}}{\text{Increase in output}}$

(C) Cost C + Rent paid to leased rent

(D) $\frac{\text{Total variable cost}}{\text{Increase in output}}$

6. If the price elasticity of barley is -0.5, it means:

(A) Barley demand will be decreased by 0.5%, if the price increased by 1%

(B) Barley demand will be increased by 0.5% if the price is increased by 1%

(C) If the price of barley is increased by 0.5%, the demand will also be increased by 0.5%

(D) Price and demand both either increased or decreased by 0.5%

7. If the petrol price fall from Rs. 10 to 8 for 1 litre and as a result the demand of petrol is increased from 5 units to 9 units. What will be the price elasticity of demand?

(A) -1 (B) -2

(C) -3 (D) -4

8. Cross elasticity of demand refers to:

(A) $\frac{\text{\% change in quantity demanded of X}}{\text{\% change in income}}$

(B) $\frac{\text{\% change in quantity demanded of X}}{\text{\% change in price}}$

(C) $\frac{\text{\% change in income}}{\text{\% change in price}}$

(D) None of these

9. Income elasticity of demand refers to:

(A) $\frac{\text{\% change in quantity demanded}}{\text{\% change in income}}$

(B) $\frac{\text{\% change in quantity demanded}}{\text{\% change in price}}$

(C) $\frac{\text{\% change in income}}{\text{\% change quantity demanded}}$

(D) None of these

10. Price elasticity of supply is equal to:

(A) $\frac{\text{\% change in quantity supplied of a commodity A}}{\text{\% change in price of A}}$

(B) $\frac{\text{\% change in demand of commodity A}}{\text{\% change in price of A}}$

(C) $\frac{\text{\% change in income}}{\text{\% change in price}}$

(D) All of these

11. The production or supply of a product will be closed by a firm if:

(A) Production cost < cost of marginal product

(B) Production cost > cost of marginal product

(C) Variable cost > production cost

(D) All of these

12. Which of the following condition is required for perfect competition in the market?

(A) More number of firms in the market

(B) Homogeneous product

(C) Free entry or exit in the market

(D) All of these

13. What will happen if the market price of a commodity is higher than normal price?

(A) Demand will be reduced and supply will be increased in the market

(B) Demand will be increased

(C) Supply will be reduced

(D) None of these

14. What will happen if the market price of a commodity is lower from its normal price?

(A) Demand of a commodity will be increased
(B) Supply will be decreased
(C) Market price will increase towards the normal price line
(D) All of these

15. Income elasticity of inferior goods is:

(A) Positive
(B) Negative
(C) Between zero to one
(D) Infinite

16. An income elasticity of necessary is:

(A) 1 (B) < 1
(C) > 1 (D) Between zero to one

17. Which of the following is a feature of monopolistic competition?

(A) Many firms
(B) Firm is price maker
(C) Products/services differentiated through promotion, quality and design
(D) All of these

18. Cyclic fluctuations in prices of agricultural commodities means:

(A) Ups and downs in the prices after certain intervals
(B) Increase in prices after certain intervals
(C) Decrease in prices after certain intervals
(D) None of these

19. The fluctuations in agricultural prices due to operation of Cobweb theorem shows:

(A) Cyclic fluctuations
(B) Long term fluctuations
(C) Seasonal fluctuations
(D) All of these

20. Long term fluctuations in agricultural prices and incomes are caused mainly due to:

(A) Change in population
(B) Opening of new markets
(C) Development of new technology
(D) All of these

Panchayati Raj

Practice Paper 1

1. The election of panchayat is normally held after a period of:

(A) 3 years (B) 4 years
(C) 5 years (D) 6 years

2. The number of tier for Panchayats in a state with population below 20 lakhs:

(A) Two (B) Three
(C) Four (D) One

3. The proportion of seat in a panchayat to be filled from women through election:

(A) Two-third (B) One-third
(C) One-fifth (D) One-fourth

4. Which of the following states have no Panchayati Raj:

(A) Karnataka (B) Bihar
(C) Orissa (D) Meghalaya

5. Panchayat Raj is a system of:

(A) Local government
(B) Local administration
(C) Local self-government
(D) None of these

6. The term Panchayat Raj was christened by:

(A) M.K. Gandhi (B) J.L. Nehru
(C) B.R. Mehta (D) M.S. Mehta

7. The Panchayat Raj institutions are intended to strengthen:

(A) Structural planning
(B) Indicative planning
(C) Multilevel planning
(D) Functional planning

8. The three tier of Panchayati Raj is:

(A) Zilla Parishad
(B) Grama Sabha
(C) Panchayat Samiti
(D) Block Panchayat

9. **The responsibility of village panchayat include:**
   (A) Agricultural production
   (B) Rural Industrial development
   (C) Maternity and child welfare
   (D) All of these

10. **The need for constitutional sanction of village panchayats was first felt by:**
   (A) Balwantrai Mehta Committee
   (B) Ashoka Mehta Committee
   (C) GVK Rao Committee
   (D) L M Singhvi Committee

11. **The Panchayati Raj institutions are intended to strengthen by:**
   (A) Structural planning
   (B) Indicative planning
   (C) Multilevel planning
   (D) Functional planning

12. **The term panchayati raj was christened by:**
   (A) M.K. Gandhi (B) J. L. Nehru
   (C) B. R. Mehta (D) M. S. Mehta

13. **Panchayati raj was introduced in India in:**
   (A) 1950 (B) 1952
   (C) 1958 (D) 1967

14. **Panchayati raj institution is formed based on:**
   (A) Ashok Mehta Committee
   (B) Hanumantha Rai Committee
   (C) Dantiwala Committee
   (D) Balwant rai Mehta Committee

15. **Panchayat raj is a system of:**
   (A) Local government
   (B) Local administration
   (C) Local self government
   (D) None of these

16. **Panchayati Raj system was adopted first in:**
   (A) Rajasthan and Madhya Pradesh
   (B) Andhra Pradesh and West Bengal
   (C) Rajasthan and Andhra Pradesh
   (D) Tamilnadu and Rajasthan

17. **The basic unit of Panchayati Raj institutions:**
   (A) Zila Parishad (B) Grama Sabha
   (C) Panchayat Samiti (D) Block Panchayat

18. **Co-operative society is basic institutions for:**
   (A) Political growth of the villagers
   (B) Social growth of the villagers
   (C) Cultural growth of the villagers
   (D) Socio-economic growth of the villagers

19. **According to Balwantrai Mehta committee, Panchayati Raj should be a**
   (A) Two tier structure of local self government bodies from village to district
   (B) Three tier structure of local self government bodies from village to district
   (C) Two tier structure of local self government bodies from village to block
   (D) Three tier structure of local self government bodies from village to block

20. **Panchayati Raj was first time started on 2nd October, 1959 in:**
   (A) Bombay (Maharastra)
   (B) Nagaur (Rajasthan)
   (C) Shimla (Himanchal Pradesh)
   (D) Patna (Bihar)

Practice Paper 2

1. **The Lakhina experiment is associated with:**
   (A) District Administration
   (B) Panchayat Raj Adminsitration
   (C) Rural Development Administration
   (D) Tribal Development Administration

2. **Panchayati Raj came after**
   (A) Pre-determined programme
   (B) Self determined programme
   (C) Fact determined programme
   (D) None of these

3. **Panchayati Raj in India was first recommended by:**
   (A) Balwant Rai Mehta Committee
   (B) Mahatma Gandhi
   (C) Kriplani committee
   (D) Pandit Jawaharlal Nehru

4. Tehsil Divas in Uttar Pradesh is organized on every month of:
(A) 1st and 2nd Monday
(B) 1st and 3rd Monday
(C) 1st and 3rd Tuesday
(D) 2nd and 4rth Tuesday

5. The concept of multipurpose village level worker was evolved from:
(A) Etawah Pilot Project
(B) Community Development Project
(C) National Extension Service
(D) Gurgaon Project

6. The first bank to launch Kisan Credit Card:
(A) Punjab National Bank
(B) State Bank of India
(C) Bank of Baroda
(D) Allahabad Bank

7. Nilokheri project was also called:
(A) Mazdoor Manzil
(B) Kisan Manzil
(C) Mazdoor Madad
(D) Kisan Madad

8. The term community development have originated in:
(A) Oxford
(B) Cambridge
(C) Ohio
(D) San Francisco

9. Firka development scheme was started by:
(A) Bihar government
(B) Bengal government
(C) Maharastra government
(D) Madras government

10. DRDA was established in:
(A) 1919 (B) 1980
(C) 1967 (D) 1947

11. The chairman of District Rural Development Agency is:
(A) Block Development Officer
(B) District Planning Officer
(C) District Panchayat President
(D) District Collector

12. The authority responsible for planning and implementation of rural development programme at state level is:
(A) Planning Commission
(B) State Development Committee
(C) Ministry of Rural Development
(D) Chief Minister

13. The chairman of State Development Committee is:
(A) Governor
(B) Chief Minister
(C) State Agriculture Minister
(D) Minister of Rural Development

14. The Gandhian approach of rural development have its basic emphasis on:
(A) Village industries
(B) Development of inner man
(C) Local self government
(D) Production improvement

15. The responsibility of village panchayat include:
(A) Agricultural production
(B) Rural Industrial development
(C) Maternity and child welfare
(D) All of these

16. The 'Year of Grama Sabha' was observed in:
(A) 1999-2000 (B) 2001-2002
(C) 2004-2005 (D) 2006-2007

17. The total number of panchayats in the country:
(A) 2.31 lakhs (B) 2.23 lakhs
(C) 2.47 lakhs (D) 2.79 lakhs

18. The intermediate level in Panchayati Raj institutions:
(A) Zila Parishad
(B) Grama Sabha
(C) Panchayat Samiti
(D) Block panchayat

19. The village Panchayati Act came into existence in:
(A) 1958 (B) 1959
(C) 1960 (D) 1961

20. The total number of panchayat at intermediate level:
(A) 6000 (B) 7000
(C) 8000 (D) 9000

Farm Management

Practice Paper 1

1. Agricultural commodities have greater price fluctuations mainly because of:
(A) Inelastic demand and supply
(B) Elastic demand and supply
(C) More control of government
(D) All of these

2. Farm management is:
(A) An art (B) A science
(C) A business (D) All of these

3. Farm layout means:
(A) Physical arrangement of farm in the form of plots, building, roads, channels etc.
(B) Planning of farm to maximize the profit
(C) Production of crops
(D) None of these

4. Farm profit is equal to:
(A) Gross revenue – cost
(B) Gross revenue – net returns
(C) Gross income – gross revenue
(D) All of these

5. Farm budget refers to:
(A) A tentative estimation of expenses and returns of a farm
(B) A blueprint of production and cost
(C) Loan and its interest
(D) All of these

6. Complete farm budget refers to:
(A) Farm budget for a entire farm as a single unit
(B) Farm budget separately for each enterprise
(C) It is a budget for a Gram Panchayat as a single unit
(D) It is a budget for whole area of the region

7. Partial farm budget refers to:
(A) Farm budget for a entire farm as a single unit
(B) Farm budget separately for each enterprise
(C) It is a budget for a Gram Panchayat as a single unit
(D) It is a budget for whole area of the region

8. Farm inventory refers to:
(A) List of items possessed by a farm on a specific date
(B) Land under cultivation
(C) Inventory of new farm
(D) All of these

9. Farm production records indicate:
(A) Financial position of farm
(B) Production in quantity
(C) Production in quality
(D) All of these

10. Which of the following is a farm production record?
(A) Livestock record
(B) Crop cultivation record
(C) Crop production record
(D) All of these

11. Mechanisation of agriculture refers to:
(A) Use of power driven machines for various farm jobs
(B) Ploughing with tractor operated plough only
(C) Crushing of sugarcane with crusher only
(D) All of these

12. Mechanisation of agriculture is required for:
(A) Better agricultural yield
(B) Multiple cropping
(C) Generation of employment
(D) All of these

13. How much farm power is available per hectare?
(A) 0.5 KW (B) 1.0 KW
(C) 1.5 KW (D) 2.0 KW

14. The wage is not paid to:
(A) Permanent labour
(B) Contract labour
(C) Family labour
(D) Casual labour

15. Level of farm mechanization is affected by:
(A) Size of holding
(B) Labour requirement
(C) Intensification of crop production
(D) All of these

16. Farm planning refers to:
(A) Reorganization and/or alteration of present farm organization and operation in order to get maximum return from limited resources
(B) Activities for crop production on farm
(C) Physical arrangement of farm resources
(D) All of these

17. The basic objective of farm planning is:
(A) To guide a farmer to move in the right direction
(B) To indicate the destination
(C) Proper use and allocation of input
(D) All of these

18. Farm price means to:
(A) The price that farmer receives for his produce
(B) Price prevailing in the market
(C) Price declared by Govt.
(D) All of these

19. Farm plan refers to:
(A) Systematic pre-thinking of future activities related to farm business
(B) It is a tentative future plan for farm business
(C) It is related to maximize the farm profit
(D) All of these

20. A farm plan can be prepared by using:
(A) Partial budgeting
(B) Complete budgeting
(C) Input-output model
(D) Break-even point

Practice Paper 2

1. The cost of production of the crop can be minimized by using economic principle:
(A) Law of diminishing return
(B) Law of opportunity cost
(C) Cost principle
(D) Principle of least cost combination

2. Principle of equi-marginal returns is applied when:
(A) The resources are unlimited
(B) The resource prices are high
(C) The prices of resources are varying
(D) The resources are limited

3. Do the T.C. and V.C. increases with increase in the output?
(A) Never
(B) Yes
(C) Partially
(D) With decrease in output

4. When a new variety of crop is introduced:
(A) Partial budgeting is required
(B) Full budgeting is a must
(C) No budgeting is needed
(D) Only planning is needed

5. The objective of preparation of calendar of operations in:
(A) To prepare a cropping scheme
(B) To calculate the cost of production
(C) To assess the need of human and bullock power
(D) To adjust farm enterprises

6. The farm plan would be based on the availability of efficiency of various factors of production:
(A) True
(B) False
(C) Partially true
(D) Incomplete statement

7. Farm budget is a tool of farm management analysis:
(A) True
(B) False
(C) Never include cost
(D) It is a futile exercise

8. Mixed farming is when:
(A) Crop production is combined with livestock enterprise and 50% : 50% income is from both
(B) Entire farm land is sown under mixed crops
(C) Only milch cattle are raised on the farm
(D) Income from milch animal is 95%

9. When marginal product is increasing at increasing rate than total product increases at a decreasing rate:
(A) Total product increases at a decreasing rate
(B) Total product increases at a increasing rate
(C) Average product decreases
(D) None of these

10. Price elasticity of supply of food products is generally:

(A) Infinite (B) Less than one
(C) More than one (D) Zero

11. One of the important fixed cost on a farm:

(A) Seed cost (B) Land rent
(C) Plant protection cost (D) Fertilizer cost

12. When marginal product is more than marginal cost, further investment on production would be:

(A) Profitable (B) Harmful
(C) Equal (D) No effect

13. The record of farm produce is entered in:

(A) Log book (B) Inventory book
(C) Farm stock register (C) Cash book

14. Price theory is a part of:

(A) Micro economics
(B) Macro economics
(C) Farm management
(D) Marketing

15. When marginal product is declining but remains more than zero, the total product will/will be:

(A) Zero
(B) Increase with increasing rate
(C) Increase with decreasing rate
(D) Initially decreasing than increasing

16. Which one of the following is not included in the factors of production:

(A) Enterprise (B) Organisation
(C) Land (D) Time

17. Which of the following is not a Consumer Right?

(A) Right to Freedom of Religion
(B) Right to Information
(C) Right to safety
(D) Right to Equality

18. Which of the following is not a feature of money?

(A) It includes currency – paper notes and coins and demand deposits
(B) It is without any use of its own
(C) It is authorized by the government of the country
(D) Gold has high liquidity so it is also included in the money

19. The modern currency is without any use of its own but still it is accepted as a medium of exchange because:

(A) The currency is authorized by the World Bank
(B) The currency is authorized by the government of the country
(C) The currency is authorized by RBI
(D) There is no other thing which can be used as a medium of exchange

20. Which of the following issues currency notes in India?

(A) RBI (B) SBI
(C) ICICI (D) President

Practice Paper 3

1. When a farm is managed by Government officials it is called a:

(A) Peasant farm
(B) Capitalistic farm
(C) Collective farm
(D) State farm

2. When the depreciation of an asset is measured by divide the first cost by its estimated life, it is called:

(A) Diminishing value method
(B) Sum of the years digit method
(C) Straight line method
(D) Revaluation methods

3. When $\frac{\Delta x1}{\Delta x2} = \frac{\Delta x2}{\Delta x1}$ then we can :

(A) Employ any combination of x1 and x2
(B) Do nothing
(C) Go on substituting x1 ad x2
(D) Employ x2 only

4. Farming system approach ensures the best use of:

(A) Land
(B) Labour
(C) Capital
(D) Land + Labour + Capital + Management

5. The chief function of money is:

(A) Reserve base of credit expansion
(B) Providing liquidity
(C) Medium of exchange
(D) None of these

6. Support prices of agricultural commodities are worked out by:
(A) Finance ministry
(B) Agricultural Price Commission
(C) Food and Agricultural ministry
(D) None of these

7. According to the law of Diminishing Return the optimum profit will be at the point where:
(A) AP > MP (B) MP > AP
(C) MP = AP (D) MC = MP

8. The best level of output is obtained at point where:
(A) MR < MC (B) MR = MC
(C) MR > MC (D) MR = AC

9. Most prevalent "system of farming" in India is:
(A) Cooperative farming
(B) Peasant farming
(C) Collective farming
(D) State farming

10. When the depreciation of an asset is measured by dividing the first cost by its estimated life, it is called:
(A) Diminishing value method
(B) Sum of years digit method
(C) Straight line method
(D) Revaluation method

11. The commission for agricultural cost and prices is located in:
(A) New Delhi (B) Hyderabad
(C) Lucknow (D) Faridabad

12. The record of farm produce is entered in:
(A) Log book
(B) Cash book
(C) Farm stock register
(D Inventory book

13. The marginal cost -curve intersects the average cost-curve when average cost is:
(A) Maximum (B) Minimum
(C) Rising (D) Falling

14. The overhead cost is also known as:
(A) Variable cost (B) Fixed cost
(C) Total cost (D) Marginal cost

15. The subsidiary service of a marketing system is:
(A) Welfare (B) Assembling
(C) Dispersion (D) Processing

16. According to law of diminishing return, the optimum [profit will be at the point where:
(A) MP = AP (B) MP > AP
(C) AP > MP (D) MC = MR

17. The vertical demand curve for a commodity shows that the demand is:
(A) Perfectly inelastic
(B) Highly elastic
(C) Unit elastic
(D) Perfectly elastic

18. The demand of agricultural product is mainly dependent on:
(A) Increase in agricultural holding
(B) Increase in income
(C) Increase in population
(D Increase in cost

19. Type of farming is based on:
(A) Law of comparative advantage
(B) Law of equi-marginal return
(C) Law of diminishing return
(D) Law of substitution

20. The ultimate aim of farm management is to:
(A) Increase gross income
(B) Optimize factor utilization
(C) Reduce total cost
(D) None of these

Practice Paper 4

1. Which one of the following is a gift of nature?
(A) Land (B) Labour
(C) Capital (D) None of these

2. The efficiency o marketing channel is determined mainly on the basis of:
(A) Price spread
(B) Difference between retail and wholesale prices
(C) Regulated market
(D) Producer's share in consumer's rupees

3. **Which one is not the function of a production manager?**
(A) Planning (B) Organisation
(C) Marketing (D) Monitoring

4. **Which system of farming does not involve farmer?**
(A) Collective farming
(B) Dry farming
(C) State farming
(D) Diversified farming

5. **Livestock insurance is included to:**
(A) Fixed cost (B) Special cost
(C) Variable cost (D) Marginal cost

6. **Farm management is a:**
(A) Inter-farm science
(B) Intra-farm science
(C) Inter-marginal farm science
(D) International farm science

7. **Which one is not a tool of farm management?**
(A) Farm planning
(B) Farm budgeting
(C) Production and cost functions
(D) Farm stock keeping

8. **One of the important fixed cost on a farm is:**
(A) Seed cost (B) Land cost
(C) Irrigation cost (D) Fertilizer cost

9. **When marginal product is at the maximum, the marginal cost is at the:**
(A) Higher point (B) Lowest point
(C) Parallel (D) No change

10. **If the marginal product is more than marginal cost than further investment in production is:**
(A) Profitable (B) Harmful
(C) No effect (D) Equal

11. **If the substitution ratio is less than the price ratio, than the cost:**
(A) Remains the same (B) Increases
(C) Decreases (D) None of these

12. **Principle of equi-marginal return is applied when:**
(A) The resources are unlimited
(B) The price of resources is high
(C) The prices of resources vary
(D) The resources are limited

13. **Which of the following is also referred to as "Opportunity cost"?**
(A) Money cost (B) Alternate cost
(C) Social cost (D) Fixed cost

14. **The law of diminishing return refers to an eventual fall in:**
(A) Marginal product of the variable factor
(B) Average income of the farm
(C) Total production of all factors
(D) Total earning of the farm

15. **Which one of the following is not correctly matched?**

Type of market	Basis of classification
(A) Unorganised market	Organization
(B) Regulated market	Control
(C) Perfect market	Competition
(D) Terminal market	Area

16. **Which of the following cost is subtracted from gross income (GI) to get the value of family labour income?**
(A) Cost A1 (B) Cost A2
(C) Cost B (D) Cost C

17. **Marketing surplus is:**

(A) Total production	Total family consumption
(B) Total consumption	Marginal consumption
(C) Total cost	Marginal cost
(D) Total production	Total cost

18. **Read the following statements and mark your answer according to code given below:**
I. For variable input, APP = MPP in the I stage
II. For variable input, APP = MPP at the end of I stage
III. For variable input, APP = MPP at the start of II stage
IV. For variable input, APP = MPP in the II stage

Answer code
(A) I and III are correct
(B) II and III are correct
(C) I and IV are correct
(D) II and IV are correct

19. The concept of farming is best explained by:
(A) Secretarial planning approach
(B) Mixed farming approach
(C) Holistic farming approach
(D) Diversified farming approach

20. When MPP is zero, TPP is at its maximum, because:
(A) Further addition to MPP is negative
(B) Further addition to MPP is positive
(C) MC = MR
(D) Elasticity of productive is more than one

Practice Paper 5

1. When marginal product increases at increasing rate, then:
(A) Total product increases at a decreasing rate
(B) Total product increases at a increasing rate
(C) Average product decreases
(D) None of these

2. When increase or decrease in the level of one product does not affect the production level of the other product, it is known as:
(A) Supplementary relationship
(B) Complementary relationship
(C) Competitive relationship
(D) None f these

3. According to A. Marshall the " law of diminishing return" is applicable to:
(A) Industry (B) Agriculture
(C) Service sector (D) All of these

4. Why people prefers to receive payments in money?
(i) Because a person holding money can easily exchange it for any commodity or service that he/she might want
(ii) Because money raises the status of a person
(A) (i) is true but (ii) is false
(B) (i) is false but (ii) is true
(C) Both (i) and (ii) are true
(D) Both (i) and (ii) are false

5. What is barter system?
(i) A system under which goods are directly exchanged without the use of money
(ii) A system under which different currencies are exchanged
(iii) A system under which banks work as mediator between buyer and seller
(A) Only (i) is true
(B) (iii) is true
(C) Both (i) and (ii) are true
(D) Both (i) and (ii) are false

6. Farm production efficiency can be measured with following overall standard?
(A) Input per unit of output
(B) Total production index
(C) Output per hectare
(D) All of these

7. Principle of equi-marginal return is also known as:
(A) Product-product relationship
(B) Output-output relationship
(C) Substitution between output/products
(D) All of these

8. Law of diminishing marginal utility states that:
(A) Utility varies with time and space
(B) A commodity generates different utilities for different people and for the same person at different times
(C) Utility of additional unit of a commodity to any consumer commodity to already consumption increases
(D) All of these

9. Principle of least-cost combination is also known as:
(A) Substitution between inputs
(B) Factor-factor relationship
(C) Rate of technical substitution
(D) All of these

10. Which of the following is not a type of production function?
(A) Linear (B) Quadratic
(C) Square root (D) Rectangular

11. Maximum profit can be achieved in agri-business when marginal return (MR) is equal to:
(A) Marginal cost
(B) Average output
(C) Total output
(D) Total cost

12. Production function illustrates the relationship between:
(A) Input levels and corresponding output
(B) Input levels and corresponding marginal input
(C) Demand and supply pattern
(D) Demand and competition

13. At optimum level of production or output, the marginal return should be:
(A) Positive
(B) Negative
(C) Zero
(D) Equal to average return

14. The maximum limit is land holding under best soil is:
(A) 10-18 acres (B) > 27 acres
(C) > 34 acres (D) None of these

15. The maximum limit of land holding if the soil is poor is:
(A) 20-22 acres (B) > 37 acres
(C) > 54 acres (D) None of these

16. What is the main provision of land consolidation act?
(A) Fixation of minimum standard area for regulating transfers
(B) Prohibition of fragmentation below standard area
(C) Filling of objectives, appeals and penalties
(D) All of these

17. The contribution of livestock to gross farm income in mixed farming is:
(A) At least 10% (B) At least 30%
(C) At least 40 % (D) More than 50%

18. Under diversified farming, the income from the single enterprise should not be:
(A) < 50% of total farm income
(B) > 50% of total farm income
(C) > 33% of total farm income
(D) None of these

19. Under specialized farming, the income from the single enterprise should not be:
(A) < 50% of total farm income
(B) > 50% of total farm income
(C) > 75% of total farm income
(D) None of these

20. Ranching refers to:
(A) Growing of crops only
(B) Growing of livestock only
(C) Growing of crops and livestock
(D) All of these

Agricultural marketing

Practice Paper 1

1. Agricultural marketing involves the operations:
(A) Movement of food and raw material from field to final consumer
(B) Handling of produce at farm
(C) Processing, grading and packing
(D) All of these

2. The main objective of agricultural marketing is:
(A) To make available agricultural produce to consumer at optimum prices
(B) To help farmers to get maximum profit from their produce
(C) To maintain optimum market prices of agricultural commodities
(D) All of these

3. Functional approach of agricultural marketing deals with:
(A) Different functions
(B) Different institutions
(C) Different commodities
(D) None of these

4. Which of the following is a facilitating function of agricultural marketing:
(A) Standardization
(B) Financing
(C) Risk bearing
(D) All of these

5. Which of the following is physical function of agricultural marketing?

(A) Buyers and sellers
(B) Transportation
(C) Processing and storage
(D) All of these

6. Which of the following is known as a middlemen?

(A) Whole sellers
(B) Retailers
(C) Commission agents
(D) All of these

7. Specialised market refers to:

(A) Sell-buy of specialized kind of a commodity
(B) Sell-buy of different kind of a commodity
(C) Communication, transport and storage play an important role
(D) All of these

8. Terminal market are found at:

(A) Large cities and sea ports
(B) Village level
(C) Town levels
(D) District level

9. Local market generally found at:

(B) Large cities and sea ports
(B) Village level
(C) Town levels
(D) District level

10. Complete competitive market should have:

(A) Large number of sellers and buyers
(B) No product differentiation
(C) Dynamic factors of production
(D) All of these

11. The market that is organized under some rules and regulations made by trade body federation etc. is known as:

(A) Regulated market
(B) Organized market
(C) Economic market
(D) Unorganized market

12. The market that is under control of government rules and regulations is called as:

(A) Regulated market
(B) Organized market
(C) Economic market
(D) Unorganized market

13. Agricultural products are characterized as:

(A) Bulky raw material
(B) Perishable
(C) Essential for life
(D) All of these

14. Agricultural production is characterized as:

(A) Annual variability
(B) Seasonal variation
(C) Varying costs of production
(D) All of these

15. A line on a graph covering all possible choices between the two commodities for a person with a given amount of money to spend is known as:

(A) Consumption line
(B) Marginality line
(C) Supply line
(D) Fixed supply line

16. Which of the following statement is correct?

(A) Marketable surplus > marketed surplus
(B) Marketable surplus < Marketed surplus
(C) Marketable surplus = marketed surplus
(D) All of these

17. Marketable surplus is affected by:

(A) Prices of agricultural produce
(B) Support prices
(C) Availability of resources and its prices
(D) All of these

18. Marketed surplus is affected by:

(A) Family size
(B) Production level
(C) Sale prices
(D) All of these

19. Which of the following government agency is responsible for marketing and distribution of agricultural produce in the country?

(A) FCI
(B) State Trading Corporation Ltd
(C) Agricultural Price Commission
(D) All of these

20. Which of the following government agency is responsible for marketing and distribution of tobacco produce in the country?

(A) FCI
(B) State Trading Corporation Ltd
(C) Tobacco Board
(D) Agricultural Price Commission

Practice Paper 2

1. Which of the following government agency is responsible for international trading of agricultural produce in the country?

(A) FCI
(B) State Trading Corporation Ltd
(C) Agricultural Price Commission
(D) All of these

2. Crop area production and productivity data are available with:

(A) Directorate of Economics and Statistics
(B) Forward Market Commission
(C) Central Storage Corporation
(D) All of these

3. Marketing margin can be defined as:

(A) Part of total cost that goes to marketing firms
(B) It is a difference between producer's selling cost and consumer's purchasing cost
(C) It is the produce of total services of marketing firms
(D) All of these

4. Absolute margin equal to:

(A) Sale price – Purchase price
(B) Sale rice + Purchase price
(C) Sale price – Total marketing cost
(D) None of these

5. Net margin is equal to:

(A) Absolute margin – absolute marketing expenditure
(B) Sale price – Purchase price
(C) Sale price – total cost
(D) Absolute margin/sale price × 100

6. Marketing margin can be calculated by:

(A) Concurrent margin method
(B) Time lag method
(C) Both (A) and (B)
(D) None of these

7. Which of the following is a first regulated market opened in India?

(A) Karanjia cotton market
(B) Brar cotton market
(C) Cashew market of Kerala
(D) None of these

8. The new name of Agricultural Price Commission is:

(A) Commission on Agricultural Cost and Prices (CACP)
(B) Commission on agricultural Prices
(C) Agricultural Prices and Cost Commission
(D) None of these

9. The CACP came into existence in year:

(A) 1982 (B) 1985
(C) 1987 (D) 1990

10. Which of the following basis is adopted in determination of prices of agricultural commodities in our country?

(A) Production cost
(B) Parity cost
(C) Prevalent market price
(D) All of these

11. Which of the following aspect is related to macro economics?

(A) Agricultural employment
(B) Instability of agriculture
(C) Tenurial system
(D) All of these

12. The principle of least-cost combination of two factors indicates that profit can be maximized at:

(A) $\frac{\Delta X2}{\Delta X2} = \frac{Px1}{Px2}$

(B) $\frac{\Delta X2}{\Delta X1} = \frac{Px2}{Px1}$

(C) $\frac{\Delta X1}{X2} = \frac{Px1}{Px2}$

(D) None of these

13. In the production possibility curve, the maximum profit will be at:

(A) $\frac{\Delta Y2}{\Delta Y1} = \frac{-Py1}{Py2}$

(B) Py2, Δ y2 = Py1, Δy1

(C) Both (A) and (B)

(D) None of these

14. Which of the following is representing Cobb-Douglas production function?

(A) $Q = KL^{\alpha} C^{\beta}$

(B) $Q = L^{\alpha} C^{\beta} M^{\gamma}$

(C) $Q = KC^{\alpha} M^{\beta}$

(D) All of these

15. Elasticity of production is equal to:

(A) $\frac{\Delta Y1}{\Delta X} \times \frac{X}{Y}$

(B) $\frac{MP}{AY}$

(C) $\frac{\text{Present change in output}}{\text{Present change in input}}$

(D) All of these

16. If MP, AP and EP represent marginal product, average product and elasticity of production, which of the following inference is correct?

(A) When MP = AP, Ep=1

(B) When MP>AP, Ep>1

(C) When MP<AP, Ep<1

(D) All of these

17. The least cost combination of inputs will be at:

(A) $\frac{\Delta X1}{\Delta X2} = \frac{Px2}{Px1}$

(B) $\frac{\Delta X1}{\Delta X2} > \frac{Px2}{Px1}$

(C) $\frac{\Delta Px1}{\Delta Px2} < \frac{Py2}{Py1}$

(D) None of these

18. The size of holding that gives full employment to the farm family and reasonable living standard is known as:

(A) Family holding

(B) Optimum holding

(C) Economic holding

(D) Non-economic holding

19. Operating holding refers to:

(A) Land area owned by a farmers

(B) Cropped area cultivated by a farmer

(C) Net land area cultivated by a farmer

(D) Land area leased in by a farmer

20. Price elasticity of demand is equal to:

(A) $\frac{\text{Percentage change in quantity demanded}}{\text{Percentage change in price}}$

(B) $\frac{\text{Percentage change in supply of quantity}}{\text{Percentage change in price}}$

(C) $\frac{\text{Percentage change in quantity demanded}}{\text{Percentage change in quantity supplied}}$

(D) All of these

Practice Paper 3

1. The income elasticity is generally:

(A) 1% increase income increase only 0.2% demand

(B) 1% increase income decreases 0.2 % demand

(C) Both (A) and (B)

(D) None of these

2. International trade is more difficult in arrange than inter regional trade within the country because:

(A) factors of production are less mobile

(B) Each country uses a different currency

(C) Government interferes with foreign trade for political reasons

(D) All of these

3. Risk and uncertainty differs in:

(A) Risk has known probability but uncertainty has no probability

(B) Risk is temporary by uncertainty is sure

(C) Risk is more fatal than uncertainty in agriculture

(D) None of these

4. **The difference of average price and minimum support price of a particular commodity is paid to each farmer if the average price is lower than minimum support price, under the scheme of:**
   (A) Crop Insurance
   (B) Forward price
   (C) Deficiency payment
   (D) All of these

5. **Labour efficiency can be increased by:**
   (A) Improving farm layout
   (B) Farm work simplification
   (C) Training of labourers
   (D) All of these

6. **Farm management refers to:**
   (A) Managing the farm
   (B) The art of managing the farm
   (C) The art of managing the farm successfully as measured by the test of profitability
   (D) All of these

7. **Farm building is important because of:**
   (A) It saves feed, both in quantity and quality
   (B) It increases safety in farm work
   (C) It helps to increase production
   (D) All of these

8. **Which of the following state is most mechanization in India?**
   (A) UP (B) Punjab
   (C) Haryana (D) Kerala

9. **Which of the following is a partial standard of measuring farm production efficiency?**
   (A) Crop yield index
   (B) Production per unit livestock
   (C) Yield per hectare of the most important crop
   (D) All of these

10. **Farm survey refers to:**
   (A) An enquiry in the composition activities and living conditions of farmers and farm
   (B) Collection of information related expenses and profit of the farms
   (C) Visit of farms
   (D) All of these

11. **Market refers to:**
   (A) A place where producers and sellers are assembled for selling and purchasing of goods
   (B) Sellers and producers have free but competitive relationship
   (C) Prices of commodities remain at normal prices
   (D) All of these

12. **Market process refers to:**
   (A) Activities through which a commodity is reached from producer to final consumer
   (B) Grading and storage of a commodity
   (C) Middlemen through which a commodity is either sold or purchase
   (D) None of these

13. **Which of the following approach useful and simple to study the agricultural marketing?**
   (A) Functional approach
   (B) Institutional approach
   (C) Commodity approach
   (D) Behavioral approach

14. **What is difference between Dalal and Commission agent?**
   (A) Commission agents have more rights than Dalal
   (B) Dalal does not have physical control of produce
   (C) Commission agents can sell the produce and charge the commission
   (D) All of these

15. **Which of the following statement is correct?**
   (A) All regulated markets should be organized markets
   (B) All organized markets should be regulated markets
   (C) All regulated markets may not be organized markets
   (D) All unorganized markets may not be regulated markets

16. **Which of the following middlemen accept the ownership of produce during the marketing process?**
   (A) Itinerant traders (B) Village traders
   (C) FCI (D) All of these

17. Which of the following agency is responsible for creating awareness about quantity control of agricultural products?
(A) Export Inspection Standard Council of India
(B) Indian Standards Institute
(C) Indian Society for Quality control
(D) All of these

18. Which of the following is related to market development?
(A) Advertisement
(B) Quality control
(C) Packing
(D) All of these

19. Which of the following is a trade practices of marketing?
(A) Personal selling
(B) Advertisement based sale
(C) Sales by inspection
(D) All of these

20. Percentage marketing margin is equal to:
(A) Sale price – purchase price
(B) Absolute margin/sale price × 100
(C) Sale price – total marketing cost
(D) None of these

Practice Paper 4

1. Agricultural Production Mandi Act was passed in UP in year:
(A) 1962 (B) 1964
(C) 1966 (D) 1968

2. Which of the following is a feature of a regulated market?
(A) Proper marketing system
(B) Proper communication of marketing information
(C) Availability of optimum grading, storage and packing facilities
(D) All of these

3. What is the objective of keeping farm records?
(A) To know annual performance of farm
(B) To know the present worth
(C) To compare the performance over a period of years
(D) All of these

4. Programme related to elimination of rural poverty is:
(A) JRY (B) RLEGP
(C) IRDP (D) All of these

5. Which of the following balances is needed in the development of underdeveloped countries?
(A) Crosswise balance
(B) Backward balance
(C) Monetary balance
(D) All of these

6. RBI was nationalized in:
(A) 1945 (B) 1947
(C) 1949 (D) 1939

7. Short term credit needs of a farmer is assessed by taking into account:
(A) Area under different crops
(B) Existing level of inputs used in crop production
(C) Recommended doses of inputs needed in crop production
(D) Recommended doses of inputs needed – (minus) existing level of inputs used in crop production

8. India's largest netionalised commercial bank is:
(A) Reserved Bank of India
(B) Central Bank of India
(C) State Bank of India
(D) Bank of India

9. Which of the following is the most effective way to overcome the defects of agricultural marketing?
(A) Cooperative marketing
(B) Forward marketing
(C) Regulated marketing
(D) Monopoly

10. Who is a consumer?
(A) Anyone who purchase goods and services from the market and pays for these
(B) Anyone who sells goods and services in the market
(C) Anyone who manufactures goods
(D) Anyone who invests in the market

11. Anyone who produce goods and services and sell these in the market is known as:

(A) Producer (B) Consumer

(C) Buyer (D) Entrepreneur

12. Under which of the following condition the market do not work in a fair manner?

(i) When producers are few and powerful whereas consumers purchase in small amounts and are scattered

(ii) When consumers are few and producers are in a large number

(A) Only (i) and (ii) (B) Only (i)

(C) Only (iii) (D) Only (ii)

13. Which of the following is an appropriate reason for the Consumer Movement?

(A) It arose out dissatisfaction of the consumers as many unfair practices were being indulged in by the sellers

(B) It arose because of spread of democracy

(C) It arose because of cheap goods and services

(D) It arose because of MNC's

14. Which of the following factor's gave birth to the Consumer Movement in India?

(i) Food shortage

(ii) Hoarding

(iii) Black marketing and adulteration

(iv) Government of India

(A) Only (i) and (ii)

(B) Only (i), (ii) and (iii)

(C) Only (iii)

(D) Only (i) and (ii)

15. The Consumer Movement in an organized form was started in:

(A) 1950s (B) 1960s

(C) 1970s (D) 1980s

16. Why is the year 1986 known as a turning point in the history of Consumer Movement?

(A) COPRA was enacted

(B) CHOPRA was enacted

(C) Policy of liberalization was followed in this year

(D) WTO was established

17. What is COPRA?

(A) Consumer Preservation Act

(B) Consumer Protection Act

(C) Consumer Protection Activities

(D) None of these

18. "We do find bad quality products in the market". Which of the following is an appropriate reason for the same?

(i) Production is being done by MNC's

(ii) Consumer Movement is still very weak

(iii) Consumer are not aware of their rights and duties

(iv) Prevailing laws to safe guard the consumers are very weak

(A) Only (i)

(B) Only (ii), (iii) and (iv)

(C) Only (iii) and (iv)

(D) Only (i) and (ii)

19. What is MRP?

(A) Minimum Retail Price

(B) Maximum Retail Price

(C) Marginal Retail Price

(D) Marginal Revenue Price

20. While buying a product from a market which of the following logo assures you of quality?

(A) ISI or Agmark (B) LTTE

(C) KCF (D) BJP

Rural development

Practice Paper 1

1. Success in rural development project depends upon:

(A) Regular training of workers
(B) Regular contacts of workers
(C) Amount of subsidy
(D) Participation of beneficiaries

2. The University Education Commission was headed by Dr. S. Radha Krishanan recommended the establishment of:

(A) Urban Universities
(B) Rural Universities
(C) Both (A) & (B)
(D) None of these

3. There are .... development blocks in India:

(A) 5000 (B) 6000
(C) 7000 (D) 8000

4. .... is a dominant social institution permeating social and economic relations

(A) Society (B) Creed
(C) Caste (D) Customs

5. Rural community differ from urban community in relation to

(A) Population density
(B) Culture
(C) Environment
(D) All of these

6. Which of the following statements about "joint family system of rural areas" is correct?

(A) Joint families are always a cause
(B) Joint families are both an impediment and a factor or rural development
(C) There are no joint families in the rural areas
(D) Joint families are the only factors of rural development

7. Rural development programme at district level is being looked after by:

(A) DRDA
(B) MP of that constituency
(C) Additional S.P.
(D) None of the above

8. Family is included under:

(A) Secondary group
(B) Primary group
(C) Tertiary group
(D) Depends upon the situation

9. Integrated Rural Development Programme was started in:

(A) 1978 (B) 1974
(C) 1975 (D) 1979

10. Zamidari system was introduced by:

(A) Mugals (B) British
(C) Dutch (D) Frenches

11. List-I List-II

List-I	List-II
(A) Theory of rent	(1) Edward H.
(B) Quasi rent	(2) Adam Smith
(C) Monopolistic competition	(3) Alfred Marshall
(D) Theory of Wealth of Nation	(4) Ricardio

Code	a	b	c	d
(A)	3	4	2	1
(B)	4	3	1	2
(C)	4	2	1	3
(D)	2	3	4	1

12. The establishment of Panchayati Raj bodies at various level fulfils:

(A) Principle of cooperation
(B) Principle of leadership
(C) Principle of participation
(D) Principle of grass root

13. RRB were set up under the recommendation of:

(A) Nariman Committee
(B) Narasimhan committee
(C) Sen Committee
(D) CRAFICARD Committee

14. The new Panchayat Raj Act (73rd Constitution amendment) was enacted all over the country from:

(A) 24th April, 1993
(B) 24th April, 1994
(C) 24th April, 1995
(D) None of these

15. Which of the following rural development project was launched before independence:
(A) Etawah Pilot Project
(B) Cooperative movement
(C) Community development
(D) National Extension Service

16. Regional Rural Bank was set up in:
(A) 1972 (B) 1973
(C) 1974 (D) 1975

17. Success of rural projects depend upon:
(A) Regular training of staff
(B) Constant visit of staff
(C) Quantum of subsidies
(D) Partnership of beneficiaries

18. Programme related to elimination of rural poverty is:
(A) JRY (B) RLEGP
(C) IRDP (D) All of these

19. Name the 'Poverty Eradication Programme' in which bank play a dominant role:
(A) MNP (B) IRDP
(C) RBY (D) JRY

20. When National Development Council was formed?
(A) August 1947 (B) August 1950
(C) August 1952 (D) August 1960

Practice Paper 2

1. Which one of the following scheme is exclusively meant for self-employment of rural youth?
(A) NREP (B) TRYSEM
(C) IRDP (D) DPAP

2. Which one of the following is not the example of social system?
(A) Village (B) Communicator
(C) Block (D) District

3. The first cooperative societies established in India were:
(A) Credit societies
(B) Marketing societies
(C) Agricultural societies
(D) Housing societies

4. In distinguished unemployment is:
(A) Marginal physical productivity becomes zero
(B) Average physical productivity becomes zero
(C) Total physical productivity becomes zero
(D) None of these

5. Social changes refer to changes in:
(A) Social structure of the society
(B) Culture of the society
(C) Adoption behavior of the people
(D) Communication network

6. Capital is the result of savings and its formation depends upon:
(A) Capacity to save
(B) Desire to save
(C) Facility to save
(D) All of these

7. Selecting 'poorest among poor first' is the approach under:
(A) Jawahar Rojgar Yojna
(B) Training and Visit
(C) Integrated Rural Development Programme
(D) National Agricultural Technology Programme

8. Which one of the following does directly affect the division of labour?
(A) Capital
(B) Climate
(C) Demand of labour
(D) Nature of work

9. The basic unit of development under IRD programme is:
(A) Village Community
(B) Development block
(C) Family
(D) District

10. An effective facilitator in the development of any society is:
(A) Communication technology
(B) Information and communication technology
(C) Information technology
(D) Rural Development

11. Rural India reflects the very essence of:
(A) Indian culture (B) Tradition
(C) Both (A) and (B) (D) None of these

12. The most serious economic problem of India are:

(A) Poverty and unemployment
(B) Unemployment and stagnation
(C) Underdevelopment and poverty
(D) Corruption and unemployment

13. The percentage of people below poverty line in India is:

(A) 25.1 (B) 27.8
(C) 29.3 (D) 21.5

14. The National Rural Employment Guarantee Act was enacted initially to cover:

(A) 175 districts (B) 200 districts
(C) 257 districts (D) 304 districts

15. "Attacking Rural Poverty" is a book written by:

(A) Paul leagans
(B) A.T. Mosher
(C) Coomb and Ahamed
(D) Henry Parker

16. The important philosophy behind community development is:

(A) Work based on felt needs
(B) Peoples participation
(C) Philosophy of self help
(D) Emancipatory training

17. The number of unemployed people living in India is around:

(A) 13.10 million (B) 9.02 million
(C) 7.98 million (D) 10.51 milion

18. Community development project was started on the basis of:

(A) Nalagarh Commission Report
(B) Royal Commission Report
(C) Ford Foundation Report
(D) GMF Enquiry Report

19. The population of India according to 2001 census is:

(A) 102.21 crores
(B) 104.56 crores
(C) 107.56 crores
(D) 102.70 crores

20. The sex ratio of India according to 2001 census is:

(A) 927 (B) 933
(C) 938 (D) 942

Practice Paper 3

1. India's population projection for 2026 is:

(A) 1.21 billion (B) 1.27 billion
(C) 23.50 billion (D) 1.40 billion

2. The literacy rate of India according to 2001 census is:

(A) 75.85 (B) 54.16
(C) 62.67 (D) 65.38

3. The chief proponent of the concept of "Empowerment"is:

(A) Alfred York (B) David Werner
(C) Paulo Fiere (D) Osgood

4. Which among the following is a correct equation:

(A) Motivation + Empowerment = Enabling
(B) Motivation + Enabling = Empowerment
(C) Enabling + Empowerment = Motivation
(D) Motivation + Empowerment + Enabling = 0

5. The transfer of authority from central government to state government implies:

(A) Decentralization
(B) Delegation
(C) Devolution
(D) Deconcentration

6. In India land reforms have been equated with total:

(A) Urban development
(B) Agrarian development
(C) Industrial development
(D) Educational development

7. The term land reforms should be used in broadest sense to serve as:

(A) Inscription (B) Prescription
(C) Description (D) Formulation

8. Ministry of Panchayat Raj was established on:

(A) 27th May, 2004
(B) 15th August, 2002
(C) 2nd October, 2001
(D) 15th August, 2006

9. The backward class commission was headed by:

(A) B.N. Roy (B) Mahatma Gandhi
(C) Kaka Kalelkar (D) None of these

10. NIRD under Ministry of Rural Development is located at:

(A) New Delhi (B) Hyderabad
(C) Lucknow (D) Chennai

11. Journal of Rural Development is published from:

(A) MANAGE
(B) Ministry of Rural Development
(C) NIRD
(D) IARI

12. Which among the following is a rural development journal:

(A) Kurukshetra (B) Yojana
(C) Gram vikash (D) Gramin Soochna

13. The number of languages in which Yojana is published:

(A) 10 (B) 11
(C) 12 (D) 13

14. International Women's day is celebrated on:

(A) 8th March (B) 8th October
(C) 21st November (D) 2nd December

15. World Habitat day is celebrated on the first Monday of:

(A) October (B) December
(C) May (D) July

16. Vikas Volunteer Vahini is sponsored by:

(A) DRDA (B) RRB's
(C) NGO's (D) NABARD

17. Rural infrastructural fund was set up by:

(A) RBI (B) NABARD
(C) IDRB (D) SBI

18. The Department of Rural Development under Government of India was started on:

(A) 1919 (B) 1928
(C) 1931 (D) 1947

19. People's plan was envisaged by:

(A) Jawaharlal Nehru
(B) Visweswrayya
(C) M.N. Roy
(D) Mahalanoibis

20. The report on All India Rural Credit Enquiry Survey was submitted on:

(A) 1947 (B) 1954
(C) 1960 (D) 1961

Practice Paper 4

1. Community Development projects were inaugurated in India on:

(A) 26th January, 1952
(B) 2nd October, 1952
(C) 25th August, 1954
(D) 8th September, 1953

2. Grow More Food enquiry committee was constituted in:

(A) 1952 (B) 1953
(C) 1955 (D) 1956

3. Grow More Food enquiry committee was headed by:

(A) V.T. Krishnamachari
(B) M.S. Swaminathan
(C) M.K. Gandhi
(D) F.L. Brayne

4. Sarvodaya Scheme was started by:

(A) R.N. Ragore (B) Vinoba Bhave
(C) M.K. Gandhi (D) F.L. Brayne

5. Sarvodaya Scheme was started in:

(A) 1945 (B) 1946
(C) 1947 (D) 1948

6. Key Village Scheme was started in:

(A) 1949 (B) 1950
(C) 1951 (D) 1952

7. Rural Reconstruction Institute at Sundarban district was started by:

(A) R.N. Tagore (B) Daniel Hamilton
(C) M.K. Gandhi (D) F.L. Brayne

8. The structure of CD blocks were taken from:

(A) Etawah Project
(B) Firka Project
(C) GMF Campaign
(D) Nilokheri Experiment

9. The technical cooperation agreement of CDP was signed with:

(A) Africa (B) France
(C) USA (D) Netherland

10. CD is sponsored by:

(A) Government agencies
(B) Private agencies
(C) Voluntary agencies
(D) Both (A) and (B)

11. The community development programme was launched in India as:
(A) Development organisation
(B) Administrative setup
(C) As a means to an end
(D) Local self government

12. "A Guide to Community Development" was written by:
(A) M. Horton (B) D. Hamilton
(C) T. Moscher (D) Ensminger

13. The CDP was planned and conceived as peoples:
(A) Planned programme
(B) Local programme
(C) Self help programme
(D) Organize programme

14. The number of community development projects implemented in India:
(A) 54 (B) 55
(C) 56 (D) 57

15. The operational area of each community development project includes:
(A) 100 villages (B) 200 villages
(C) 300 villages (D) 400 villages

16. Under community development project each village level worker is appointed for:
(A) 1-5 villages
(B) 5-10 villages
(C) 10-15 villages
(D) 15-20 villages

17. "Extension Education in Community Development" is published by:
(A) Ministry of Planning
(B) Ministry of Food and Agriculture
(C) Ministry of Rural Development
(D) Ministry of Information and Broadcasting

18. National extension service was started in:
(A) 1952 (B) 1953
(C) 1954 (D) 1955

19. NES was considered basically as a:
(A) Method (B) Agency
(C) Project (D) Plan

20. Community Development Blocks was started in:
(A) 1952 (B) 1953
(C) 1954 (D) 1955

Practice Paper 4

1. A community development project have an operational area of:
(A) 400-500 square miles
(B) 600-700 square miles
(C) 700-800 square miles
(D) 800-1000 square miles

2. The entire nation was covered under CD programme by the year:
(A) 1957 (B) 1958
(C) 1959 (D) 1960

3. The committee constituted to review CD and NES projects was headed by:
(A) Ashok Mehta
(B) Hanumantha Rao
(C) O.P. Dahama
(D) Balwant Ray Mehta

4. The origin of the word democracy is from:
(A) Roman (B) Sanskrit
(C) Greek (D) Persian

5. The word "democracy" means:
(A) Peoples participation
(B) Peoples rule
(C) Peoples government
(D) Peoples judgement

6. The process of dispersing decision making closer to the point of service or action:
(A) Decentralization
(B) Delegation
(C) Devotion
(D) Deconcentration

7. The constitutional amendment act was passed in:
(A) 1996 (B) 2001
(C) 1992 (D) 1997

8. The election to panchayat is normally held after a period of:
(A) 3 years (B) 4 years
(C) 5 years (D) 6 years

9. The number of tiers for Panchayat in a state with population below 20 lakhs:
(A) Two (B) Three
(C) Four (D) Five

10. The proportion of seat in a panchayat to be filled from women through election:

(A) Two-third (B) One-third
(C) One-fifth (D) One-fourth

11. Which of the following states have no Panchayati Raj:

(A) Karnataka (B) Bihar
(C) Orissa (D) Meghalaya

12. The 73rd amendment act does not apply to the states of:

(A) Nagaland (B) Mizoram
(C) Meghalaya (D) All of these

13. The total number of district in India:

(A) 256 (B) 594
(C) 488 (D) 604

14. The responsibility of village panchayat include:

(A) Agricultural production
(B) Rural industrial development
(C) Maternity and child welfare
(D) All of these

15. The pioneer state in India to bring the whole of it under democratic decentralization:

(A) Uttar Pradesh
(B) Bihar
(C) Rajasthan
(D) Madhya Pradesh

16. The constitutional 74th amendment act relates to:

(A) Urban local bodies
(B) Election to Gram Sabha
(C) Free primary education
(D) Reservation in panchayat

17. Which of the following is a committee on block level planning:

(A) Hanumantha Rao Committee
(B) Dantewada Committee
(C) G.V.K. Rao Committee
(D) L.M. Singhvi Committee

18. Which of the following is a committee on district planning:

(A) Hanumantha Rao Committee
(B) Diwakar Committee
(C) Dantewada Committee
(D) Santhanam Committee

19. Ashok Mehta Committee recommended the establishment of:

(A) Mandal Panchayat
(B) Nagar Panchayat
(C) Panchayat Samiti
(D) Gram Panchayat

20. .... is also called : "package progamme":

(A) IRDP (B) IADP
(C) CDP (D) NES

Practice Paper 6

1. Ford foundation was invited to India on:

(A) 1958 (B) 1959
(C) 1960 (D) 1961

2. "India's food crisis and step to meet it" is a:

(A) Book on food situation in India
(B) Report of food foundation team
(C) Formula to solve food problems
(D) A slogan for campaign

3. Farmers can get short term loan from:

(A) Money lenders
(B) Land Development Banks
(C) Government
(D) All of these

4. What is the main principle of good system of credit to agriculture?

(A) Repayment capacity
(B) Risk bearing capacity
(C) Estimation of returns to credit
(D) All of these

5. The lead Bank Scheme was started in 1969 on the recommendation of:

(A) Nariman Committee
(B) All India Rural Credit Survey Committee
(C) Committee on Cooperative Credit
(D) Expert group of RBI on credit scheme

6. Abolition of bounded labour was recommended by:

(A) Dhebar Commission
(B) National Agriculture Commission
(C) Planning Commission
(D) Labour Commission

7. The bounded labour practiced was abolished in India by the notification made on:

(A) January 26, 1975
(B) August 15, 1975
(C) October 2, 1975
(D) October 25, 1975

8. The programme started based n the report of ford foundation:

(A) CDP (B) IADP
(C) IRDP (D) Etawah project

9. The IADP was launched in:

(A) 1960 kharif (B) 1960 rabi
(C) 1961 kharif (D 1961 rabi

10. "Betting on the strong horse" is a slogan associated with:

(A) IRDP (B) HYVP
(C) IADP (D) SFMA

11. Linking farm plan with credit is a characteristic feature of:

(A) IRDP (B) IADP
(C) CDP (D) Etawah Project

12. CDP differs from IADP as the former have:

(A) Area based approach
(B) Decentralized administration
(C) Degree of investment
(D) Multi purpose approach

13. The extension approach used under IADP was:

(A) Proscriptive (B) Prescriptive
(C) Participatory (D) Commercial

14. The IAAP was launched in:

(A) 1962 (B) 1963
(C) 1964 (D) 1965

15. High Yielding Variety Programme was started in:

(A) 1962 (B) 1963
(C) 1964 (D) 1965

16. High yielding technologies are:

(A) Resource positive
(B) Resource negative
(C) Resource neutral
(D) None of these

17. Farm and Home unit of Akashvani was started in connection with:

(A) IAAP (B) HYVP
(C) IRDP (D) CDP

18. Radio Rural Forum was started in:

(A) 1963 (B) 1964
(C) 1965 (D) 1966

19. Broadcasting of rural programme in television started in:

(A) 1965 (B) 1966
(C) 1967 (D) 1968

20. Which of the following is not a key feature of T&V system:

(A) Professionalism
(B) Time bound work
(C) Single line of command
(D) Farming System Research

Practice Paper 7

1. The base level extension functionary in T&V system:

(A) BDO (B) VLW
(C) AEO (D) SDAO

2. The first state in India to introduce Training and Visit system was:

(A) Punjab (B) Rajasthan
(C) Tamilnadu (D) Bihar

3. The ratio of extension agent to farmer in a typical T &V system is:

(A) 1-250 (B) 1-500
(C) 1-800 (D) 1-1500

4. T&V system of extension can be characterized as:

(A) Management system
(B) Supervisory system
(C) Integrated system
(D) Broad based system

5. Small Farmers Development Agency was set up in:

(A) 1969-70 (B) 1970-71
(C) 1971-72 (D) 1972-73

6. The basic approach followed by SFDA:

(A) Employment approach
(B) Integrated approach
(C) Welfare approach
(D) Target group approach

7. **SFDA is meant for the farmers having land holding of:**
(A) 1-2 hectare (B) 0-1 hectare
(C) 1-5 hectare (D) 2-5 hectare

8. **SFDA-MFAL is associated with:**
(A) Venkatpaiah committee
(B) Venkataraman committee
(C) Chelliah committee
(D) None of these

9. **An agricultural labourer earns more than of his wage from agriculture:**
(A) 33.33 per cent (B) 40.00 per cent
(C) 50.00 per cent (D) 66.66 per cent

10. **A welfare programme for the rural poor:**
(A) Krishi Vigyan Kendra
(B) Lab to Land
(C) National Demostration
(D) Minikit trials

11. **The Ganga Kalyan Yojna is for:**
(A) Consumption purpose of farmers
(B) Irrigation for small and marginal farmers
(C) Credit support scheme for rural women
(D) Credit support scheme for small farmers

12. **Intensive Cattle Development Project was launched in:**
(A) 1963-64 (B) 1964-65
(C) 1965-66 (D) 1966-67

13. **The rural development programme launched on the basis of area development approach:**
(A) DPAP (B) MWP
(C) RWP (D) IRDP

14. **Desert Development Programme was launched on:**
(A) 1977-78 (B) 1968-69
(C) 1971-72 (D) 1956-57

15. **Tribal Area Development Programme was started in the year:**
(A) 1970-71 (B) 1983-84
(C) 1984-85 (D) 1985-86

16. **Integrated Wasteland Development Programme was launched on:**
(A) 1989-90 (B) 1968-69
(C) 1971-72 (D) 1956-57

17. **National Food Security Mission covers all except:**
(A) Rice (B) Wheat
(C) Pulses (D) Sugarcane

18. **Annual growth rate in agriculture sector expected during the eleventh plan:**
(A) 8 per cent (B) 6 per cent
(C) 4 per cent (D) 3 per cent

19. **Planning commission was setup in:**
(A) 1949 (B) 1950
(C) 1951 (D) 1952

20. **The first chairman of planning commission was:**
(A) Dr. Rajendra Prasad
(B) Dr. S. Radhakrishnan
(C) Pt Jawaharlal Nehru
(D) Dr. B.R. Ambedkar

Practice Paper 8

1. **Planning commission is a:**
(A) Statutory body
(B) Advisory body
(C) Autonomous body
(D) Elected body

2. **Which one of the following is not an objective of India's economic planning:**
(A) Population growth
(B) Industrial growth
(C) Self reliance
(D) Poverty alleviation

3. **The long term objective of development in India has been:**
(A) High economic growth
(B) Economic self reliance
(C) Social justice
(D) All of these

4. **The National Development Council gets its administrative support from:**
(A) Parliament
(B) Planning commission
(C) Finance commission
(D) Union cabinet

5. **The Planning commission of India is a:**
(A) Constitutional body
(B) Statutory body
(C) Non-statutory body
(D) Autonomous body

6. **Who is not a member in the National Development council:**
(A) Prime Minister
(B) President
(C) Chief Ministers
(D) Cabinet Ministers

7. **Who was the first deputy caiman of planning commission of India?**
(A) Jawahrlal Nehru
(B) Rajendra Prasad
(C) V.T. Krishnamachari
(D) C. Rajagopalachari

8. **The main objective of National Development Council is:**
(A) To promote common economic policies in all vital spears
(B) To mobilize the efforts and resources of the nation in support of plan
(C) To ensure balanced and rapid development of all part of the country
(D) To secure cooperation of states in the execution of the plan

9. **The book entitled"Planned Economy for India" was the attempt of:**
(A) Mahalanobis
(B) Viswesvarayya
(C) Jawaharlal Nehru
(D) M.N. Roy

10. **The highest policy making body in matters relating to national planning is the:**
(A) Planning commission
(B) Finance commission
(C) Interstate council
(D) National Development Council

11. **The National Development Council was set up in:**
(A) 1950 (B) 1951
(C) 1952 (D) 1953

12. **Reviewing the implementation of national plans from time to time is the function of:**
(A) Research Programmes Committee
(B) Union Council of Ministers
(C) Planning commission
(D) National Development Council

13. **The Programme Evaluation Organization is an integral part of the:**
(A) National Development Council
(B) Central Secretariat
(C) Department of Public Enterprises
(D) Planning Commission

14. **The plan implementation and its impact evaluation is done by:**
(A) NDC
(B) Union cabinet
(C) PEO
(D) Planning commission

15. **Which of the following is not a function of planning commission in India?**
(A) Formulation of a plan
(B) Execution of development plans
(C) Appraisal of plan progress
(D) Policy and administrative recommendation

16. **The rolling plan concept in national planning was introduced by:**
(A) Congress government
(B) UDF government
(C) Janata Party government
(D) NDA government

17. **The credit for coining and applying the concept of indicative planning goes to:**
(A) USSR (B) USA
(C) Britain (D) France

18. **Economic planning is a subject included under the:**
(A) Union List
(B) State List
(C) Concurrent List
(D) Unspecified List

19. **In 1938 the National Planning Committee was set up under the chairmanship of:**
(A) Jawaharlal Nehru
(B) Vallabhbhai Patel
(C) Moulana Abdul Kalam Azad
(D) Mohammad Ali Jinnah

20. **Encyclopedia of social work in India is published by:**
(A) NIRD
(B) Planning commission
(C) ICSSR
(D) UNESCO

Practice Paper 9

1. **The first five year plan was submitted to the parliament in:**
(A) 1950 (B) 1951
(C) 1952 (D) 1953

2. **Harrod-Domar model was adopted in:**
(A) First five year plan
(B) Second five year plan
(C) Third five year plan
(D) Fourth five year plan

3. **Removal of poverty was the foremost objective of which five year plan:**
(A) Fourth (B) Fifth
(C) Sixth (D) Seventh

4. **Which decade of planning has been the best for agricultural growth:**
(A) Fifties (B) Sixties
(C) Seventies (D) Eighties

5. **Till which plan employment strategy has been growth linked:**
(A) Third plan (B) Fourth plan
(C) Fifth plan (D) Sixth plan

6. **Command Area Development Programme was launched during:**
(A) Fifth five year plan
(B) Sixth five year plan
(C) Seventh five year plan
(D) Eight five year plan

7. **The opening of Indian economy and signing of WTO happened during:**
(A) Seventh five year plan
(B) Eight five year plan
(C) Ninth five year plan
(D) Tenth five year plan

8. **Perspective plan is drawn out for a period of:**
(A) One year (B) 2 to 3 years
(C) 5 to 10 years (D) 15-20 years

9. **The term "Hindu rate of growth" was coined by:**
(A) J.N. Bhagwathi
(B) K.N. Raj
(C) Raj Krishna
(D) Mahalanoibis

10. **Atainment of economic self reliance and removal of poverty were the main objectives of:**
(A) First five year plan
(B) Fourth five year plan
(C) Fifth five year plan
(D) Sixth five year plan

11. **Which one of the following did the Mahalonobis model lay greater emphasis:**
(A) Agriculture
(B) Heavy industries
(C) Power generation
(D) Science and technology

12. **India achieved self reliance in food production during:**
(A) Fourth five year plan
(B) Fifth five year plan
(C) Sixth five year plan
(D) Seventh five year plan

13. **Micro level planning is an example for:**
(A) Imperative planning
(B) Five year planning
(C) Regional planning
(D) Target group planning

14. **Planning in India as it is at present can be termed as:**
(A) State planning
(B) Centralized planning
(C) Decentralized planning
(D) Indicative planning

15. **Indicative planning started from:**
(A) Eight five year plan
(B) Seventh five year plan
(C) Sixth five year plan
(D) Fifth five year plan

16. **Five year plan is a:**
(A) Perspective plan
(B) Medium term plan
(C) Adhoc plan
(D) Contingent plan

17. **Why was plan holiday necessary in the national planning:**
(A) First Indo-Pakistan war
(B) Second Indo-Pakistan war
(C) Third Indo-Pakistan war
(D) Indo-China war

18. The highest priority of the first five year plan was given for:
(A) Agriculture and irrigation
(B) Medium and large industry
(C) Unemployment and poverty
(D) Communication and education

19. The goal of socialistic pattern of society was presented before the national in the:
(A) First five year plan
(B) Second five year plan
(C) Third five year plan
(D) Fourth five year plan

20. The concept of mixed economy in planning was introduced during:
(A) Second five year plan
(B) Fifth five year plan
(C) Seventh five year plan
(D) Ninth five year plan

Practice Paper 10

1. A rolling plan is a plan for:
(A) One year
(B) Two year
(C) Three year
(D) Year to year basis

2. The idea of family planning as a concept of population control was widely popularized during:
(A) Fifth five year plan
(B) Sixth five year plan
(C) Seventh five year plan
(D) Eight five year plan

3. The current five year plan is:
(A) Tenth (B) Eleventh
(C) Twelfth (D) Thirteenth

4. The tenth five year plan ended on:
(A) 2007 (B) 2008
(C) 2009 (D) 2010

5. The eleventh five year plan ended on:
(A) 2009 (B) 2010
(C) 2011 (D) 2012

6. The GDP growth envisaged in the tenth plan is:
(A) 8.10 per cent (B) 8.50 per cent
(C) 9.0 per cent (D) 10.0 per cent

7. The percentage of reducing in poverty envisaged at the end of tenth five year plan is:
(A) 4 per cent (B) 5 per cent
(C) 6 per cent (D) 7 per cent

8. The approximate number of new employment opportunities envisaged in the tenth five year plan is:
(A) 50 million (B) 60 million
(C) 70 million (D) 80 million

9. The allocation of funds for rural development programme in the tenth plan is:
(A) 77474 crores (B) 82276 crores
(C) 85123 crores (D) 94000 crores

10. IRDP was introduced in India on:
(A) 1977-78 (B) 1978-79
(C) 1979-80 (D) 1980-81

11. The basic approach followed in IRDP is:
(A) Individual based
(B) Group based
(C) Community based
(D) Family based

12. IRDP was conceived essentially as a:
(A) Adult education programme
(B) Anti-poverty programme
(C) Anti-boarding programme
(D) Fisheries programme

13. The basic unit of development under IRDP is:
(A) District (B) Village
(C) CD blocks (D) Family

14. Which of the following is not a characteristic of IRDP:
(A) Unidimensional in nature
(B) Adopt a holistic approach
(C) Calls for higher coordination
(D) Poverty alleviation programme

15. PM krishi samman nidhi yojna launched in:
(A) 2020 (B) 2021
(C) 2000 (D) 2019

16. TRYSEM was introduced in the year:
(A) 1978 (B) 1979
(C) 1979 (D) 1980

17. Training under TRYSEM is mainly given for:
(A) Young scientists
(B) Rural Youth
(C) Young students
(D) Young women

18. DWCRA was implemented in:
(A) 1982 (B) 1979
(C) 1978 (D) 1977

19. The sill requirement for the members of DWCRA was met through:
(A) SFDA (B) IRDP
(C) NREP (D) TRYSEM

20. The programme aimed at creating income generating assets through a mix of bank credit and government subsidies:
(A) SGSY (B) NREP
(C) SITRA (D) JGSY

Answers Key

Priciples of Agricultural Economics

■ Practice Paper 1

Q. No.	Ans	Q. No.	Ans	Q. No.	Ans	Q. No.	Ans	Q. No.	Ans	Q. No.	Ans
1.	A	2.	D	3.	A	4.	B	5.	C	6.	C
7.	A	8.	D	9.	B	10.	D	11.	A	12.	C
13.	A	14.	C	15.	B	16.	C	17.	A	18.	B
19.	A	20.	A								

■ Practice Paper 2

Q. No.	Ans	Q. No.	Ans	Q. No.	Ans	Q. No.	Ans	Q. No.	Ans	Q. No.	Ans
1.	B	2.	C	3.	B	4.	B	5.	A	6.	A
7.	B	8.	B	9.	A	10.	C	11.	B	12.	A
13.	B	14.	C	15.	A	16.	C	17.	B	18.	A
19.	D	20.	C								

■ Practice Paper 3

Q. No.	Ans	Q. No.	Ans	Q. No.	Ans	Q. No.	Ans	Q. No.	Ans	Q. No.	Ans
1.	A	2.	A	3.	B	4.	C	5.	C	6.	B
7.	D	8.	B	9.	A	10.	C	11.	A	12.	C
13.	D	14.	B	15.	D	16.	C	17.	A	18.	B
19.	A	20.	C								

■ Practice Paper 4

Q. No.	Ans	Q. No.	Ans	Q. No.	Ans	Q. No.	Ans	Q. No.	Ans	Q. No.	Ans
1.	D	2.	A	3.	B	4.	A	5.	A	6.	D
7.	A	8.	B	9.	B	10.	D	11.	A	12.	C
13.	B	14.	A	15.	B	16.	A	17.	B	18.	C
19.	D	20.	C								

■ Practice Paper 5

Q. No.	Ans	Q. No.	Ans	Q. No.	Ans	Q. No.	Ans	Q. No.	Ans	Q. No.	Ans
1.	A	2.	C	3.	A	4.	C	5.	D	6.	A
7.	D	8.	A	9.	B	10.	A	11.	B	12.	C
13.	A	14.	A	15.	A	16.	B	17.	C	18.	D
19.	A	20.	C								

■ Practice Paper 6

Q. No.	Ans	Q. No.	Ans	Q. No.	Ans	Q. No.	Ans	Q. No.	Ans	Q. No.	Ans
1.	D	2.	A	3.	B	4.	B	5.	C	6.	C
7.	B	8.	A	9.	B	10.	C	11.	D	12.	A
13.	A	14.	C	15.	A	16.	D	17.	A	18.	B
19.	A	20.	B								

■ Practice Paper 7

Q. No.	Ans	Q. No.	Ans	Q. No.	Ans	Q. No.	Ans	Q. No.	Ans	Q. No.	Ans
1.	A	2.	B	3.	A	4.	B	5.	B	6.	A
7.	C	8.	A	9.	B	10.	A	11.	B	12.	C
13.	A	14.	A	15.	A	16.	B	17.	A	18.	A
19.	B	20.	C								

■ Practice Paper 8

Q. No.	Ans	Q. No.	Ans	Q. No.	Ans	Q. No.	Ans	Q. No.	Ans	Q. No.	Ans
1.	A	2.	B	3.	A	4.	C	5.	B	6.	B
7.	D	8.	C	9.	A	10.	A	11.	A	12.	C
13.	C	14.	D	15.	D	16.	A	17.	D	18.	B
19.	C	20.	B								

■ Practice Paper 9

Q. No.	Ans	Q. No.	Ans	Q. No.	Ans	Q. No.	Ans	Q. No.	Ans	Q. No.	Ans
1.	B	2.	A	3.	A	4.	A	5.	D	6.	A
7.	A	8.	B	9.	A	10.	B	11.	A	12.	A
13.	B	14.	A	15.	D	16.	A	17.	A	18.	C
19.	A	20.	B								

■ Practice Paper 10

Q. No.	Ans	Q. No.	Ans	Q. No.	Ans	Q. No.	Ans	Q. No.	Ans	Q. No.	Ans
1.	C	2.	B	3.	A	4.	B	5.	B	6.	B
7.	D	8.	A	9.	A	10.	A	11.	B	12.	D
13.	A	14.	D	15.	B	16.	D	17.	D	18.	A
19.	A	20.	D								

Panchayati Raj

■ Practice Paper 1

Q. No.	Ans	Q. No.	Ans	Q. No.	Ans	Q. No.	Ans	Q. No.	Ans	Q. No.	Ans
1.	C	2.	A	3.	B	4.	D	5.	C	6.	B
7.	C	8.	A	9.	D	10.	D	11.	C	12.	B
13.	C	14.	D	15.	C	16.	C	17.	B	18.	D
19.	B	20.	B								

■ Practice Paper 2

Q. No.	Ans	Q. No.	Ans	Q. No.	Ans	Q. No.	Ans	Q. No.	Ans	Q. No.	Ans
1.	B	2.	C	3.	A	4.	C	5.	A	6.	B
7.	A	8.	B	9.	D	10.	B	11.	D	12.	B
13.	B	14.	A	15.	D	16.	A	17.	A	18.	C
19.	A	20.	A								

Farm Management

■ Practice Paper 1

Q. No.	Ans	Q. No.	Ans	Q. No.	Ans	Q. No.	Ans	Q. No.	Ans	Q. No.	Ans
1.	A	2.	D	3.	A	4.	A	5.	A	6.	A
7.	B	8.	A	9.	B	10.	D	11.	A	12.	D
13.	D	14.	C	15.	D	16.	A	17.	D	18.	D
19.	D	20.	C								

■ Practice Paper 2

Q. No.	Ans	Q. No.	Ans	Q. No.	Ans	Q. No.	Ans	Q. No.	Ans	Q. No.	Ans
1.	D	2.	D	3.	B	4.	A	5.	A	6.	A
7.	A	8.	A	9.	B	10.	B	11.	B	12.	A
13.	B	14.	A	15.	B	16.	D	17.	A	18.	D
19.	B	20.	A								

■ Practice Paper 3

Q. No.	Ans	Q. No.	Ans	Q. No.	Ans	Q. No.	Ans	Q. No.	Ans	Q. No.	Ans
1.	D	2.	C	3.	B	4.	D	5.	B	6.	B
7.	A	8.	C	9.	B	10.	C	11.	A	12.	D
13.	B	14.	B	15.	D	16.	A	17.	A	18.	C
19.	A	20.	B								

■ Practice Paper 4

Q. No.	Ans	Q. No.	Ans	Q. No.	Ans	Q. No.	Ans	Q. No.	Ans	Q. No.	Ans
1.	A	2.	C	3.	D	4.	C	5.	A	6.	B
7.	A	8.	B	9.	B	10.	A	11.	C	12.	D
13.	B	14.	A	15.	D	16.	C	17.	A	18.	B
19.	D	20.	D								

■ Practice Paper 5

Q. No.	Ans	Q. No.	Ans	Q. No.	Ans	Q. No.	Ans	Q. No.	Ans	Q. No.	Ans
1.	A	2.	A	3.	B	4.	A	5.	A	6.	D
7.	D	8.	C	9.	D	10.	D	11.	A	12.	A
13.	C	14.	A	15.	C	16.	D	17.	A	18.	A
19.	B	20.	B								

Agricultural Marketing

■ Practice Paper 1

Q. No.	Ans	Q. No.	Ans	Q. No.	Ans	Q. No.	Ans	Q. No.	Ans	Q. No.	Ans
1.	D	2.	D	3.	A	4.	D	5.	D	6.	D
7.	A	8.	A	9.	B	10.	D	11.	B	12.	A
13.	D	14.	D	15.	D	16.	B	17.	D	18.	D
19.	A	20.	C								

■ Practice Paper 2

Q. No.	Ans	Q. No.	Ans	Q. No.	Ans	Q. No.	Ans	Q. No.	Ans	Q. No.	Ans
1.	B	2.	A	3.	D	4.	A	5.	A	6.	C
7.	A	8.	A	9.	C	10.	D	11.	D	12.	A
13.	C	14.	A	15.	D	16.	D	17.	A	18.	C
19.	C	20.	C								

■ Practice Paper 3

Q. No.	Ans	Q. No.	Ans	Q. No.	Ans	Q. No.	Ans	Q. No.	Ans	Q. No.	Ans
1.	A	2.	D	3.	A	4.	C	5.	D	6.	D
7.	D	8.	B	9.	D	10.	A	11.	D	12.	A
13.	A	14.	D	15.	A	16.	D	17.	D	18.	D
19.	D	20.	B								

■ Practice Paper 4

Q. No.	Ans	Q. No.	Ans	Q. No.	Ans	Q. No.	Ans	Q. No.	Ans	Q. No.	Ans
1.	B	2.	D	3.	D	4.	D	5.	D	6.	C
7.	A	8.	C	9.	C	10.	A	11.	A	12.	B

Q. No.	Ans	Q. No.	Ans	Q. No.	Ans	Q. No.	Ans	Q. No.	Ans	Q. No.	Ans
13.	A	14.	B	15.	B	16.	A	17.	B	18.	B
19.	B	20.	A								

Rural Development

■ Practice Paper 1

Q. No.	Ans	Q. No.	Ans	Q. No.	Ans	Q. No.	Ans	Q. No.	Ans	Q. No.	Ans
1.	D	2.	B	3.	A	4.	C	5.	D	6.	B
7.	A	8.	B	9.	A	10.	B	11.	B	12.	D
13.	B	14.	A	15.	B	16.	D	17.	D	18.	D
19.	B	20.	C								

■ Practice Paper 2

Q. No.	Ans	Q. No.	Ans	Q. No.	Ans	Q. No.	Ans	Q. No.	Ans	Q. No.	Ans
1.	B	2.	B	3.	A	4.	B	5.	A	6.	D
7.	C	8.	D	9.	C	10.	B	11.	C	12.	A
13.	B	14.	B	15.	C	16.	C	17.	A	18.	D
19.	D	20.	B								

■ Practice Paper 3

Q. No.	Ans	Q. No.	Ans	Q. No.	Ans	Q. No.	Ans	Q. No.	Ans	Q. No.	Ans
1.	D	2.	D	3.	B	4.	B	5.	C	6.	B
7.	B	8.	A	9.	C	10.	B	11.	C	12.	A
13.	D	14.	A	15.	A	16.	D	17.	B	18.	C
19.	C	20.	B								

■ Practice Paper 4

Q. No.	Ans	Q. No.	Ans	Q. No.	Ans	Q. No.	Ans	Q. No.	Ans	Q. No.	Ans
1.	B	2.	A	3.	A	4.	B	5.	D	6.	D
7.	B	8.	A	9.	C	10.	D	11.	C	12.	D
13.	C	14.	B	15.	C	16.	•	17.	B	18.	B
19.	B	20.	C								

■ Practice Paper 5

Q. No.	Ans	Q. No.	Ans	Q. No.	Ans	Q. No.	Ans	Q. No.	Ans	Q. No.	Ans
1.	A	2.	B	3.	D	4.	C	5.	B	6.	A
7.	C	8.	C	9.	A	10.	B	11.	D	12.	D
13.	D	14.	D	15.	C	16.	A	17.	B	18.	A
19.	A	20.	B								

■ Practice Paper 6

Q. No.	Ans	Q. No.	Ans	Q. No.	Ans	Q. No.	Ans	Q. No.	Ans	Q. No.	Ans
1.	B	2.	B	3.	D	4.	D	5.	A	6.	A
7.	D	8.	B	9.	A	10.	C	11.	B	12.	D
13.	B	14.	C	15.	C	16.	A	17.	B	18.	D
19.	C	20.	D								

■ Practice Paper 7

Q. No.	Ans	Q. No.	Ans	Q. No.	Ans	Q. No.	Ans	Q. No.	Ans	Q. No.	Ans
1.	B	2.	B	3.	C	4.	A	5.	B	6.	D
7.	A	8.	A	9.	C	10.	B	11.	B	12.	B
13.	A	14.	A	15.	A	16.	A	17.	D	18.	C
19.	B	20.	C								

■ Practice Paper 8

Q. No.	Ans	Q. No.	Ans	Q. No.	Ans	Q. No.	Ans	Q. No.	Ans	Q. No.	Ans
1.	C	2.	A	3.	D	4.	B	5.	A	6.	B
7.	C	8.	D	9.	B	10.	D	11.	C	12.	D
13.	D	14.	C	15.	B	16.	C	17.	D	18.	C
19.	A	20.	B								

■ Practice Paper 9

Q. No.	Ans	Q. No.	Ans	Q. No.	Ans	Q. No.	Ans	Q. No.	Ans	Q. No.	Ans
1.	B	2.	A	3,	B	4.	B	5.	C	6.	A
7.	B	8.	D	9.	C	10.	C	11.	B	12.	A
13.	C	14.	D	15.	A	16.	B	17.	A	18.	A
19.	B	20.	A								

■ Practice Paper 10

Q. No.	Ans	Q. No.	Ans	Q. No.	Ans	Q. No.	Ans	Q. No.	Ans	Q. No.	Ans
1.	D	2.	B	3.	C	4.	A	5.	D	6.	A
7.	B	8.	A	9.	D	10.	A	11.	D	12.	B
13.	C	14.	A	15.	A	16.	B	17.	B	18.	A
19.	D	20.	A								

PART - 3

ICAR-JRF QUESTION PAPERS OF 2020

Agronomy

1. **Which one of the following is a ceremonial weeds?**
(A) *Cynodon dactylon*
(B) *Melilotus alba*
(C) *Chorchorus capsularis*
(D) *Mentha spicata*

2. **Match list I with list II:**

List I	List II
1. Safflower	i. The South West United State & Mexico
2. Sunflower	ii. Indian sub continent
3. Soybean	iii. The Meditarrenian and the Pursian Gulf
4. Green gram	iv. China

Choose the correct answer from the options given below:
(A) 1-ii, 2-iii, 3-i, 4-iv
(B) 1-i, 2-iii, 3-iv, 4-ii
(C) 1-iii, 2-i, 3-iv, 4-ii
(D) 1-iii, 2-iv, 3-ii, 4-i

3. **For the analysis of variance of data from plot sampling in a CRD with 't' treatments, 'Y' replications and 'S' sampling units per plot, sampling error will be:**
(A) (r – 1) (s – 1) (B) rt (s – 1)
(C) r(st – 1) (D) t(r- 1) (s – 1)

4. **Identify the correct order of the following methods of irrigation as per their water use efficiency from the maximum to minimum efficiency:**
(A) Continuous flooding > flooding > cabilgation > bubbler
(B) Bubbler > cablegation > flooding > continuous flooding
(C) Cablegation > flooding > bubbler > continuous flooding
(D) Bubbler > flooding >continuous flooding > cablegation

5. **The important biological control of insect for *Parthenium hysterophorus* in India is:**
(A) *Zygogramma bicolorata*
(B) *Teleonemia scrupulosa*
(C) *Sida latifolia*
(D) *Orsioliella javanica*

6. **Which one of the following is NOT a quality parameter of rice grain?**
(A) Milling quality (B) Cooking quality
(C) Eating quality (D) Curing quality

7. **What will be the error d.f. for a RBD field experiment with 8 treatments and 3 replications?**
(A) 5 (B) 11
(C) 14 (D) 24

8. **Triose phosphate is formed in the reduction step of:**
(A) Hatch and Slack cycle
(B) Calvin cycle
(C) Oxidative photosynthetic carbon cycle
(D) CAM

9. **Match list I with list II:**

List I	List II
1. PAA	i. Pluviometer
2. Canopy temperature	ii. Quantum sensor
3. Rainfall	iii. Wind vane
4. Wind direction	iv. IR-thermometer

Choose the correct answer from the options given below:
(A) 1-i, 2-iii, 3-iv, 4-ii
(B) 1-i, 2-iv, 3-ii, 4-iii
(C) 1-ii, 2-iv, 3-i, 4-iii
(D) 1-ii, 2-i, 3-iv, 4-iii

10. Ozone concentration in the atmosphere is reduced by:

(A) CO_2 and O_2 (B) H_2S and CH_4
(C) CFC and Cl_2 (D) N_2 and N_2O

11. Under dryland farming situation which practice is most prominent?

(A) Single cropping
(B) Double cropping
(C) Relay cropping
(D) Multiple cropping

12. The highest number of resistant weeds are associated with which group of herbicides?

(A) Dinitroanilines (B) Bipyridiliums
(C) Synthetic auxins (D) Trizines

13. Match list I with list II:

List I	List II
1. Stomata closing	i. Silicon
2. Film forming	ii. PMA
3. Reflectant	iii. Cycocel
4. Growth retardant	iv. Kaolin

Choose the correct answer from the options given below:

(A) 1-ii, 2-i, 3-iv, 4-iii
(B) 1-vi, 2-i, 3-ii, 4-iii
(C) 1-iv, 2-i, 3-iii, 4-ii
(D) 1-i, 2-iii, 3-ii, 4-iv

14. Useful part of plantago (Psyllium) is:

1. Root 2. Shoot
3. Seed husk 4. Leaves

Choose the correct answer from the options given below:

(A) 3 only (B) 1, 3 and 4 only
(C) 1 and 3 only (D) 1 and 2 only

15. Weeds propagated through tubers:

(A) *Cyperus rotundus*
(B) *Cirsium arvens*
(C) *Allium vineale*
(D) *Cynodon dactylon*

16. Given below are two statements:

Statement I: Ammonium sulphate can supply nitrogen, calcium and sulphate to crops

Statement II: Ammonium sulphate is the most suited N fertilizer for acidic soils

In light of the above statements, choose the most appropriate answer from the options given below:

(A) Both statement I and statement II are correct
(B) Both statement I and statement II are incorrect
(C) Statement I is correct but statement II is incorrect
(D) Statement I is incorrect but statement II is correct

17. Post emergence herbicide for weed control in rice is:

(A) Butachlor
(B) Bispyribac sodium
(C) Thiobencarb
(D) Nitrofen

18. Given below are two statements:

Statement I: Circulation of air over tropics is dominated by Hardley cell and Walker cell

Statement II: Hardley cell is associated with N-S direction and walker cell is associated with E-W direction

In light of the above statements, choose the most appropriate answer from the options given below:

(A) Both statement I and statement II are correct
(B) Both statement I and statement II are incorrect
(C) Statement I is correct but statement II is incorrect
(D) Statement I is incorrect but statement II is correct

19. Given below are two statements:

Statement I: Seedling blight, color rot, stem rot and root rot are the important fungal diseases of jute crop

Statement II: Anthrac nose is a major bacterial disease of jute

In light of the above statements, choose the most appropriate answer from the options given below:

(A) Both statement I and statement II are correct

(B) Both statement I and statement II are incorrect

(C) Statement I is correct but statement II is incorrect

(D) Statement I is incorrect but statement II is correct

20. A weed with short lived blue coloured flowers is:

(A) *Erigeron Canadensis*

(B) *Echinochloa colona*

(C) *Avena ludoviciana*

(D) *Commelina benghalensis*

21. In which year did the ICAR-National Bureau of Soil Survey and Land Use Planning (ICAR-NBSSS&LP) categorizes India in to different agroecological regions (AER) for the first time:

(A) 1996 (B) 1997

(C) 1998 (D) 1999

22. Match list I with list II:

List I	List II
1. Phototropism	i. Bending towards soil
2. Geotropism	ii. Response to day length
3. Thigmotropism	iii. Bending towards light
4. Photoperiodism	iv. Touch me not

Choose the correct answer from the options given below:

(A) 1-i, 2-ii, 3-iv, 4-iii

(B) 1-ii, 2-iii, 3-i, 4-iv

(C) 1-iv, 2-ii, 3-iii, 4-i

(D) 1-iii, 2-i, 3-iv, 4-ii

23. The principle objective of alley cropping is:

(A) To get green and palatable fodder from hedge rows and reasonable yield of intercrop

(B) To maximize crop yield

(C) To control wind erosion

(D) To increased soil fertility

24. Chronological development in commercial use of herbicide is:

1. Triazines
2. Phenoxyalkanoic acids
3. Dinitroanilines
4. Sulfonylureas
5. Phenoxy (aryloxy) propionates

Choose the correct answer from the options given below:

(A) 5 – 2 – 1 – 3 – 4

(B) 1 – 2 – 3 – 4 – 5

(C) 2 – 1 – 4 – 5 – 3

(D) 2 – 3 – 1 – 5 – 4

25. Given below are two statements:

Statement I: Waterlogging cause injury to crops due to low oxygen content and accumulation of hydrocarbon gases, hydrogen sulphide and carbon dioxide

Statement II: Ethanol production and activity of dehydrogenase decreases in roots of waterlogged plants.

In light of the above statements, choose the most appropriate answer from the options given below:

(A) Both statement I and statement II are correct

(B) Both statement I and statement II are incorrect

(C) Statement I is correct but statement II is incorrect

(D) Statement I is incorrect but statement II is correct

26. Given below are two statements:

Statement I: Crop sequences involving cereals and pulses are commonly followed under the irrigated agro-ecosystem in the southern region of Indian sub-continents.

Statement II: One of the prominent cropping sequences followed in the irrigated agro-ecosystem of southern India is the rice-rice – green gram

In light of the above statements, choose the most appropriate answer from the options given below:

(A) Both statement I and statement II are correct
(B) Both statement I and statement II are incorrect
(C) Statement I is correct but statement II is incorrect
(D) Statement I is incorrect but statement II is correct

27. The technique of growing plants with heir roots suspended in air without soil is called:
(A) Phytotronics (B) Aeroponics
(C) Hydroponics (D) Aerosols

28. Which of the following causes a resource influencing competition in a weed crop:
(A) pH
(B) Bulk density
(C) Water holding capacity
(D) Phosphorus

29. Chronology in the grass - weed growth stages is:
1. Booting
2. Tillering
3. Anthesis
4. Dough
5. Seedling

Choose the correct answer from the options given below:
(A) 5 – 2 – 1 – 3 – 4
(B) 1 – 2 – 3 – 4 – 5
(C) 4 – 1 – 2 – 5 – 3
(D) 5 – 4 – 3 – 2 – 1

30. Monocropping of tobacco is always discouraged as:
1. It leads to the development of pest and disease complex
2. It decreases soil fertility
3. It reduces germination of crops
4. It leads to crop failure due to water stress
5. Curing become difficult

In light of the above statements, choose the most appropriate answer from the options given below:
(A) 1 and 2 only (B) 1 and 3 only
(C) 1 only (D) 4 and 5 only

31. The point at which CO_2 uptake exactly balances CO_2 release is called:
(A) Quantum yield
(B) Solar tracking
(C) Light compensation point
(D) Threshold point

32. Given below are two statements:
Statement I: Two continuous watersheds are mostly similar in characteristics
Statement II: Each watershed unit is a well defined topographically delineated area with a distinct boundary.

In light of the above statements, choose the most appropriate answer from the options given below:
(A) Both statement I and statement II are correct
(B) Both statement I and statement II are incorrect
(C) Statement I is correct but statement II is incorrect
(D) Statement I is incorrect but statement II is correct

33. Match list I with list II:

List I	List II
1. Protein legend interaction	i. IRGA
2. Rate of photosynthesis	ii. Planimeter
3. Transpiration rate	iii. Isothermal titration calorimetry
4. Leaf area	iv. Porometer

Choose the correct answer from the options given below:
(A) 1-iii, 2-i, 3-iv, 4-ii
(B) 1-i, 2-iv, 3-iii, 4-ii
(C) 1-ii, 2-iii, 3-iv, 4-i
(D) 1-i, 2-iii, 3-ii, 4-iv

34. Desmotubules are identified in which of the following organelles:
(A) Golgibodies
(B) Endoplasmic reticulum
(C) Cytoskeleton
(D) Plasmodesmata

35. Calculate the plant population of sugarcane in 1 ha area of planted in paired row spacing of 60/120 cm and the number of canes in 4 meter row length is 50 meter:

(A) 135000 (B) 138888
(C) 139888 (D) 140888

36. Given below are two statements:

Statement I: A remote sensing system that possess only a sensor and depends on an external source to irradiate the target to be called an active remote sensing system.

Statement II: INSAT series of satellites are solar-orbitory satellites

In light of the above statements, choose the most appropriate answer from the options given below:

(A) Both statement I and statement II are correct
(B) Both statement I and statement II are incorrect
(C) Statement I is correct but statement II is incorrect
(D) Statement I is incorrect but statement II is correct

37. Given below are two statements:

Statement I: A species is generally defined as a uniform inter breeding population spread over time and space

Statement II: A commonly is a group of population of different species in a given area.

In light of the above statements, choose the most appropriate answer from the options given below:

(A) Both statement I and statement II are correct
(B) Both statement I and statement II are incorrect
(C) Statement I is correct but statement II is incorrect
(D) Statement I is incorrect but statement II is correct

38. Diara lands comes under:

(A) Salt affected areas
(B) High rainfall areas
(C) Flood prone areas
(D) Arid areas

39. Arranged the following major events in dryland agriculture research in chronological order of their establishment:

(1) ICRISAT (2) AICRPDA
(3) CRPs (4) KVK
(5) CRIDA

Choose the correct answer from the options given below:

(A) 2 – 1 – 3 – 4 – 5
(B) 1 – 2 – 3 – 4 – 5
(C) 2 – 3 – 1 – 5 – 4
(D) 3 – 4 – 5 – 1 – 2

40. Moisture availability index of 0.34 to 0.67 indicates which of the following class:

(A) Very deficient
(B) Moderate deficient
(C) Excessive moisture
(D) Adequate moisture

41. Plastids that contain high concentration of carotenoid pigments rather than chlorophyll pigments are:

(A) Leucoplasts
(B) Chemoplasts
(C) Amyloplasts
(D) Chloroplasts

42. Rate of transpiration in cop plant is maximum during:

(A) Morning
(B) Little before noon
(C) Afternoon
(D) Evening

43. Which of the following weeds is a holo root parasite

(A) Loranthus (B) Striga
(C) Orobanche (D) Cuscuta

44. A wheat crop growing in a 40 hectare area was irrigated with 10 cumecs of water for 6 hours. Water stores in the root zone was 30 cm. What is the field application efficiency?

(A) 20.55% (B) 44.44%
(C) 50.35% (D) 55.55%

45. Given below are two statements:

Statement I: Potassium is accumulated passively by both the cytosel and vacuole when extra cellular K^+ concentration are very low in which case it is taken up actively.

Statement II: Nitrogen is an important element that participates in the opening and closing of stomata.

In light of the above statements, choose the most appropriate answer from the options given below:

(A) Both statement I and statement II are correct

(B) Both statement I and statement II are incorrect

(C) Statement I is correct but statement II is incorrect

(D) Statement I is incorrect but statement II is correct

46. What will be the effect of tillage on the thermal conductivity of soil:

(A) Remain unchanged

(B) Decreases

(C) Increases

(D) First increases than decreases

47. The concentration of macronutrients in dry plant tissue is:

(A) Less than 0.1 per cent

(B) More than 0.1 per cent

(C) Less than 0.01 per cent

(D) More than 0.01 per cent

48. Which among the following soil resource indicators gives the highest priority for evaluating agricultural sustainability:

(A) Weed intensity

(B) Soil pH

(C) Water table depth

(D) Soil organic matter

49. Nutrient index approach of soil fertility evaluation was proposed by:

(A) Liebig

(B) Parker

(C) Mehlich

(D) Piper

50. Given below are two statements:

Statement I: ATP synthesize produces ATP as protons diffuse back through it from the lumen in to the stroma.

Statement II: Malate and aspartate are carboxylation products of the C_3 cycle.

In light of the above statements, choose the most appropriate answer from the options given below:

(A) Both statement I and statement II are correct

(B) Both statement I and statement II are incorrect

(C) Statement I is correct but statement II is incorrect

(D) Statement I is incorrect but statement II is correct

51. Which among the following is a commonly used method for determining soil moisture constant under laboratory conditions:

(A) Tensiometer

(B) Nutrient probe

(C) Gypsum block

(D) Pressure plate membrane apparatus

52. Given below are two statements:

Statement I: If the ozone concentration decreases in the atmosphere, more UV radiation will reach the earth's surface.

Statement II: The adverse effect of 1^0C in temperature can be compensated by increasing CO_2 concentration upto 600 ppm

In light of the above statements, choose the most appropriate answer from the options given below:

(A) Both statement I and statement II are correct

(B) Both statement I and statement II are incorrect

(C) Statement I is correct but statement II is incorrect

(D) Statement I is incorrect but statement II is correct

53. CAM plants are adapted to which type of environment:

(A) Humid (B) Perhumid

(C) Arid (D) Monsoon

54. Parthenium Awareness Week in India during the month of August each year is coordinated/organised by:

(A) ICAR-DWR Jabalpur

(B) ICAR-IARI, New Delhi

(C) ICAR-CSSRI, Kernal

(D) ICAR-IIFSR, Modipuram

55. Determining the volume of HCl required to prepare 0.6 N HCl solution (given specific gravity of HCl = 1.18 g/ml, purity percentage = 35%)

(A) 6.50 ml (B) 53.03 ml

(C) 73.83 ml (D) 65.10 ml

56. Match list I with list II:

List I	**List II**
1. Feroz Shah Tuglaq's rule	i. Eastern canal project
2. Mughal rule	ii. Ganga canal project
3. Rangila Mohammad Shah rule	iii. Western Yamuna canal
4. Sir Aukland Tenure	iv. Ground water resources

Choose the correct answer from the options given below:

(A) 1-i, 2-ii, 3-iv, 4-iii

(B) 1-iii, 2-iv, 3-i, 4-ii

(C) 1-ii, 2-iii, 3-i, 4-iv

(D) 1-iv, 2-ii, 3-iii, 4-i

57. Which of the following weeds has a cup shaped corolla:

(A) *Avena ludoviciana*

(B) *Convolvulus arvens*

(C) *Chenopodium arvensis*

(D) *Cyperus rotundus*

58. Which of the following irrigation methods is most suitable for the wheat crop:

(A) Sprinkler (B) Drip

(C) Bubbler (D) Drun kit

59. Match list I with list II:

List I (Year)	**List II (Pulse production in million tons)**
1. 2014-15	i. 17.15
2. 2015-16	ii. 16.35
3. 2016-17	iii. 25.41
4. 2017-18	iv. 23.13

Choose the correct answer from the options given below:

(A) 1-i, 2-ii, 3-iv, 4-iii

(B) 1-i, 2-ii, 3-iii, 4-iv

(C) 1-iv, 2-iii, 3-i, 4-ii

(D) 1-iv, 2-ii, 3-iii, 4-i

60. Awned varieties give more yield under drought conditions as compared to awnless varieties and the contribution of awn in photosynthesis to grain is:

(A) 5% (B) 10%

(C) 12% (D) 15%

61. Given below are two statements:

Statement I: Single super phosphate, double super phosphate and triple super phosphate are mono calcium phosphates

Statement II: Monocalcium phosphates are water soluble and thus have high leaching losses

In light of the above statements, choose the most appropriate answer from the options given below:

(A) Both statement I and statement II are correct

(B) Both statement I and statement II are incorrect

(C) Statement I is correct but statement II is incorrect

(D) Statement I is incorrect but statement II is correct

62. Which of the following is not zymogen?

(A) Pepsinogen

(B) Trypsinogen

(C) Procarboxy peptidase

(D) Pyrrolidine

63. Match list I with list II:

List I (Disease)	List II (Control measure)
1. Blast of foxtail millet	i. Wettable sulphur (0.3%)
2. Rust of foxtail millet	ii. Mencozeb (0.2%)
3. Downey mildew of foxtail millet	iii. Carboxin (1 g)
4. Head smut of kodo millet	iv. Ridomil (MZ)

Choose the correct answer from the options given below:

(A) 1-ii, 2-i, 3-iv, 4-iii
(B) 1-i, 2-ii, 3-iv, 4-iii
(C) 1-ii, 2-iii, 3-iv, 4-i
(D) 1-iii, 2-iv, 3-i, 4-ii

64. The NAVSTAR global Positioning System is maintained by:

(A) Japan space Agency
(B) Indian Space Research Organization
(C) USA Department of Defence
(D) European Space Agency

65. Given below are two statements:

Statement I: Nutrient that are important in energy storage reactions are often present in plant tissues

Statement II: Some virus induced plant diseases may produce symptoms similar to those of nutrient deficiencies

In light of the above statements, choose the most appropriate answer from the options given below:

(A) Both statement I and statement II are correct
(B) Both statement I and statement II are incorrect
(C) Statement I is correct but statement II is incorrect
(D) Statement I is incorrect but statement I is correct

66. First operational Indian Remote Sensing Satellite - IRS-1A was launched in the year:

(A) 1988 (B) 1989
(C) 1990 (D) 1991

67. Arrange the sequence of steps involved in the procedure for contribution of ANOVA in Latin Square Design:

1. Arrange the raw data according to their rows and column designation with corresponding treatments specified for each observation.
2. Outline analysis of variance table.
3. Compute F value
4. Compute DF for each source of variation, CF and the various sum of square and MSS.
5. Compute CV

Choose the correct answer from the options given below:

(A) 1 – 2 – 4 – 3 – 5 (B) 1 – 4 – 2 – 3 – 5
(C) 2 – 1 – 4 – 5 – 3 (D) 1 – 2 – 3 – 4 – 5

68. How many stereoisomers does a ketopentose have?

(A) 4 (B) 6
(C) 8 (D) 10

69. Given below are two statements:

Statement I: A heat unit is the departure from the mean daily temperature below the maximum temperature.

Statement II: Q 10 is the ratio of the rate of a process at a certain temperature to that at a temperature 10 °C lower.

In light of the above statements, choose the most appropriate answer from the options given below:

(A) Both statement I and statement II are correct
(B) Both statement I and statement II are incorrect
(C) Statement I is correct but statement II is incorrect
(D) Statement I is incorrect but statement II is correct

70. Which of the following indicates a situation of hidden hunger in plants?

(A) Plant expresses nutrient deficiency symptoms when a nutrient is deficient
(B) Plant registers severe yield reduction without expressing deficiency symptoms.

(C) Plant nutrient is in sufficient quantity but seems to be deficient.
(D) Plant is unable to express symptoms due to other elements.

71. What amount of urea, SSP and MOP is required for application in the wheat crop at the rate of 120, 64 and 30 kg N, P_2O_5 and K_2O/ha?
(A) 261 : 400 and 50 kg Urea, SSP and MOP
(B) 130 : 200 and 50 kg Urea, SSP and MOP
(C) 261 : 200 and 25 kg Urea, SSP and MOP
(D) 261 : 100 and 25 kg Urea, SSP and MOP

72. What is the amount of Topik 10% WP required for 4000 m^2 wheat field if the rate of application of clodinafop propargyl is 60 g/ha?
(A) 120 g (B) 240 g
(C) 480 g (D) 600 g

73. Match list I with list II:

List I	List II
1. Law of minimum	i. Mitscherlich
2. Dy/dx=(A-y)C	ii. Spillman
3. Y = m(1 – R)	iii. Baule
4. Percent yield	iv. Liebig

Choose the correct answer from the options given below:
(A) 1-iii, 2-ii, 3-iv, 4-i
(B) 1-iv, 2-i, 3-ii, 4-iii
(C) 1-iv, 2-iii, 3-i, 4-ii
(D) 1-iii, 2-ii, 3-i, 4-iv

74. Calculate the water conveyance efficiency of water is delivered to the farm is 84 mm and water delivered from the source is 140 mm.
(A) 50% (B) 60%
(C) 70% (D) 80%

75. Branching and pod formation are the most critical stages for irrigation in:
(1) Cereal crops
(2) Pulse crops
(3) Pulse and forage crops
(4) Medicinal crops
(5) Aromatic crops

Choose the correct answer from the options given below:
(A) 2 only (B) 2 and 3 only
(C) 1and 4 only (D) 4 and 5 only

76. The most commonly used method of herbicide application in India is:
(A) Spraying
(B) Sand application
(C) Cut and stump
(D) Band placement

77. Calculate the plant population of maize in 6000 m^2 area, if it is sown at spacing of 60 cm × 25 cm.
(A) 40000 plants
(B) 50000 plants
(C) 60000 plants
(D) 100000 plants

78. In irrigated salt affected soils, under raised bed planting system where should the seedlings be planted:
(A) At top of the bed
(B) At mid slope of the bed
(C) At centre of the furrow
(D) Seedling can't be transplanted

79. The mode of herbicide action wherein the weed plants are bleached while it caused by which of the following herbicides:
(A) Atrazine
(B) Clodinafop propargyl
(C) Pendimethalin
(D) Tembotrione

80. Match list I with list II:

List I	List II
1. Oats	i. NB 21
2. Berseem	ii. African tall
3. Maize	iii. Kent
4. Napier	iv. Mescavi

Choose the correct answer from the options given below:
(A) 1-i, 2-ii, 3-iii, 4-iv
(B) 1-i, 2-iii, 3-iv, 4-ii
(C) 1-iii, 2-iv, 3-ii, 4-i
(D) 1-iii, 2-ii, 3-iv, 4-i

81. NDVI of a crop can be measured by using;

(A) Leaf colour chart

(B) Chlorophyll meter

(C) Green seeker

(D) Canopy analyser

82. Each turn of ß-oxidation spiral generates two reduced coenzymes in the form of:

(A) Two NADH moles

(B) Two $FADH_2$ moles

(C) One NADH and one $FADH_2$ moles

(D) One $NADH_2$ and one GTP moles

83. When an abnormal failure of rainfall occurs the type of drought is referred as:

(A) Seasonal drought

(B) Contingent drought

(C) Invisible drought

(D) Atmospheric drought

84. Maggots enter the stem and feed on the growing point of the central shoot and the attacked tiller is converted into a shinning tubular structure called 'silver shoot'. It is due to which of the following pests?

(A) Rice Gandhi bug

(B) Rice root weevil

(C) Gall midge

(D) Rice hispa

85. Which of the following does not represent the use of rice husk:

(A) Source of plant nutrients

(B) Fuel

(C) Substrate for production of biofertilizers

(D) Direct human consumption

86. Which one of the following is NOT a variety of chickpea:

(A) SCS – 3 (B) PBG – 1

(C) GNG – 1581 (D) IPM 2 – 3

87. Which of the following statement regarding shifting cultivation is NOT true:

(A) Initially, it was low input and sustainable land use system

(B) Now, it is not sustainable due to the drastic increase in the gestation period (shifting cycle)

(C) Now, it is not sustainable due to the drastic decrease in the gestation period (Shifting cycle).

(D) It causes erosion and land degradation

88. Given below are two statements:

Statement I: Ginning percentage, colour, trash, fibre quality and fibre length are the important commercial value parameters of jute.

Statement II: Ribboning, retting and extraction are the quality factors of cotton.

In light of the above statements, choose the most appropriate answer from the options given below:

(A) Both statement I and statement II are correct

(B) Both statement I and statement II are incorrect

(C) Statement I is correct but statement II is incorrect

(D) Statement I is incorrect but statement II is correct

89. NIASM is located at:

(A) Baramati (B) Nagpur

(C) Ranchi (D) Jaipur

90. Match list I with list II:

List I (Watersheds)	List II (State)
1. Sukhomajri	i. Uttarakhand
2. Bunga	ii. Punjab
3. Relmajra	iii. Chandigarh
4. Fakot	iv. Haryana

Choose the correct answer from the options given below:

(A) 1-iii, 2-iv, 3-ii, 4-i

(B) 1-ii, 2-iv, 3-i, 4-iii

(C) 1-iii, 2-ii, 3-i, 4-iv

(D) 1-iv, 2-iii, 3-ii, 4-i

91. The growth rate which is also called as efficiency index:

(A) Aggresivity

(B) Relative growth rate

(C) Net assimilation rate

(D) Leaf area duration

92. Given below are two statements:

Statement I: India ranks Ist in the global market of essential oils of aromatic plants ranges from aromatic plants

Statement II: Production of essential oils of aromatic plants ranges from 22000 to 23000 t/year.

In light of the above statements, choose the most appropriate answer from the options given below:

(A) Both statement I and statement II are correct

(B) Both statement I and statement II are incorrect

(C) Statement I is correct but statement II is incorrect

(D) Statement I is incorrect but statement II is correct

93. Benzoquinones and quinines are:

(A) Slow release N fertilizers

(B) Nitrification inhibitors

(C) Urease inhibitors

(D) All of these

94. The science that deals with the aspects of meteorology having direct relevance to agriculture is called:

(A) Atmospheric science

(B) Environment science

(C) Agrometeorology

(D) Earth science

95. Given below are two statements:

Statement I: A vector in GIS system in the graphic or visual data model

Statement II: ARC-INFO is a GIS software

In light of the above statements, choose the most appropriate answer from the options given below:

(A) Both statement I and statement II are correct

(B) Both statement I and statement II are incorrect

(C) Statement I is correct but statement II is incorrect

(D) Statement I is incorrect but statement II is correct

96. Given below are two statements:

Statement I: Conservation agriculture involves no – tillage or drastically reduced tillage but biological tillage is allowed.

Statement II: During the initial years of conservation agriculture, there is a heavy infestation of weeds but it decreases with passing years.

In light of the above statements, choose the most appropriate answer from the options given below:

(A) Both statement I and statement II are correct

(B) Both statement I and statement II are incorrect

(C) Statement I is correct but statement II is incorrect

(D) Statement I is incorrect but statement II is correct

97. Which among the following phenomenon directly affects crop growth and yield?

(A) Competition for nitrogen

(B) Competition for water

(C) Competition for carbon dioxide

(D) Allelopathy

99. Which of the following is NOT a 1:1 type of clay mineral?

(A) Kaolinite (B) Halloysite

(C) Nontronite (D) Chlorite

100. The important event in the early history of agriculture started in 4400 BC:

1. Domestication of sheep
2. Wheat and barley cultivation
3. Maize cultivation
4. Potato cultivation
5. Rice cultivation

Choose the correct answer from the options given below:

(A) 3 only

(B) 1 and 2 only

(C) 3 and 4 only

(D) 1 and 5 only

101. Match list I with list II:

List I (method of determining) Hydraulic conductivity (k)	List II (expressed)
1. Augor hole method of determining Hydraulic conductivity (k)	i. $4k(b^2 - a^2)/q$
2. Pipe cavity method of determining Field hydraulic conductivity (k)	ii. $52300a^2 \log 10(Y0 - Y1)/Bt$
3. Constant head permeameter for Determining hydraulic conductivity in A laboratory	iii. Πr^2 in $(Y0-Y1)/[s(t2 - t1)]$
4. Spacing between drains (s)	iv. $qxL/[A \times t \times \Delta H]$

Choose the correct answer from the options given below:

(A) 1-iv, 2-iii, 3-i, 4-ii
(B) 1-iv, 2-ii, 3-iii, 4-i
(C) 1-ii, 2-iii, 3-iv, 4-i
(D) 1-iii, 2-ii, 3-i, 4-iv

101. A solution contains 1/10000 moles/1H$^+$ ions. What is the concentration of hydroxyl ions in this solution?

(A) 10^{-10} moles/litre (B) 10^{-4} moles/litre
(C) 10^{10} moles/litre (D) 10^{4} moles/litre

102. Match list I with list II:

List I	List II
1. Blossom end rot	i. Cd
2. Itai-Itai disease	ii. Ca
3. Marsh spot of bean	iii. Mo
4. Whiptail disease of cauliflower	iv. Mn

Choose the correct answer from the options given below:

(A) 1-iv, 2-ii, 3-iii, 4-i
(B) 1-ii, 2-i, 3-iv, 4-iii
(C) 1-ii, 2-iv, 3-iii, 4-i
(D) 1-ii, 2-iv, 3-i, 4-iii

103. The fertilizer phosphate use efficiency under optimum condition is:

(A) 10-15% (B) 15-20%
(C) 20-25% (D) 25-30%

104. Match list I with list II:

List I	List II
1. Finger millet	i. *Panicum miliaceum*
2. Barnyard millet	ii. *Paspalum scorbiculatum*
3. Kodo millet	iii. *Eleusine corocana*
4. Proso millet	iv. *Echinocloa frumentocea*

Choose the correct answer from the options given below:

(A) 1-iii, 2-iv, 3-i, 4-ii
(B) 1-ii, 2-iii, 3-iv, 4-i
(C) 1-iii, 2-iv, 3-ii, 4-i
(D) 1-ii, 2-iii, 3-i, 4-iv

105. Which of the following is released during conversion of succinate to fumarate in TCA cycle?

(A) CO_2
(B) $FADH_2$
(C) H_2O
(D) ATP

106. ICAR-Central Research Institute for Dryland Agriculture is located at:

(A) Jodhpur
(B) Hyderabad
(C) Rajahmundry
(D) Bikaner

107. The effective post emergence control of *Ageratum houstoniamum* can be activated with:

(A) Atrazin @ 3.5 kg/ha
(B) Pendimethalin @ 1.5 kg/ha
(C) Alachlor @ 2.0 kg/ha
(D) Thiobencarb @ 1.0 kg/ha

108. Given below are two statements:

Statement I: Organic manure contain a low amount of nutrients and are applied in large quantities.

Statement II: Nutrient fractions in organic manures are inorganic nature only.

In light of the above statements, choose the most appropriate answer from the options given below:

(A) Both statement I and statement II are correct

(B) Both statement I and statement II are incorrect

(C) Statement I is correct but statement II is incorrect

(D) Statement I is incorrect but statement II is correct

109. What is the unit of SAR (Sodium absorption ratio)

(A) moles/litre (B) me/litre

(C) $(me/litre)^{0.5}$ (D) $(moles/litre)^{0.5}$

110. In soil plaque method of soil fertility assessment developed by Winogradsky, which microorganism is studied?

(A) Azotobacter

(B) Aspergillus

(C) Both Azotobacter and Aspergillus

(D) Pseudomonas

111. A person got his soil sample analysed and received a test report monitoring soil pH 8.3, ESP 12% and electrical conductivity 5.2 dsm^{-1}. What type of soil it is?

(A) Acid soil

(B) Saline soil

(C) Alkaline soil

(D) Saline-alkaline soil

112. Maximum emission of radiation from the sun takes place at the wavelength:

(A) 0.45 μm (B) 0.55 μm

(C) 0.90 μm (D) 1.0 μm

113. Electron transport inhibition in PS-II (photosystem II) for herbicidal action is observed in:

(A) Nitriles (B) Ureas

(C) Triazines (D) All of these

114. The formulation of 2, 4-D which is observed faster by plants is:

(A) Ethyl ester (B) Na-salt

(C) Amine-salt (D) Isopropyl amine

115. Given below are two statements:

Statement I: As per IMD investigation, drought in Assam is very rare and occurs once in 15 years

Statement II: In Gujarat and Kashmir, its period of occurance is once in eight years whereas in West Bengal and Bihar once in ten years.

In light of the above statements, choose the most appropriate answer from the options given below:

(A) Both statement I and statement II are correct

(B) Both statement I and statement II are incorrect

(C) Statement I is correct but statement II is incorrect

(D) Statement I is incorrect but statement I is correct

116. Match list I with list II:

List I	List II
1. Parallel system	i. Junction box
2. Grouping system	ii. Bedding system
3. Special drains	iii. Mixed system
4. Surface drains	iv. Grid system

Choose the correct answer from the options given below:

(A) 1-iv, 2-iii, 3-i, 4-ii

(B) 1-iii, 2-ii, 3-i, 4-iv

(C) 1-i, 2-ii, 3-iii, 4-iv

(D) 1-iv, 2-ii, 3-iii, 4-i

117. Match list I with list II:

List I (Crop)	List II (Water requirement)
1. Rice	i. 400 mm
2. Wheat	ii. 300 mm
3. Ground nut	iii. 1200 mm
4. Pulses	iv. 600 mm

Choose the correct answer from the options given below:

(A) 1-ii, 2-iv, 3-i, 4-iii

(B) 1-iii, 2-iv, 3-ii, 4-i

(C) 1-ii, 2-iii, 3-iv, 4-i

(D) 1-iii, 2-iv, 3-i, 4-ii

118. The first confirmed occurance of herbicide resistant strain of a weed species in US was:

(A) ALS inhibitors (B) Bipyridillium
(C) Glycines (D) Triazines

119. An amino acid induced during moisture stress condition and also known as stress indicator is:

(A) Serine (B) Tryptophan
(C) Glycine (D) Proline

120. Chemiosmotic theory was first put forward by:

(A) Peter Mitchell (B) Paul Boyer
(C) John Walker (D) Julius Sachs

Answers Key

Q. No.	Ans	Q. No.	Ans	Q. No.	Ans	Q. No.	Ans	Q. No.	Ans	Q. No.	Ans
1.	A	2.	B	3.	A	4.	A	5.	A	6.	D
7.	C	8.	B	9.	C	10.	C	11.	A	12.	D
13.	A	14.	A	15.	A	16.	B	17.	B	18.	A
19.	C	20.	D	21.	D	22.	D	23.	A	24.	D
25.	C	26.	A	27.	B	28.	D	29.	A	30.	A
31.	C	32.	D	33.	A	34.	D	35.	B	36.	B
37.	A	38.	C	39.	A	40.	B	41.	B	42.	B
43.	C	44.	D	45.	C	46.	B	47.	B	48.	D
49.	B	50.	C	51.	D	52.	A	53.	C	54.	A
55.	B	56.	B	57.	B	58.	A	59.	A	60.	C
61.	C	62.	D	63.	A	64.	C	65.	A	66.	A
67.	A	68.	A	69.	D	70.	B	71.	A	72.	B
73.	B	74.	B	75.	A	76.	A	77.	A	78.	B
79.	D	80.	C	81.	C	82.	C	83.	B	84.	C
85.	D	86.	D	87.	B	88.	B	89.	A	90.	A
91.	B	92.	A	93.	C	94.	C	95.	A	96.	A
97.	A	98.	C	99.	C	100.	A	101.	A	102.	B
103.	B	104.	C	105.	B	106.	B	107.	A	108.	C
109.	C	110.	A	111.	B	112.	A	113.	D	114.	A
115.	B	116.	A	117.	D	118.	D	119.	D	120.	A

Plant Biotechnology

1. **Which type of RNA molecules make up more than 80% of the total RNA in the cell:**
   - rRNA
2. **The H-O-H bond angle in water is:**
   - 104.5°
3. **The chromosome complement of 2n - 2 is:**
   - Nullisomy
4. **The growth rate which is also called as efficiency index is:**
   - Relative growth rate
5. **Which of the following is a omega - 9 fatty acid:**
   - Erusic acid
6. **The enzyme involved in removal of RNA primer in DNA replication process is:**
   - Exonuleases
7. **Which one of the following cannot be consider as a climacteric fruit:**
   - Citrus
8. **Chromosomes are relaxed to the maximum at:**
   - telophase
9. **Energy bond in double helix structure of DNA which are responsible for holding the two strands together are:**
   - Hydrogen bonds
10. **Which one of the following elements can cause necrosis of the young meristematic region, when present in insufficient amounts for plant physiological processes:**
    - Calcium
11. **Which of the following nucleotides has been shown to be functioning as a local hormone to control various physiological functions in humans:**
    - Guanosine
12. **A double haploid (DH) plant is:**
    - Homozygous (get after 6 generations and after interbreeding)
13. **In vegetative tissues and seeds, phosphorus is stored in the form of:**
    - Phytin
14. **Which Of the following sugars enters in the glycolysis via leloir pathway:**
    - Glactose
15. **The most active metabolite of vitamin-D which helps in the regulation of calcium and phosphorus metabolism is:**
    - 1,25 – dihydroxy Cholecalciferol
16. **The term 'biological yield' is given by:**
    - Nichiporovich (also gave economic yield)
17. **The first step of calvin cycle in C4 plant is:**
    - Carboxylation
18. **Which one of the following plants does not belong to the family of CAM plants:**
    - Sugarcane
19. **Which one of the following vitamin is required for the conversion of di-phosphate (duDP) to deoxy thymidine di-phosphate (dtDP):**
    - Thymin
20. **In photosystem (PS-II) the pigment molecules absorb maximum light at:**
    - 700 nm
21. **Tumore cells can be identified in PET scan as sites of unusually high 2-[^{18}F] floro-2-depxyglucose (FDG) accumulation, as:**
    - Glucose occur at a high rate of cancer cell rather than normal cells
22. **How many protons are released on the cytoplasmic side of inner mitochondrial membrane by complete for every two protons taken from its matrix side:**
    - 4

23. An enzyme involved in initiation of DNA synthesis during replication is:

– RNA primase

24. Scientist who had given name mitochondria:

– Carl benda

25. The gene that expresses all the time and keeps up cellular process are called:

– House keeping genes

26. The country that deregulated 'golden rice' recently to address the deficiency of vitamin A is:

– Philippines

27. The type of proteins that are involved in higher order coiling of DNA post-nucleosome stage are:

– Histone proteins

28. Polyploid with multiple complete genome derives from two or more different species is:

– Allopolyploid

29. The process that operates in a diametrically opposite direction of CO_2 fixation leading to a loss of photo-synthetically fixed CO_2 is:

– Photorespiration

30. Primary role of post-transcriptional modification such as 5′ capping and 3-poly A trailing of mRNA is:

– Precision of translation

31. Insulin is an oligo-saccharide of great nutritional significance, it is formed from which one of the following monosachharide:

– Fructose

32. Which one of the following rhizobial nod proteins is activated by flavonoids, secreted by the roots of the legume plants and induce transcription of other nod genes after its activation:

– nod-D

33. Which of the following post-translational modification of the proteins may tag them for degradation in 26s proteosome:

Poly-ubiquitination

34. A nucleic acid molecule is a linear polymer in which the monomers nucleotides are linked together by:

– Phosphodiester bond

35. Which of the following E.coli DNA polymerases possesses both 5′ and 3′ exonuclleolytic activity and is thus its own proofreader and editor:

– DNA polymerase – I

36. Which of the following is not a ketone body:

– Isobutyryl CoA

37. The most abundant protein found in biosphere is:

– Rubisco

38. For thinking in apple, the growth regulator used commercially is:

– NAA

39. Which of the following statement is true about the D-glucose and D-talose:

– They are diestomers not primers

40. Resitance development by some weeds specific weedicide due to:

– gene mutation

41. A molecule marker that samples genome in a random and genome wide manner is:

– RAPD

42. Which one of the following organelles contain enzymes that help in converting stored fatty acids into sugars in germinating seeds:

– Glyoxysomes

43. The plant cellular organelles that contributes to the extra chromosomal inheritance are:

– Mitochondria and chloroplast

44. In plant cells non-pigmented plastids are called:

– Leucoplast

45. An example of crop where the reserve food material in seeds is stored in endosperm is:

– Castor

46. The identification of specific DNA sequence by southern blotting involves the following precedures

Step 1 : DNA digestion
Step 2 : Gel electrophoresis
Step 3 : Blotting
Step 4 : Probe labelling
Step 5 : Hybridization & Washing
Step 6 : Detection

47. The natural made of exchange of genetic material in bacteria is:

– Conjugation

48. A protein has eight cysteine residues. How many possible ways are there to make four disulfide bridges in its structure:

– 105

49. The bioinformatics tools used to identify microsatellites (repeated motifs) in a genome is:

– MISA

50. Correcting a genetic disorder by the way of introduction of normal genes in place of missing or defective one is reflected as:

– Gene therapy

51. The instrument used for the measurement of water potential of plant tissue is:

– Pressure chamber

52. The water potential of pure water is always:

– 0.0 Mpa

53. The activity of enzyme pyruvate dehydrogenase is regulated by which one of the following modification:

– Phosphorylation/dephosphorylation

54. The crop that flowers when the day exceeds a critical duration of 24 hrs cycle are called:

– long day plant

55. Nobel prize was awarded to fire and melo for the discovery of:

– RNA interference

56. Cryopreservation of plant material using liquid nitrogen (-196°C) for conservation eventualities is a:

– Long term strategy

57. Nucleotides differ from nucleotide with respect to:

– Phosphate group

58. The unit area of leaf area index (LAI) is:

– Dimension less

59. The empirical formula of the chlorophyll molecules is:

– $C_{55}H_{72}O_5N_4Mg$

60. A polyketide that inhibits the enzyme ATP synthase by blocking the movement of protones through its F_0 unit is:

– Oligomycin

61. The common bio assay used to detect the hormone brassinosteriods is:

– dwarf rice lamina inclination bio assay

62. The crop plant with highest amount of water requirement is:

– sugarcane (1500-3000 L)

63. Some plants have solar tracking mechanism and such mechanisms of bending towards sunlight is called:

– Heliotropism

64. The pH of 10^{-8} M HCl solution will be:

– 6.96 (near 7)

65. Which one of the following nutrient element remains in ionic form and is the main cation in establishing plant cell turgor:

– Potassium (K^+)

66. Which of the following reagents cannot be used for the determination of the N-terminus of the protein:

– Hydrazine

67. Which one of the sugar is non-reducing

– Sucrose, Trehalose

68. Which of the following are bioinformatics databases:

– SWISS-PROT, Gene Bank, ENTREZ

69. The crop plant tolerant to available iron in the soil is:

– Brinjal

70. The place or site of oxidative phosphorylation in plant cell is:

– Inner mitochondrial membrane

71. Match the following

List I	List II
1. National Institute for Rural Development	i. Hyderabad
2. Central Agricultural University	ii. Imphal
3. ICAR-Sugarcane Breeding Institute	iii. Coimbatore
4. ICAR-National Institute of Biotic Stress Management	iv. Raipur

72. Oxidation of one glucose molecule under anaerabic respiration gives:

– 2 ATP

73. The vector that could be used in two different hosts are called:

– Shuttle vector

74. The vasicular mycorrhiza genus which plays an important role in P-cycling and its uptake:

– Glomus

75. A cluster of genes under the control of a single promoter is called:

– Operon

76. Calvin's group used for following plant to identify the intermediates in C3 pathways:

– Alfalfa

77. The transfer DNA (T-DNA) portion of Ti plasmid contains genes for:

– Hormone and opines

78. An example of a model crop in biology is:

– Wheat/tobacco

79. Light is the signal that controls stomatal movement in plants. The stomatal opening is stimulated by which light:

– Red light

80. Urease catalyzes hydrolysis of urea to CO_2 and NH_4. Which of the following element is an essential part of urease:

– Nickel

81. The viruses that infects bacteria are called:

– Bacteriophages

82. Transgenic papaya variety 'Rainbow' is resistant to:

– PRSV

83. The DNA replication in prokaryotes is of:

– Semi-conservative type

84. The reaction NO_3-NAD(P)H + H^+ + 2e → NO_2 + NAD(P) + H_2O is catalysed by the enzyme:

– Nitrate reductase

85. The epigenetic modification, which regulates gene expression is:

– Spiling of hnRNA

86. Match the following:

List I	List II
1. Vid A	i. Beta-glucuronidase
2. Cry IAC	ii. Insecticidal gene
3. Bt Brinjal	iii. Fruit and shoot borer resistant
4. DREB gene	iv. Drought tolerant

87. The water quality is categorized 'excellent' for irrigation when the chloride concentration is

_ less than or equal to 4

88. The biosynthesis pathway of chlorophyll begins with the compound:

– Glutamic acid

89. The α - helical structure of proteins are result of formation of hydrogen bonds between which of the following amino acids in its primary structure:

– nth and n+ 4th

90. The peptide containing which of the following amino acid are well suited for buffering the physiological pH:

– Histidine

91. The structure of amino acids in a protein during trasnlation process is determined by:

– m – RNA codons

92. The fatty acids are converted into which of the following volatile derivatives for their separation by gas liquid chromatography (GLC):

– Fatty acid methyl esters (FAMEs)

93. The plant hormone that activated many defense reponses is:

– Jasmonic acid

94. Which one of the following plant hormones can induce flowering in pineapple:

– Ethylene

95. The first commercialized transgenic crop/variety was:

– Bt cotton

96. The most abundant compound that is transported through sieve tubes is:

– Sucrose

97. ^{35}S isotope is commonly used in experiments involving which of the following biomolecules:

– Nucleic acid

98. A bio-molecular complex composed of four peptides two identical light chains and two identical heavy chains is called:

– Antibody

99. The 14-3-3 proteins are involved in the post -translational regulation of which one of the following enzymes of nitrogen metabolism:

– Nitrate reductase

100. The data base used by the ENTREZ for bibliographic search is:

– PubMed

101. Which one of the following crop genotype exhibits the dimorphic structure of lead called Krantz anatomy

– Sugarcane

102. The in-vitro 'slow growth culture' is a method used for:

– Conservation off plant genetic resources

103. The symbol ' ' in the T C stem - loop of the t-RNA denotes which one of the following:

– Pseudouridine

104. Majority of the t-RNA molecules have the following sequence at their 3' end:

– CCA-3'-OH

105. Which one of the following hormone functions as a 'morphogan' during embryogenesis:

– Cytokinin

Plant Sciences

1. **The hormone needed for cell division during germination of seed as:**
   (A) Gibbrellin
   (B) Abscisic acid
   (C) Cytokinin
   (D) Ethylin

2. **Seed dormancy can be detected by:**
   (A) Vigour test
   (B) Germination count
   (C) Tetrazolium test
   (D) Purity test

3. **Two genes are linked or not can be ascertained by:**
   (A) F-test (B) t-test
   (C) Chi-square test (D) Z-test

4. **Cell wall of archae bacteria do not contain:**
   (A) Proteins
   (B) Glycoproteins
   (C) Polysaccharides
   (D) Peptidoglycans

5. **Given below are two statements:**

   Statement I: In bacterial transformation a segment of naked DNA from the surroundings passes through the cell wall and lasma membrane of a recipient cell

   Statement II: Only a few competent recipient cells can take up the DNA from the surroundings

   In light of the above statements, choose the most appropriate answer from the options given below:
   (A) Both statement I and statement II are correct
   (B) Both statement I and statement II are incorrect
   (C) Statement I is correct but statement II is incorrect
   (D) Statement I is incorrect but statement Ii is correct

6. **The first product to be tagged with GI in India is:**
   (A) Assam tea (B) Nilgiri tea
   (C) Darjeeling tea (D) Organic tea

7. **Match list I and list II:**

List I	List II
1. Rose weed	i. *Shorea robusta*
2. Sal	ii. *Diopyros ebenum*
3. Ebeny	iii. *Dalbergia latifolia*
4. Teak	iv. *Tectona grandis*

   Choose the correct answer from the options given below:
   (A) 1-ii, 2-iii, 3-iv, 4-i
   (B) 1-iv, 2-ii, 3-iii, 4-i
   (C) 1-i, 2-ii, 3-iii,4-iv
   (D) 1-iii, 2-i, 3-ii,4-iv

8. **Which one of the following bacteria pathogenic genes is classified under gram - negative aerobic and microaerophilic rods:**
   (A) Acidovorer (B) Erwinia
   (C) Pantoea (D) Entrobactor

9. **Folian plant pathogens are pigmented in nature and cause crop losses in major staple crops in the world as:**
   1. Pigment help them to tolerate high rediation
   2. They cause diseases only on monocots
   3. They cause diseases only on dicots
   4. They move from one plant to another plant with the help of wind
   5. They cause polycyclic diseases

   Choose the correct answer from the options given below:
   (A) 1, 2 and 3 only
   (B) 2, 3 and 4 only
   (C) 3, 4 and 5 only
   (D) 1, 4 and 5 only

10. Given below are two statements:

Statement I: Kosambi's mapping function considers the absence of interferance (I)

Statement II: In light of the above statements, choose the most appropriate answer from the options given below:

(A) Both statement I and statement II are correct

(B) Both statement I and statement II are incorrect

(C) Statement I is correct but statement II is incorrect

(D) Statement I is incorrect but statement Ii is correct

11. Match list I and list II:

List I	List II
1. Chitin	i. Archaebacteria
2. Pseudomurein	ii. Gram positive bacteria
3. O- antigens	iii. Fungal cell wall
4. Teichain acid	iv. Gram negative bacteria

Choose the correct answer from the options given below:

(A) 1-iii, 2-i, 3-ii, 4-iv

(B) 1-i, 2-iii, 3-ii, 4-iv

(C) 1-i, 2-iii, 3-iv,4-ii

(D) 1-iii, 2-i, 3-iv,4-ii

12. Zein, a special kind of protein is obtained from:

(A) Ramie (*Boehmeria nivea*)

(B) Guar (*Cyamopsis tetragonoloba*)

(C) Grasspea (*Lathyrus sativus*)

(D) Maize (*Zea mays*)

13. Given below are two statements:

Statement I: Cleistogamy and chamogony favours self fertilization.

Statement II : Male sterility and self incompatibility do not favour cross fertilization.

In light of the above statements, choose the most appropriate answer from the options given below:

(A) Both statement I and statement II are correct

(B) Both statement I and statement II are incorrect

(C) Statement I is correct but statement II is incorrect

(D) Statement I is incorrect but statement II is correct

14. Match list I and list II:

List I (Organism)	List II (Specific activities)
1. Mycorrhizae	i. Autotropic nitrogen fixation
2. Symbiosome	ii. Microaerophilic conditions
3. Azospirillum	iii. Symbiotic nitrogen fixation
4. Nitrosomonas	iv. Phosphorus nutrition

Choose the correct answer from the options given below:

(A) 1-i, 2-ii, 3-iv, 4-iii (B) 1-iv, 2-iii, 3-ii, 4-i

(C) 1-iv, 2-ii, 3-iii,4-i (D) 1-iv, 2-i, 3-iii,4-ii

15. Given below are two statements:

Statement I: Mera Gaon mera Gaurav scheme under extension education was launched by Prime Minister Sh. Narendra Modi with the theme Lab to Land by asking scientist to adopt villages and remain in touch with them providing technical information to farmer and other aspects.

Statement II: The scheme mera gaon mera gaurav functions with the help of Krishi Vigyan Kendra (KVKs) and Agriculture Technology Management Agency (ATMA).

In light of the above statements, choose the most appropriate answer from the options given below:

(A) Both statement I and statement II are correct

(B) Both statement I and statement II are incorrect

(C) Statement I is correct but statement II is incorrect

(D) Statement I is incorrect but statement II is correct

16. Translocation of carbohydrates in plants usually occurs in the form of:

(A) Glucose (B) Maltose

(C) Starch (D) Sucrose

17. Transfer of a portion of DNA from one bacterium to another by a bacteriophage is called:

(A) Transformation

(B) Conjugation

(C) Transduction

(D) Translocation

18. In which of the following seed storage conditions the longivity of the seeds would be maximum:

(A) Normal ambient storage

(B) Moisture proof storage

(C) Dehumidified with low temperature (5-8°) storage

(D) Dehumidified storage only

19. Which of the following stages of cell division is primarily arrested by treating with colchicine used in polyploidization of plants:

(A) Anaphase (B) Telophase

(C) Interphase (D) Metaphase

20. Given below are two statements:

Statement I: Light reaction takes place in the stroma.

Statement II: Dark reaction takes place in grama.

In light of the above statements, choose the most appropriate answer from the options given below:

(A) Both statement I and statement II are correct

(B) Both statement I and statement II are incorrect

(C) Statement I is correct but statement II is incorrect

(D) Statement I is incorrect but statement II is correct

21. End products of glycolysis is:

(A) Acetyl CoA

(B) Pyruvic acid

(C) Glucose-1-phosphate

(D) Fructose -1-phosphate

22. Which one of the following viruses is transmitted by pollen:

(A) Tospovirus

(B) Tobravirus

(C) Prunus nicrotic ringspot virus

(D) Papaya ringspot virus

23. With regard to the cultivation of groundnut which of the following statements are correct:

1. Groundnut is a warm season crop requiring abundant sunshine
2. Well drained sandy loams are most favourable for growing groundnut
3. Groundnut is propagated through whole pods, decorticated seed or from seedlings
4. Groundnut is widely grown in coastal and deltaic regions
5. Groundnut takes six to nine months to mature

Choose the correct answer from the options given below:

(A) 1, 2 and 3 only

(B) 2, 3 and 4 only

(C) 3,4 and 5 only

(D) 1, 4 and 5 only

24. A cultivar developed through population improvement approach can be maintained indefinately with proper identity through:

(A) Pureline selection

(B) Random mating in isolation

(C) Selfing in isolation

(D) Top crossing

25. The gene A-B-D are present in one chromosomes in the coupling phase of linkage. One of the double cross-over gametes has A-b-D gene combinations. the gene which was in the middle of the sequence is:

(A) A

(B) B

(C) D

(D) Cannot determine

26. Match list I and list II:

List I (Theory)	List II (Scientist who proposed)
1. Cell theory	i. Clarence P Oliver
2. Chromosomal theory of inheritance	ii. Scheiden, Schwann & Virchow
3. Divisibility of gene	iii. Vernon Ingram
4. One gene one polypeptide	iv. Walter Sutton & Theodor Boveri

Choose the correct answer from the options given below:

(A) 1-ii, 2-i, 3-iv, 4-iii
(B) 1-ii, 2-iv, 3-i, 4-iii
(C) 1-iv, 2-iii, 3-ii,4-i
(D) 1-iii, 2-ii, 3-i,4-iv

27. In a sexually reproducing organism, the chromosome number in the progeny does not double even after fertilization. The reason is that the male and female gametes contains:

(A) Quarter set of somatic chromosome
(B) Half set of somatic chromosome
(C) Full set of somatic chromosomes
(D) Inhibitor that destroys the extra chromosomes

28. Given below are two statements:

Statement I: Higher the degree of cross polination in crops, more will be the isolation distance requirement

Statement II: Self pollinated crops need more isolation distance like cross pollinated crops.

In light of the above statements, choose the most appropriate answer from the options given below:

(A) Both statement I and statement II are correct
(B) Both statement I and statement II are incorrect
(C) Statement I is correct but statement II is incorrect
(D) Statement I is incorrect but statement II is correct

29. Match list I and list II:

List I	List II
1. Breeder seed	i. No tag are used
2. Nucleus seed	ii. Golden brown
3. Certified seed	iii. Azure blue
4. Foundation seed	iv. White

Choose the correct answer from the options given below:

(A) 1-iii, 2-i, 3-ii, 4-iv
(B) 1-i, 2-ii, 3-iii, 4-iv
(C) 1-ii, 2-i, 3-iii,4-iv
(D) 1-iv, 2-i, 3-ii,4-iii

30. The tag for breeder seed issued by:

(A) State Seed Certification Agency
(B) Central Seed Certification Agency
(C) Concerned Breeder of the variety
(D) Independent organization operated under the state government

31. Arrange the operations in seed processing in correct order:

(1) Cleaning and grading
(2) Drying
(3) Seed treatment
(4) Seed testing
(5) Bagging and labelling

Choose the correct answer from the options given below:

(A) 2, 1, 4, 3, 5
(B) 1, 2, 3, 4, 5
(C) 1, 4, 3, 2, 5
(D) 3, 4, 2, 1, 5

32. Match list I and list II:

List I	List II
1. White rust	i. Puccinia steriiformis
2. Black rust	ii. Albugo candida
3. Yellow rust	iii. Cephaleuros parasiticus
4. Red rust	iv. Puccinia tritici

Choose the correct answer from the options given below:

(A) 1-i, 2-ii, 3-iii, 4-iv
(B) 1-ii, 2-iv, 3-i, 4-iii
(C) 1-iv, 2-iii, 3-ii,4-i
(D) 1-iii, 2-ii, 3-i,4-iv

33. Which disease is greatly stimulated interest in the cause the control of plant diseases:

(A) Downy mildew of grapevine
(B) Early blight of potato
(C) Late blight of potato
(D) Wheat rust disease

34. In which of the following disease, the symptoms is spatially separated from the pathogens niche:

(A) Crown gall
(B) Die back
(C) Leaf spot
(D) Blight

35. Given below are two statements:

Statement I: The process of CTC in tea stands for crushing, tearing and curling

Statement II: The process of CTC in tea stands for curling, tearing and crushing

In light of the above statements, choose the most appropriate answer from the options given below:

(A) Both statement I and statement II are correct
(B) Both statement I and statement II are incorrect
(C) Statement I is correct but statement II is incorrect
(D) Statement I is incorrect but statement II is correct

36. The genome of Triticale is designated as:

(A) AABB (B) AABBCC
(C) AABBDD (D) AABBRR

37. Stock 6 is a haploid inducer line of:

(A) Wheat (B) Maize
(C) Rice (D) Sorghum

38. Conjugation between two bacteria is mediated through:

(A) F-plasmid (B) R-plasmid
(C) C-plasmid (D) D-plasmid

39. Naked arci can be observed in which ascomycete fungus:

(A) Erysiphe (B) Claviceps
(C) Taphrina (D) Uncinula

40. Water potential values of biological systems are always:

(A) Positive
(B) Negative
(C) Neither positive nor negative
(D) Neutral

41. Given below are two statements:

Statement I: Witches broom symptoms are caused by phytoplasm

Statement II: Nutrient deficiency also causes witchs broom in crop plants.

In light of the above statements, choose the most appropriate answer from the options given below:

(A) Both statement I and statement II are correct
(B) Both statement I and statement II are incorrect
(C) Statement I is correct but statement II is incorrect
(D) Statement I is incorrect but statement II is correct

42. Scientific name of oil palm is:

(A) *Olea europea*
(B) *Elaeis guineensis*
(C) *Ricinus communis*
(D) *Sesamum indicum*

43. Which one of the following is NOT an introduced variety od wheat crop:

(A) Turkey (B) Marquis
(C) Federation (D) Latisail

44. In which host - patho system does not the Gene-for-Gene hypothesis operate:

(A) Host-Nectrotrophic interaction
(B) Host-Seprophytic interaction
(C) Host-Biotrophic interaction
(D) Host-obligate Saprophytic interaction

45. Which of the following is NOT a non-sense codon:

(A) UAG (B) UAA
(C) AUG (D) UGA

46. Minimum standard of seed germination (%) for seed certification in maize is:

(A) 80 (B) 85
(C) 90 (D) 75

47. Which one of the following is a biotrophic oomycetes:

(A) *Phytophthora capsici*

(B) *Pythium ultimum*

(C) *Phytophthora ramorum*

(D) *Plasmopora viticola*

48. Match list I and list II:

List I	List II
1. Baiting	i. Nematodes
2. Ooze test	ii. Bacteria
3. Elisa	iii. Fungi
4. Baermann-funnel technique	iv. Virus

Choose the correct answer from the options given below:

(A) 1-i, 2-ii, 3-iii, 4-iv (B) 1-iii, 2-ii, 3-iv, 4-i

(C) 1-iv, 2-iii, 3-ii,4-i (D) 1-ii, 2-i, 3-iii,4-iv

49. The National Gene Bank Facility to conserve plant genetic resources in India located at:

(A) Chennai (Tamilnadu)

(B) Ahmedabad (Gujarat)

(C) New Delhi (Delhi)

(D) Hyderabad (Andhra Pradesh)

50. Toxic shock syndron through which bacteria:

(A) *Bacillus anthracis*

(B) *Staphylococcus aureus*

(C) *Streptococcus pyogenes*

(D) *Enterococcus foecalis*

51. Given below are two statements:

Statement I: Many viral diseases are transmitted by insects vectors.

Statement II: Few insects trasmit bacterial disease.

In light of the above statements, choose the most appropriate answer from the options given below:

(A) Both statement I and statement II are correct

(B) Both statement I and statement II are incorrect

(C) Statement I is correct but statement II is incorrect

(D) Statement I is incorrect but statement II is correct

52. A crop variety enjoys legal protection in India, provided it is :

(A) Developed by a breeder in a government organization

(B) Registered with the PVR&FR Authority

(C) Cultivated in a larger area

(D) Released by any Variety Release Committee

53. The possible number of triplet codons from 6 different types of nucleotides in the DNA:

(A) $(6)^3$ (B) $(3)^6$

(C) $(6)^6$ (D) $(3)^3$

54. How many linkage groups are present in maize (Zea mays L):

(A) 10 (B) 20

(C) 12 (D) 13

55. Given below are two statements:

Statement I: Nitrogenase enzyme is sensitive to molecular oxygen.

StatementII: Bacteria of genus Azospirillum fix nitrogen under microaerophillic conditions.

In light of the above statements, choose the most appropriate answer from the options given below:

(A) Both statement I and statement II are correct

(B) Both statement I and statement II are incorrect

(C) Statement I is correct but statement II is incorrect

(D) Statement I is incorrect but statement II is correct

56. With respect to chickpea, which of the following statement is INCORRECT:

(A) Chickpea is a rabi crop

(B) Heavy rains after sowing or at flowering and fruiting stage is useful

(C) The crop cannot frost at any stage of growth

(D) Chickpea is well adapted to arid and semiarid regions and a cool and dryclimate

57. Which of the following listed items do NOT come under the purview of the National Plant Protection Organization (NPPO) under the International Plant Protection Convention (IPPC):

1. Surveillance and inspection
2. Controlling pests (example administering treatments, preventing spread, disinfection or disinfectant
3. Conduction of pest risk analysis
4. Conservation and regeneration of plant genetic resources
5. Multiplication of plant genetic resources

Choose the correct answer from the options given below:

(A) 1, 2 only (B) 2, 3 only
(C) 1, 5 only (D) 4, 5 only

58. Polycyclic diseases are usually caused by pathogens adapted:

(A) on foliar plant parts
(B) in soil
(C) in rhizosphere
(D) in phloem

59. Given below are two statements:

Statement I: Hybrids are heterzygousn and homogenus

Statement II: Purelines are heterozygous and heterogenous

In light of the above statements, choose the most appropriate answer from the options given below:

(A) Both statement I and statement II are correct
(B) Both statement I and statement II are incorrect
(C) Statement I is correct but statement II is incorrect
(D) Statement I is incorrect but statement II is correct

60. As per Seymour, a unit of mutation (muton) and unit of recombination (recon), respectively represent:

1. Single gene and any pair of genes
2. Single nucleotide and any adjacent pairs of nucleotide
3. Single gene and any homologous pair of chmosomes
4. Single nucleotide and any single gene
5. Deletion mutants and any cross-over genes

Choose the correct answer from the options given below:

(A) 1 only (B) 2 only
(C) 1, 2, 3 only (D) 3, 4, 5 only

61.. When two self - incompatible but cross compatible parental lines are interplanted, the seeds harvested from both the self-incompatible lines will be:

(A) Composite seed (B) Selfed seed
(C) Hybrid seed (D) Inbred seed

62. A type of mutualistic relationship in which individual of one species lives in close association with individual of another species is called:

(A) Symbiosis (B) Parasitism
(C) Syntrophism (D) Commensation

63. Fractional sterilization is also called:

(A) Pasteurization
(B) Tyndallization
(C) Fractionation
(D) Dry heat sterilization

64. Which of the following statements is NOT true for Koch's postulates:

1. A specific organism can always be found in association with a given disease.
2. The organism can be isolated from a diseases animal and grown in oure culture in the laboratory.
3. The pure culture will produce the disease when inoclated into a resistant animal
4. The pure culture will not produce the disease when inoculated into a susceptible animal.
5. It is possible to recover the organism in pure culture from the expperimentally infected animal.

Choose the correct answer from the options given below:

(A) 3 and 4 only (B) 3, 4 and 5 only
(C) 3 only (D) 3 and 5 only

65. An organism has haploid chromosome number 7. The probability that a sperm produced by it will contain all the 7 chromosomes where centromeres were derived from the maternal homologue is:

(A) $(1/2)^1$ (B) $(1/2)^3$
(C) $(1/2)^5$ (D) $(1/2)^7$

66. What is the role of antibiotics in disease management:

1. Prevent the growth of pathogen
2. Prevent the development of pathogens
3. Accelerated the growth of pathogens
4. Accerelate the development of pathogens

Choose the correct answer from the options given below:

(A) 1, 4 only (B) 2, 3 only
(C) 1, 2 only (D) 3, 4 only

67. .... is phyto resistance used by:

(A) Cotton (B) Potato
(C) Pepper (D) Pea

68. Match list I and list II:

List I	List II
1. *Fusarium oxysporum*	i. Microsclerotia
2. *Macrophomina phaselolina*	ii. Clamydospores
3. *Bipolaris sarokiniana*	iii. Thick-walled conidia
4. *Ralstonia solanacearum*	iv. Naked cells

Choose the correct answer from the options given below:

(A) 1-iii, 2-i, 3-ii, 4-iv
(B) 1-iv, 2-i, 3-iii, 4-ii
(C) 1-ii, 2-i, 3-iii,4-iv
(D) 1-i, 2-ii, 3-iii,4-iv

69. Arranged the correct sequence:

1. World Trade Organisation-Trade Related Aspect of Intelectual Property Rights (WTO-TRIPS)
2. Convention on Biological Diversity (CBD)
3. Nagoya Protocol on Access and Befifit Sharing (NP)
4. Geographical Indication of Goods (Registration and Protection), GI Act
5. International Treaty on Plant Genetic Resources for Food and Agriculture (ITPGRFA)

Choose the correct answer from the options given below:

(A) 2, 4, 1, 5, 3 (B) 1, 2, 4, 5, 3
(C) 5, 3, 2, 1, 4 (D) 4, 2, 1, 5, 3

70. Match list I and list II:

List I	List II
1. Tumor	i. Xanthomonas
2. Wilt	ii. Pectobacterium
3. Spot	iii. Pantoea
4. Rot	iv. Agrobacterium

Choose the correct answer from the options given below:

(A) 1-iv, 2-iii, 3-i, 4-ii
(B) 1-iv, 2-ii, 3-iii, 4-i
(C) 1-iii, 2-ii, 3-iv,4-i
(D) 1-i, 2-iv, 3-ii,4-iii

71. Isolation distance required for production of the certified seed of maize composite varieties:

(A) 200 m (B) 300 m
(C) 100 m (D) 50 m

72. Who among the following scientist is considered as Father of Green Revolution in India:

(A) Dr. B.P. Pal
(B) Dr. R.S. Paroda
(C) Dr. M.S. Swaminathan
(D) Dr. Gurudev S. Khush

73. Which of the following crops are condidered to have originated in South Americal centre:

(A) Potato, Tomato, Tobacco, Rice
(B) Potato, Tomato, Tobacco, Maize
(C) Potato, Tomato, Sugarcane, Wheat
(D) Potato, Tomato, Soybean, Brinjal

74. Cell theory was put forward by:

(A) Singh and Nicolsan
(B) Leeuwenhock
(C) Schleiden and Schwan
(D) Calvin and Malvin

75. Which of the following plant cell organelles is considered as semi-autonomous?

(A) Endoplasmic reticulum
(B) Mitochondria
(C) Golgibody complex
(D) Lysosomes

76. The allelo combination of the variety 1 is AAbbCCddEE and has the phenotype of 6 yield units. The allelic combination of variety 2 is aaBBccDDee and has 4 yield units. What would be the yield unit and heterosis (over the better parent) unit of the F1(if there is complete dominance)?

(A) 6 and 4 (B) 10 and 6
(C) 10 and 4 (D) 4 and 4

77. To estimate the genetic diversity of germplasm collection, a technique involving multi variate analysis based on ordination giving graphical representation is:

(A) D^2 analysis
(B) Principal component analysis
(C) Cluster analysis
(D) Linkage analysis

78. The yield of inbred A is 40 q/ha, inbred B is 20 q/ha, check hybrid yield is 45 q/ha and the yield of hybrid (AxB) is 50 q/ha. The statndard heterosis (%) is:

(A) 25% (B) 11.1%
(C) 66.6% (D) 50%

79. Match list I and list II:

List I (Organelle)	**List II** (Function)
1. Mitochondria	i. Packaging protein, lipid and phospholipid for transport to other cells
2. Endoplasminc reticulum	ii. Oxidation of carbohydrate, lipid and protein
3. Lysosomes	iii. Digestion of intra cellular substances and foreign material
4. Golgi complex	iv. Provide passage for mRNA from nucleus to cytoplasm

Choose the correct answer from the options given below:

(A) 1-ii, 2-i, 3-iii, 4-iv
(B) 1-ii, 2-iv, 3-iii, 4-i
(C) 1-ii, 2-iv, 3-i,4-iii
(D) 1-i, 2-ii, 3-iii,4-iv

80. Which of the statements below do NOT stand true about the convention on Biological Diversity (CBD)?

1. CBD deals with all biological resources
2. One of the objectives of CBD is the conservation and sustainable use of the component of biodiversity
3. The CBD affirms the sovereign rights of the nation over its genetic resources
4. CBD is an instrument for homonizing phytosanitary measures in international trade
5. CBD provides a model for securing protection for plant breeder rights for plant varieties

Choose the correct answer from the options given below:

(A) 1, 2 only (B) 2, 3 only
(C) 3, 4 only (D) 4, 5 only

81. Which of the following fungi is:

(A) Erysiphe (B) Uncinula
(C) Podosphaera (D) Giberella

82. Identify the correct statement(s) about speed Breeding:

1. It is targetted to reduce the breeding cycle of a crop
2. It manipulates photoperiod and temperature to enhance cop maturity
3. It is done in open field condition
4. It was initially used to grow crops in the space station
5. It increases both vegetable and reproductive phases of the plants

Choose the correct answer from the options given below:

(A) 1, 2, 3 only
(B) 2, 3, 4 only
(C) 3, 4, 5 only
(D) 1, 2, 4 only

83. What are the genotypes that will be included in the trial for the estimation of heterobeltiosis?

(A) F1 hybrid developed and the parental lines

(B) F1 hybrid and commercially released hybrid (check)

(C) F1 hybrid developed only

(D) Parental lines and commercially released hybrid (check)

84. Certain plant pathogens do not ive at the site of the symptom or symptomate plant niche:

1. They only transfer virulent factors to host
2. Plant attracts the virulence factors by its chemical signals
3. This kind of pathogen survive in the rhizosphere of the infected plant
4. Virulence factors can be cultured on media
5. Infected plant can be recovered with time

Choose the correct answer from the options given below:

(A) 2, 3, 5 only

(B) 3, 4, 5 only

(C) 1, 3, 4 only

(D) 1, 2, 3 only

85. Match list I and list II:

List I	List II
1. Dithiocarbamate	i. Metalaxyl
2. Oxathin	ii. Fesctyl-al
3. Acylalanine	iii. Mancozed
4. Organophosphate	iv. Carboxin

Choose the correct answer from the options given below:

(A) 1-iv, 2-iii, 3-ii, 4-i

(B) 1-i, 2-ii, 3-iii, 4-iv

(C) 1-iii, 2-iv, 3-ii,4-i

(D) 1-iii, 2-iv, 3-i,4-ii

86. Maximum assimilation of nitrate takes place in which part of the plant:

(A) Stem (B) Root

(C) Leaves (D) Fruit

87. Which one of the following species has an inflorescence of terminal digitate spikes of spikelets:

(A) *Setaria italica*

(B) *Eleusine caracana*

(C) *Panicum sumatrense*

(D) *Panicum miliaceum*

88. Which of the following statements are correct with respect to fibres or fibre yielding plants:

1. Vegetable fibres are mainly composed of cellulose
2. Vegetable fibres are mainly composed of proteins
3. Fibre durability is conditioned by the diameter of the fibre or fibrous aggregate
4. The flexible wall of cotton fibre contains no lignin

Choose the correct answer from the options given below:

(A) 1, 2, 3 only

(B) 1, 3, 4 only

(C) 2, 3, 4 only

(D) 1, 2, 4 only

89. The degree of compost maturity can be assessed by:

(A) Seed germination test only

(B) Infrared technique only

(C) Moisture content only

(D) Both seed germination test and infrared technique

90. Starting from interior to exterior, arrange the following in correct order:

1. Periplasmic stage
2. Nucleus
3. Plasma membrane
4. Cell wall
5. Endoplasmic reticulum

Choose the correct answer from the options given below:

(A) 2, 5, 1, 3, 4

(B) 2, 5, 3, 4, 1

(C) 2, 5, 3, 1, 4

(D) 2, 1, 5, 3, 4

91. Identify the correct order of activities followed for testing and release of crop varieties:

1. Multi location testing
2. Station trial
3. Variety identification
4. Release and notification
5. Seed multiplication and distribution

Choose the correct answer from the options given below:

(A) 1, 2, 3, 5, 4 (B) 2, 1, 3, 4, 5
(C) 2, 3, 4, 5, 1 (D) 3, 4, 1, 5, 2

92. Match list I and list II:

List I (Organization)	**List II** (Location)
1. International Water Management Institute	i. Lima, Peru
2. International Rice Research Institute	ii. Ibadan, Nigeria
3. International Institute of Tropical Agriculture	iii. Columbia, Sri Lanka
4. International Centre for Potato	iv. Los Banos, Philippines

Choose the correct answer from the options given below:

(A) 1-iii, 2-iv, 3-i, 4-ii
(B) 1-iii 2-i, 3-iv, 4-ii
(C) 1-iii, 2-iv, 3-ii,4-i
(D) 1-ii, 2-iii, 3-iv,4-i

93. In the Harshey and Chase experiment, ^{32}P and ^{35}S were used because:

1. Both P and S are absent in RNA
2. P and S both are present in DNA and protein
3. P is present in DNA and S is present in protein only
4. Selecting labelling of DNA and protein can be done ^{22}P and ^{35}S
5. Both are easily available in gasseous form

Choose the correct answer from the options given below:

(A) 1 and 5 only (B) 3 and 4 only
(C) 2 and 5 only (D) 3 and 5 only

94. Match list I and list II:

List I (Plant)	**List II** (Product)
1. *Hevea brassilensis*	i. Henna
2. *Achras zapota*	ii. Tanin
3. *Lawsonia inermis*	iii. Rubber
4. *Schinopsis lorentzii*	iv. Chewing gum

Choose the correct answer from the options given below:

(A) 1-iii, 2-iv, 3-i, 4-ii (B) 1-i, 2-iii, 3-ii, 4-iv
(C) 1-ii, 2-i, 3-iii,4-iv (D) 1-iv, 2-ii, 3-iii,4-i

95. Match list I and list II:

List I (Crop)	**List II** (Edible plant)
1. Cauliflower	i. Root
2. Chilli	ii. Leaves
3. Carrot	iii. Inflorescence
4. Spinach	iv. Fruits

Choose the correct answer from the options given below:

(A) 1-i, 2-ii, 3-iii, 4-iv
(B) 1-iii, 2-iv, 3-i, 4-ii
(C) 1-ii, 2-i, 3-iii,4-iv
(D) 1-i, 2-iii, 3-iv,4-ii

96. In a crop, genes C and P are required for the development of colour in the flower. The white flower are produced in the either or both the genes. what will be the percentage of coloured flowers in the off-spring of the cross CcPP x ccPp?

(A) 50% (B) 25%
(C) 75% (D) 100%

97. Which of the following statements are true with respect to tomato:

1. Tomato plant are not sensitive to frost
2. Typically tomato seeds are first sown in a nursery and transported later in to the field
3. Tomato is a native crop of Asia
4. Under normal conditions, tomato assumes spreading habit
5. Tomato is a annual crop

Choose the correct answer from the options given below:

(A) 1, 3, 5 only (B) 2, 4, 5 only
(C) 4, 1, 3 only (D) 1, 2, 4 only

98. What are the probability when both parents with bloom type 'A' and normal vision (not colour-blind) produce their Ist son, who has blood type 'O' and is colour blind:

(A) ½ (B) ¼
(C) 1/8 (D) 1/16

99. Chief functions of leaves:

(A) Gasseous exchange and photosynthesis
(B) Respiration and photosynthesis
(C) Respiration and Digestion
(D) Respiration and gaseous exchange

100. In a cross (AaBbCCDdEE) × (AABbCcDdEe), what is the probability that any perticular gamete from the female parent will be AbCdE:

(A) ¼ (B) 1/6
(C) 1/16 (D) 1/8

101. Biogas production from cattle manure involves which of the following types of bacteria:

1. Hydrolytic 2. Methanogenic
3. Transitional

Choose the correct answer from the options given below:

(A) 1 only (B) 2 only
(C) 3 only (D) 1, 2 and 3 only

102. Match list I and list II:

List I	List II
1. *Magnaporthe oryzae*	i. Stomata
2. *Puccinia tritici*	ii. Hydathodes
3. *Xanthomonas axonopodis*	iii. Wound
4. *Agrobacterium tunefaciens*	iv. Direct penetration

Choose the correct answer from the options given below:

(A) 1-iv, 2-i, 3-ii, 4-iii
(B) 1-i, 2-ii, 3-iii, 4-iv
(C) 1-iv, 2-iii, 3-ii,4-i
(D) 1-i, 2-iv, 3-ii,4-iii

103. Which of the following is a symbiotic nitrogen fixer?

(A) Clostridium (B) Bradyrhizobium
(C) Azospirillum (D) Azatobacter

104. Given below are two statements:

Statement I: Leaf curl is one of the symptoms of viral disease

Statement II: Fungi can also cause leaf curl symptoms

In light of the above statements, choose the most appropriate answer from the options given below:

(A) Both statement I and statement II are correct
(B) Both statement I and statement II are incorrect
(C) Statement I is correct but statement II is incorrect
(D) Statement I is incorrect but statement II is correct

105. The purity standard of certified seeds of carrot is:

(A) 95% (B) 98%
(C) 97% (D) 70%

106. Which of the following is an INCORRECT statement about hybrids:

(A) Can be produced only in cross pollinated crop species
(B) Exploits both GCA and SCA components
(C) Are highly homogenous
(D) Genetic constitution changes over generations

107. Given below are two statements:

Statement I: Xanthomonas causes disease on fibre plants parts

Statement II: Xanthomonas can cause vascular wilt in rice

In light of the above statements, choose the most appropriate answer from the options given below:

(A) Both statement I and statement II are correct
(B) Both statement I and statement II are incorrect
(C) Statement I is correct but statement II is incorrect
(D) Statement I is incorrect but statement II is correct

108. Given below are two statements:

Statement I: Spatial and temporal transmission of plant pathogens are in general guided by asexual and sexual spores

Statement II: Bacterial pathogen contan asexual cells

In light of the above statements, choose the most appropriate answer from the options given below:

(A) Both statement I and statement II are correct

(B) Both statement I and statement II are incorrect

(C) Statement I is correct but statement II is incorrect

(D) Statement I is incorrect but statement II is correct

109. Given below are two statements:

Statement I: Potato is essentially a crop of cool moist regions and grows best in a climate where cool nights alternate with warm days during the period of tuber formation

Statement II: Irrigation or rainfall at or just before the initiation of tuber formation will result in decreases number of tuber sets

In light of the above statements, choose the most appropriate answer from the options given below:

(A) Both statement I and statement II are correct

(B) Both statement I and statement II are incorrect

(C) Statement I is correct but statement II is incorrect

(D) Statement I is incorrect but statement II is correct

110. Match list I and list II:

List I	**List II**
1. Incineration	i. Culture media
2. Sodium hypochlorite	ii. Sterilization of inoculation needle
3. Moist heat sterilization	iii. Surface sterilization of seeds
4. Filter sterilization	iv. Thermo labile liquids

Choose the correct answer from the options given below:

(A) 1-ii, 2-iv, 3-iii, 4-i

(B) 1-iv, 2-ii, 3-iii, 4-i

(C) 1-ii, 2-iii, 3-i,4-iv

(D) 1-iii, 2-i, 3-iv,4-ii

111. Irradiation of UV light causes:

(A) Dimerization of purines

(B) Dimerization of pyrimidines

(C) Dimerization of both purines and pyrimidines

(D) Dimerization of either purines or pyrimidines

112. Given below are two statements:

Statement I: When one dominant allele of one gene and recessive allele of another genes are present on one chromosome and their apposite alleles are present on the homologous chromosomes, the arrangement is called Cis-heterozygous

Statement II: When the dominent alleles of two genes are present on one chromosome and their corresponding recessive alleles are present on the homologous chromosome, the arrangement is called Cis-heterozygous.

In light of the above statements, choose the most appropriate answer from the options given below:

(A) Both statement I and statement II are correct

(B) Both statement I and statement II are incorrect

(C) Statement I is correct but statement II is incorrect

(D) Statement I is incorrect but statement II is correct

113. Vertical resistance is useful in crop improvement programme for imparting resistance.

1. Because of their monogenic nature
2. Because of their polygenic nature
3. The pathogen can break their resistance by involving new race
4. The pathogen cannot break their resistance
5. Vertical resistance will work against biotrophic pathogens

Choose the correct answer from the options given below:

(A) 1, 2, 3 only (B) 2, 3, 4 only
(C) 1, 3, 5 only (D) 3, 4, 5 only

114. Match list I and list II:

List I	List II
1. Double cross hybrid	i. Standard heterosis
2. F1 hybrid	ii. Mid-parent heterosis
3. Average heterosis	iii. F1(A×B) × F1(C×D)
4. Economic heterosis	iv. Single cross hybrid

Choose the correct answer from the options given below:

(A) 1-ii, 2-i, 3-iv, 4-iii
(B) 1-iii, 2-iv, 3-ii, 4-i
(C) 1-iii, 2-ii, 3-iv,4-i
(D) 1-iv, 2-i, 3-iii,4-ii

115. Given below are two statements:

Statement I: Cis-trans test can assign mutant alleles to the same or different genes but cannot estimate the map distance

Statement II: Recombination frequency can be used to estimate map distance between mutant luci

In light of the above statements, choose the most appropriate answer from the options given below:

(A) Both statement I and statement II are correct
(B) Both statement I and statement II are incorrect
(C) Statement I is correct but statement II is incorrect
(D) Statement I is incorrect but statement II is correct

116. Match list I and list II:

List I (Genotype)	List II (Population)
1. In bred lines	i. Homozygous and heterogenous
2. 3-way cross hybrid	ii. Homozygous and homogenous
3. Single cross hybrid	iii. Heterozygous and heterogenous
4. Multiline	iv. Heterozygous and homogenous

Choose the correct answer from the options given below:

(A) 1-i, 2-iv, 3-iii, 4-ii
(B) 1-ii, 2-iii, 3-i, 4-iv
(C) 1-iii, 2-iv, 3-ii,4-i
(D) 1-ii, 2-iii, 3-iv,4-i

117. Which of the following vascular wilt is transmitted by beetle?

(A) Bacterial wilt of tomato
(B) Fungal wilt of banana
(C) Bacterial wilt of banana
(D) Fungal wilt of date palm

118. At the minimum, how many years of multilocation testing is required for a conventionally bred genotype to be released as a new variety:

(A) 2 years (B) 3 years
(C) 4 years (D) 5 years

119. Given below are two statements:

Statement I: Active collection in the seed genebank is conserved for a medium period of 2-10 years. they are regularly used for research and distribution to bonafide users.

Statement II: Base collection in the seed gene bank are conserved for a long duration, 15 years or more for safety and prosperity.

In light of the above statements, choose the most appropriate answer from the options given below:

(A) Both statement I and statement II are correct

(B) Both statement I and statement II are incorrect

(C) Statement I is correct but statement II is incorrect

(D) Statement I is incorrect but statement II is correct

120. Given below are two statements:

Statement I: In breeding depression is the loss of vigour and fertility due to homozygosity of deleterious recessive alleles.

Statement II: In breeding depression is due to loss of vigour and fertility due to heterzygosity of recessive alleles.

In light of the above statements, choose the most appropriate answer from the options given below:

(A) Both statement I and statement II are correct

(B) Both statement I and statement II are incorrect

(C) Statement I is correct but statement II is incorrect

(D) Statement I is incorrect but statement II is correct

Answers Key

Q. No.	Ans	Q. No.	Ans	Q. No.	Ans	Q. No.	Ans	Q. No.	Ans	Q. No.	Ans
1.	C	2.	C	3.	C	4.	D	5.	A	6.	C
7.	D	8.	A	9.	D	10.	B	11.	D	12.	D
13.	C	14.	B	15.	A	16.	D	17.	C	18.	C
19.	D	20.	B	21.	B	22.	C	23.	A	24.	B
25.	B	26.	B	27.	B	28.	A	29.	C	30.	C
31.	A	32.	B	33.	C	34.	A	35.	C	36.	D
37.	B	38.	A	39.	C	40.	B	41.	A	42.	B
43.	D	44.	C	45.	C	46.	C	47.	D	48.	B
49.	C	50.	B	51.	A	52.	B	53.	A	54.	A
55.	A	56.	B	57.	D	58.	A	59.	C	60.	B
61.	C	62.	A	63.	B	64.	A	65.	D	66.	D
67.	B	68.	C	69.	B	70.	A	71.	A	72.	C
73.	B	74.	C	75.	B	76.	C	77.	B	78.	B
79.	B	80.	D	81.	D	82.	D	83.	A	84.	D
85.	D	86.	B	87.	B	88.	B	89.	D	90.	C
91.	B	92.	C	93.	B	94.	A	95.	B	96.	A
97.	B	98.	C	99.	A	100.	D	101.	D	102.	A
103.	B	104.	A	105.	A	106.	A	107.	A	108.	A
109.	C	110.	C	111.	B	112.	D	113.	•	114.	B
115.	A	116.	D	117.	D	118.	B	119.	A	120.	C

Soil Science

1. **What is C % N ratio of normal soil?**
   (A) 5 : 1 (B) 10 : 1
   (C) 15 : 1 (D) 20 : 1
2. **The WHO standard of drinking water is:**
   (A) 20 mg NO_3^--N/L
   (B) 10 mg NO_3^--N/L
   (C) 30 mg NO_3^--N/L
   (D) 50 mg NO_3^--N/L
3. **DDT has stable and high persistance in:**
   (A) Plant only
   (B) Soil only
   (C) Water only
   (D) Plant, Soil and water
4. **If the mean annual soil temperature is 8 – 15^{0}C and the difference between mean annual summer and mean annual winter temperature is <5^{0}C, then the soil temperature regime is known as**
   (A) Mesic (B) Isomesic
   (C) Thermic (D) Isothemic
5. **Under saturated condition, the soil water potential is:**
   (A) Always negative
   (B) Either positive or negative
   (C) Always positive
   (D) May be zero or negative
6. **Which of the following is NOT a suborder of entisols?**
   (A) Aqualfs (B) Fluvents
   (C) Orthents (D) Arents
7. **Process of intermixing of soil horizons by external factor is called:**
   (A) Argillation (B) Horizonation
   (C) Gleization (D) Pedoturbation
8. **Soil heat flux is equal to**
   (A) dT/dZ (B) k dZ/dT
   (C) k dT/dZ (D) DZ/DT
9. **The unconsolidated material above the parent material is called:**
   (A) Profile (B) Regolith
   (C) Solum (D) Pedon
10. **Soil pH is measured by making soil : water suspension in the ratio of:**
    (A) 1 : 10 (B) 1 : 5
    (C) 1: 2.5 (D) 1 : 1
11. **Soil heat flux is upward during:**
    (A) Noon (B) Afternoon
    (C) Forenoon (D) Midnight
12. **Natural aggregate of soil are called:**
    (A) Clods (B) Peds
    (C) Clusters (D) Flocules
13. **Which master horizon of soil is known as the zone of illuviation?**
    (A) O horizon (B) A horizon
    (C) B horizon (D) C horizon
14. **Water requirement of rice is:**
    (A) 25 cm (B) 50 cm
    (C) 120 cm (D) 200 cm
15. **Which one of the following is the soil order for old alluvial soils?**
    (A) Inceptisols (B) Entisols
    (C) Histosols (D) Molisols
16. **According ot Ministry of Water Resources, the waterlogged areas are the land areas having:**
    (A) Water within 2 m from land surface
    (B) Water table between 2 – 3 m from land surface
    (C) Water table below 3 m from land surface
    (D) Standing water column
17. **Which soft ware is used from spatial data analysis?**
    (A) GIS (B) GPS
    (C) IRS (D) GSM
18. **Weight of one ha furrow slice of root zone depth soil is about:**
    (A) 2.2×10^6 kg (B) 1.5×10^6 kg
    (C) 20×10^6 kg (D) 15×10^6 kg

19. Which of the following is a man made epipedon?

(A) Albic (B) Ochric

(C) Anthropic (D) Mollic

20. Infrared thermometer works on the principle of which of the following radiation law?

(A) Beer's law

(B) Stefen Boltzman's law

(C) Wien's law

(D) Kirchoff's law

21. The essentiality of molybdenum was dicovered by:

(A) S.L. Sommer and CP Lipman

(B) D I Arnon and P R Stout

(C) C. S. Spreengle

(D) G Gris and P H Brown

22. Illuviation is the process of:

(A) Accumulation (B) Leaching

(C) Transportation (D) Weathering

23. Which nutrient element imparts indirect drought resistance:

(A) K (B) Ca

(C) Mg (D) S

24. The spectral bands used in Greenness vegetation index are:

1. Red 2. Blue 3. Green

4. NIR 5. MIR

Choose the correct answer from the options given below:

(A) 1, 4, 3 and 5 only

(B) 1, 2 and 3 only

(C) 2, 3 and 4 only

(D) 1, 2, 3 and 4 only

25. The parent material constituted by the wind blown transported material consisting of silt:

(A) Alluvium (B) Colluvium

(C) Loess (D) Dune

26. Transformed Vegetation Index is expressed as:

(A) $\sqrt{NDVI}$ (B) $(NDVI)^2$

(C) $\sqrt{GVI}$ (D) $\sqrt{0.5 + NDVI}$

27. In general, longer the plant is in contact with nitrate, there would be:

(A) Constant accumulation of nitrate reserve

(B) Greater nitrogen fixation

(C) No effect on nitrate accumulation

(D) Reduced denitrification

28. The formula for Ethane is:

(A) CH_4 (B) CH_3CH_3

(C) $CH_3CH_2CH_2CH_3$ (D) $CH_3(CH_2)_4CH_3$

29. The common table sugar is:

(A) Glucose (B) Fructose

(C) Sucrose (D) Maltose

30. Biological control of wind erosion is:

(A) Strip cropping

(B) Contour bunding

(C) Graded bunding

(D) Ridge and furrow system

31. Most limiting factor for crop cultivation in desert areas is:

(A) Water only

(B) Temperature only

(C) Water and temperature

(D) Solar radiation

32. Comparatively stable fraction of the soil organic matter, left after the decomposition of major portion of added plant and animal residue is called:

(A) Podzol (B) Mineralization

(C) Humus (D) Immobilization

33. Geohydrological unit of land which drains through a common point is called:

(A) Pond (B) Waterbody

(C) Watershed (D) Rainshed

34. Lime is required for reclamation of:

(A) Saline soil (B) Sodic soil

(C) Acid soil (D) Acid sulphate soil

35. Chroma variables in determination of soil colour is represented by:

(A) Spectral region of the colour

(B) Whiteness of the colour

(C) Relative brightness of the colour

(D) Relative purity of the colour

36. The critical stages in rice with respect to irrigation are:
(A) Grow root initiation and flowering
(B) Sprouting and panicle initiation
(C) Early tillering and flowering
(D) Ripenining and maturity

37. The process of enrichment of surface water bodies with nutrients is called:
(A) Eutrophication
(B) Oligotrophication
(C) Putrefaction
(D) Nutrification

38. Bench terrace sloping outwards are effective only in:
(A) High rainfall area
(B) Low rainfall area
(C) Medium rainfall area
(D) Heavy rainfall area

39. The chemical used to remove organic matter from the soil during mechanical analysis is:
(A) $CaCl_2$ (B) HCl
(C) NaOH (D) H_2O_2

40. The major factors controlling the leaching loss of nitrate are:
(1) Concentration of nitrate in soil profile at the time of leaching
(2) Quantity of water passing through the soil profile
(3) Waterlogging capacity of soil

Select the correct answer from the options given below:
(A) 1 only
(B) 2 only
(C) Both 1 and 2
(D) 3 only

41. The potentially renewable resource is:
(A) Petroleum (B) Fossile soil
(C) Fertile soil (D) Coal

42. Total nitrogen in soil is measured by:
(A) Kjeldahl method
(B) Walkley and Black method
(C) Colorimeter method
(D) Hydrometer method

43. The building units for oligosaccharides and polysaccharides are:
(A) Monosaccharides
(B) Glyceraldehyde
(C) Dihdroxyacetone
(D) Trioses

44. Which of the following diagnostic surface horizons is not common in India?
(A) Mollic (B) Anthropic
(C) Ochric (D) Umbric

45. Rhizosphere is closely related to:
(A) Soil structure
(B) Soil texture
(C) Plant root system
(D) Wilting range

46. Match list I and list II:

List I	List II
1. Spectral resolution	i. 22 Days
2. Spatial resolution	ii. 8 bits
3. Radiometric resolution	iii. 110 nm
4. Temporal resolution	iv. 36 m

Choose the correct answer from the options given below:
(A) 1-iv, 2-ii, 3-iii, 4-i
(B) 1-iii, 2-iv, 3-ii, 4-i
(C) 1-iii, 2-iv, 3-i,4-ii
(D) 1-iii, 2-ii, 3-i,4-iv

47. The enzymes which catalyze the interconversion of aldose and ketose sugers are:
(A) Isomerases (B) Epimerases
(C) Carboxylases (D) Transferases

48. Flame photometry is based on the principle of:
(A) Absorption spectroscopy
(B) Emission spectroscopy
(C) ESR spectroscopy
(D) IR soectroscopy

49. Which one of the following organisms in NOT free living?
(A) Azotobacter (B) Cyanobacteria
(C) Rhodospirillum (D) Frankia

50. Match list I and list II:

List I	List II
1. *In situ* measurement	i. Textural classification chart
2. USDA	ii. Wilting range Arrangement of soil particles
3. Radiometric resolution	iii. Soil moisture
4. Above 15 bar	iv. Soil structure

Choose the correct answer from the options given below:

(A) 1-i, 2-ii, 3-iii, 4-iv
(B) 1-iv, 2-iii, 3-ii, 4-i
(C) 1-iii, 2-i, 3-iv,4-ii
(D) 1-iii, 2-iv, 3-i,4-ii

51. Which one of the following is the active factor of soil formation?

(A) Climate (B) Relief
(C) Topography (D) Time

52. Absence of air in soil limits the activity of:

(A) Microorganisms only
(B) Mesoorganisms only
(C) Microorganisms and affects oxidation process
(D) Pollutants only

53. The most imoportant process of chemical weathering process of rocks is:

(A) Solution (B) Hydration
(C) Hydrolysis (D) Oxidation

54. Amount of boron uptake may be higher in:

(A) Organic soil
(B) Acidic soil
(C) Acidic soil of humid region
(D) Clayey soil

55. Match list I and list II:

List I	List II
1. Soil quality	i. Removal of fertile layer
2. Water pollution	ii. Point and non point source
3. Soil erosion	iii. Agricultural crops
4. Land capability	iv. Soil health

Choose the correct answer from the options given below:

(A) 1-iv, 2-iii, 3-i, 4-iii
(B) 1-ii, 2-iv, 3-iii, 4-i
(C) 1-ii, 2-iii, 3-iv, 4-i
(D) 1-iii, 2-ii, 3-iv,4-i

56. Match list I and list II:

List I	List II
1. Robinson	i. Homozygous and heterogenous
2. Yoder	ii. Homozygous and homogenous
3. Munsell	iii. Heterozygous and heterogenous
4. Atterburg	iv. Heterozygous and homogenous

Choose the correct answer from the options given below:

(A) 1-iii, 2-iv, 3-i, 4-ii
(B) 1-iii, 2-ii, 3-i, 4-iv
(C) 1-ii, 2-i, 3-iv,4-iii
(D) 1-ii, 2-iii, 3-i,4-iv

57. Which gas is emitted from the soil due to rice cultivation?

(A) N_2 (B) C_2H_2
(C) CH_4 (D) SO_2

58. The first step in aggregate formation is:

(A) Cementation
(B) Dispersion
(C) Flocculation
(D) Compaction

59. The instrument used for estmating consumptive use of water is:

(A) Anemometer
(B) Piezometer
(C) Lysimeter
(D) Hydrometer

60. Deficiency of vitamin A causes:

(A) Night blindness (B) Scurvy disease
(C) Anaemia (D) Beri-Beri

61. Orientation of leaves and other parts of plant towards sunlight is called:

(A) Photoperiodism (B) Phototropism
(C) Long day plants (D) Short day plants

62. Sodium Adsorption Ratio can be expressed as:

(A) $\dfrac{Na^+}{\sqrt{\dfrac{Ca^{++} + Mg^{++}}{2}}}$

(B) $\dfrac{\sqrt{Na^+}}{\dfrac{Ca^{++} + Mg^{++}}{2}}$

(C) $\dfrac{Na^+}{\sqrt{Ca^{++} + Mg^{++}}}$

(D) $\dfrac{Na^+}{\dfrac{Ca^{++} + Mg^{++}}{2}}$

63. The potential difference between the fixed part and freely mobile portion of the solution side of double layer in soil colloids is known as:

(A) Total water potential
(B) Theta potential
(C) Gravitational potential
(D) Zeta potential

64. Emissivity of black body is:

(A) 0.50 (B) 0.25
(C) 1.00 (D) 0.75

65. Water erosion permitting crop is:

(A) Red gram (B) Black gram
(C) Soybean (D) Cowpea

66. If the soil temperature at 5 cm depth is 20°C then what will be its value in degree Farenheat?

(A) 52 °F (B) 86 °F
(C) 68 °F (D) 98 °F

67. GPS needs atleat .... signals from satellites for providing geographic location:

(A) 1 (B) 3
(C) 2 (D) 4

68. RADAR works in which of the following EMR bonds?

(A) Visible (B) Radiowave
(C) Microwave (D) Infrared

69. Phosphorus in DAP:

(A) 25% (B) 60%
(C) 46% (D) 18%

70. Which one of the following physical processes is responsible for sol heat loss?

(A) Infiltration (B) Transpiration
(C) Evaporation (D) Consensation

71. Instrumental method of analysis of pesticide residue include chromatographic techniques such as:

(1) TLC (2) GLC (3) HPLC

Choose the correct answer from the options given below:

(A) Both 1 and 2 (B) All 1, 2 and 3
(C) 3 only (D) Both 2 and 3

72. The rate of decomposing organic matter is affected negatively by the presence of:

(A) Protein (B) Fat
(C) Lignin
(D) Water soluble compounds

73. Soil moisture characteristic curve represents the relationship between:

(A) Soil texture and soil moisture content
(B) Soil moisture and soil moisture content
(C) Available soil moisture and soil moisture content
(D) Soil moisture tension and soil moisture content

74. Biofertilizers are:

(A) Organic matter
(B) Green manure
(C) FYM
(D) Culture of microorganisms

75. Decomposition of algae and mycrophytes:

(A) Leads to reduced nitrogen content
(B) Increases dissolved oxygen
(C) Enriches water quality
(D) Reduced dissolved oxygen, leads to fish mortalitty and impairs water quality

76. Clay particles in soil have the diameter of

(A) <0.002 mm (B) 0.02 – 0.002 mm
(C) 0.02 – 0.2 mm (D) 2.0 – 0.20 mm

77. Which one of the following is an aerobic non-symbiotic nitrogen fixing bacteria:

(A) Azotobacter
(B) Clostridium
(C) Rhizobium
(D) Beizerinkia

78. In general lower the latitude of a place, the:

(A) More the solar energy it receives
(B) Lesser the solar energy, it receives
(C) The more windy is the place
(D) The climate is cold

79. Match list I and list II:

List I	List II
1. Thomas way	i. pH
2. Thomas Graham	ii. pF
3. Sorenson	iii. Colloids
4. Schofield	iv. CEC

Choose the correct answer from the options given below:

(A) 1-iv, 2-ii, 3-i, 4-iii
(B) 1-iii, 2-iv, 3-i, 4-ii
(C) 1-iii, 2-iv, 3-ii, 4-i
(D) 1-iv, 2-iii, 3-i,4-ii

80. The On-Farm Development (OFD) activities undertaken under Command Area Development include:

1. Construction of field channels and drains
2. Reclamation of waterlogged areas
3. Renovation and rehabilitation of minor irrigation tanks

Choose the correct answer from the options given below:

(A) 1 only (B) Both 2 and 3
(C) 3 only (D) All 1, 2 and 3

81. The albedo is the highest in:

(A) Water surface (B) Bare soil
(C) Cropped area (D) Ice

82. Which one of the following is 2:1 non expanding clay minerals?

(A) Kaolinite (B) Vermiculite
(C) Montmorillonite (D) Illite

83. Stoke's law suggests that velocity of fall of particles is directly proportional to the:

(A) Density of the medium
(B) Square of the density of the medium
(C) Radius of the particle
(D) Square of the radius of the particle

84. Shale on metamorphosis changes to:

(A) Quartizite (B) Marble
(C) Slate (D) Basalt

85. Match list I and list II:

List I	List II
1. Splash erosion	i. Wind erosion
2. Saltation	ii. Rainfall factor
3. Erosibility	iii. Impact force required to detach soil particle
4. Dipersion ratio	iv. Impact of falling rain drops

Choose the correct answer from the options given below:

(A) 1-i, 2-ii, 3-iv, 4-iii
(B) 1-iv, 2-i, 3-ii, 4-iii
(C) 1-ii, 2-iii, 3-iv, 4-i
(D) 1-iii, 2-iv, 3-i,4-ii

86. Regur is used to identify which of the following soils?

(A) Lateritic soil (B) Red soil
(C) Black soil (D) Acid soil

87. The ratio of the total mass of soil solids to the total volume of soil solids is called:

(A) Bulk density (B) Porosity
(C) Particle density (D) Plasticity

88. The consumption of water containing high level of nitrate -N can lead to:

(A) Methohemoglobianemia in infants
(B) Night blindness
(C) Colour blindness
(D) Scurby disease

89. Which instrument is used for the determination of K content in soil:

(A) UV-VIS spectrophotometer
(B) Atomic absorption spectrophotometer
(C) Flame photometer
(D) Calorimeter

90. Available water for plant indicates water held in:

(A) 0.33 – 15 bar
(B) 0.1 – 0.2 bar
(C) Above 15 bar
(D) Below 0.1 bar

91. Specific agricultural practices used to slow down the water runoff include:

1. Contour tillage
2. Strip cropping
3. Terracing

Choose the correct answer from the options given below:

(A) Both 1 and 2
(B) Both 2 and 3
(C) All 1, 2, and 3
(D) Both 1and 3

92. Environmental Protection Act was enacted in the year:

(A) 1988 (B) 1987
(C) 1986 (D) 1985

93. The most abundant element in the earth's crust is:

(A) Oxygen (B) Silica
(C) Aluminium (D) Iron

94. Keen's box is used to determine which of the following properties?

(A) Water holding capacity
(B) Soil temperature
(C) Soil permeability
(D) Soil moisture tension

95. The conversion of a sol to a gel is called:

(A) Solation (B) Gelation
(C) Filtration (D) Precipitation

96. Which of the following is NOT associated with podzolization?

(A) Removal of Fe and Al
(B) Removal of silica
(C) Removal of humus
(D) Acidic condition

97. If a satellite completes revolution in orbit two times a day, then angular velocity is:

(A) 180°/day (B) 360°/day
(C) 720°/day (D) 90°/day

98. The correct order of wind erosion process is:

(A) Surface creep, Saltation, Suspension
(B) Saltation, Surface creep, Suspension
(C) Suspension, Surface creep, Saltation
(D) Saltation, Suspension, Surface creep

99. Hydrometer is used in soil analysis to determine:

(A) Soil moisture content
(B) Soil hydraulic conductivity
(C) Soil structure
(D) Soil texture

100. The equation to estimate soil loss by wind erosion is E = IRKFCWDB, in which K stands for:

(A) Soil textured class factor
(B) Surface roughness factor
(C) Surface cover factor
(D) Wind barrier factor

101. Under what condition, denitrification is a major mechanism of nitrogen loss from the soil:

(A) Well-drained soil
(B) Fallow soils
(C) Submerged soil
(D) Pasture land soil

102. Working range of tensiometer for determination of soil moisture is:

(A) 0-15 bar (B) 0-0.8 bar
(C) 0.8-15 bar (D) 0-0.3 bar

103. Soils of these textured class have minimum water holding capacity:

(A) Sandy soil (B) Clay soil
(C) Loam soil (D) Clay loam soil

104. The Resource Conservation Technologies inclde:

(1) Zero or minimum tillage
(2) Direct seeding
(3) Bed planting with residue mulch

Choose the correct answer from the options given below:

(A) All 1, 2 and 3
(B) Both 1 and 2
(C) 1 only
(D) Both 2 and 3

105. Match list I and list II:

List I	List II
1. Gleization	i. Desilication
2. Laterization	ii. Process of decomposition of orgnic matter
3. Humification	iii. Concretions of Fe and Mn
4. Podzolization	iv. Removal of oxides of Fe and Al

Choose the correct answer from the options given below:

(A) 1-iii, 2-i, 3-ii, 4-iv
(B) 1-ii, 2-i, 3-iv, 4-iii
(C) 1-iv, 2-i, 3-ii, 4-iii
(D) 1-i, 2-ii, 3-iii,4-iv

106. Black cotton soil is commonly represented in order of:

(A) Alfisols (B) Vertisols
(C) Entrisols (D) Aridisols

107. The number of epipedon commonly found in India is:

(A) 3 (B) 8
(C) 9 (D) 13

108. A surface horizon produced by continuous cropping is known as:

(A) Plaggen epipedon
(B) Ochric epipedon
(C) Umbric epipedon
(D) Anthropic epipedon

109. Which of the following best represents the pH range of sodic soils P?

(A) 8-10 (B) <8.5
(C) >8.5 (D) 6.0-8.0

110. Which of the following is related with in-situ moisture conservation:

(A) Fertigation (B) Weeding
(C) Mulching (D) Seeding

111. The predominant mechanism of gasseous exchange in soil is:

(A) Mass flow (B) Contact exchange
(C) Ion exchange (D) Diffusion

112. Biotite is the example of which type of rock forming mineral:

(A) Orthosilicate (B) Iososilicate
(C) Tectosilicate (D) Phyllosilicate

113. The highest category of soil classification system is:

(A) Series (B) Type
(C) Great group (D) Order

114. The K beaing mineral dominated in soil is:

(A) Olivine (B) horneblende
(C) Micas (D) Zincon

115. Phosphorus availability maximum at pH:

(A) 6.5-7.5 (B) 5.5-6.5
(C) 7.5-8.5 (D) 8.5-9.5

116. The drought tolerant leguminous green manure are:

(A) *Crotolaria juncea and Tephrosia purpurea*
(B) *Vigna trilabola and Lens culinaris*
(C) *Trifolium alexandrinum and Cymopsis tetragonoloba*
(D) *Vigna unguiculata and Macrolyloma uniflorum*

117. 'K' in Darcy's law ($q=K\ \Delta A/L$) is represented by:

(A) Flux density
(B) Hydraulic head
(C) Hydraulic gradient
(D) Hydraulic conductivity

118. Soil temperature at 5 cm depth is higher in:

(A) Crop field
(B) Bare field
(C) Wet field
(D) Ploughed field

119. The make up of soil according to proportion of sand, silt and clay is known as:

(A) Soil texture (B) Soil structure
(C) Soil horizon (D) Soil health

120. For the restoration of highly degraded areas, plantation of which of the following is recommended:

(A) Tree seedlings
(B) Grasses
(C) Fat growing tree seedlings
(D) Nitrogen-fixing tree seedlings

Answers Key

Q. No.	Ans	Q. No.	Ans	Q. No.	Ans	Q. No.	Ans	Q. No.	Ans	Q. No.	Ans
1.	B	**2.**	B	**3.**	C	**4.**	B	**5.**	D	**6.**	A
7.	D	**8.**	C	**9.**	B	**10.**	C	**11.**	D	**12.**	B
13.	C	**14.**	C	**15.**	A	**16.**	A	**17.**	A	**18.**	A
19.	C	**20.**	B	**21.**	B	**22.**	A	**23.**	A	**24.**	D
25.	C	**26.**	D	**27.**	A	**28.**	B	**29.**	C	**30.**	A
31.	A	**32.**	C	**33.**	C	**34.**	C	**35.**	D	**36.**	C
37.	A	**38.**	B	**39.**	D	**40.**	C	**41.**	C	**42.**	A
43.	A	**44.**	B	**45.**	C	**46.**	B	**47.**	A	**48.**	B
49.	D	**50.**	C	**51.**	A	**52.**	C	**53.**	C	**54.**	A
55.	A	**56.**	A	**57.**	C	**58.**	C	**59.**	C	**60.**	A
61.	B	**62.**	A	**63.**	D	**64.**	C	**65.**	A	**66.**	C
67.	B	**68.**	C	**69.**	C	**70.**	C	**71.**	B	**72.**	C
73.	D	**74.**	D	**75.**	D	**76.**	A	**77.**	A	**78.**	A
79.	D	**80.**	D	**81.**	D	**82.**	D	**83.**	D	**84.**	C
85.	B	**86.**	C	**87.**	C	**88.**	A	**89.**	B	**90.**	A
91.	C	**92.**	C	**93.**	A	**94.**	A	**95.**	B	**96.**	B
97.	C	**98.**	D	**99.**	D	**100.**	B	**101.**	C	**102.**	B
103.	A	**104.**	A	**105.**	A	**106.**	B	**107.**	A	**108.**	A
109.	C	**110.**	C	**111.**	D	**112.**	D	**113.**	D	**114.**	C
115.	C	**116.**	A	**117.**	D	**118.**	B	**119.**	A	**120.**	B

Agricultural Extension

1. **Which one of the following countries was the partner country of 34th Surajkhund International Craft Fair held in Haryana:**
   (A) China (B) Nepal
   (C) Bhutan (D) Uzbekistan

2. **The term 'Delimitation' which was in news recently is associated with which process:**
   (A) Fixing the limit of income tax
   (B) Fixing limits of affordable housing
   (C) Fixing limits of territorial constituencies
   (D) Fixing limits of chairpersons salary

3. **Who was the first administrators - statesment to attempt planning as a means for economic development?**
   (A) C. Rajagopalachari
   (B) V.T. Krishnamachari
   (C) M. Viswasvarayya
   (D) Sir C.P. Ramaswami Aiyyer

4. **Economic liberalization in India started with:**
   (A) Sustainable charges in industrial licensing policy
   (B) The convertibility of the Indian rupee
   (C) Significant reduction of tax rates
   (D) Changes in procedurial formalities for FDI

5. **Which type of N-fertilizer is suitable for waterlogged soils:**
   (A) Zinc phosphate
   (B) Ammonium sulphate
   (C) Calcium
   (D) Potassium nitrate

6. **Which one of the following countries is the main rival of India in tea export:**
   (A) China (B) Japan
   (C) Germany (D) Sri Lanka

7. **Which of the following are the two most important wheat growing zones of the country:**
   (1) Ganga-Satluj plains in the north-west
   (2) Deltaic region
   (3) Plains of north-east
   (4) Black soil regions of the Deccan

   Choose the correct answer from the options given below:
   (A) Only 1 and 2 (B) Only 2 and 3
   (C) Only 3 and 4 (D) Only 1 and 3

8. **Which one of the following is a global association of nations of the world to help cooperation in international law, security, economic development and social equity?**
   (A) United Nations Organization
   (B) World Bank
   (C) International Monetary Fund
   (D) World Health Organization

9. **Which of the following would be deficient if one finds high somatic cell count in the cow's milk:**
   (A) Butter fat
   (B) Whey protein
   (C) Cascin
   (D) Trace mineral

10. **A farmer wishes to start a 100 sq. m. rectangular vegetable garden. Since he has only 30 m barbed wire, he fences three sides of the garden letting his house compound wall act as the fourth side fencing. What is the dimension of the garden:**
    (A) 15 m × 6.67 m
    (B) 20 m × 5.0 m
    (C) 30 m × 3.33 m
    (D) 40 m × 2.5 m

11. **Under the Ganga Action Plan Phase II, pollution abatement works are being taken up in how many towns:**
    (A) 57 (B) 65
    (C) 52 (D) 69

12. Indian agriculture is typically characterized as:
(A) Land surplus, labour scarce economy
(B) Land surplus, labour surplus economy
(C) Land scarce, labour surplus economy
(D) Land scarce, labour scarce economy

13. Given below are two statements:
Statement A: MSP is declared by the government every year to provide incentives to the farmers for raising the production of crops.
Statement B: Buffer stock is created to distribute food grains in the deficit areas and among the poorer strata of the society at the lower price than the market price .
In light of the above statements, choose the most appropriate answer from the options given below:
(A) Both statement A and statement B are correct.
(B) Both statement A and statement B are incorrect
(C) Statement A is correct but statement B is incorrect
(D) Statement A is incorrect but statement B is correct

14. Amazon rainforest span over nine countries. In August 2019, in which country did a deadly fire broke in the Amazon forest:
(A) Brazil (B) Venezuela
(C) Colombia (D) Argentina

15. Which one of the following crops is a commercial crop in Punjab and Haryana but is a subsistance crop in Orissa?
(A) Tea (B) Cotton
(C) Rubber (D) Rice

16. Where is the headquarter of European Union?
(A) Geneva, Switzerland
(B) Washington DC, USA
(C) New York, US
(D) Brussels, Belgium

17. The famous Bailadila range of hills in the Bastar distict of Chattisgarh is famous for its:
(A) Bauxite deposits
(B) Manganese deposits
(C) Copper deposits
(D) High grade haematite deposits

18. Which one of the following statement is incorrect:
(A) The highest number of pigs in the world are found inIndia
(B) Assam has the largest number of pork producing pigs
(C) Uttar Pradesh provides largest amount of pork
(D) Pork constitutes about 7.6 % of the total meat produced in India

19. Which industry consume more rubber:
(A) Auto tyres and tubes
(B) Foot wear
(C) Dipped goods
(D) Latex foam

20. Nobel laurate Abhijit Vinayak Benerjee was conferred the honorary Doctor of Letters (D. Litt.) by which university?
(A) University of Delhi
(B) University of Kolkata
(C) University of Mumbai
(D) University of Hyderabad

21. Which one of the following is an information highway for sharing Rice Knowledge through the latest ICT tolls including mobile telephony?
(A) Research Knowledge Management Portal
(B) Research Knowledge Messaging Portal
(C) Rice Knowledge Management Portal
(D) Research Knowledge Management Plateform

22. The unique method of appointing a 'village Guide' to act as a link between administration and people was introduced by:
(A) Albert Mayer
(B) Spencer Hatch
(C) Rabindranath Tagore
(D) S.K. Dey

23. Mr Gupta wants to become good extension professional for this, what kind of public speaking style should he have:

(1) Ceremonial speaking style
(2) Demostrative speaking style
(3) Informative speaking style
(4) Submissive speaking style

Choose the correct answer from the options given below

(A) 1 only
(B) 1 and 2 onlt
(C) 2 and 3 only
(D) 3 and 4 only

24. Arrange the following extension methods in an increasing order based on the cost:

1. Telephone call
2. Circular letter
3. Radio talk
4. Video film
5. Newspaper article

Choose the correct answer from the options given below:

1, 2, 3, 4, 5 (B) 2, 1, 3, 4, 5
1, 2, 3, 5, 4 (D) 2, 1, 5, 3, 4

25. Which one of the following is the practice of comparing other cultural practices with one's own and considering other cultural practices as inferior:

(A) Xenocentrism
(B) Egocentrism
(C) Cultural lag
(D) Ethnocentrism

26. Models are which type of aids:

(A) Display (B) Presentation
(C) Traditional (D) Projected

27. Shri Harikishan Singh, Block Technology Manager wants to promote 'wider row spacing in sugarcane' among farmers. As an extension researcher, which combination of extension methods would be most appropriate during the decision stage of the adoption process:

(A) Field day + Compaign + Extension literature
(B) Circular letter + Compaign + Kisan mela
(C) Personal letter + Exhibition + Group discussion
(D) Circular letter + Kisan gosthi + Whatsapp message

28. In designing a poster, the abbrebiation ABC refers to:

(A) Attractive, Bold, Complete
(B) Apealing, Better, Complete
(C) Attractive, Brevity, Clear
(D) Attractive, Brief, Complex

29. Match list I with list II:

List I	List II
1. Common IQ test	i. Defeur
2. Leadership Quotient test	ii. Paul Peterson
3. Socio Linguistic theory	iii. Arthur Paul
4. Theories of Mass Communication	iv. Hymes

Choose the correct answer from the options given below:

(A) 1-iv, 2-iii, 3-ii, 4-i
(B) 1-ii, 2-iii, 3-i, 4-iv
(C) 1-ii, 2-iii, 3-iv, 4-i
(D) 1-i, 2-ii, 3-iii, 4-iv

30. Which one of the following methods of analysis is known as the queen of statistics:

(A) Comparartive analysis
(B) Factor analysis
(C) Regression analysis
(D) Cluster analysis

31. Given below are two statements, one is labelled as Assertion (A) and the is labelled as Reason (R)

Asertion (A) – As per the NSSO survey, the youth of India are not interested in agriculture because it is not remunerative.

Reason (R) – ICAR has initiated a programme called ARYA to attract rural youth towards agriculture

In light of the above statements, choose the correct answer from the option given below:

(A) Both A and R are true and R is the correct explanation of A
(B) Both A and R are true but R is NOT correct explanation of A
(C) A is true but R is false
(D) A is false but R is true

32. **Match list I and list II:**

List I	List II
1. Pannel	i. Emphasis is on creativity and to encourage new thinking
2. Symposium	ii. Diagnostic training group
3. T – Group	iii. Discussion of a single subject by experts
4. Brainstorming	iv. A series of speeches by experts

Choose the correct answer from the options given below:

(A) 1-iv, 2-iii, 3-ii,4-i
(B) 1-iii,2-iv, 3-i, 4-ii
(C) 1-i, 2-iii, 3-ii, 4-iv
(D) 1-iii, 2-iv, 3-ii, 4-i

33. **Arrange the following ICAR projects in the order of their implementation year, from the earliest to the latest:**

1. NARP 2. NAHEP
3. Farmers first 4. NAIP
5. NATP

Choose the correct answer from the options given below

(A) 1, 5, 4, 2, 3 (B) 1, 5, 4, 3, 2
(C) 2, 4, 5, 1, 3 (D) 3, 4, 5, 2, 1

34. **If it is violated, authority is under mined disciplne is zeopardized order disturbed and stability threatened refers mainly to the principle of:**

(A) Hierarchy (B) Span of control
(C) Coordination (D) Unity of control

35. **Match list I with list II:**

Match I	Match II
1. Entropy	i. Amount of information that could be added or deleted in a channel
2. Redundancy	ii. One's degree of freedom or choice to select a message
3. Noise	iii. Amount of information that can be trasmitted per unit time
4. Channel capacity	iv. Any unintended message/ information in the communication process

Choose the correct answer from the options given below:

(A) 1-i, 2-iv, 3-iii, 4-ii
(B) 1-iv, 2-ii, 3-i, 4-iii
(C) 1-iii, 2-i, 3-ii, 4-iv
(D) 1-ii, 2-i, 3-iv, 4-iii

36. **The concept of farmers friend' represents, which one of the following combinations:**

(A) Formal and cosmopolite
(B) Informal and localite
(C) Formal and localite
(D) Informal and cosmopolite

37. **The letter 'A' in PLA stand for:**

(A) Assessment
(B) Analysis
(C) Action
(D) Appraisal

38. **WYSIWYG is the display of a document on screen as it actually WYSIWYG is an acronym for:**

(A) What You Say is What You Get
(B) What You See is What You Get
(C) What You Save is What You Get
(D) What You Suggest is What You Get

39. Which one of the following is a forecasting process framework based on the result of multiple round of quetionnaires sent to a panel of experts?

(A) Brainstorming
(B) Buzz session
(C) Panel Discussion
(D) Delphi technique

40. The practice of Technology Assessment and Refinement as used in programmes like IVLP (Institution Village Linkage Programme) was based on which principle of extension?

(A) Whole Family Principle
(B) Principle of Adaptability
(C) Principle of Learning by Doing
(D) Principle of Satisfaction

41. Which one of the following is the path finder for need based researchers?

(A) Receivers observations
(B) Receivers feedback
(C) Source feedback
(D) Message of the communication process

42. A systematic integrated and planned approach for improving enterprise effectiveness is known as:

(A) Managerial grid
(B) Organisational development
(C) Management development
(D) Executive development

43. Match list I with list II:

Data could be converted into information using five (S) main processes (Leibowitz, 2003)

List I	List II
1. Condensation	i. Assigning a type
2. Contextualization	ii. Data is processed Calculation
3. Calculation	iii. Items are summerized
4. Categorization	iv. Purpose of data is understood

Choose the correct answer from the options given below:

(A) 1-ii. 2-iii, 3-iv, 4-i
(B) 1-iii, 2-iv, 3-i, 4-ii
(C) 1-ii, 2-iv, 3-iii, 4-i
(D) 1-iii, 2-iv, 3-ii, 4-i

44. Given below are two statements:

Statement I: e-Readiness encompasses basic elements viz., Infrastructure, Accessibility, affordability, Skill and Policy

Statement II: Extension Service Delivery Mechanism to be e-Ready, should have e-Ready extension educationist, e-Ready extensions service provider and e-Ready farmer

In light of the above statements, choose the most appropriate answer from the options given below:

(A) Both statement A and statement B are correct.
(B) Both statement A and statement B are incorrect
(C) Statement A is correct but statement B is incorrect
(D) Statement A is incorrect but statement B is correct

46. Given below are two statements:

Statement I: Pradhan Mantry Phasal Bima Yojna (PMFBY) aims at providing financial support to farmers suffering from crop loss/damage due to unforeseen events

Statemet II: Ensuring the flow of credit to the agricultural sector, this will contribute to food for society, crop diversification and enhancing the growth of the agriculture sector.

In light of the above statements, choose the most appropriate answer from the options given below:

(A) Both statement A and statement B are correct.
(B) Both statement A and statement B are incorrect
(C) Statement A is correct but statement B is incorrect
(D) Statement A is incorrect but statement B is correct

47. Bhoomi project was implemented in:

(A) Karnataka (B) Andhra Pradesh

(C) Madhya Pradesh (D) Rajasthan

48. The person who satisfies the needs of people and is willing to take the risk and go after unnoticed opportunities is known as:

(A) Intrapreneur (B) Social enterpreneur

(C) Entepreneur (D) Businessman

49. Which one of the following programmes was based on the report titled "India's Food Crisis & Steps to meet it":

(A) IADP (B) IVLP

(C) HYVP (D) LLP

50. 'The whole is greater than the sum of individuals' is the theory associated with:

(A) Pavlov (B) Thorndike

(C) Gestait (D) Woodworth

51. Management refers to activities involved in which one of the following four general functions:

(A) Planning, Training, Loading and Controlling

(B) Planning, Leading, Controlling and Coordinating

(C) Planning, Leading, Controlling and Evaluating

(D) Planning, Evaluating, Loading and Controlling

52. Given below are two statements:

Statement I: Virtual Extension and Research Network (VERCON) was experimental in Egypt.

Statement II: VERCON was established by ICRISAT.

In light of the above statements, choose the most appropriate answer from the options given below:

(A) Both statement A and statement B are correct.

(B) Both statement A and statement B are incorrect

(C) Statement A is correct but statement B is incorrect

(D) Statement A is incorrect but statement B is correct

53. Which type of trainee has maximum potential for effective transfer of learning in the back home situation?

(A) Early convert (B) Missionery

(C) Linear critic (D) Expatriate

54. Which one of the following extension programmes used management principles for the first time:

(A) Community Development Programme

(B) Intensive Agricultural Area Programme

(C) Intergrated Rural Development Programme

(D) Training and Visit System

55. Given below are two statements:

Statement I : Theory 'Z' by William Quchi proposed that interpersonal skills are need for group interaction. Yet, despite the emphasis on group discussion making the responsibility remains with the individual.

Statement II : Japanese manageent style emphasizes collective responsibility for effective management.

In light of the above statements, choose the most appropriate answer from the options given below:

(A) Both statement A and statement B are correct.

(B) Both statement A and statement B are incorrect

(C) Statement A is correct but statement B is incorrect

(D) Statement A is incorrect but statement B is correct

56. Given below are two statements:

Statement I : Rural radio forums are a combintion of mass media and interpersonal

Statement II : Media forum were originally developed in Egypt

In light of the above statements, choose the most appropriate answer from the options given below:

(A) Both statement A and statement B are correct.

(B) Both statement A and statement B are incorrect

(C) Statement A is correct but statement B is incorrect
(D) Statement A is incorrect but statement B is correct

57. In ABCDEF of training for extension professionals, 'D' stands for:

(A) Difficult (B) Diversity
(C) Diligence (D) Decorum

58. Given below are two statements:

Statement I : PERT and CPM are techniques of project management useful in the basic managerial functions of planning, scheduling and control.

Statement II : PERT is a technique of planning and control of both time and costs. CPM is a method to control time only.

In light of the above statements, choose the most appropriate answer from the options given below:

(A) Both statement A and statement B are correct.
(B) Both statement A and statement B are incorrect
(C) Statement A is correct but statement B is incorrect
(D) Statement A is incorrect but statement B is correct

59. Given below are two statement:

Statement I : It is not possible to convert truely discontinuous variables to a continuous variable.

Statement II: It is always possible to convert a continuous variables to polytomics/dichotomics

In light of the above statements, choose the most appropriate answer from the options given below:

(A) Both statement A and statement B are correct.
(B) Both statement A and statement B are incorrect
(C) Statement A is correct but statement B is incorrect
(D) Statement A is incorrect but statement B is correct

60. Which one of the following emphasized the development of digitized audio visual content not only in the regional language but also in regional dialects?

(A) Digital green (B) Plantix
(C) agMoocs (D) Digital blue

61. You have developed a tewt to measure farmer's attitude towards natural farming when this test was used by 25 other research scholars, their results are consistent and measured (empirically) farmers attitude towards natural farming in this scenario, your test is known to be:

(A) Reliable but not valid
(B) Unreliable but not valid
(C) Neither reliable not valid
(D) Both reliable and valid

62. In the context of the meaning of extension, the term 'AVFKLARUNG' is used for enlightment. From which language is the term derived:

(A) Dutch (B) German
(C) Spanish (D) Korean

63. "Hanging over the stick to an insider" relates to which one of the following methods:

(A) Participatory Learning and Action (PLA)
(B) Farming Systems Research and Extension (FSRE)
(C) Rapid Rural Appraisal (RRA)
(D) Participatary Rural Appraisal (PRA)

64. Match list I with list II:

List I	List II
1. Oculesits	i. Body language
2. Chronemics	ii. Space and distance
3. Kinesics	iii. Time
4. Proxemics	iv. Eye contact

Choose the correct answer from the options given below:

(A) 1-iv, 2-iii, 3-i,4-ii
(B) 1-i, 2-ii, 3-iii, 4-iv
(C) 1-iii, 2-i,3-ii,4-iv
(D) 1-ii,2-iv, 3-iii, 4-i

65. Sharing of information through whatsapp group to farmers sequentially, over a period of cropping season, may be classified as which method of communication:

(A) Individual method
(B) Group method
(C) Mass method
(D) Non-projeted method

66. You are newly appointed as an Agricultural Officer and you have been assigned the job of 'gaining first hand knowledge' goodwill and creating confidence among farmers of a village. Which extension method would you use for the purpose:

(A) Mobile app
(B) Farm and Home visit
(C) Compaign
(D) Radio talk

67. The four submission under National Mission on Agricultural Extension and Technology (NMAET) are:

(A) Seed, Mechanisation, Extension and Irrigation
(B) Mechanisation, Irrigation, Fertilizer and Extension
(C) Seed, Fertilizer, Mechanisation and Markeing
(D) Extension, Seed, Plant protection and Mechanization

68. Given below are two statements, one is labelled as Assertion A and the other is labelled as Reason R:

Assertion A: Result demonstrations are used to show relative worthh of a technology over another existing technology in a farmer's field. Result demonstration is an individual method but the field day is a group method.

Reason R: Result demonstration may or may not include field days . When neighbouring farmers participate in field days, it become a group method.

In light of the above statements, choose the correct annswer from the option given below:

(A) Both A and R are true and R is the correct explanation of A
(B) Both A and R are true but R is NOT correct explanation of A
(C) A is true but R is false
(D) A is false but R is true

69. Given below are two statements:

Statement I: Directorate of Knowledge Management (DKMA) is the official publication wing of ICAR

Statement II: DKMA works under the administrative control of the Agricultural Extension Division of ICAR

In light of the above statements, choose the most appropriate answer from the options given below:

(A) Both statement A and statement B are correct
(B) Both statement A and statement B are incorrect
(C) Statement A is correct but statement B is incorrect
(D) Statement A is incorrect but statement B is correct

70. Given below are two statements:

Statement I: In recent times, timely and accurate traceability of agricultural products and actvities is the supply chain has become a new factor in agricultural eextension.

Statement II: Distributed Aledger Technologies (DLT) can be used for traceability in agricultural extension.

In light of the above statements, choose the most appropriate answer from the options given below:

(A) Both statement A and statement B are correct.
(B) Both statement A and statement B are incorrect
(C) Statement A is correct but statement B is incorrect
(D) Statement A is incorrect but statement B is correct

71. Which of the following are correct about the Innovation Decision Process:

(1) Mass media channels are relatively more important at processing stage

(2) Interpersonal channels are more important at the knowledge stage

(3) Cosmopolite channels are relatively more important at the knowledge stage

(4) Localite channels are relatively more important at the persuasion stage

(5) Cosmopolite channels are relatively more important at the persuation stage

Choose the correct answer from the options given below:

(A) 1 and 2 only

(B) 2 and 3 only

(B) 3 and 4 only

(D) 4 and 5 only

72. Women's Empowerment in Agriculture Index (WEAI) was developed by IFPRI, Oxford Poverty and Human Development Initiative (OPH) and USAID in 2012. According to this, how many domains are identified to measure the women empowerment of agriculture:

(A) 4 (B) 5

(C) 6 (D) 7

73. Which one of the following is a 'non parametric statistical test' for two way analysis of variance:

(A) Krushkal Wallis test

(B) Chi-square test

(C) Friedman test

(D) Mann Whitney test

74. In research 'objectivity' refers to:

(A) Bias free (B) Multiplicity

(C) Suggestion (D) Opinion

75. Given below are two sttements:

Statement I: Reliability refers to the degree to which a scale provides consistent results when repeated measurements are made.

Statement II: Validity implies to the extent to which a research instrument measures what is intended to measure.

In light of the above statements, choose the most appropriate answer from the options given below:

(A) Both statement A and statement B are correct.

(B) Both statement A and statement B are incorrect

(C) Statement A is correct but statement B is incorrect

(D) Statement A is incorrect but statement B is correct

76. Which one of the folowing is the actual physical product of source encoder?

(A) Channel (B) Treatment

(C) Receiver (D) Message

77. In any organisation, whenever a problem can't be solved or a decision made without pooling the authority of two or more managers, the authority is known as:

(A) Autocratic authority

(B) Split authority

(C) Splintered authority

(D) Non-distributed authority

78. Given below are two statements:

Statement I: All valid scales are supposed to be reliable

Statement II: All reliable scales are supposed to be valid

In light of the above statements, choose the most appropriate answer from the options given below:

(A) Both statement A and statement B are correct.

(B) Both statement A and statement B are incorrect

(C) Statement A is correct but statement B is incorrect

(D) Statement A is incorrect but statement B is correct

79. Given below are two statements:

Statement I: Free rein leadership is the one which uses his/her power very little and depends largely on subordinates

Statement II: Benevolent autocratic leader is the one who listens to followers before making a decision but makes the decision of his/her own.

In light of the above statements, choose the most appropriate answer from the options given below:

(A) Both statement A and statement B are correct.

(B) Both statement A and statement B are incorrect

(C) Statement A is correct but statement B is incorrect

(D) Statement A is incorrect but statement B is correct

80. Given below are two statements:

Statement I: In management, Espirit de corps is popularity known as 'in union there is strength'

Statement II: Individual interest is more important than organizational interest.

In light of the above statements, choose the most appropriate answer from the options given below:

(A) Both statement A and statement B are correct.

(B) Both statement A and statement B are incorrect

(C) Statement A is correct but statement B is incorrect

(D) Statement A is incorrect but statement B is correct

81. In recent year, direct marketing by farmers has gained importance, which one of the following is NOT a direct marketing program:

(A) Rythu Bazar in Andhra Pradesh

(B) Apni mandi in Punjab

(C) Uzavar Sinthaigil in Tamilnadu

(D) Apno Bazar in West Bengal.

82. Given below are two statements:

Statement I: Product, Price, Promotion and Place together makes the marketing mix

Statement II: Fifth 'P' of the marketing mix is 'Packaging'

In light of the above statements, choose the most appropriate answer from the options given below:

(A) Both statement A and statement B are correct.

(B) Both statement A and statement B are incorrect

(C) Statement A is correct but statement B is incorrect

(D) Statement A is incorrect but statement B is correct

83. A statutory body created by an act of parliament concerning village/cottage industries:

(A) KVIC (B) SIDBI

(C) SMEDA (D) EDP

84. Given below are two statements:

Statement I: The term 'management by objectives' was first used by Peter E. Drucker

Statement II: The book The Effective Executive was written by Peter E. Drucker.

In light of the above statements, choose the most appropriate answer from the options given below:

(A) Both statement A and statement B are correct.

(B) Both statement A and statement B are incorrect

(C) Statement A is correct but statement B is incorrect

(D) Statement A is incorrect but statement B is correct

85. Classification of adopters into varius categories is based on which of the following:

(A) Mean and variance

(B) Mean and median

(C) Median and standard deviation

(D) Mean and standard deviation

86. Given below are two statements, one is labelled as Assertion A and the other is labeller as Reason R:

Assertion A: 'Startup India' and 'Standup India' are the key initiatives of Government of India for promotion of enterpreneurship.

Reason R: Startup India provides relaxation in certain rules and regulations for startup companies and standup India on potential enterpreneurs belonging to SC and ST communities

In light of the above statements, choose the correct answer from the option given below:

(A) Both A and R are true and R is the correct explanation of A
(B) Both A and R are true but R is NOT correct explanation of A
(C) A is true but R is false
(D) A is false but R is true

87. Given below are two statements:

Statement I: MOOCS (Massive Online Open Courses) provide a learning plateform to a range of learners in asynchonous mode.

Statement II: Asynchronous mode of learning is an advantage for both teacher and learner, where they can share their opinion, ask questions and get response but not in real time.

In light of the above statements, choose the most appropriate answer from the options given below:

(A) Both statement A and statement B are correct.
(B) Both statement A and statement B are incorrect
(C) Statement A is correct but statement B is incorrect
(D) Statement A is incorrect but statement B is correct

88. In Punjab, the indiscriminate sinking of shallow tubewells in a limited area resulted in lowering of the watertable. Such behaviour can be termed as:

(A) Social marketing (B) Over adoption
(C) Under adoption (D) Dissonance

89. Given below are two statements:

Statement I: Theory X and Y were proposed by Mc Gregor.

Statement II: Theory X states average human being has on inherent 'dislike of work'. Theory Y states that people will exercise self direction and self control.

In light of the above statements, choose the most appropriate answer from the options given below:

(A) Both statement A and statement B are correct.
(B) Both statement A and statement B are incorrect
(C) Statement A is correct but statement B is incorrect
(D) Statement A is incorrect but statement B is correct

90. The method of research which solves an on doing problem having immediate relevance to society is called:

(A) Explanatory research
(B) Descriptive research
(C) Action research
(D) Applied research

91. Thurstone equal appeaing scale of attitude is an example of measurement at which level:

(A) Ordinal (B) Nominal
(C) Interval (D) Ratio

92. Thematic analysis is generally used for:

(A) Delineating construct
(B) Defining concept
(C) Testing hypothesis
(D) Report writing

93. An abstract formed from generalization from a particular is called:

(A) Concept (B) Construct
(C) Variable (D) Theory

94. Given below are two statements:

Statement I: Rejecting the null hypothesis when it is true is called Type II error

Statement II: Accepting the null hypothesis when it is NOT true is called Type I error

In light of the above statements, choose the most appropriate answer from the options given below:

(A) Both statement A and statement B are correct.
(B) Both statement A and statement B are incorrect

(C) Statement A is correct but statement B is incorrect

(D) Statement A is incorrect but statement B is correct

95. Given below are two statements:

Statement I: APEDA is the national level agency for promoting export of agricultural and processed food products.

Statement II: APEDA is under the administrative juridiction of the Union Minister of Agriculture and Farmers Welfare.

In light of the above statements, choose the most appropriate answer from the options given below:

(A) Both statement A and statement B are correct.

(B) Both statement A and statement B are incorrect

(C) Statement A is correct but statement B is incorrect

(D) Statement A is incorrect but statement B is correct

96. Match list I with list II:

List I	List II
1. Venturesome	i. 13.5%
2. Deliberate	ii. 16.0%
3. Traditional	iii. 2.5%
4. Respectable	iv. 34.0%

Choose the correct answer from the options given below:

(A) 1-iii, 2-iv, 3-i, 4-ii

(B) 1-iv, 2-iii, 3-i, 4-ii

(C) 1-ii, 2-iv, 3-i, 4-iii

(D) 1-iii, 2-iv, 3-ii, 4-i

97. Given below are two statements:

Statement I: Management Information System (MIS) focussed on semi-structural decision requiring managerial judgement

Statement II: Decision Support System (DSS) focus on structural and routine decision

In light of the above statements, choose the most appropriate answer from the options given below:

(A) Both statement A and statement B are correct.

(B) Both statement A and statement B are incorrect

(C) Statement A is correct but statement B is incorrect

(D) Statement A is incorrect but statement B is correct

98. Which of the following statements are correct

1. Contingency theory of leadership was proposed by Fred E. Fielder
2. Path and Goal leadership theory was proposed by Victor Vroom
3. Normative leadership theory was proposed by Robert H. House.
4. Path and Goal theory states leader should make goals and clarity the path to subordinates.
5. Normative theory offers guidelines on low decision ought to be made in specific situation.

Choose the correct answer from the options given below:

(A) 1, 2 and 3 only

(B) 2, 3 and 4 only

(B) 1, 4 and 5 only

(D) 3 and 5 only

99. Which one of the following is included in criteial of Roger's theory of diffusion:

(A) Pro-Innovation bias and individual blame bias

(B) The extent of extension workers promotional efforts is too high

(C) The attributes of innovation are not defined properly

(D) It did not focus on how to reduce innovation decision period.

100. CMAT means the condidates managerial:

(A) Attitude (B) Ability

(C) Aptitude (D) Aspiration

101. A farmer has considered the adoption of drip irrigation in his farm (he even undertook a trial) but then the decision not to adopt it. Such a decision is called:

(A) Discontinuation (B) Dissonance

(C) Passive rejection (D) Active rejection

102. The new management practices in integrated pest management (IPM) are difficult to understood for farmers. Which one of the following innovation attribute is influencing diffusion/ adoption of IPM among the farmers:

(A) Complexity (B) Compatibility

(C) Observability (D) Trialability

103. Identify the statement which is correct with respect to the Panchayati Raj system:

(A) It was recommended by Balwant Rai Mehta Committee

(B) Originally S.L. Mehta Committee was appointed to review CDP and NES

(C) It was recommended by V.T. Krishmachari Committee

(D) It was recommended by Mohan Singh Mehta Committee

104. Teacher made test is usually used to measure which one of the following domains of behaviour:

(A) Attitude (B) Skill

(C) Knowledge (D) Aptitude

105. Which one of the following is considered as the seed for enterpreneurship development programme:

(A) Microlab experiment

(B) ACABC

(C) Kakinada experiment

(D) Nilokheri experiment

106. Match the list I with list II:

Match I	**Match II**
1. Adoption quotient	i. Gabrial Tarde
2. Preservative innovation	ii. Ryan and Gross
3. Diffusion theory	iii. Rogers
4. Diffusion in hybrid seed corn	iv. Chattopadhayay

Choose the correct answer from the options given below:

(A) 1-iv, 2-iii, 3-i, 4-ii

(B) 1-iv, 2-iii, 3-ii, 4-i

(C) 1-iii,2-iv, 3-i, 4-ii

(D) 1-iii, 2-iv, 3-ii, 4-i

107. Central Institute of Women and Agricultuure (CIWA) was established in Bhuvneshwar, which is directly under the control of:

(A) DDG (Crop Science)

(B) DDG (Agricultural Education)

(C) DDG (Agricultural Extension)

(D) DDG (NRM)

108. Given below are two statements:

Statement I: The managerial grid was proposed by Robert R. Blake and Jane Mouton

Statement II: The managerial grid consists of two dimesions, concern for production and concern for people.

In light of the above statements, choose the most appropriate answer from the options given below:

(A) Both statement A and statement B are correct.

(B) Both statement A and statement B are incorrect

(C) Statement A is correct but statement B is incorrect

(D) Statement A is incorrect but statement B is correct

109. The abbreviation IMRAD is commonly used in:

(A) Research Report Writing

(B) Popular writing

(C) Writing for Radio

(D) Writing business reports

110. Which one of the following is an example of covert behaviour:

(A) Replacement discontinuation

(B) Active rejection

(C) Disenchantment discontinuation

(D) Symbolic adoption

111. Which one of the following are the process barriers of communication:

(1) Feedback barrier (2)Decoding barriet

(3) Receiver barrier (4) Encoding barrier

Choose the correct answer from the options given below

(A) 1 only (B) 1 and 2 only

(C) 1, 2 and 3 only (D) All of the above

112. Given below are two statements:

Statement I: Ohio state leadership studies were proposed by Ralph M. Stogdim.

Statement II: Ohio state leadership studies determine the relationship between concern for people and concern for production.

In light of the above statements, choose the most appropriate answer from the options given below:

(A) Both statement A and statement B are correct.

(B) Both statement A and statement B are incorrect

(C) Statement A is correct but statement B is incorrect

(D) Statement A is incorrect but statement B is correct

113. Given below are two statements.

Statement I: Sex refers to psychological, social and cultural differences between males and females, whereas gender refers to bilogical difference in the body.

Statement II: Sex refers to the physical differences in the body, whereas gender refers to the socio-cultural differences between males and females

In light of the above statements, choose the most appropriate answer from the options given below:

(A) Both statement A and statement B are correct.

(B) Both statement A and statement B are incorrect

(C) Statement A is correct but statement B is incorrect

(D) Statement A is incorrect but statement B is correct

114. All the domestic support measures, considered to distort production and trade are grouped under:

(A) Green Box (B) Red Box

(C) Amber Box (D) Black Box

115. In the context of management theories, which of the following statements are correct:

(1) Principles of scientific management were prposed by Fredrick Winslow Tayler

(2) Gnatt Chart is the forerunner of a modern technique like PERT

(3) Henry Fayal is known as the father of modern management theory

(4) 7 S approach was proposed by Taylor

(5) Father of inductrial psychology is Walter Dill Scott

Choose the correct answer from the options given below:

(A) 1, 2, 3 and 4 only

(B) 1, 3, 4 and 5 only

(C) 1, 2 and 3 only

(D) 2, 3 and 4 only

116. SIDBI is the first bank set up exclusively to provide funds to which one of the following types of enterprises:

(A) Medium scale

(B) Large scale

(C) Small scale

(D) Corporate

117. Estabishment in 2010, the united Nations entity for 'Gender Equity and Empowerment of Women' is known as:

(A) UN Gender Pro

(B) UN Equity

(C) UN Women

(D) UN Empowerment Fund

118. Which one of the following grants Intellectual Property Rights to plant breeders, researchers and farmers which have developed new/extent plant varieties

(A) TRIPS (B) TRIMS

(C) PBRFRA (D) PPVFRA

119. Arrange the following stages of startup funding from earlier to later

1. Initial Public Offering (IPO)
2. Seed Funding/capital
3. Mezzanine funding and bridge loam
4. Angel invester funding
5. Venture capital funding

Choose the correct answer from the options given below:

(A) 5, 4, 3, 2, 1 (B) 2, 4, 5, 3, 1
(C) 1, 2, 3, 4, 5 (D) 1, 5, 1, 3, 4

120. .... is the employer whereas a .... is an employee in an enterprise.

(A) Manage, Subordinate
(B) CEO, worker
(C) Enterpreneur, Manager
(D) Intrapreneur, CEO

Answers Key

Q. No.	Ans	Q. No.	Ans	Q. No.	Ans	Q. No.	Ans	Q. No.	Ans	Q. No.	Ans
1.	D	2.	C	3.	C	4.	D	5.	B	6.	D
7.	D	8.	A	9.	C	10.	B	11.	A	12.	C
13.	A	14.	A	15.	A	16.	D	17.	D	18.	A
19.	A	20.	B	21.		22.	C	23.	A	24.	C
25.	D	26.	A	27.	B	28.	C	29.	C	30.	B
31.	A	32.	D	33.	B	34.	D	35.	D	36.	B
37.	C	38.	B	39.	D	40.	B	41.	B	42.	B
43.	D	44.	D	45.	A	46.	A	47.	A	48.	B
49.	A	50.	B	51.	B	52.	C	53.	C	54.	D
55.	A	56.	C	57.	C	58.	C	59.	A	60.	A
61.	D	62.	B	63.	D	64.	A	65.	B	66.	B
67.	D	68.	A	69.	C	70.	A	71.	C	72.	B
73.	C	74.	A	75.	A	76.	D	77.	C	78.	C
79.	A	80.	C	81.	D	82.	C	83.	A	84.	A
85.	C	86.	A	87.	A	88.	B	89.	A	90.	C
91.	C	92.	A	93.	A	94.	B	95.	C	96.	D
97.	B	98.	C	99.	A	100.	C	101.	D	102.	A
103.	A	104.	C	105.	C	106.	A	107.	B	108.	A
109.	A	110.	D	111.	D	112.	C	113.	D	114.	C
115.	C	116.	C	117.	C	118.	D	119.	B	120.	C

Entomology/Pathology

1. **Organic matter content (%) of most Indian soils on a dry soil basis varies between:**
   (A) 3-4 (B) 0.2-0.5
   (C) 5-6 (D) 0.5-2.0
2. **Degree of freedom is related to:**
   (A) Number of observations in a set
   (B) Number of independent observations in a set
   (C) A clear hypothesis under set
   (D) Number of dependant obsrvations in a set
3. **The most effective means of control 'Molya disease' of wheat is:**
   (A) Flooding of crop
   (B) Hot water treatment
   (C) Crop rotation
   (D) Lindane spray
4. **Under natural conditions, *Drosicha mangiferae* lays ovisec inside the:**
   (A) Leaf curls
   (B) Stem crevices
   (C) Soil cracks
   (D) Inflorescence of plant host
5. **The most suitable design for an experiment involving a varying number of tillage and nitrogen levels is:**
   (A) RBD (B) Split plot
   (C) Strip plot (D) CRD
6. **Which of the following stored grain pests is incapable of feeding on whole grain:**
   (A) Rice weevil
   (B) Red flour beetle
   (C) Khapra beetle
   (D) Lesser grain borer
7. **Potassium is obserbed by plants in the ionic form as:**
   (A) K (B) K^+
   (C) K^{++} (D) K_2O
8. **Given below are two statements.**
   Statement I: Straw berry is propagated through suckers.
   Statement II: Large cardamom is propagated through suckers.
   In light of the above statements, choose the most appropriate answer from the options given below:
   (A) Both statement A and statement B are correct.
   (B) Both statement A and statement B are incorrect
   (C) Statement A is correct but statement B is incorrect
   (D) Statement A is incorrect but statement B is correct
9. **Match the list I with list II:**

Match I	Match II
1. Ethylene	i. Apical dominance
2. Gibberellin	ii. Stomatal closing
3. Abscisic acid	iii. Genetical dwarfism
4. Auxin	iv. Fruit ripening

   Choose the correct answer from the options given below:
   (A) 1–iv, 2–iii, 3–ii, 4-i
   (B) 1-iv, 2-i, 3-ii, 4-iii
   (C) 1-ii,2-iii, 3-iv, 4-i
   (D) 1-i, 2-iv, 3-iii, 4-ii
10. **Which one of the following is known as 'Emergent weed'?**
    (A) Typha (B) Pistia
    (C) Cyperus (D) Hydrilla
11. **Separation of old cuticle from epidermis is called:**
    (A) Ecdysis (B) Apolysis
    (C) Stadium (D) Exuviae
12. **Which one of the following crop is the most tolerant of soil sodicity?**
    (A) Pearlmillet (B) Cotton
    (C) Barley (D) Rice

13. Inbreeding in cross polLinated crops increase due to:

(A) Homozygosity (B) Heterozygosity
(C) Polyploidy (D) Population means

14. Seed gall of wheat is caused by:

(A) Insect (B) Fungus
(C) Nematodes (D) Mite

15. Match the list I with list II:

Match I	Match II
1. Sugarcane top borer	i. *Tetranychus cinnabarinus*
2. EDB	ii. *Tryporyza nivella*
3. Telenomus	iii. Grain fumigant
4. Mite	iv. Insect bioagent

Choose the correct answer from the options given below:

(A) 1–iii, 2–ii, 3–i, 4-iv
(B) 1-iv, 2-i, 3-ii, 4-iii
(C) 1-ii,2-iii, 3-iv, 4-i
(D) 1-i, 2-iv, 3-iii, 4-ii

16. Which one of the following is a genetically modified crop widely grown in India:

(A) Potato (B) Brinjal
(C) Cotton (D) Maize

17. Given below are two statements.

Statement I: Royal jelly is secreted by honey bee

Statement II: Drones can't string

In light of the above statements, choose the most appropriate answer from the options given below:

(A) Both statement A and statement B are correct.
(B) Both statement A and statement B are incorrect
(C) Statement A is correct but statement B is incorrect
(D) Statement A is incorrect but statement B is correct

18. The length of silk thread in a cocoon of *Bombyx mori* ranges from:

(A) 900 – 1000 m (B) 100-150 m
(C) 200-250 m (D) 100 – 350 m

19. 'Flacheric disease' of silk worm is caused by:

(A) Virus (B) Bacterium
(C) Fungus (D) Nematode

20. Which one of the following is an insect growth regulator:

(A) Carbothion (B) Diflubenzuron
(C) Benzimidozoles (D) Diafenthiuron

21. Directorate of Plant Protection and Storage (DPPQS) is located at:

(A) Faridabad (B) Dehraun
(C) Bhopal (D) Hyderabad

22. Axillary sclerites are absent in:

(A) Coleoptera and Hymenoptera
(B) Hemiptera and Orthoptera
(C) Psocoptera and Lepidoptera
(D) Odonata and Ephemeroptera

23. The nitrogen ontent (%) in farm yard manure is:

(A) 0.1 (B) 1.5
(C) 2.0 (D) 0.5

24. Which one of the following schemes is exclusively meant for self employment of rural youth:

(A) NREP (B) TRYSEM
(C) DPAP (D) IRDP

25. The insecticidal property of DDT was discovered by Paul Muller in this year:

(A) 1874 (B) 1032
(C) 1939 (D) 1942

26. Given below are two statements.

Statement I: RCH 2 is NOT Bt cotton

Statement II: MECH 12 is a Bt cotton

In light of the above statements, choose the most appropriate answer from the options given below:

(A) Both statement A and statement B are correct.
(B) Both statement A and statement B are incorrect
(C) Statement A is correct but statement B is incorrect
(D) Statement A is incorrect but statement B is correct

27. Which one among the following is the oldest:

(A) PAU, Ludhiana (Punjab)
(B) GBPUAT, Pant nagar (Uttarakhand)
(C) BCKVV, Mohanpur (WB)
(D) KAU, Thrissur (Kerala)

28. Which one of the following is an internal grain feeder?

(A) Pulse beetle
(B) Khapra beetle
(C) Rice meal moth
(D) Indian meal moth

29. Zero tillage is successful in India in the cropping systems of:

(A) Rice crop in monocropping
(B) Wheat crop in rice-wheat
(C) Wheat crop in monocropping
(D) Rice crop in rice-wheat

30. Match the list I with list II:

Match I	Match II
1. Pulse beetle	i. *Lasioderma serricorne*
2. Rice weevil	ii. *Callosobruchus chinensis*
3. Tobacco beetle	iii. *Sitophilus oryzae*
4. Drug store beetle	iv. *Stegobium panicum*

Choose the correct answer from the options given below:

(A) 1–iii, 2–ii, 3–i, 4-iv
(B) 1-iv, 2-iii, 3-ii, 4-i
(C) 1-ii,2-iii, 3-i, 4-iv
(D) 1-ii, 2-i, 3-iv, 4-iii

31. Match the list I with list II:

Match I	Match II
1. G.B. Pant University of Agriculture and Technology	i. Uttar Pradesh
2. Central Agricultural University	ii. Maharastra
3. Gandhi Krishi Vigyan Kendra University of Agriculture Sciences	iii. Uttarakhand
4. Dr. Punjabrao Deshmukh Krishi Vidyapeeth	iv. Kernataka

Choose the correct answer from the options given below:

(A) 1–ii, 2–i, 3–iv, 4-iii
(B) 1-iv, 2-i, 3-ii, 4-iii
(C) 1-iii,2-i, 3-iv, 4-ii
(D) 1-iv, 2-ii, 3-iii, 4-i

32. False root -knot nematode is:

(A) *Nacobhus* sp.
(B) *Meloidogyne* sp.
(C) *Hirschmaniella* sp.
(D) *Criconema* sp.

33. Which one of the following is a carbamate:

(A) Thiometon
(B) Tetradifon
(C) Lamda-cyhalothrin
(D) Methomyl

34. What is full form of NABARD:

(A) National Bank for Agriculture and Rural Development
(B) National Bank for Agricuture and Research Development
(C) National Bank for Apiculture and Research Development
(D) National Bank for Aquaculture and Research Development

35. Match the list I with list II:

Match I	Match II
1. Chlorinated Hydrocarbons	i. Dichlovas
2. Organo phosphates	ii. Propoxur
3. Carbamates	iii. DDT
4. Pyrethroids	iv. Fenvalerate

Choose the correct answer from the options given below:

(A) 1-iii, 2-ii, 3-iv, 4-i
(B) 1-iv, 2-iii, 3-ii, 4-i
(C) 1-ii, 2-iii, 3-iv, 4-i
(D) 1-iii, 2-i, 3-ii, 4-iv

36. Given below are two statements.

Statement I: Author of the book Silent Spring in David Watson

Statement II: The main idea of the book Silent Spring is genetically modified organism

In light of the above statements, choose the most appropriate answer from the options given below:

(A) Both statement A and statement B are correct.

(B) Both statement A and statement B are incorrect

(C) Statement A is correct but statement B is incorrect

(D) Statement A is incorrect but statement B is correct

37. How much pesticide spray solution is required to apply 12 litres of 50 EC insecticide formulation if the recommended application rate is 0.5%:

(A) 1.2 litre (B) 120 ml

(C) 24 ml (D) 12 litre

38. Which one of the following possesses contact, systemic and fumigant actions:

(A) Chloropyriphos (B) Dichlorvos

(C) Ethion (D) Phorate

39. Chemically chitin is:

(A) Nitrogenous polysaccharide

(B) Polypeptide

(C) Halogenated polysaccharide

(D) Saturated fatty acid

40. How many major agroclimatic zones are there in India:

(A) 16 (B) 18

(C) 17 (D) 15

41. The resistant structure produced by *Fusarium* sp. is:

(A) Sclerotia (B) Stroma

(C) Conidia (D) Chlamydospores

42. Given below are two statements.

Statement I: Mahatma Gandhi National Rural Employment Guarantee Act is related to rural development

Statement II: Jawahar Rojgar Yojna is related to urban – rural development.

In light of the above statements, choose the most appropriate answer from the options given below:

(A) Both statement A and statement B are correct.

(B) Both statement A and statement B are incorrect

(C) Statement A is correct but statement B is incorrect

(D) Statement A is incorrect but statement B is correct

43. Resurgence of whitefly is locally associated with:

(A) Excessive use of phosphamidon

(B) Excessive use of Lindane

(C) Excessive use of Triazophos

(D) Indiscriminate use of Synthetic pyrethroids

44. Which one of the following mineral elements plays an important role in biological nitrogen fixation:

(A) Copper

(B) Zinc

(C) Boron

(D) Molybdenum

45. Given below are two statements.

Statement I: Mosquito, Aedes aegyptii is a vector of dengue fever, chikungunya and yellow fever virus.

Statement II: The disease rickettsia is spread by house fly.

In light of the above statements, choose the most appropriate answer from the options given below:

(A) Both statement A and statement B are correct.

(B) Both statement A and statement B are incorrect

(C) Statement A is correct but statement B is incorrect

(D) Statement A is incorrect but statement B is correct

46. Which of the following enhances seed germinatin:

(A) IBA (B) GA

(C) NAA (D) BA

47. Bird lice belong to the order:

(A) Mallophaga (B) Sihuneulata
(C) Psocoptera (D) Thysanoptera

48. The micro element available in soil with alkaline pH is:

(A) Zinc (B) Copper
(C) Iron (D) Molybdenum

49. The mechanism favouring maize as an allogamous crop species is:

(A) Dicling (B) Dichogamy
(C) Apogamy (D) Autogamy

50. Chromosomal theory of inheritance was proposed by:

(A) Jacob and Monad
(B) Mendel and strass berger
(C) Sutton and Boveri
(D) Muller and Jacob

51. Given below are two statements.

Statement I: The female of lac insect has true eyes, legs and wings.

Statement II: The tussar silkworm feeds on caster.

In light of the above statements, choose the most appropriate answer from the options given below:

(A) Both statement A and statement B are correct.
(B) Both statement A and statement B are incorrect
(C) Statement A is correct but statement B is incorrect
(D) Statement A is incorrect but statement B is correct

52. Mat nursery is related to:

(A) Onion (B) Rice
(C) Tobacco (D) Chillies

53. Which one of the following was domesticated durig the year 7500 BC:

(A) Rice (B) Wheat
(C) Barley (D) Maize

54. Match the list I with list II:

Match I	Match II
1. Vector of spike disease	i. *Nephotettix virescens*
2. Sesamum phyllody	ii. *Jassus indicus*
3. Rice yellow dwarf	iii. *Hishimonus phycitis*
4. Little leaf of brinjal	iv. *Orosius albicinetus*

Choose the correct answer from the options given below:

(A) 1–iii, 2–ii, 3–i, 4-iv
(B) 1-iv, 2-iii, 3-ii, 4-i
(C) 1-ii,2-iii, 3-iv, 4-i
(D) 1-ii, 2-iv, 3-i, 4-ii

55. The standard Bombyx mori loose egg pack contains:

(A) 40000 eggs
(B) 30000 eggs
(C) 20000 eggs
(D) 10000 eggs

56. Which one of the following is NOT a cultural method:

(A) Tillage
(B) Light trap
(C) Resistant variety
(D) Pruning and thining

57. Which one of the following is NOT a part of the insect leg:

(A) Coxa
(B) Trochanter
(C) Scape
(D) Tibia

58. Johnstan organs is absent in:

(A) Collembola
(B) Diplura
(C) Both (A) and (B)
(D) Protura

59. Rosette disease in groundnut is caused by:

(A) Fungus
(B) Bacterium
(C) Virus
(D) Mealybug

60. Given below are two statements.

Statement I: Zinc deficiency is common in calcareous soil.

Statement II: Zinc availability to crops is higher in alkaline soil.

In light of the above statements, choose the most appropriate answer from the options given below:

(A) Both statement A and statement B are correct.

(B) Both statement A and statement B are incorrect

(C) Statement A is correct but statement B is incorrect

(D) Statement A is incorrect but statement B is correct

61. The word Agronomy was derived from:

(A) Spanish (B) English

(C) Latin (D) Greek

62. Which one of the followig is NOT a herbicide:

(A) Diuron (B) Butachlor

(C) Proclain (D) Amiben

63. The acronym GEAC stands for:

(A) Government Engineering Approval Committee

(B) Global Engineering Approval Committee

(C) Genetically Engineering Appraisal Committee

(D) Genetically Engineering Approval Committee

64. Origin of silk gland in silk worm is:

(A) Ectodermal (B) Endodermal

(C) Mesodermal (D) Epidermal

65. Which one of the following is popularly known as boneless meat?

(A) Safflower (B) Potato

(C) Soybean (D) Wheat

66. Match the list I with list II:

List I	List II
1. Nosema disease	i. *Acarapsis woodi*
2. Isle of weight	ii. *Streptococcus* sp.
3. American fuelbrood	iii. *Bacillus pluton*
4. European fuel wood	iv. *Nosema species*

Choose the correct answer from the options given below:

(A) 1–iv, 2–i, 3–iii, 4-ii

(B) 1-iv, 2-iii, 3-ii, 4-i

(C) 1-ii, 2-i, 3-iv, 4-iii

(D) 1-iii, 2-i, 3-ii, 4-iv

67. Match the list I with list II:

List I	List II
1. Arkil	i. Potato
2. Langra	ii. Pea
3. Hara madhu	iii. Mango
4. Kufri jyoti	iv. Melon

Choose the correct answer from the options given below:

(A) 1–iii, 2–ii, 3–i, 4-iv

(B) 1-iv, 2-i, 3-ii, 4-iii

(C) 1-ii, 2-iii, 3-iv, 4-i

(D) 1-i, 2-iv, 3-iii, 4-ii

68. Which one of the following has the lowest circular transpiration:

(A) Heliothus (B) Sorghum

(C) Oputia (D) Hordeum

69. Which one of the following is a herbicide:

(A) Simazine (B) Parathion

(C) Phorate (D) Malathion

70. In termite, 'Royal Jelly' is produced by:

(A) Mandibular glands of queen

(B) Lateral pharyngeal glands of worker

(C) Lateral pharyngeal glands of queen

(D) Well-developed mandibles of soliers

71. Class insecta belongs to the sub soldiers:

(A) Uniramia (B) Pycnogonida

(C) Xiphosura (D) Diplopoda

72. Cotton leaf curl virus diseaase is transmitted by:

(A) Cotton aphid (B) White fly

(C) Thrips (D) Mite

73. Match the list I with list II:

List I	List II
1. ICAR-NBAIR	i. Nagpur
2. ICAR-IARI	ii. Bangaluru
3. ICAR-CCRI	iii. New Delhi
4. ICAR-NBAIM	iv. Mau

Choose the correct answer from the options given below:

(A) 1–iii, 2–ii, 3–i, 4-iv
(B) 1-iv, 2-i, 3-ii, 4-iii
(C) 1-ii, 2-iii, 3-i, 4-iv
(D) 1-iii, 2-iv, 3-ii, 4-i

74. The insecticide NOT banned for application in crop pest management is:

(A) DDT (B) Aldrin
(C) Chlorpyriphos (D) BHC

75. The entomopathogenic nematode used in the microbial control of insect pest is:

(A) *Radophilus similis*
(B) *Globodera rostochinensis*
(C) *Neoplactna carpocapsae*
(D) *Platylenchus culnus*

76. Eggs are laid in an ootheca by:

(A) Silverfish (B) Gass hopper
(C) Mole cricket (D) Cockroach

77. Snow flea belong to:

1. Siphonaptera
2. Thysanura
3. Colllembola
4. Thysanoptera

Choose the correct answer from the options given below:

(A) Both 1 and 2 (B) Both 1 and 3
(C) 3 only (D) Both 2 and 4

78. In insecta, the origin of foregut is:

(A) Ectodermal
(B) Mesodermal
(C) Endoderml
(D) Meso and endodermal

79. Scansorial or clinging type of legs are found in:

(A) Bed bugs (B) Mites
(C) Ticks (D) Head lice

80. Which one of the following is the sugarcane root borer:

(A) *Sesamia inferens*
(B) *Scirpophaga nivella*
(C) *Scirpophaga incertulas*
(D) *Emmalocera nivella*

81. The degree to which a chemical is poisonous to the insect is its:

(A) Kill value (B) Lethal dose
(C) Hazard rating (D) Toxicity

82. Complete metamorphosis occurs in:

(A) Gradd hoppers (B) Leaf hoppers
(C) Cockroaches (D) Moths

83. Given below are two statements.

Statement I: Turnip belongs to cole crop.

Statement II: Brocoli belongs to cole crop.

In light of the above statements, choose the most appropriate answer from the options given below:

(A) Both statement A and statement B are correct
(B) Both statement A and statement B are incorrect
(C) Statement A is correct but statement B is incorrect
(D) Statement A is incorrect but statement B is correct

84. The use of legislative restrictions to control pests is called:

(A) Plant protection (B) Plant quarantine
(C) Plant law (D) Plant Enclosure

85. Which one of the following is a physical method of nematode control:

(A) Heat only
(B) Irradiation only
(C) Osmotic pressure only
(D) All of the above

86. Trichogrammatids are parasitic on insects at which stage:

(A) Egg (B) Larva
(C) Pupa (D) Adult

87. If a linear relation exists between the variable X and Y, then the coeffiient of correlation between them will be:

(A) One (B) Infinite
(C) Zero (D) Two

88. From which part of chrysanthemum cinerarifolium, pyrethrum is extracted:
(A) Leaf (B) Flower
(C) Stem (D) Root

89. Resin glands in lac insect are present in:
(A) Mouth parts
(B) Brething pores
(C) All over the cuticle
(D) Anus and hindgut

90. Match the list I with list II:

Match I	Match II
1. International pest	i. *Chilo partellus*
2. Dead heart	ii. *Schistocerca gregaria*
3. Spinosad	iii. JH analogue
4. Altocid	iv. Microbial origin

Choose the correct answer from the options given below:
(A) 1–iii, 2–ii, 3–i, 4-iv
(B) 1-iv, 2-i, 3-ii, 4-iii
(C) 1-ii, 2-iii, 3-iv, 4-i
(D) 1-ii, 2-i, 3-iv, 4-iii

91. What is the full form of ATARI:
(A) Agriculture Technology Applied Research Institute
(B) Agriculture Technology Application Research Institute
(C) Agriculture Technology Apex Research Institute
(D) Agriculture Technology Animal Research Institute

92. Molya disease of wheat is most severe in:
(A) Rajasthan
(B) Gujarat
(C) Haryana
(D) Punjab

93. Retting is process associated with:
(A) Cotton (B) Jute
(C) Sugarcane (D) Opium

94. Insect hind wings are modified into knob like structure called 'halteres' in:
(A) Mustard sawfly (B) Fruit fly
(C) White fly (D) Damselfly

95. A top cross test is done to evaluate the performance of:
(A) Inbred line
(B) Single cross hybrid
(C) Pureline
(D) Double cross hybrid

96. Which one of the following is NOT a suitable method for irrigating fruit crops:
(A) Drip irrigation (B) Border strip
(C) Ring basin (D) Checkbasin

97. Given below are two statements.
Statement I: Trition x 100 is used as an emulsifier.
Statement II: Protein hydrolysate is used as an attractant against fruit fly.
In light of the above statements, choose the most appropriate answer from the options given below:
(A) Both statement A and statement B are correct
(B) Both statement A and statement B are incorrect
(C) Statement A is correct but statement B is incorrect
(D) Statement A is incorrect but statement B is correct

98. "Ufra disease' mainly occurs in:
(A) Wheat (B) Maize
(C) Paddy (D) Barley

99. Repetitive growing of same sole crop on the same field is known as:
(A) Mono cropping
(B) Sequential cropping
(C) Relay cropping
(D) Multiple cropping

100. Which nematode produces 'Nemawool':
(A) *Aphelenchus* sp.
(B) Anguina agrostic
(C) Ditylenchus dipsaci
(D) Rotylenchus reniformis

101. Geniculate type of antennae are present in:
(A) Honey bee (B) Grass hopper
(C) Termite (D) House fly

102. Infective stage of Entomo-pathogenic nematodes is:

(A) IJ 3 (B) IJ2

(C) Pre-adult male (D) Pre-adult female

103. In India, the first Plant Quentine Act was passed in the year:

(A) 1912 (B) 1910

(C) 1914 (D) 1920

104. First use of DDT resistance against glabally was reported in the year:

(A) 1947 (B) 1937

(C) 1957 (D) 1967

105. Sexual spores of fungi are:

(A) Chlamydospores (B) Sporangiospores

(C) Zoospores (D) Zygospores

106. Cartap hydrochloride was isolated for the first time from:

(A) Actinomycetes (B) Bacteria

(C) Annelid (D) Protozoa

107. Agronomic nutrient use efficiency is calculated by:

(A) Grain yield/amount of nutrient applied

(B) Grain yield in treated plot – grain yield in control/amount of nutrient applied

(C) Grain yield in treated plot – nutrient uptake in control plot / amount of nutrient applied

(D) Grain yield in treated plot – grain yield in control/nutrient uptake in treated plot – nutrient uptake in control plot

108. The test of significance of differences between two means in small samples is calculated by:

(A) x^2 (Chi-square test)

(B) Z- test

(C) t – test (D) F-test

109. Nematode reproductive system is:

(A) Bilaterally symmetrical

(B) Triradially symmetrical

(C) Tetraradially symmetrical

(D) Asymmetrical

110. First generation insecticide era was between the years:

(A) 1920-1928 (B) 1939-1942

(C) 1969-1973 (D) 1980-1984

111. Most India Bt cotton hybrid contains the genetic material:

(A) Cry 1 Ab (B) Cry 1 Ac

(C) Cry 1 3A (D) Cry 4 gene

112. The insecticide Act in India was passed in the year:

(A) 1971 (B) 1959

(C) 1968 (D) 1985

113. Which among the following has higher irrigation efficiency than other:

(A) Drip irrigation

(B) Flooding

(C) Channel irigation

(D) Sprinkler irrigation

114. The downward entry of water into the soil is termed as:

(A) Percolation (B) Infiltration

(C) Permeability (D) Seepage

115. Match the list I with list II:

Match I	Match II
1. ICAR - CIPHET	i. Hyderabad
2. ICAR - NAARM	ii. Ludhiana
3. ICAR – NBSS&LUP	iii. Lucknow
4. ICAR - IISR	iv. Nagpur

Choose the correct answer from the options given below:

(A) 1–iii, 2–ii, 3–i, 4-iv

(B) 1-iv, 2-i, 3-ii, 4-iii

(C) 1-ii, 2-iii, 3-iv, 4-i

(D) 1-ii, 2-i, 3-iv, 4-iii

116. Water potential of pure water is equal to:

(A) -0.0 (B) -1.0

(C) -1.5 (D) -0.5

117. World wide resistance to inorganic pesticide was first reported in:

(A) Scale insect (B) Mealy bug

(C) Moth (D) Housefly

118. Which one of the following is a rice cyst nematode:

(A) *Heterodera oryzae*

(B) *Heterodera graminis*

(C) *Heterodera oryzicola*

(D) *Heterodera major*

119. Juvenile hormone is secreted by:

(A) *Corpora allata*

(B) *Corpora cardiaca*

(C) Neurosecretary brain cell

(D) Salivary gland

120. Mendel did not face the problem of linkage, however, seven characters studied by him were located on how many chromosomes:

(A) 7 (B) 4

(C) 5 (D) 3

Answers Key

Q. No.	Ans	Q. No.	Ans	Q. No.	Ans	Q. No.	Ans	Q. No.	Ans	Q. No.	Ans
1.	B	2.	B	3.	C	4.	C	5.	C	6.	B
7.	B	8.	A	9.	A	10.	A	11.	B	12.	C
13.	A	14.	C	15.	C	16.	A	17.	A	18.	A
19.	B	20.	B	21.	A	22.	D	23.	D	24.	B
25.	B	26.	D	27.	B	28.	B	29.	B	30.	C
31.	C	32.	A	33.	D	34.	A	35.	D	36.	B
37.	B	38.	A	39.	A	40.	D	41.	D	42.	C
43.	D	44.	D	45.	C	46.	B	47.	A	48.	D
49.	B	50.	C	51.	B	52.	B	53.	C	54.	D
55.	C	56.	B	57.	C	58.	C	59.	C	60.	C
61.	D	62.	C	63.	D	64.	A	65.	C	66.	A
67.	A	68.	C	69.	A	70.	B	71.	A	72.	B
73.	C	74.	C	75.	C	76.	D	77.	C	78.	A
79.	D	80.	D	81.	C	82.	D	83.	D	84.	B
85.	D	86.	A	87.	A	88.	B	89.	C	90.	D
91.	B	92.	C	93.	B	94.	B	95.	B	96.	B
97.	A	98.	C	99.	A	100.	C	101.	A	102.	B
103.	C	104.	A	105.	D	106.	C	107.	B	108.	C
109.	D	110.	B	111.	B	112.	C	113.	A	114.	B
115.	D	116.	A	117.	A	118.	C	119.	A	120.	B

Horticulture

1. **Flame of the forest is botanically known as:**
   (A) *Collistemon lanceolous*
   (B) *Casia fistula*
   (C) *Butea monosperma*
   (D) *Cassia javanica*
2. **Given below are two statements.**
   Statement I: Phyllody and witches Broom disease of cucurbits is caused by phytoplasma.
   Statement II: This phytoplasma is transmitted by leaf hopper.
   In light of the above statements, choose the most appropriate answer from the options given below:
   (A) Both statement A and statement B are correct
   (B) Both statement A and statement B are incorrect
   (C) Statement A is correct but statement B is incorrect
   (D) Statement A is incorrect but statement B is correct
3. **Pusa aditi is a newly developed variety of:**
   (A) Mango (B) Guava
   (C) Grapes (D) Banana
4. **Andro monoceous form is found in which spice crop:**
   (A) Coriender (B) Black pepper
   (C) All spice (D) Nut meg
5. **Holleyhock and sweet pea are suitable annual for:**
   (A) Hanging basket
   (B) Screening purpose
   (C) Oil extraction
   (D) Rockery
6. **The most common sex form of topioca is:**
   (A) Monoecious (B) Dioecious
   (C) Bisexual (D) Gynoecious
7. **Tree banding is used for the control of**
   (A) *Drosicha mangifera*
   (B) *Clumetia transversa*
   (C) *Batocera rufomaculata*
   (D) *Eublemma versicola*
8. **Multiple disease resistant variety of tomato is:**
   (A) Arka Samrat (B) Arka Rakshak
   (C) Arka vardan (D) Arka Rohini
9. **Nutritional value of legume vegetable is reduced due to the presence of which of the following inhibitors:**
   (A) Trypsin (B) Cystine
   (C) Tyrosine (D Tryptophan
10. **Which of the following crop is most prone to red spider mite:**
    (A) Carnation (B) Tuberose
    (C) Gladiolus (D) Calendula
11. **Vesicular arbuscular mycorrhiza are:**
    (A) Free living organisms
    (B) Obligate symbiont
    (C) Facultative symbiont
    (D) Obligate parasite
12. **Zinc is responsible for the biosynthesis of which plant auxin:**
    (A) NAA (B) IBA
    (C) IAA (D) 2, 4-D
13. **Match the list I with list II:**

Match I	**Match II**
1. Chrysanthemum	i. Birbal Sahani
2. Carnation	ii. Top secret
3. Rose	iii. Arka agni
4. Marigold	iv. Dona

Choose the correct answer from the options given below:
(A) 1–iv, 2–iii, 3–i, 4-ii
(B) 1-i, 2-iv, 3-ii, 4-iii
(C) 1-ii, 2-iii, 3-iv, 4-i
(D) 1-iii, 2-iv, 3-i, 4-ii

14. **Match the list I with list II:**

Match I	Match II
1. Berry elongation in grapes	i. IBA
2. Single sigmoid curve	ii. GA_3
3. Induce rooting	iii. *Mallus domestica*
4. Veraison stage in grape	iv. Onset of ripenng

Choose the correct answer from the options given below:

(A) 1–ii, 2–iii, 3–iv, 4-i
(B) 1-iii, 2-iv, 3-ii, 4-i
(C) 1-i, 2-ii, 3-iv, 4-iii
(D) 1-ii, 2-iii, 3-i, 4-iv

15. **Which of the following is a winter annual:**

(A) Balsam (B) Zinnia
(C) Snapdragon (D) Kochia

16. **Given below are two statements.**

Statement I: Cytokinin promotes chloroplast development and chlorophyll synthesis.

Statement II: Cytokinins delays senescence of cut flowers.

In light of the above statements, choose the most appropriate answer from the options given below:

(A) Both statement A and statement B are correct
(B) Both statement A and statement B are incorrect
(C) Statement A is correct but statement B is incorrect
(D) Statement A is incorrect but statement B is correct

17. **Which crop does NOT have a chilling requirement for flowering?**

(A) Cabbage (B) Tulip
(C) Beet leaf (D) Beet root

18. **Ty - 1 gene conferringTYLCV resistance in tomato has been derived from:**

(A) *Solanum habrochites*
(B) *Solanum chmieleskii*
(C) *Solanum chilense*
(D) *Solanum peruvianum*

19. **Kashi Lalima is the latest variety of:**

(A) Okra (B) Carrot
(C) Tomato (D) Radish

20. **Match the list I with list II:**

Match I	Match II
1. Granulation	i. Mango
2. Black tip	ii. Citrus
3. Fruit cracking	iii. Grapes
4. Pink brush	iv. Tomato

Choose the correct answer from the options given below:

(A) 1–ii, 2-i, 3-iv, 4-iii
(B) 1-ii, 2-i, 3-iii, 4-iv
(C) 1-iii, 2-i, 3-iv, 4-ii
(D) 1-iii, 2-i, 3-ii, 4-iv

21. **The export of which one of the following Basmati rice varieties earns the highest share of foreign exchange of total rice exports:**

(A) Pusa 1121
(B) Improved Pusa basmati
(C) Pusa Basmati 1
(D) Pusa 1509

22. **Which one of the following is a pre-emergence herbicide**

1. Atrazine
2. Pendmethalin
3. Glyphosate
4. 2,4-D

Choose the correct answer from the options given below:

(A) 1 only (B) 2 and 3 only
(C) 1 and 2 only (D) 1 and 4 only

23. **The botanical term of banana inflorescence is:**

(A) Thyrse (B) Catkin
(C) Hypanthodium (D) Spadix

24. **Which of the following is salinity tolerance root stock of mango:**

(A) Vellai Kollumban
(B) Rumani
(C) Korukkam
(D) Olour

25. Given below are two statements.

Statement I: Leaf curl of tomato caused by TOLCNDV is a serious disease of North India

Statement II: TOLCNDV is a monopartite virus.

In light of the above statements, choose the most appropriate answer from the options given below:

(A) Both statement A and statement B are correct
(B) Both statement A and statement B are incorrect
(C) Statement A is correct but statement B is incorrect
(D) Statement A is incorrect but statement B is correct

26. Given below are two statements.

Statement I: The rice variety Pusa 1460 was developed by marker assested selection (MAS).

Statement II: The genes for bacterial blight resistance have been introgressed in Pusa Basmati 1

In light of the above statements, choose the most appropriate answer from the options given below:

(A) Both statement A and statement B are correct
(B) Both statement A and statement B are incorrect
(C) Statement A is correct but statement B is incorrect
(D) Statement A is incorrect but statement B is correct

27. Melting flesh peaches mesocarp undergo softening at:

(A) First stage of ripening
(B) Mid stage of ripening
(C) Last stage of ripening
(D) Senescence phase

28. Cole crops are considered to have evolved from which of the following common ancester:

(A) *Brassica oleracea var. capitata*
(B) *Brassica oleracea var. sylvestris*
(C) *Brassica oleracea var. italica*
(D) *Brassica oleracea var. acephala*

29. In which of the following temperate fruit crop, the double growth corve is followed:

(A) Apple (B) Pear
(C) Peach (D) Plum

30. Gulkand is prepared by mixing sugar and rose petals in the ratio:

(A) 3 : 1 (B) 1 : 1
(C) 1 : 4 (D) 1 : 3

31. Removal of field heat of flower by keeping them at lower temperature is known as:

(A) Startification (B) Hardening
(C) Pulsing (D) Pre-cooling

32. Which of the following is foliage annual:

(A) Kochia (B) Ice-plant
(C) Marigold (D) Zinia

33. The preservation of cell, tissue and organs of plant under ultra low temperature (-196°C) is called:

(A) Hyperbolic preservation
(B) Liquid preservation
(C) Hypobolic preservation
(D) Cryo preservation

34. Girdling is generally practiced in case of:

(A) Mango (B) Grapes
(C) Papaya (D) Guava

35. The progenator of cultivated cucumber is:

(A) *Cucumis humifractus*
(B) *Cucumis hytivus*
(C) *Cucumis metuliferus*
(D) Cucumis sativus var hardwickii

36. Rani Sahiba is a variety of:

(A) Rose (B) Chrysanthemum
(C) Orchid (D) Carnation

37. Weedy or semi woody perennial plants with little or no trunk are known as:

(A) Herb (B) Shrub
(C) Tree (D) Annual

38. Match the list I with list II:

Match I	Match II
1. Diamond black moth	i. *Plasmodiophora brassicae*
2. Club rot	ii. *Plutella xylostella*
3. Aphids	iii. *Pernospora parastica*
4. Downey mildew	iv. *Myzus persicae*

Choose the correct answer from the options given below:

(A) 1–ii, 2–i, 3–iv, 4-iii
(B) 1-i, 2-ii, 3-iii, 4-iv
(C) 1-ii, 2-iii, 3-i, 4-iv
(D) 1-iii, 2-ii, 3-i, 4-iv

39. The 'Pd$_1$' gene from Allium roylei conferring resistance to downey mildew is located on:

(A) Chromosome 3 (B) Chromosome 5
(C) Chromosome 6 (D) Chromosome 8

40. Coconut toddy is obtained by topping the:

(A) Unopened spadix
(B) Unopened spathe
(C) Opened spathe
(D) Opened spadix

41. The symbol for the gene controlling bitter free cucumber fruits is:

(A) bi/bi (B) Bi/Bi
(C) Bi/bi (D) B/b

42. The crop having largest area under genetically modified seeds in India is:

(A) Brinjal (B) Soybean
(C) Cotton (D) Tomato

43. The most suitable vegetable to be grown in saline soil is:

(A) Cabbage (B) Water melon
(C) Potato (D) Beet

44. Match the list I with list II:

Match I	Match II
1. Pusa Arunima	i. Alphanso × Baganpalli
2. Ratna	ii. Neelam × Alphanso
3. Sindhu	iii. Ratna × Alphanso
4. Arka puneet	iv. Amrapali × Sensation

Choose the correct answer from the options given below:

(A) 1–iv, 2–ii, 3–iii, 4-i
(B) 1-iv, 2-ii, 3-i, 4-iii
(C) 1-iv, 2-iii, 3-ii, 4-i
(D) 1-ii, 2-iv, 3-iii, 4-i

45. Match the list I with list II:

Match I	Match II
1. Khaira disease of rice	i. Copper
2. Brown heart in radish	ii. Chloride
3. Herring bone disease of tomato	iii. Zn deficiency
4. Stem melansis disease of wheat	iv. Phosphorus

Choose the correct answer from the options given below:

(A) 1-iii, 2–iv, 3-ii, 4-i
(B) 1-ii, 2-iii, 3-i, 4-iv
(C) 1-iii, 2-ii, 3-i, 4-iv
(D) 1-iv, 2-iii, 3-i, 4-ii

46. Rashtrapati Bhawan Garden was designed by:

(A) Robert Brown (B) Edwin lutyens
(C) Robinson (D) G.S. Randhawa

47. If crosses were made with 8 parents in the diallel mating system (excluding reciprocals), what would be total number of F$_1$s?

(A) 56 (B) 64
(C) 32 (D) 28

48. Seed albumin gene AmA1 was cloned from:

(A) *Amaranthus tricolor*
(B) *Amaranthus tristis*
(C) *Amaranthus dubius*
(D) *Amaranthus hypocondriacus*

49. Match the list I with list II:

Match I	Match II
1. Lycopene	i. Beet root
2. Beta cyanin	ii. Carrot
3. Beta carotene	iii. Chilli
4. Capsanthin	iv. Water melon

Choose the correct answer from the options given below:

(A) 1–i, 2-ii, 3-iii, 4-iv
(B) 1-iv, 2-i, 3-ii, 4-iii
(C) 1-ii, 2-iii, 3-i, 4-iv
(D) 1-iii, 2-iv, 3-i, 4-ii

50. 2,4-Dichlorophenoxy acetic acid (2,4-D) is a:

1. Growth regulator
2. Mutagene
3. Gasseous hormone
4. Herbicide

Choose the correct answer from the options given below:

(A) 1 and 2 only (B) 2 and 3 only
(C) 3 and 4 only (D) 1 and 4 only

51. Which one of the following is known as'Mood food':

(A) Water melon (B) Musk melon
(C) Red cherry (D) Date palm

52. Tulip bulbs must be stored/exposed to the following environmental conditions for quality flower initiation:

(A) 10-12^{0}C for 4-6 weeks
(B) 7-9^{0}C for 6-8 weeks
(C) 12-14^{0}C for 5-6 weeks
(D) 9-11^{0}C for 4-6 weeks

53. One year old apple trees consisting of a single stem is known as:

(A) Whip (B) Vertigo
(C) Feather (D) Monarch

54. Some of the important characteristics of predators are:

1. They feed on prey smaller than themselves
2. Lady bird beetles are not being used for managing mealy bugs and scale insects
3. They are free living and require seceral preys to complete their life cycle
4. They feed on prey biggen than themselves

Choose the correct answer from the options given below:

(A) 1 and 3 only (B) 2 and 3 only
(C) 1 and 4 only (D) 3 and 4 only

55. Given below are two statements.

Statement I: Tuberose is susceptible to root-knot nematode.

Statement II: Root-knot nematode in tuberose can be minimized or controlled by crop rotation with marigold.

In light of the above statements, choose the most appropriate answer from the options given below:

(A) Both statement A and statement B are correct
(B) Both statement A and statement B are incorrect
(C) Statement A is correct but statement B is incorrect
(D) Statement A is incorrect but statement B is correct

56. The first bi-specific origin of banana was suggested by:

(A) Lineaus (B) Simmonds
(C) Cheesman (D) Kurz

57. The stage of inflorescence emergence in pineapple is:

(A) Piping (B) Red heart
(C) Rosette (D) Sorose

58. Directorate of Floricultural Research is located at:

(A) Pune (B) New Delhi
(C) Lucknow (D) Bangaluru

59. Given below are two statements.

Statement I: Rangpur lime (*Citrus limonia*) is considered as most suitable rootstock for Nagpur mandarin.

Statement II: This root stock does not have resistance to gummosis and root rot.

In light of the above statements, choose the most appropriate answer from the options given below:

(A) Both statement A and statement B are correct
(B) Both statement A and statement B are incorrect
(C) Statement A is correct but statement B is incorrect
(D) Statement A is incorrect but statement B is correct

60. When female papaya plant is intercrossed hermaphrodite, the ratio of female and hermaphrodite in progeny will be:
(A) 1 : 2 (B) 1 : 1
(C) 1 : 3 (D) 2 : 1

61. Which one of the following wild relatives of tomato is the most important donor for drought resistance:
(A) *Solanum pimpinellifollum*
(B) *Solanum cheesmanii*
(C) *Solanum pennellii*
(D) *Solanum chmielewskii*

62. Mango is subjected to hot water treatment before export to mango:
(A) Fruit fly (B) Viruses
(C) Anthracnose (D) Spongy tissue

63. Which of the following is most commonly used rooting hormone:
(A) BAP (B) IBA
(C) ABA (D) GA3

64. Which of the following states is the largest producer of annual flowers seeds:
(A) Punjab (B) Karnataka
(C) Tamilnadu (D) Kerala

65. Match the list I with list II:

Match I	Match II
1. Monoceous	i. Ridge gourd (Satputia)
2. Dioceous	ii. Cucumber
3. Gynoeceous	iii. Pumkin
4. Hermaphrodite	iv. Pointed gourd

Choose the correct answer from the options given below:
(A) 1-iii, 2–iv, 3-ii, 4-i
(B) 1-i, 2-ii, 3-iii, 4-iv
(C) 1-ii, 2-iii, 3-iv, 4-i
(D) 1-iv, 2-i, 3-ii, 4-iii

66. Gradually loss of variability from germplasm of a crop is known as:
(A) Pleiotropy
(B) Mutation
(C) Inbreeding depression
(D) Genetic erosion

67. Seed crop of coriander should be harvested when:
(A) 10% turn yellow (B) 30% turn yellow
(C) 50% turn yellow (D) 70% turn yellow

68. Which of the following are informal style gardens?
1. English 2. Persian
3. Mughal 4. Japanese
Choose the correct answer from the options given below:
(A) 1 and 3 only (B) 1 and 4 only
(C) 1 and 2 only (D) 3 and 4 only

69. The optimum average daily temperature for wheat sowing is:
(A) 22-23°C (B) 15-20°C
(C) 25-27°C (D) 28-30°C

70. Which of the following hedge is most suitable for landscaping in the arid region?
(A) *Lawsonia inermis*
(B) *Mussaendra frondosa*
(C) *Eranthemum bicolor*
(D) *Galphimia gracilis*

71. Maturation of male and female repoductive organs of a bisexual flower at different times is known as:
(A) Dichogamy (B) Cicliny
(C) Cliestogamy (D) Chasmogamy

72. Which of the following is propagated by suckers:
(A) Tuberose (B) Gladiolus
(C) Gaillardia (D) Chrysanthemum

73. You want to cultivate onion in North India whose harvest should coincide with a higher price in November-December. Which onion variety will you choose to grow:
(A) Agrifound Light Red
(B) Agrifound Dark Red
(C) Nasik Red
(D) Pusa Red

74. ZW-20 which expresses the coat protein (CP) gene is a transgenic line of:
(A) Cucumber (B) Summer squash
(C) Water melon (D) Musk melon

75. Suppressed climateric fruit is related to:

(A) Apple (B) Banana
(C) Peach (D) Plum

76. Match the list I with list II:

Match I	Match II
1. Bitter pit	i. Zinc
2. Little leaf	ii. Calcium
3. Fruit cracking	iii. Potassium
4. Marginal burning of leaf	iv. Boron

Choose the correct answer from the options given below:

(A) 1–i, 2-ii, 3-iii, 4-iv
(B) 1-ii, 2-i, 3-iv, 4-iii
(C) 1-ii, 2-i, 3-iii, 4-iv
(D) 1-iii, 2-ii, 3-i, 4-iv

77. Application of Remote Sensing and GIS based technologies are essential for:

(A) Integrated farming
(B) Internsive farming
(C) Precision farming
(D) Mixed farming

78. Which one of the following is a coloured seeded variety of grape:

(A) Sonaka (B) Manik chaman
(C) Tas-A-Ganesh (D) Banglore blue

79. In which system of training trees are trained to form a V shape:

(A) Vertical Centre (B) Kniffin chaman
(C) Tatura trellis (D) Spindle bush

80. Given below are two statements.

Statement I: A guava wilt resistant root stock have been developed by crossing *Psidium mole* x *Psidium guajava.*

Statement II: Root rot nematodes (Meloidogyne) has become a serious threat to guava as it causes a reduction in plant vigour and wilting of leaves

In light of the above statements, choose the most appropriate answer from the options given below:

(A) Both statement A and statement B are correct
(B) Both statement A and statement B are incorrect
(C) Statement A is correct but statement B is incorrect
(D) Statement A is incorrect but statement B is correct

81. The most commercially viable technique for propagation of cashew nut is:

(A) Shoot tip grafting
(B) Soft wood grafting
(C) Hardwood grafting
(D) Epicotyl grafting

82. Citrus rootstock susceptible for phytophthora and Tristeza are:

1. Alemov 2. Rough lemon
3. Savage 4. X-639

Choose the correct answer from the options given below:

(A) 1 and 2 only (B) 2 and 4 only
(C) 3 and 4 only (D) 1 and 3 only

83. Solasodine is present in:

(A) Tomato (B) Brinjal
(C) Garlic (D) Onion

84. In epicotyl grafting of mango, callusing stage for a graft onion extends from:

(A) 10-30 days (B) 30-60 days
(C) 60-90 days (D) 90-120 days

85. Which one of the following is an attached method of grafting:

(A) Notch grafting
(B) Inarching
(C) Whip grafting
(D) Side grafting

86. Which Vitis species does NOT have tendrils opposite to the third leaf:

(A) *Vitis champini*
(B) *Vitis rupistris*
(C) *Vitis vinifera*
(D) *Vitis labrusca*

87. Which is considered as calcifuge crop:

(A) Mango (B) Banana
(C) Citrus (D) Guava

88. Which one of the following fruits has the highest TSS:

(A) Mango (B) Grape
(C) Jamun (D) Bael

89. Match the list I with list II:

Match I	Match II
1. Cut spray type	i. Snow ball
2. Cut standard type	ii. Ravi kisan
3. Garland purpose type	iii. Birbal Sahani
4. Pot mums type	iv. Pusa Sona

Choose the correct answer from the options given below:

(A) 1-ii, 2-i, 3–iv, 4-iii
(B) 1-ii, 2-i, 3-iii, 4-iv
(C) 1-iii, 2-i, 3-ii, 4-iv
(D) 1-iv, 2-i, 3-iii, 4-ii

90. Which of the following is a temperate orchid:

(A) Cymbidium (B) Oncidium
(C) Vanda (D) Aerids

91. The local Indian tradtonal drink 'Kanji' is:

(A) A fermented spiced beverage made from the juice of black carrots
(B) A nonfermented spiced beverage made from the juice of black carrots
(C) Fermented juice of beetroot
(D) Fermented juice of ornge carrot

92. Given below are two statements.

Statement I: Pine apple is separated from other genera of family Bromeliaceae on the basis of the syncarpous type of fruits.

Statement II: Its juice contains a very low amount of a digestive enzyme bromalin.

In light of the above statements, choose the most appropriate answer from the options given below:

(A) Both statement A and statement B are correct
(B) Both statement A and statement B are incorrect
(C) Statement A is correct but statement B is incorrect
(D) Statement A is incorrect but statement B is correct

93. Which of the following is warm season turf grass

(A) Zoysin (B) Rye
(C) Kentucky blue (D) Creeping bent

94. Given below are two statements.

Statement I: Scald is a storage disorder of apple which usually affects the skin only.

Statement II: It can be identified by brownish streaks radiating into flesh from the core.

In light of the above statements, choose the most appropriate answer from the options given below:

(A) Both statement A and statement B are correct
(B) Both statement A and statement B are incorrect
(C) Statement A is correct but statement B is incorrect
(D) Statement A is incorrect but statement B is correct

95. Which among the following is the richest source of carbohydrates

(A) Date palm (B) Bael
(C) Banana (D) Custard apple

96. The number of genotypes in a F2 population of digenic cross would be:

(A) 4 (B) 9
(C) 16 (D) 8

97. Which was the first hybrid Tea Rose developed in the world:

(A) La France (B) Pixie
(C) Gold star (D) Queen Elizabeth

98. The stone fruit, the period of suspended growth coincide with:

(A) Fall of sepals (B) Final sweet
(C) Pith softening (D) Pith hardening

99. Apple cultivar having triploid chromosomes are

1. Golden Delicious 2. Gravenslein
3. Janagold 4. Melntoh

Choose the correct answer from the options given below:

(A) 1 and 2 only (B) 2 and 3 only
(C) 3 and 4 only (D) 1 and 4 only

100. Pithiness is the physiological disorder of which one of the following vegetables:

(A) Radish (B) Tomato
(C) Cucumber (D) Sweet potato

101. Fenugreek seeds contain:

(A) Solanine (B) Totenine
(C) Solasodine (D) Diosgenin

102. The ability of a plant cell to regenerate is known as:

(A) Apomixis (B) Totipotency
(C) Parthenogenesis (D) Budding

103. The ancester of cultivated sweet potato is:

(A) *Ipomea triloba* (B) *Ipomea batatas*
(C) *Ipomea aquatica* (D) *Ipomea trifida*

104. Saucerkraut is prepared from:

(A) Cucumber (B) Cabbage
(C) Tomato (D) Carrot

105. Which of the following chemicals is used for doubling of chromosomes:

(A) MH (B) CCC
(C) NAA (D) Colchicine

106. Which direction should be preferred for the planting of windbreak to protest orchards from strong winds:

(A) South – west (B) North – East
(C) South – East (D) North – west

107. MDU - 1 is a variety of chilli developed through:

(A) Pureline selection
(B) Mutation breeding
(C) Mass selection
(D) Pedigree method

108. The permitted sulphur dioxide limit for sweetness in Ready to Serve Beverage (RTS) is:

(A) 50 ppm (B) 70 ppm
(C) 100 ppm (D) 200 ppm

109. Given below are two statements.

Statement I: Gynoecious line of bitter gourd has been developed in India.

Statement II: Hybrids developed by use of gynoecious line will have predominently gynoecious sex form.

In light of the above statements, choose the most appropriate answer from the options given below:

(A) Both statement A and statement B are correct
(B) Both statement A and statement B are incorrect
(C) Statement A is correct but statement B is incorrect
(D) Statement A is incorrect but statement B is correct

110. Grafting in vegetable crops is useful for managening:

(A) Fusarium wilt (B) Bud necrosis virus
(C) Mites (D) Downey mildew

111. Match the list I with list II:

Match I	Match II
1. Scab resistance	i. M 9
2. Dwarfing root stock	ii. Shalimar apple 1
3. Low chilling	iii. Anna
4. Woody apple aphid resistance	iv. Merton 793

Choose the correct answer from the options given below:

(A) 1-ii, 2–i, 3-iv, 4-iii
(B) 1-ii, 2-i, 3-iii, 4-iv
(C) 1-iii, 2-i, 3-ii, 4-iv
(D) 1-iv, 2-i, 3-iii, 4-ii

112. Bending is a specific technique being practiced in:

(A) Rose (B) Carnation
(C) Tube rose (D) Gerbera

113. Given below are two statements.

Statement I: Whiptail is a physiological disorder of cole crops caused by a deficiency of molybdenum.

Statement II: It occurs frequently in saline and alkaline soils.

In light of the above statements, choose the most appropriate answer from the options given below:

(A) Both statement A and statement B are correct
(B) Both statement A and statement B are incorrect
(C) Statement A is correct but statement B is incorrect
(D) Statement A is incorrect but statement B is correct

114. The parentage of muskmelon cultivar, Punjab hybrid is:

(A) MS-2 × Durgapu Madhu
(B) MS-2 × Hara Madhu
(C) MS-1 × Hara Madhu
(D) MS-1 × Durga Madhu

115. The most common method of propagation for phalsa (*Grewia subinaequalis*) is:

(A) Seeds
(B) Hard wood cutting
(C) Soft wood grafting
(D) Budding

116. Which one of the following is a high yielding variety of black pepper:

(A) Sreekara (B) Suprabha
(C) Suroma (D) Suranjana

117. National Botanical Research Institute is located at:

(A) New Delhi (B) Lucknow
(C) Vadodara (D) Chennai

118. Art of training plants in to different shapes of birds, animals, umbrellas etc are known as:

(A) Topiary (B) Edges
(C) Standard shrubs (D) Baradari

119. Which fruit is known as "Miracle fruit of China"?

(A) Kiwi fruit (B) Mango
(C) Litchi (D) Loquat

120. Generally seedless watermelon is:

(A) Diploid (B) Triploid
(C) Tetraploid (D) Hexaploid

Answers Key

Q. No.	Ans	Q. No.	Ans	Q. No.	Ans	Q. No.	Ans	Q. No.	Ans	Q. No.	Ans
1.	C	2.	B	3.	C	4.	A	5.	B	6.	A
7.	A	8.	B	9.	A	10.	A	11.	B	12.	C
13.	A	14.	D	15.	C	16.	A	17.	C	18.	C
19.	A	20.	B	21.	A	22.	C	23.	A	24.	C
25.	C	26.	A	27.	C	28.	B	29.	C	30.	B
31.	D	32.	A	33.	D	34.	B	35.	D	36.	A
37.	A	38.	A	39.	A	40.	A	41.	A	42.	D
43.	D	44.	A	45.	A	46.	B	47.	D	48.	D
49.	B	50.	D	51.	A	52.	B	53.	A	54.	A
55.	A	56.	D	57.	B	58.	A	59.	C	60.	B
61.	C	62.	A	63.	B	64.	A	65.	A	66.	D
67.	C	68.	B	69.	A	70.	A	71.	A	72.	D
73.	B	74.	B	75.	D	76.	B	77.	C	78.	D
79.	C	80.	A	81.	B	82.	A	83.	B	84.	A
85.	B	86.	C	87.	B	88.	D	89.	C	90.	A
91.	A	92.	C	93.	A	94.	A	95.	A	96.	B
97.	A	98.	D	99.	B	100.	A	101.	D	102.	B
103.	D	104.	B	105.	D	106.	B	107.	B	108.	B
109.	A	110.	A	111.	B	112.	A	113.	C	114.	C
115.	A	116.	A	117.	B	118.	A	119.	A	120.	B